T0318639

Bioremediation of Pollutants

Bioremediation of Pollutants

From Genetic Engineering to Genome Engineering

Edited by

Vimal Chandra Pandey

Department of Environmental Science, Babasaheb Bhimrao Ambedkar University, Lucknow, India

Vijai Singh

Department of Biosciences, Indrashil University, Rajpur, India

Elsevier
Radarweg 29, PO Box 211, 1000 AE Amsterdam, Netherlands
The Boulevard, Langford Lane, Kidlington, Oxford OX5 1GB, United Kingdom
50 Hampshire Street, 5th Floor, Cambridge, MA 02139, United States

British Library Cataloguing-in-Publication Data
A catalogue record for this book is available from the British Library

Library of Congress Cataloging-in-Publication Data
A catalog record for this book is available from the Library of Congress

ISBN: 978-0-12-819025-8

Publisher: Joe Hayton
Acquisitions Editor: Marisa LaFleur
Editorial Project Manager: Lena Sparks
Production Project Manager: Bharatwaj Varatharajan
Cover Designer: Mark Rogers

Typeset by MPS Limited, Chennai, India

ELSEVIER
Book Aid International
Working together to grow libraries in developing countries
www.elsevier.com • www.bookaid.org

Dedication

Dedicated to our beloved families

Contents

... *Acinetobacter baumannii* was ... sites (Paul et al., 2005). ... for life, but they can be haz- ... become a significant environmental hazard ... Petroleum hydrocarbon degraders include *Yokenella* sp.

List of contributors

Aditya Vikram Agarwal Department of Biochemistry, University of Lucknow, Lucknow, India; DST—Center for Policy Research, Babasaheb Bhimrao Ambedkar University, Lucknow, India

Gargi Bhattacharjee Department of Biosciences, School of Science, Indrashil University, Rajpur, Mehsana, Gujarat, India

Doongar R. Chaudhary Biotechnology and Phycology Division, CSIR-Central Salt and Marine Chemicals Research Institute, Bhavnagar, India

Madhubanti Chaudhuri Microbiology Laboratory, Department of Botany, University of Calcutta, Kolkata, India

Ankita Chaurasia Centre for Bioinformatics, Institute of Interdisciplinary Sciences (IIDS), University of Allahabad, Allahabad, India

Swati Dahiya Department of Civil Engineering, Indian Institute of Technology, Roorkee, India

Shivika Datta Department of Zoology, Doaba College, Jalandhar, India

Daljeet Singh Dhanjal Department of Biotechnology, Lovely Professional University, Phagwara, India

Molka Feki Tounsi Centre of Biotechnology of Sfax, University of Sfax, Sfax, Tunisia

Khalid Muzamil Gani Institute for Water and Wastewater Technology, Durban University of Technology, Durban, South Africa

Nisarg Gohil Department of Biosciences, School of Science, Indrashil University, Rajpur, Mehsana, Gujarat, India

Harish Department of Botany, Mohanlal Sukhadia University, Udaipur, India

Monu Jariyal Regional Centre of Organic Farming, Ministry of Agriculture & Farmers Welfare, Government of India, Ghaziabad, India

Shweta Jha Plant Functional Genomics Lab, Biotechnology Unit, Department of Botany (UGC-CAS), J.N.V. University, Jodhpur, India

Krunal Joshi Department of Biosciences, School of Science, Indrashil University, Rajpur, Mehsana, Gujarat, India

Madhu Kamle Department of Forestry, North Eastern Regional Institute of Science and Technology, Nirjuli, India

Jitendra Kumar Bangalore Bio Innovation Centre, Helix Biotech Park, Electronics City Phase 1, Bengaluru, Karnataka, India

Pradeep Kumar Department of Forestry, North Eastern Regional Institute of Science and Technology, Nirjuli, India

Priya Ranjan Kumar Department of Biotechnology, IMS Engineering College, Ghaziabad, India

Tarun Kumar Department of Civil Engineering, Greater Noida Institute of Technology, Greater Noida, Uttar Pradesh, India

Vijay Kumar Regional Ayurveda Research Institute for Drug Development, Gwalior, India

Alka Kumari Biotechnology and Phycology Division, CSIR-Central Salt and Marine Chemicals Research Institute, Bhavnagar, India

Indra Mani Department of Microbiology, Gargi College, University of Delhi, New Delhi, India

Santosh Kumar Mishra Department of Biotechnology, IMS Engineering College, Ghaziabad, India

Vachaspati Mishra Bangalore Bio Innovation Centre, Helix Biotech Park, Electronics City Phase 1, Bengaluru, Karnataka, India; Alberta Plant Health Lab, Crop Diversification Centre North, Alberta Agriculture and Forestry, Edmonton, AB, Canada

Nihal Mohammed Department of Biology, College of Science, Mosul University, Mosul, Iraq

Arun Kumar Pal Department of Molecular and Cellular Engineering, JIBB, Sam Higginbottom University of Agriculture, Technology and Sciences, Prayagraj, India

Siddhartha Pandey Department of Civil Engineering, Chalapathi Institute of Technology, Guntur, India

Vimal Chandra Pandey Department of Environmental Science, Babasaheb Bhimrao Ambedkar University, Lucknow, India

Sumya Pathak Department of Biochemistry, University of Lucknow, Lucknow, India; DST—Center for Policy Research, Babasaheb Bhimrao Ambedkar University, Lucknow, India

Gaurav Sanghvi Department of Microbiology, Marwadi University, Rajkot, India

Pallavi Saxena Department of Botany, Mohanlal Sukhadia University, Udaipur, India

Deepansh Sharma Amity Institute of Microbial Technology, Amity University, Jaipur, India

Iti Sharma Advance Technology Development Centre, Indian Institute of Technology, Kharagpur, India

Pooja Sharma Department of Environmental Microbiology, School for Environmental Sciences, B.B. Ambedkar Central University, Lucknow, India

Amit Kumar Singh Department of Biochemistry, University of Allahabad, Allahabad, India

Joginder Singh Department of Biotechnology, Lovely Professional University, Phagwara, India

Jyotsna Singh Department of Molecular and Cellular Engineering, JIBB, Sam Higginbottom University of Agriculture, Technology and Sciences, Prayagraj, India

Nitin Kumar Singh Department of Environmental Science and Engineering, Marwadi University, Rajkot, India

Rana Pratap Singh Department of Civil Engineering, Katihar Engineering College, Katihar, India

Ravi Kant Singh Amity Institute of Biotechnology, Amity University Chhattisgarh, Raipur, India

Satyender Singh Regional Advanced Water Testing Laboratory, Mohali, India

Simranjeet Singh Department of Biotechnology, Lovely Professional University, Phagwara, India; Punjab Biotechnology Incubators, Mohali, India; Regional Advanced Water Testing Laboratory, Mohali, India

Surendra Pratap Singh Department of Botany, D.A.V. College, Chhatrapati Shahu Ji Maharaj University (CSJM University), Kanpur, India

Vijai Singh Department of Biosciences, School of Science, Indrashil University, Rajpur, Mehsana, Gujarat, India

Ramendra Soni Department of Molecular and Cellular Engineering, JIBB, Sam Higginbottom University of Agriculture, Technology and Sciences, Prayagraj, India

Shubhi Srivastava Microbiology Laboratory, Department of Botany, University of Calcutta, Kolkata, India

Arti Thanki Department of Environmental Science and Engineering, Marwadi University, Rajkot, India

Heykel Trabelsi Micalis Institute, INRA, AgroParisTech, Paris-Saclay University, Jouy-en-Josas, France

Pooja Tripathi Department of Computational Biology and Bioinformatics, JIBB, Sam Higginbottom University of Agriculture, Technology and Sciences, Prayagraj, India

Vijay Tripathi Department of Molecular and Cellular Engineering, JIBB, Sam Higginbottom University of Agriculture, Technology and Sciences, Prayagraj, India

Sachin Vaidh Department of Biological Sciences and Biotechnology, Institute of Advanced Research, Gandhinagar, India

S. Veeranna Bangalore Bio Innovation Centre, Helix Biotech Park, Electronics City Phase 1, Bengaluru, Karnataka, India

Balasubramanian Velramar Amity Institute of Biotechnology, Amity University Chhattisgarh, Raipur, India

Gajendra Singh Vishwakarma Department of Biological Sciences and Biotechnology, Institute of Advanced Research, Gandhinagar, India

Manish Yadav Central Mine Planning and Design Institute Limited, Bhubaneswar, India

Suman Yadav APS University, Rewa, India

Tara Chand Yadav Department of Biotechnology, Indian Institute of Technology, Roorkee, India

About the editors

Dr. Vimal Chandra Pandey is currently a CSIR-Senior Research Associate (CSIR-Pool Scientist) in the Department of Environmental Science at Babasaheb Bhimrao Ambedkar University, Lucknow, India. He also worked as a Consultant at Council of Science and Technology, Uttar Pradesh (CSTUP) and a DST-Young Scientist at the Plant Ecology and Environmental Science Division, CSIR-National Botanical Research Institute (CSIR-NBRI), Lucknow, India. He is well recognized internationally in the field of phytomanagement of fly ash/polluted sites through low-input green technologies. His main research focuses on phytoremediation, revegetation, and restoration of heavy metal-polluted sites and waste dumpsites (fly ash, red mud, etc.) through valuable nonedible plants with minimum inputs, least risk, and low maintenance. He is a recipient of a number of awards, honors, and fellowships such as the CSTUP-Young Scientist Award, the DST SERB-Young Scientist Award, the UGC-Dr. DS Kothari Postdoctoral Fellowship, and the CSIR-Senior Research Associateship and a member (MNASc) of National Academy of Sciences, India (NASI), a commission member of IUCN-CEM Ecosystem Restoration and a member of the BECT's Editorial Board. He has written, cowritten, edited more than 90 publications, including research and review articles in refereed journals (53 peer-reviewed), book chapters (35), and books (3 published by Elsevier with several more forthcoming). His high quality research papers have good citation and have been cited worldwide by researchers within short time. He is also serving as a subject expert reviewer for several journals from Elsevier, Springer, Wiley, Taylor & Francis, etc. Email address: vimalcpandey@gmail.com, ORCID iD: https://orcid.org/0000-0003-2250-6726, Google Scholar: https://scholar.google.co.in/citations?user = B-5sDCoAAAAJ&hl.

Dr. Vijai Singh is an Associate Professor in the Department of Biosciences, School of Science at Indrashil University, Rajpur, Mehsana, Gujarat, India. Before this, he was a postdoctoral fellow at the Institute of Systems and Synthetic Biology, France and School of Energy and Chemical Engineering at Ulsan National Institute of Science and Technology, South Korea. He has served as an Assistant Professor in the Department of Biotechnology at Invertis University, Bareilly India and Department of Biological Sciences and Biotechnology, Institute of Advanced Research, Gandhinagar, India. He received his Ph.D. in Biotechnology from Dr. APJ Abdul Kalam Technical University/ICAR-National Bureau of Fish Genetic Resources, Lucknow, India with a research focus on the development of molecular and immunoassay for *Aeromonas hydrophila*. He has designed and characterized several synthetic oscillators, gene networks, lycopene pathway, MAGE, and

CRISPR-Cas systems in *Escherichia coli*. He has more than 8 years of research and teaching experience in synthetic biology, microbiology, metabolic engineering, and industrial microbiology. His laboratory focuses on the construction of a novel biosynthetic pathway for production of pigments, chemicals, and biofuels. In addition, his laboratory works on developing a CRISPR-based platform for disease diagnosis and eradication of MDR pathogens. He has published 74 articles, 25 chapters, and 3 books. He serves as a member of the editorial board and reviewer of many peer-reviewed journals.

Foreword

I am pleased to accept the invitation from Dr. Vimal Chandra Pandey and Dr. Vijai Singh to provide some introductory statements to *Bioremediation of Pollutants: From Genetic Engineering to Genome Engineering*. This is a timely volume on the rapidly evolving field of bioremediation using genetic engineering and genome engineering approaches. It provides insights into sustainable and potential strategies in remediation of a wide range of pollutants.

The book explores discussions on molecular biology tools for developing transgenic plants, root engineering, CRISPR-assisted genome editing for phytoremediation, genetically improved microbes, metagenomics, synthetic biology, microbial indicators, biosensors, and biofilms as well as biosurfactants. Eminent scientists, professors, and researchers from across the world working in the area of phytoremediation and microbial remediation have contributed to this book. This is a unique compilation on recent findings and strategies on the genetic engineering and genome engineering in phytoremediation and bioremediation methodologies.

It is a great pleasure for me to announce this book will be useful source for doctoral students, researchers, faculties, and scientists in academia and other industries. Besides, the funding agencies including public and private, policymakers, social activists, and stakeholders will also get a clear notion on the road traveled so far and the future roadmap on improvement of plant and microbes for remediation of pollutants by genetic engineering and genome editing approaches. Overall, this book is highly informative, timely, and demands wide-ranging readership to learn about these promising and effective strategies and their success case studies.

I am happy to recognize the valuable efforts of Dr. Vimal Chandra Pandey and Dr. Vijai Singh, who together created an excellent book on bioremediation using genetic engineering and genome editing approaches.

Prabodh K. Trivedi
Director
CSIR-Central Institute of Medicinal and Aromatic
Plants (CSIR-CIMAP), Lucknow, India

Preface

Environmental pollution is a serious issue and challenge worldwide. It is caused by the uncontrolled use of man-made chemicals, pesticide, plastic, phenolic compounds, industrial waste, and/or nondegradable materials. These pollutants cause a wide range of diseases including infertility, blindness, genetic diseases, cancer, and can also impoverish soil and leave detrimental effects on air and water quality. These pollutants are also transferred through the food chain and may have pernicious repercussions on plants and animals. A pressing need has arisen to develop technologies, methods, or identify plants or microorganisms that can be used to remove or reduce the pollutants in an ecofriendly and cost-effective manner. Many of these are currently used to treat pollutants from soil, water, surface materials by altering the environmental conditions and stimulating the microbes or plants grown for degradation or reduction of pollutants. A few of the bioremediation technologies that allow for removal or reduction of pollutants include phytoremediation, mycoremediation, bioleaching, landforming, bioreactor, composting, and bioaugmentation.

A number of plants and microorganisms have the natural ability to degrade or reduce pollutants owing to the biosynthetic pathways present in their genome or plasmids. Bioremediation is also used for degradation of pollutants with heavy metals including cadmium, chromium, lead, and uranium which cannot be biodegraded. However, bioremediation can potentially reduce the mobility of these materials in the subsurface, reducing the potential of human and environmental exposure. This book offers knowledge for removal or reduction of pollutants using several microorganisms and plants with various approaches including metagenomics, bioreactor, molecular biology tools, microbial indicators, biosurfactants, biofilm, genetically modified organisms, engineered fungi and bacteria, synthetic biology tools, and genome editing CRISPR-Cas9 technology. This book also covers a wide range of areas such as transgenic plants, increasing the biomass, use of genetic engineering, and genome editing technology for rapid phytoremediation of pollutants.

This book is a compilation of 21 chapters written by eminent scientists worldwide. We sincerely hope that the present book will strengthen scientific understanding to researchers, students, scientists, stakeholders, policymakers, and environmentalists. We believe that this is a single volume that offers a wide range of areas of bioremediation with the use of enriching literacy text, clarity, in-depth knowledge, and coverage. We also believe that this book can serve as a primer for students, researchers, microbiologists, plant scientists, environmental scientists, policymakers, and regulatory agencies in bioremediation of pollutants through genetic engineering and genome engineering approaches.

Vimal Chandra Pandey and Vijai Singh

Acknowledgments

We sincerely wish to thank Candice Janco and Marisa LaFleur (Acquisitions Editors), Lena Sparks (Editorial Project Manager), Swapna Praveen (Copyrights Coordinator), and Bharatwaj Varatharajan (Production Project Manager) from Elsevier for their excellent support, guidance, and coordination of this fascinating project. We would like to thank all the authors for their excellent chapter contributions. We greatly appreciate support from reviewers for their time and expertise in reviewing the chapters. Special thanks go to Prof. Prabodh K. Trivedi, Director, CSIR-Central Institute of Medicinal and Aromatic Plants (CSIR-CIMAP), Lucknow, India for providing the "Foreword" of this book. We are aware that even despite our best efforts, the first version always comes with some errors that may have crept in this compilation. We are happy to receive positive feedback from readers to improve future volumes.

Part I

Phytoremediation

Phytoremediation—a holistic approach for remediation of heavy metals and metalloids

1

Sumya Pathak[1,2], *Aditya Vikram Agarwal*[1,2] *and*
Vimal Chandra Pandey[3,*]
[1]Department of Biochemistry, University of Lucknow, Lucknow, India, [2]DST—Center for Policy Research, Babasaheb Bhimrao Ambedkar University, Lucknow, India, [3]Department of Environmental Science, Babasaheb Bhimrao Ambedkar University, Lucknow, India
*Corresponding author

1.1 Introduction

During the last century, rapid urbanization and industrialization have acutely disturbed our environmental matrices globally. Unregulated disposal of pollutants, generated via various anthropogenic activities like mining, smelting, burning of fossil fuels, use of fertilizers, military operations, and sewage have casted an irreversible negative effect on all forms of life on Earth including humans (Cristaldi et al., 2017). These environmental pollutants in the form of solid, liquid, and gaseous wastes enter the food chain through various routes of exposure and have become the major cause of increasing health issues across the globe (Muthusaravanan et al., 2018). Uncontrolled production and accumulation of environmental pollutants have critically hampered the soil health too, through soil degradation processes including erosion, salinization, and heavy-metal contamination (da Conceição Gomes et al., 2016).

Elements of the periodic table are classified as heavy metals (HMs) if their density is more than 5 g/cm^3. The majority of the transition elements, including copper, lead, zinc, mercury, arsenic, and cadmium, belong to HMs (Laghlimi et al., 2015). HMs naturally exist in the Earth's crust and are an intrinsic part of soil in many parts of the world with their profile (qualitative and quantitative measures) varying from one place to another. Some HMs (copper, manganese, zinc, etc.) have biological importance acting as micronutrients for most organisms, however, others (cadmium, lead, mercury, and arsenic) are toxic in nature with no known biological relevance (Luo et al., 2016).

The geogenic existence of HMs has never been reported to pose concern for the environment or human health. However, man-made (anthropogenic) activities like the use of agro-chemicals, over-exploitation of underground water, and waste from utensil industries have directed toward magnification of HM accumulation in soil, resulting in deleterious consequences (Clemens, 2006). It has been reported that over recent decades the global discharge of HMs into the environment has reached

Bioremediation of Pollutants. DOI: https://doi.org/10.1016/B978-0-12-819025-8.00001-6

22,000 metric ton for cadmium, 939,000 metric ton for copper, 783,000 metric ton for lead, and 1,350,000 metric ton for zinc (de Mello-Farias et al., 2011). This magnitude of enormous discharge results in accumulation of HMs in the agricultural soils and water resources, which eventually pose a threat to human health, due to potential risk of their entry into the food chain (Sarwar et al., 2017). HM contamination of soil has now turned out to be a worldwide menace not just because of unchecked production of metallic waste but also due to its immutable characteristics, persistence, and biomagnifications.

It has been recently brought forth that approximately 10 million people around the world suffer from health issues due to HM pollution in soil (Shakoor et al., 2013). China stands among the worst hit countries with HM contamination of soil. Studies report that more than 15% farmland and agricultural land area in the country have become unusable due to accumulation of HM contamination, exceeding far beyond the limit of environmental quality standards of soil. It has been found out that cadmium tops the chart of metallic contaminants (7%) in soil samples across China. Next to China, soil contamination has become an alarming issue to the European Union with 3.5 million potentially contaminated and 0.5 million highly contaminated sites seeking immediate remediation measures. Several European countries including Germany, Italy, Spain, Denmark, France, Slovakia, and many more have all been affected by soil contamination. In 2012 it was identified that approximately 600,000 ha of brown field sites in America are contaminated with HMs, giving this kind of pollution a cosmopolitan status (Mahar et al., 2016).

Restoration of sites contaminated with these hazardous and persistent pollutants requires cost effective and environmental-friendly ways of remediation. In the last two decades immense efforts have been made toward development of a range of soil-cleaning approaches which are based on physical, chemical, or biological technologies and are further divided into two categories: in situ or ex situ (Lim et al., 2014). Of these, the conventional physicochemical approaches involve soil excavation and transfer to land-filling, washing, extraction using acids, and immobilization of HMs by addition of chemicals, like limestone and EDTA (a chelating agent), to pull down their further spread in the environment (Clemens, 2006). Some of the recent and upcoming approaches include vapor-extraction, thermal desorption, and ion exchange followed by dumping waste products in landfills.

In spite of numerous conventional and recent approaches available, financial and technical challenges have made HM remediation from soil, a cumbersome process. Major limitations of these approaches include labor-intensive procedures, generation of secondary HM pollution during migration, reduction of soil fertility, destruction of natural soil micro-fauna, generation of voluminous hazardous sludge and high operational costs, restricting their exhaustive utilization in developing countries (Ali et al., 2013). According to Tsao (2003), the annual cost of global remediation efforts are between 25 and 50 billion US dollars, out of which approximately 6–8 billion US dollars are used in the United States alone.

In pursuit for an economically, environmentally, and technically feasible approach toward remediation of HMs from soil, biological remediation (bioremediation) has gained much attention in the last decade, due to its ecofriendly nature and has been considered as one of the most adequate approaches available (da Conceição Gomes et al., 2016). This approach includes varied techniques including biodegradation, bio-venting, bio-leaching, bio-augmentation, bio-filtration, bio-stimulation, and phytoremediation (Muthusaravanan et al., 2018).

Phytoremediation, also known as phytocleaning and phytocorrection, is a plant-based technology, which exploits various natural abilities of plants namely selective uptake, translocation, accumulation, and degradation of contaminants, for restoring contaminated land as well as water resources. This relatively recent approach has exhibited enormous potential by employing naturally existing or genetically engineered plants and is being considered a green alternative solution against the global hazard of HM pollution (Leguizamo et al., 2017; Pandey and Bajpai, 2019).

1.2 Heavy metals and metalloids

Environmental pollutants have been divided into two major groups, that is, organic and inorganic pollutants. The organic pollutants include halogenated hydrocarbons, polycyclic aromatic hydrocarbons, and nitroaromatics whereas HMs and metalloids are responsible for most of the inorganic pollutions. HMs, being part of the Earth's crust, are geologically found in trace quantities (less than 1 g/kg) in the soil environment (Wuana and Okieimen, 2011). Pedogenetic processes causing weathering of indigenous and sedimentary rocks lead to the existence of HMs in soil, which are rarely toxic and have also been known to be used toward human welfare (Jaishankar et al., 2014). Anthropogenic activities stand out as the leading cause of their exponential increase and categorization as pollutants for the environment (Table 1.1). Coal-based thermal power station is one of the most important anthropogenic sources that pollutes environment with a number of metal(loid)s such as Pb, As, Hg, Co, Ni, Zn, Cd, and Cr (Pandey et al., 2009, 2011).

HMs exist in various forms in the soil including free metal ions, soluble metal complexes, carbonate and silicate minerals, which make them more bioavailable for the environment (Lasat, 1999). The redox properties and chemically similar complexes of these pollutants allow them to be easily transported, compartmentalized, and integrated into cellular milieu (Bañuelos et al., 2015). HMs displace essential metals from their natural binding sites on proteins thereby disturbing the homeostasis of the cell. They accumulate in the tissue of organisms (bioaccumulation) and enter the food chain, thereby affecting other organisms that are not directly involved, leading to biomagnifications of the pollutant (Pandey et al., 2015).

1.3 Bioremediation

Bioremediation offers an environmentally sustainable way to biologically remove, degrade, or immobilize a pollutant, under controlled ambience, so as to bring down its levels below the defined regulatory limits. It employs living organisms by

Table 1.1 Sources, toxicological effects, and permissible limits of some heavy metals and metalloids.

Metals	Source	Forms of existence	Effect on plant	Effect on animal	EPA[a] limit (ppm)
Arsenic (As)	Rock weathering, mining	Inorganic and methylated forms	Oxidative stress, physiological disorders	Skin cancer, respiratory disorders	0.01
Cadmium (Cd)	Refinery, fertilizers	Marine phosphates	Reduced seed germination	Prostate cancer, testicular atrophy	5.0
Chromium (Cr)	Paints, textile	Ferrochromites	Chlorosis, reduced growth	Lung cancer, liver disease	0.1
Copper (Cu)	Polishing, plating	Sulfides and chalcocite	Oxidative stress	Kidney damage, metabolic disorder	1.3
Mercury (Hg)	Volcanic eruptions, batteries	Elemental, inorganic and methylated form	Genotoxic effects	Ataxia, blindness	2.0
Lead (Pb)	Electroplating, mining	Organic forms	Inhibited enzyme activity	Neuronal damage, nephropathy	15
Nickel (Ni)	Porcelin enamelling, silver refinery	Inorganic salts of iron and sulfur	Reduced nutrient uptake	Lung and nasal cancer	0.2
Selenium (Se)	Coal combustion, mining	Inorganic salts of sodium	Altered protein properties	Endocrine system dysfunction	50

[a]EPA, Environmental Protection Agency.

Source: Peralta-Videa, J.R., Lopez, M.L., Narayan, M., Saupe, G., Gardea-Torresdey, J., 2009. The biochemistry of environmental heavy metal uptake by plants: implications for the food chain. Int. J. Biochem. Cell Biol. 41, 1665–1677; Tchounwou, P.B., Yedjou, C.G., Patlolla, A.K., Sutton, D.J., 2012. Heavy metal toxicity and the environment. In: Molecular, Clinical and Environmental Toxicology. Springer, pp. 133–164; Dixit, R., Malaviya, D., Pandiyan, K., Singh, U., Sahu, A., Shukla, R., et al., 2015. Bioremediation of heavy metals from soil and aquatic environment: an overview of principles and criteria of fundamental processes. Sustainability 7, 2189–2212; Ayangbenro, A., Babalola, O., 2017. A new strategy for heavy metal polluted environments: a review of microbial biosorbents. Int. J. Environ. Res. Public Health 14, 94.

exploiting their natural ability, thereby reclaiming HM contaminated soil without impacting the environment. In simplified terms, bioremediation utilizes special microorganisms (bacteria and fungi) and plants that work together to breakdown pollutants present in soil or water bodies (Vijayalakshmi et al., 2018). Bioremediation can be performed by either treating the contaminants directly on the site (*in situ*) or by collecting, followed by treating them somewhere else (*ex situ*) (Ramachandran et al., 2013). Some of remediation mechanisms involved are extracellular-complexation, precipitation, leaching, intracellular-accumulation, and biosorption. Moreover, siderophores and biosurfactants produced by specialized microorganisms, act as chelating and complex forming agents respectively, thereby reducing the solubility and bioavailability of HMs from bioaccumulation (Cristaldi et al., 2017).

1.4 Phytoremediation

In this type of bioremediation approach, contaminants along with essential nutrients, are taken up and sequestered inside plants at the contaminated site. Although, a systematic economic analysis is unavailable (Wan et al., 2016), implementation of this approach has been rising considerably due to the possibility of low-cost remediation. In the past two decades enormous efforts have been made in accessing terrestrial and aquatic plant species for their bio-accumulative potential and remediation capabilities under various hazardous conditions. Phytoremediation can be further categorized according to the mechanism involved in HM uptake, which includes phytoextraction, phytofiltration, phytostabilization, phytovolatilization, and phytotransformation (Halder and Ghosh, 2014; Pandey and Bajpai, 2019). Several potential plant species (i.e., *Ipomea carnea*, *Azolla caroliniana*, *Ricinus communis*, *Typha latifolia*, *Saccharum munja*, and *Thelypterys dentata*) have been identified to remediate metal(loid)s from fly ash polluted sites or fly ash dumpsites (Pandey, 2012a,b, 2013; Pandey et al., 2012, 2014; Kumari et al., 2013).

1.5 Strategies for phytoremediation

Phytoextraction is one of the most useful techniques of phytoremediation approach and involves the removal of metals and metalloids from polluted soils through their extraction and accumulation in harvestable tissues of the plant. However, success of this technique depends critically on the properties of the metal contaminants and the soil they are present in, for example, the redox state of contaminant and bioavailability in soil (Ali et al., 2013). Phytostabilization, another useful technique of phytoremediation, runs on the mechanism of converting metal pollutants to less bioavailable forms, thereby preventing their migration from the contamination site. As the pollutant becomes stabilized in the plant rhizosphere, chances of erosion, leaching, and run-off of pollutants are minimized (Tamburini et al., 2017). Phytovolatilization, a contentious technique, works by transferring the metalloids from one environment (soil) to another environment

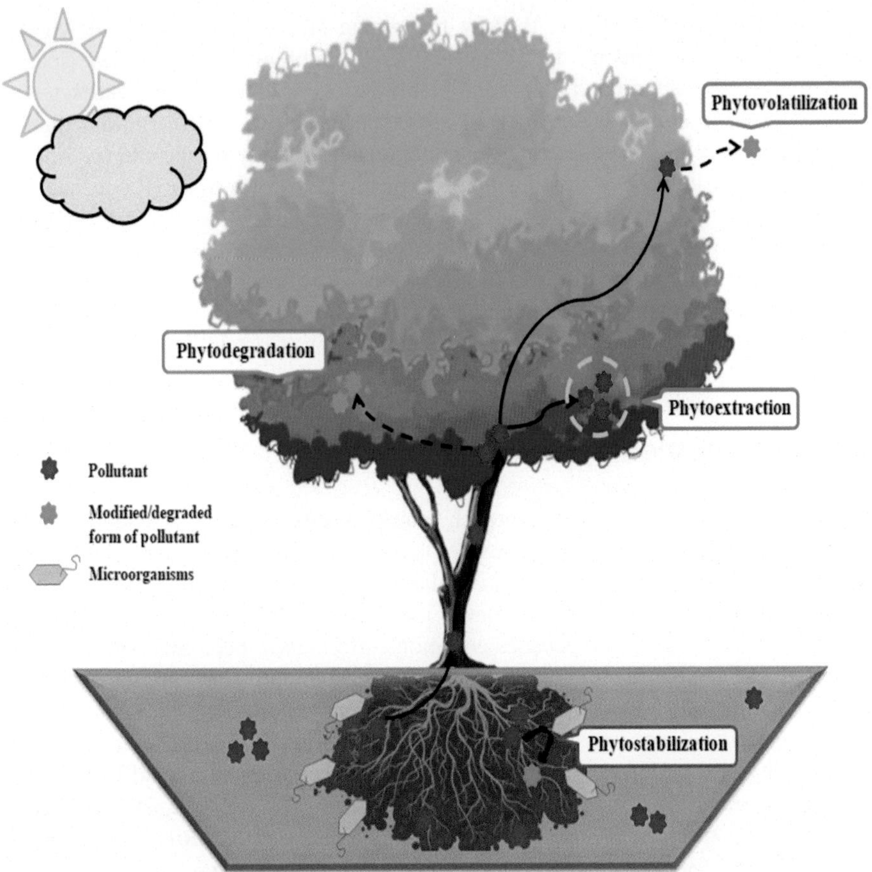

Figure 1.1 Schematic representation of various phytoremediation strategies.

(atmosphere) through converting the pollutants into volatile forms along with exchanging other necessary gases by the plant (Sarma, 2011). Phytotransformation, also termed phytodegradation, results in immobilization, deactivation, or degradation of contaminants through chemical modifications by enzymatic activity inside the plant root or shoot. This mechanism of remediation requires a comparatively longer duration of time and is many times assisted with bacterial-, yeast-, and fungal-based degradation processes in the soil (da Conceição Gomes et al., 2016) (Fig. 1.1).

1.6 Biological mechanism of heavy-metal phytoremediation

The exposure of plants to pollutants triggers detoxification strategies in the forms of diverse cellular and molecular perturbations. These strategies involve different plant parts and proceed through a more or less common pathway of uptake, transport, sequestration, and detoxification of pollutants (Pilon-Smits and Freeman, 2006). Moreover, plant–microbe interactions in the rhizosphere as well as plant-chelator-based sequestrations also form part of the biological machinery. As HMs are found in the soil, uptake of these contaminants by plants is the primary task for remediation process. For enhanced uptake of metals, plants release nutrients, organic materials, oxygen, and moisture in the soil. This not only alters the pH, osmotic potential, and redox potential but also creates a rich environment for microbial activity thereby enhancing the tolerance of plants toward pollutants and at the same time making the pollutant bioavailable for uptake by the plant (Sharma et al., 2013). HMs enter the root cells via two routes, that is, apoplastic and symplastic pathways marking the first step in phytoremediation (Salt and Rauser, 1995). Being in direct physical contact with the soil, root cells are the primary site of HM accumulation.

Next, transport of HMs involve specialized transmembrane transporters which carry out the process of moving metals inside various intracellular compartments as well as translocating them across cells to maintain homeostasis (Colangelo and Guerinot, 2006; Tangahu et al., 2011). It has been found that some plant species can carry out the initial uptake process, many times faster than other plant species, and are termed as hyperaccumulators. Although no dedicated transporters for HMs have been identified in the plant system, it is anticipated that enhanced uptake by hyperaccumulators is due to the overexpression of metal transporters (Krämer, 2010). Further, the HMs are exported from roots to aerial parts of the plant through symplastic movement and transpiration pull inside xylem vessels.

Once they reach the aerial parts of the plant, the pollutants are destined toward sequestration and detoxification. The sequestration and detoxification properties, acquired during evolution, are survival strategies of plants against toxic and detrimental effects of HMs accumulated in different plant parts. In a plant cell, the cell wall, vacuoles, and Golgi bodies form major sites of HM sequestration. Accumulation in the cell wall reduces entry of metal inside cytoplasm which decreases their metabolic interaction and further translocation. Some of the cell wall components, namely suberin and pectin, bind to HMs and sequester them in the cell wall (Chen et al., 2013). Sequestration inside vacuoles and Golgi apparatus are other means of restricting HM-mediated enzyme inactivation and other biochemical disturbances inside cells. Detoxification mechanisms also include metal chelating agents and antioxidants which act as effective means of reducing metal-induced toxicity in the cytoplasm of plant cells. Several transcriptional and translational modifications are also part of detoxification strategies at the molecular level inside plant cells (Rodriguez-Hernandez et al., 2015).

1.7 Factors affecting heavy-metal phytoremediation

Having read about various biological mechanisms, it seems evident that phytoremediation critically depends on the bioavailability of the metal for the plant, that is, phytoavailability. Different components involved in phytoremediation, including the pollutants, soil, microorganisms, plant species, and the downstream cascade of events together control this complex phenomenon of phytoavailability of HMs.

Soil properties like soil pH, texture, electrical conductivity, and the presence of organic matter directly influence the plant uptake by affecting the solubilization of metal in soil solution. The chemical form/speciation of metals in soil has a significant impact on their mobility toward plant uptake. Depending on their mobility, HMs have been classified into four groups, (1) weakly soluble in soil, (2) relatively easily absorbed through roots but sparsely transported to shoots, (3) easily absorbed and transported to shoots, and (4) potential risk to the food chain. Environmental conditions have been found to have a profound effect on accumulation mechanisms. Several reports reveal substantial influence of higher temperature, and components of sedimentary rocks (lignite and lime) in the soil, on phytoavailability. The organic soil matter also plays a significant role in metal uptake by altering the cation exchange capacity and sorption capabilities of soil. Comparative studies suggest that the addition of organic matter (compost) in soil containing HMs reduces the availability of metals for the plant (Laghlimi et al., 2015).

Rhizospheric conditions are another important factor for HM phytoavailability. Root exudates including phytosiderophores (metal solubilizing compounds), enzymes, and H^+ released by roots into the soil enhance metal solubility in the soil. Similarly, rhizospheric microorganisms have been well characterized for their ability to enhance HM bioavailability and uptake by plants (da Conceição Gomes et al., 2016). Finally, various characteristics of plant such as its growth rate in local conditions, root architecture and surface transporters, xylem loading efficiency, ability to detoxify the contaminant, and ease of maintenance in field, largely determine the uptake and accumulation of HMs. Phytoavailability of HMs therefore is critical for phytoremediation and is dependent on several diverse factors which determine the success or failure of the technique.

1.8 Plants used for phytoremediation

Selection of plant species for the process of phytoremediation is based on their ability to tolerate, accumulate, stabilize and degrade the contaminant of interest like HMs and metalloids. In the last few decades, about 500 species of plants have been identified with traits of accumulating extraordinary high levels of HMs in aerial parts. These plants termed as hyperaccumulators are being extensively exploited for HM remediation through phytomining and phytoextraction. Hyperaccumulators are capable of accumulating up to 500 times higher concentrations of HMs in different plant parts as compared to that present in soil. Herbaceous plants are the most preferred hyperaccumulating species due to their intrinsic properties of rapid growth, high accumulation capacity for HM, and adaptation on varying kind of soils (Cappa and Pilon-Smits, 2014). However, due to limited biomass, total HM accumulated in aerial parts of herbaceous plants remains very limited. Fast growing woody plants, in spite of being not-so-good hyperaccumulators as compared to herbaceous plant, offer another suitable option for effective phytoremediation due to their large aboveground biomass and extensive root system. Comparative studies have demonstrated greater advantage of using fast growing woody plants over hyperaccumulating herbaceous plants for remediating HMs from contaminated soils (Luo et al., 2016). Some studies have suggested the use of mixed vegetation, that is, combined use of herds, shrubs and trees, for sustained remediation programs in which specialized species play role at specific stage of remediation process (Laghlimi et al., 2015).

1.9 Enhancing phytoremediation

Traditional phytoremediation approaches propose a feasible way to remove HMs from soil by engaging naturally available mechanisms; however the technique has a number of limitations too. Hyperaccumulating plants need a long standing time to uptake HMs from soil and are able to remediate a limited amount of contamination available in the vicinity of their rhizosphere. Moreover, use of hyperaccumulating plants is restricted only for remediation purposes in low and moderately contaminated soils and not at highly contaminated locations. Further, the unexplored spectrum of food chain contamination due to mismanaged agronomic practices and plant predators, impose a serious threat to traditional phytoremediation techniques (Luo et al., 2016; Mahar et al., 2016). This pressing need for alternative approaches that have fast and large scale utility, led the scientific community to look into other chemical and biological possibilities. Consequently a number of recent advancements have been made toward enhancing phytoremediation techniques, which are discussed below.

1.9.1 Enhancement using chemicals

Synthetic chelating agents and surfactants like Ethylenediaminetetraacetic acid (EDTA), Ethylenebis(oxyethylenenitrilo)tetraacetic acid (EGTA), 1,2-cyclohexylenedinitrilotetraacetic acid (CDTA), Diethylenetriaminepentaacetic acid (DTPA), and citric acid have been shown to enhance the absorption rate of HMs by making them more bioavailable for plants. An increase in the mobility of metals in soil due to these chemicals, compensates for the relatively low absorption capacity of non-hyperaccumulators thereby improving their overall uptake potential and utility. As of now the use of chemical agents is restricted due to a number of related issues including large expenses, environmental hazards, and plant toxicity (Souza et al., 2013).

1.9.2 Enhancement using agronomic techniques

Various agronomic techniques like using soil additives, adopting of specific cultivation practices, and employing energy crops, are other means of improving phytoremediation. Soil additives like fertilizers, biochars, and inoculation with microbes, are known to increase the plant growth and biomass which tends toward increased uptake and accumulation of HMs from soil (Liu et al., 2013). Land farming practices devised for improved light and air exposure to soil are another means of escalating plant growth. Energy crops also offer a means of enhancing the phytoremediation practices. These have been identified as perennial crops which can be used for phytoremediation purposes as well as bioenergy production. Some of the examples of energy crops are *Populous* and *Jatropha* (Bauddh and Singh, 2012).

1.9.3 Enhancement using transgenic plants

A lot of scientific efforts have gone toward plants that can generate huge biomass under hazardous soil conditions and also uptake, translocate, sequester, and detoxify HMs in their aerial parts. Genetic engineering offers a completely unexplored horizon for enhancing the capability of plants toward remediation of HMs from soil. Altering the genetic structure of hyperaccumulator plants by over-expressing native genes/introducing foreign genes to enhance their tolerance and accumulation capabilities has turned out to be a promising strategy toward improving phytoremediation (Cherian and Oliveira, 2005). In the past decade, understanding regulatory control of uptake and translocation, identification of transporter proteins and metal-chelators, as well as unique genes from other organisms through molecular biology tools has empowered scientists to develop effective and economic transgenic plants which are better equipped for phytoremediation (Kotrba et al., 2009; Fasani et al., 2018).

1.10 Advantages and disadvantages of phytoremediation

A number of studies suggest phytoremediation as the best alternative for removing contaminants from soil, however there are certain issues that need to be taken care of before embarking on the approach. Numerous advantages of using phytoremediation approaches have been discussed in literature including its inexpensive installation, capability to treat large area, generation of recyclable metal-enriched plant product, prevention against soil erosion, and its aesthetic character making it acceptable for society (Pandey and Bajpai, 2019; Pandey and Souza-Alonso, 2019). However, it is also equally important to analyze the concurrent disadvantages related with the approach. It has been indicated through a number of studies that phytoremediation approaches show typical limitations that create bottlenecks in the success of these measures. The prime limitation of phytoremediation is the incomplete removal of contaminants from soil because plant roots can only uptake contaminants present in their vicinity and cannot reach the deep levels of soil. Another major limitation of phytoremediation is the long duration of time needed by plants to extract contaminants from soil (Leguizamo et al., 2017). Moreover, it has been observed in many cases that soil contains multiple contaminants at the same time whereas the plants employed for remediation in such soil may not be tolerant to or hyperaccumulator for all the contaminants thereby making the process ineffective (de Mello-Farias et al., 2011; Muthusaravanan et al., 2018).

1.11 Conclusion

Anthropogenic accumulation of HMs in soil is a serious environmental concern worldwide, demanding immediate implementation of effective remediation efforts. Conventional physiochemical remediation methods are fast and efficient, however they are expensive and they also cause environmental hazards if used at large scales. Phytoremediation strategies are quite recent and show an unprecedented potential for removing HMs and metalloids contaminations from soil. Phytoremediation is considered a "green technology" as it does not employ harmful substances and can be economically utilized at large scales for long durations (Laghlimi et al., 2015).

Hyperaccumulator plants have been demonstrated to effectively extract HMs and metalloids from soil, however their dependence on various environmental factors like soil pH, moisture, micro-fauna, and type of metal species cripples down the efficiency. This drawback calls for larger efforts toward understanding of mechanisms behind uptake, accumulation, and detoxification processes inside plants. Moreover, efforts need to be made to identify critical interactions between rhizospheric components, that is, roots, microbes, soil, and HM contaminants, so as to make the uptake process more efficient (Sarwar et al., 2017). The transgenic approach for enhancing phytoremediation, proposes a completely different dimension of research where characterization, transformation, and overexpression of key genes involved in HM hyperaccumulation, can be used to maximize remediation process. Along with collective endeavors toward effective and efficient phytoremediation techniques, it is of equal importance to manage the disposal strategies of enriched toxicants as well as, critical assessment of cost−benefit ratio keeping in mind the social aspects related to it.

Acknowledgments

SP and AVA acknowledge DST for NPDF as well as STI-PDF grant. Financial assistance given to Dr. V.C. Pandey under the Scientist's Pool Scheme [Pool No. 13 (8931-A)/2017] by the Council of Scientific and Industrial Research, Government of India is gratefully acknowledged.

Conflict of interest

The authors declare no conflicts of interest.

References

Ali, H., Khan, E., Sajad, M.A., 2013. Phytoremediation of heavy metals—concepts and applications. Chemosphere 91, 869–881.

Ayangbenro, A., Babalola, O., 2017. A new strategy for heavy metal polluted environments: a review of microbial biosorbents. Int. J. Environ. Res. Public Health 14, 94.

Bañuelos, G.S., Arroyo, I., Pickering, I.J., Yang, S.I., Freeman, J.L., 2015. Selenium biofortification of broccoli and carrots grown in soil amended with Se-enriched hyperaccumulator *Stanleya pinnata*. Food Chem. 166, 603–608.

Bauddh, K., Singh, R.P., 2012. Cadmium tolerance and its phytoremediation by two oil yielding plants *Ricinus communis* (L.) and *Brassica juncea* (L.) from the contaminated soil. Int. J. Phytoremediat. 14, 772–785.

Cappa, J.J., Pilon-Smits, E.A., 2014. Evolutionary aspects of elemental hyperaccumulation. Planta 239, 267–275.

Chen, G., Liu, Y., Wang, R., Zhang, J., Owens, G., 2013. Cadmium adsorption by willow root: the role of cell walls and their subfractions. Environ. Sci. Pollut. Res. 20, 5665–5672.

Cherian, S., Oliveira, M.M., 2005. Transgenic plants in phytoremediation: recent advances and new possibilities. Environ. Sci. Technol. 39, 9377–9390.

Clemens, S., 2006. Toxic metal accumulation, responses to exposure and mechanisms of tolerance in plants. Biochimie 88, 1707–1719.

Colangelo, E.P., Guerinot, M.L., 2006. Put the metal to the petal: metal uptake and transport throughout plants. Curr. Opin. Plant Biol. 9, 322–330.

Cristaldi, A., Conti, G.O., Jho, E.H., Zuccarello, P., Grasso, A., Copat, C., et al., 2017. Phytoremediation of contaminated soils by heavy metals and PAHs. A brief review. Environ. Technol. Innov. 8, 309–326.

da Conceição Gomes, M.A., Hauser-Davis, R.A., de Souza, A.N., Vitória, A.P., 2016. Metal phytoremediation: general strategies, genetically modified plants and applications in metal nanoparticle contamination. Ecotoxicol. Environ. Saf. 134, 133–147.

de Mello-Farias, P.C., Chaves, A.L.S., Lencina, C.L., 2011. Transgenic plants for enhanced phytoremediation—physiological studies. Genetic Transformation. IntechOpen.

Dixit, R., Malaviya, D., Pandiyan, K., Singh, U., Sahu, A., Shukla, R., et al., 2015. Bioremediation of heavy metals from soil and aquatic environment: an overview of principles and criteria of fundamental processes. Sustainability 7, 2189–2212.

Fasani, E., Manara, A., Martini, F., Furini, A., DalCorso, G., 2018. The potential of genetic engineering of plants for the remediation of soils contaminated with heavy metals. Plant Cell Environ. 41, 1201–1232.

Halder, S., Ghosh, S., 2014. Wetland macrophytes in purification of water. Int. J. Environ. Sci. 5, 432–437.

Jaishankar, M., Tseten, T., Anbalagan, N., Mathew, B.B., Beeregowda, K.N., 2014. Toxicity, mechanism and health effects of some heavy metals. Interdiscip. Toxicol. 7, 60–72.

Kotrba, P., Najmanova, J., Macek, T., Ruml, T., Mackova, M., 2009. Genetically modified plants in phytoremediation of heavy metal and metalloid soil and sediment pollution. Biotechnol. Adv. 27, 799–810.

Krämer, U., 2010. Metal hyperaccumulation in plants. Annu. Rev. Plant Biol. 61, 517–534.

Kumari, A., Pandey, V.C., Rai, U.N., 2013. Feasibility of fern *Thelypteris dentata* for revegetation of coal fly ash landfills. J. Geochem. Explor. 128, 147–152.

Laghlimi, M., Baghdad, B., El Hadi, H., Bouabdli, A., 2015. Phytoremediation mechanisms of heavy metal contaminated soils: a review. Open J. Ecol. 5, 375.

Lasat, M., 1999. Phytoextraction of metals from contaminated soil: a review of plant/soil/metal interaction and assessment of pertinent agronomic issues. J. Hazard Subst. Res. 2, 5.

Leguizamo, M.A.O., Gómez, W.D.F., Sarmiento, M.C.G., 2017. Native herbaceous plant species with potential use in phytoremediation of heavy metals, spotlight on wetlands—a review. Chemosphere 168, 1230–1247.

Lim, K., Shukor, M., Wasoh, H., 2014. Physical, chemical, and biological methods for the removal of arsenic compounds. Biomed. Res. Int. 2014.

Liu, X., Zhang, A., Ji, C., Joseph, S., Bian, R., Li, L., et al., 2013. Biochar's effect on crop productivity and the dependence on experimental conditions—a meta-analysis of literature data. Plant Soil 373, 583–594.

Luo, Z.-B., He, J., Polle, A., Rennenberg, H., 2016. Heavy metal accumulation and signal transduction in herbaceous and woody plants: paving the way for enhancing phytoremediation efficiency. Biotechnol. Adv. 34, 1131–1148.

Mahar, A., Wang, P., Ali, A., Awasthi, M.K., Lahori, A.H., Wang, Q., et al., 2016. Challenges and opportunities in the phytoremediation of heavy metals contaminated soils: a review. Ecotoxicol. Environ. Saf. 126, 111–121.

Muthusaravanan, S., Sivarajasekar, N., Vivek, J., Paramasivan, T., Naushad, M., Prakashmaran, J., et al., 2018. Phytoremediation of heavy metals: mechanisms, methods and enhancements. Environ. Chem. Lett. 16, 1339–1359.

Pandey, V.C., 2012a. Phytoremediation of heavy metals from fly ash pond by *Azolla caroliniana*. Ecotoxicol. Environ. Saf. 82, 8–12.

Pandey, V.C., 2012b. Invasive species based efficient green technology for phytoremediation of fly ash deposits. J. Geochem. Explor. 123, 13–18.

Pandey, V.C., 2013. Suitability of *Ricinus communis* L. cultivation for phytoremediation of fly ash disposal sites. Ecol. Eng. 57, 336–341.

Pandey, V.C., Bajpai, O., 2019. Phytoremediation: from theory towards practice. In: Pandey, V.C., Bauddh, K. (Eds.), Phytomanagement of Polluted Sites. Elsevier, Amsterdam, pp. 1–49. Available from: https://doi.org/10.1016/B978-0-12-813912-7.00001-6.

Pandey, V.C., Souza-Alonso, P., 2019. Market opportunities in sustainable phytoremediation. In: Pandey, V.C., Bauddh, K. (Eds.), Phytomanagement of Polluted Sites. Elsevier, Amsterdam, pp. 51–82. Available from: https://doi.org/10.1016/B978-0-12-813912-7.00002-8.

Pandey, V.C., Abhilash, P.C., Singh, N., 2009. The Indian perspective of utilizing fly ash in phytoremediation, phytomanagement and biomass production. J. Environ. Manag. 90, 2943–2958.

Pandey, V.C., Singh, J.S., Singh, R.P., Singh, N., Yunus, M., 2011. Arsenic hazards in coal fly ash and its fate in Indian scenario. Resour. Conserv. Recy. 55, 819–835.

Pandey, V.C., Singh, K., Singh, R.P., Singh, B., 2012. Naturally growing *Saccharum munja* on the fly ash lagoons: a potential ecological engineer for the revegetation and stabilization. Ecol. Eng. 40, 95–99.

Pandey, V.C., Singh, N., Singh, R.P., Singh, D.P., 2014. Rhizoremediation potential of spontaneously grown *Typha latifolia* on fly ash basins: study from the field. Ecol. Eng. 71, 722–727.

Pandey, V.C., Pandey, D.N., Singh, N., 2015. Sustainable phytoremediation based on naturally colonizing and economically valuable plants. J. Clean. Prod. 86, 37–39.

Peralta-Videa, J.R., Lopez, M.L., Narayan, M., Saupe, G., Gardea-Torresdey, J., 2009. The biochemistry of environmental heavy metal uptake by plants: implications for the food chain. Int. J. Biochem. Cell Biol. 41, 1665–1677.

Pilon-Smits, E.A., Freeman, J.L., 2006. Environmental cleanup using plants: biotechnological advances and ecological considerations. Front. Ecol. Environ. 4, 203–210.

Ramachandran, P., Sundharam, R., Palaniyappan, J., Munusamy, A.P., 2013. Potential process implicated in bioremediation of textile effluents: a review. Adv. Appl. Sci. Res. 4, 131–145.

Rodriguez-Hernandez, M., Bonifas, I., Alfaro-De la Torre, M., Flores-Flores, J., Bañuelos-Hernández, B., Patiño-Rodríguez, O., 2015. Increased accumulation of cadmium and lead under Ca and Fe deficiency in *Typha latifolia*: a study of two pore channel (TPC1) gene responses. Environ. Exp. Bot. 115, 38–48.

Salt, D.E., Rauser, W.E., 1995. MgATP-dependent transport of phytochelatins across the tonoplast of oat roots. Plant Physiol. 107, 1293–1301.

Sarma, H., 2011. Metal hyperaccumulation in plants: a review focusing on phytoremediation technology. J. Environ. Sci. Technol. 4, 118–138.

Sarwar, N., Imran, M., Shaheen, M.R., Ishaque, W., Kamran, M.A., Matloob, A., et al., 2017. Phytoremediation strategies for soils contaminated with heavy metals: modifications and future perspectives. Chemosphere 171, 710–721.

Shakoor, M.B., Ali, S., Farid, M., Farooq, M.A., Tauqeer, H.M., Iftikhar, U., et al., 2013. Heavy metal pollution, a global problem and its remediation by chemically enhanced phytoremediation: a review. J. Bio. Env. Sci. 3, 12–20.

Sharma, R., Sharma, K., Singh, N., Kumar, A., 2013. Rhizosphere biology of aquatic microbes in order to access their bioremediation potential along with different aquatic macrophytes. Recent Res. Sci. Technol. 5.

Souza, L.A., Piotto, F.A., Nogueirol, R.C., Azevedo, R.A., 2013. Use of non-hyperaccumulator plant species for the phytoextraction of heavy metals using chelating agents. Sci. Agric. 70, 290–295.

Tamburini, E., Sergi, S., Serreli, L., Bacchetta, G., Milia, S., Cappai, G., et al., 2017. Bioaugmentation-assisted phytostabilisation of abandoned mine sites in south west Sardinia. Bull. Environ. Contam. Toxicol. 98, 310–316.

Tangahu, B.V., Abdullah, S., Rozaimah, S., Basri, H., Idris, M., Anuar, N., et al., 2011. A review on heavy metals (As, Pb, and Hg) uptake by plants through phytoremediation. Int. J. Chem. Eng. 2011.

Tchounwou, P.B., Yedjou, C.G., Patlolla, A.K., Sutton, D.J., 2012. Heavy metal toxicity and the environment. Molecular, clinical and environmental toxicology. Springer, pp. 133–164.

Tsao, D.T., 2003. Overview of phytotechnologies. Phytoremediation. Springer, pp. 1–50.

Vijayalakshmi, V., Senthilkumar, P., Mophin-Kani, K., Sivamani, S., Sivarajasekar, N., Vasantharaj, S., 2018. Bio-degradation of bisphenol A by *Pseudomonas aeruginosa* PAb1 isolated from effluent of thermal paper industry: kinetic modeling and process optimization. J. Radiat. Res. Appl. Sci. 11, 56–65.

Wan, X., Lei, M., Chen, T., 2016. Cost–benefit calculation of phytoremediation technology for heavy-metal-contaminated soil. Sci. Total Environ. 563, 796–802.

Wuana, R.A., Okieimen, F.E., 2011. Heavy metals in contaminated soils: a review of sources, chemistry, risks and best available strategies for remediation. ISRN Ecol. 2011.

Role of potential native weeds and grasses for phytoremediation of endocrine-disrupting pollutants discharged from pulp paper industry waste

2

Pooja Sharma[1], *Surendra Pratap Singh*[2,*], *Siddhartha Pandey*[3], *Arti Thanki*[4] *and Nitin Kumar Singh*[4]

[1]Department of Environmental Microbiology, School for Environmental Sciences, B.B. Ambedkar Central University, Lucknow, India, [2]Department of Botany, D.A.V. College, Chhatrapati Shahu Ji Maharaj University (CSJM University), Kanpur, India, [3]Department of Civil Engineering, Chalapathi Institute of Technology, Guntur, India, [4]Department of Environmental Science and Engineering, Marwadi Education Foundations Group of Institutions, Rajkot, India

*Corresponding author.

2.1 Introduction

Due to rapid industrialization of, that is, pulp paper industry, distillery industry, tannery industry development discharge, there is a huge amount of wastewater and solid waste laden with different residual organic and inorganic pollutants along with heavy metals (Hewitt et al., 2008). Among these industries, the pulp paper industry has been ranked sixth in the world, in terms of its pollution potential. During the manufacturing process of the pulp paper industry, a significant amount of hazardous and toxic elements is generated (Ugurlu et al., 2007). Approximately, 411 million tons of paper is generated worldwide in 2017, and contribution is dominated mainly by the United Nations, European Union, India, Brazil, and China (Demirel and Altin, 2017). The Environmental Protection Agency of the United States also reported that over 250 million tons of waste is discharged every year, and out of that ~30% is contributed only by the pulp paper industry, especially after secondary treatment. More specifically, ~0.4 tons of pulp paper manufacturing industrial waste is generated per each ton of paper produced (Toczylowska-Maminska, 2017). As per recent statistics, around 6 million tons of paper are produced from 515 paper mills in India. Various agro and forest origin raw materials such as wheat, rice, bagasse, and rice straw, etc. are used for these mills. Beside these materials, significant amounts of various commercial chemicals (organic/inorganic solvents, sodium hydroxide, and chlorine compounds, etc.) are also used in

Bioremediation of Pollutants. DOI: https://doi.org/10.1016/B978-0-12-819025-8.00002-8

different processing stages of paper and pulp mills. The major processing stages which contribute pollutants to secondary effluents, are debarking, washing of pulp, and bleaching, etc. (Ali and Sreekrishnan, 2001). These wastes are laden with an enormous high quantity of heavy metals and other hazardous pollutants, thus may cause severe pollution to the environment, ecosystem, and living organisms. These pollutants are also reported to be responsible for mutagenicity and carcinogenicity in humans and animals. Moreover, these substances have a tendency to accumulate in the soil, water, and plant environment (Sarwar et al., 2010). The discharged wastewaters from the pulp paper industry are rich in color and have an unpleasant smell due to the presence of lignin ions and chlorinated compounds. Such compounds include Biological oxygen demand (BOD) and Chemical oxygen demand (COD) exerting substances, suspended solids, tannins, resin acids, sulfur compounds, and lignin, etc. (Pokhrel and Viraraghavan, 2004). Previous investigations revealed the presence of more than 250 types of chemicals in pulp paper industry waste, which also include resin acids and sterols. At present, although various treatment technologies have been proven effective for wide variety of wastewater in the context of pulp paper industrial waste treatment systems, some pollutants still pass through and consequently are released into the environment. In the last few decades, phytoremediation techniques have emerged as one of the best technologies for the management of potential toxic elements of pulp paper industry waste contaminated sites (Gupta and Sinha, 2006; Mani and Kumar, 2014). In this direction, the luxurious growth of some native plants on disposal site of pulp paper industry revealed the potential for phytoextraction of pollutants and bioremediation of complex hazard compounds containing metals and persistent organic pollutants (POPs) laden sludges (Chandra et al., 2017). Some of the limiting constraints in sustainable applications of these plants include mobility of trace metals, their specific forms, and binding potential with other organic copollutants. Besides, the condition of soil and plant growth is also among the pollutant regulating factors in such systems (Rosselli et al., 2003). Recent research highlighted that native plants and/or plant grown on waste disposal sites may play a significant role in phytoremediation of pulp and paper wastes. Such plants are reported to have the capability to grow in a harsh environment, even in the presence of hazardous pollutants with appreciable survival and reproduction. These evidences have led to the exploration of potential native plants which could sustain in the presence of hazardous pollutants, that is, metals, organic/inorganic elements, and POPs, etc. (Shu et al., 2002; McGrath and Zhao, 2003). A literature review revealed that native weeds and grasses (cattail, *Typha latifolia*; common reed, *Phragmites australis*; and motha, *Cyperus malaccensis*), grown in a complex organometallic pollutant environment, have the potential ability to accumulate heavy metals in their tissues (Ye et al., 2001; Deng et al., 2004). Experimental investigations of secondary effluents, collected from the pulp paper industry, revealed that these streams are rich in residual organic pollutants. The commonly observed compounds including 4-isopropoxybutyric acid, butane-1-ol, hexahydropyrrole (1,2A)pyrazine-1,4-dione, 6-chlorohexanoic acid, 1-chlorooctadane,

1,2-benzene carboxylic acid, 2-hydroxymethylcyclopropane carboxylic acid, 2-methoxyphenol phthalic anhydride, 2,6-dimethoxyphenol 2-methoxy-4-ethylphenol, 3-allyl-6-methoxyphenol, and 2-methoxy-4(1-propenyl) (Chandra and Singh, 2012; Chandra et al., 2017, 2018). Besides the above substances, furans and some methylated compounds were reported to be appear in secondary effluents of the pulp paper manufacturing industry. The major contributors of these substances include bleaching or pulping stage of industrial wastewater treatment system (Chandra et al., 2017, 2018). Other pollutants like POPs such as hexadecanoic acid, tetradecanoic acid, pentadecanoic acid, and hexadecane are also identified from the sludge of the pulp paper industry. These are basically phytosterols of plants and are detected from humic substances also (Reveille et al., 2003). Further, these compounds are listed under endocrine-disrupting chemicals (EDCs), as per the US Environmental Protection Agency and Endocrine Disruptor Screening Program (USEPA and EDSP) (2012). In this chapter, attempts were made to analyze the toxicity effects of paper pulp industry waste, which contain EDCs, carcinogenic, and mutagenic compounds.

2.2 Physicochemical analysis of wastewater of pulp paper industry

The wastewater of the paper pulp industry contains a high amount of potentially toxic pollutants along with heavy metals. The physicochemical properties of pulp paper industry waste showed a high concentration of Total suspended solids (TSS), Total dissolved solids (TDS), COD, and BOD. These values were observed to be well above the acceptable limits. Typical characteristics of such secondary effluents are presented in Table 2.1.

These waste streams also contain varying concentrations of ions such as sodium, potassium, and chloride, etc. Some environmentally hazardous and accumulating heavy metals like Fe, Zn Cu, Cr, Cd, Mn, and Ni also release different unit processes of the manufacturing plants. All these parameters were drastically reduced after potential bacterial treatment in the bio-stimulation and bioaugmentation process. Some potential plants, that is, *Parthenium* sp., *Alternanthera* sp., *Cannabis sativa*, *P. australis*, and *Cynodon dactylon* are reported for high accumulation and remediation of pulp and paper industry waste. The combined residual effect of pulp paper waste raises the alkalinity of waste streams which may be linked with the presence of bicarbonates, carbonates, and hydroxides. These compounds are used at various stages of processing for pH adjustment (Tiwari et al., 2013). The presence of chlorolignins, cellulose, and lignin fibers material was also observed in discharged secondary effluents of pulp and paper industries. Other than these compounds, some salts and chloride are released from these industries, which are expected to be released due to byproduct reaction of sodium sulfide during bleaching and pulping process.

Table 2.1 Physico-chemical analysis of discharged pulp paper industry waste after secondary treatment.

SN	Parameters	Pulp paper waste values (mean)	Permissible limit (EPA 2002)
1	pH	8.1 ± 7.0	5–9
2	Color	2510 ± 2324	Dark Brown
4	Total solid (mg L^{-1})	671 ± 588	-
5	TDS (mg L^{-1})	561 ± 620	30
7	TSS (mg L^{-1})	58 ± 63	35
8	COD (mg L^{-1})	17,991 ± 16,548	120
9	BOD (mg L^{-1})	6159 ± 6258	40
10	Total phenols (mg L^{-1})	489 ± 527	0.50
11	Total nitrogen (mg L^{-1})	246 ± 254	143
12	Sulphate (mg L^{-1})	1954 ± 1896	250
13	Phosphorus (mg L^{-1})	364 ± 268	2001
14	Cl^{-} (mg L-1)	2.44 ± 3.45	1500
15	Na^{+} (mg L-1)	71 ± 79	200
16	K^{+} (mg L-1)	7.6 ± 8.3	–
17	Lignin (ppm)	47,540 ± 48,952	–
18	Chlorophenol (mg L^{-1})	203 ± 209	3.0
Heavy metals (mg L^{-1})			
19	Fe	68.54 ± 71.54	2.00
20	Zn	13.91 ± 14.59	2.00
21	Cu	2.19 ± 2.28	0.50
22	Cr	2.31 ± 2.68	0.05
23	Cd	0.255 ± 0.548	0.01
24	Mn	11.2 ± 12.5	0.20
25	Ni	3.31 ± 3.54	0.10

All the values are means of triplicate (n = 3) ± SD. Unit of all parameters are in mgl^{-1} except pH, and color (Co-Pt. Unit) (Chandra et al., 2017; 2018).

2.3 Endocrine-disrupting pollutants from pulp paper industry waste

Pollutants released from pulp paper industry waste contain several residual organic pollutants and EDCs compounds, even after secondary treatment. These are well-known carcinogenic, mutagenic, and androgenic in nature compounds which are considered to be one of the highest discharged pollutants from industries. Moreover, the category of pollutants, that is, organic, inorganic, and gaseous pollutants is based on the pulp paper industry waste and completely depends on the particular industry product as well as its raw material. In particular, waste production depends mainly on the pulping and bleaching stages which can be considered as the main sources of environmental pollution. With respect to the POPs, these substances are possibly generated either during the alkali pulping with Na_2S or through bacterial biotransformation during the treatment of wastewater. Along with these, plant origin fatty acids, that is, tetradecanoic acid and hexadecanoic acid, have also been reported as humic substances in such kind of wastes (Reveille et al., 2003). US Environmental Protection Agency and Endocrine Disruptor Screening Program (USEPA and EDSP) (2012) also listed these compounds under the category of EDCs. Other detected compounds in such waste include sulfurous acid, phthalic acid, which are also included in the EDC category. Research studies revealed that in chloroform extracted samples, several phenolic and nonphenolic compounds, that is, decane 1-bromo-2-methyl, pentadecanone, 2-pentadecanone, benzene dicarboxylic acid dibutyl phthalate, 1-decanol-2-hexyl, hexadecanoic acid, trimethylsilyl ester, β-sitosterol trimethylsilyl ether, pentadecanoic acid, ethyl ester, 2-methyl-4-keto-2-pentane-2-of 1TMS, octadecenoic acid, and trimethylsilyl ester are also present in such wastewater. Typical quantities of these compounds are shown in Table 2.2.

Wastewater discharged from the pulp paper industry is a major problem today, hence there is a need for urgent detoxification of these pollutants laden streams. In the paper making and pulping process, a huge quantity of fresh water, varying in the range of 5–300 m^3/ton of pulp products, is consumed. In general, the average size of the mill produces about 2000 m^3 of effluent per day. A research conducted in France reported that ~30% of the industrial pollution of surface water sources is contributed by the wood, pulp, and paper industries. These industries contribute 13%, 10%, and 4%, respectively, of discharged manganese, cadmium, and arsenic. It is also reported that more than 500 organic compounds have been identified in secondary effluents. These concerns led to the exploration of more reliable and robust technologies to remove a large number of pollutants. Some of the identified carcinogenic compounds include 9,12-octadecadienoic acid (*Z,Z*)-2,3-dihydroxy propyl ester, tetradecanoic acid methyl ester, and 2-methoxy phenol. The conventional biological processes do not explicitly reduce the complex organometallic compound of pulp paper industry sludge, as these processes are very slow. In this regard, some hyperaccumulators plants are discussed by Chandra et al. (2017) for detoxification of various pollutants, that is, *Solanum nigrum*, *Rumex dentatus*, *Ranunculus sceleratus*, *Alternanthera philoxeroides*, *Parthenium dermatitis*, *C. Sativa*, *Phragmites communis*, and *Ricinus* which are highly accumulative for heavy metals. Various chromatographic methods, that is, Gas chromatography–mass spectrometry (GC-MS), Liquid chromatography-mass spectrometry (LC-MS), High Performance Liquid Chromatography (HPLC) analysis, are mainly used for detection and identification of organic pollutants present in pulp paper industry wastes.

2.3.1 Lignin and chlorophenolic compounds

Historically, the pulp paper industry has been a major consumer of natural resources, that is, fossil fuels, electricity for energy, wood, and water; consequently, also a major contributor to environmental pollutant discharges. The commonly associated pollutants of pulp and paper mill effluents include pesticides, phenols, lignin, fatty acids, resin acid, lignin, and other chlorinated compounds. These

Table 2.2 Identified persistent organic pollutants by Gas chromatography–mass spectrometry (GC-MS) in pulp paper industry waste after secondary treatment.

Identified compounds	Toxicity
2-Butoxyethanol	Aquatic toxicity
Thymol-TMS	Aquatic toxicity
2,3,6-Trimethyl phenol	Data not available
Benzoic acid, trimethylsilyl ester	Data not available
Citral	Aquatic toxicity
2′,6′-Dihydroxyacetophenone, bis(trimethylsilyl) ether	Unknown
2-Methoxyphenyl or guaiacol	Data not available
Phenol, 2,6-dimethoxy	Unknown
Pthalatic anhydride	Data not available
9-Decanoic acid, trimethylsilyl ester	Unknown
Benzyldehyde, 4-(acetyloxy)-3-methoxy	Unknown
Octadecanoic acid, trimethylsilyl ester or stearic acid	EDCs
1,2-Benzenedicarboxylic acid, bis(2-ethylhexyl) ester	Unknown
Acetic acid [(trimethylsilyl)oxy]	Unknown
Methoxy cinnamic acid	EDCs
9,12-Octadecadienoic acid, (2-phenyl-1,3-dioxolan-4-yl) methyl ester trans	EDCs
n-Pentadecanoic acid, trimethylsilyl ester	Unknown
2,6-bis[trimethylsilyl]-3,4-dimethylphosphinine	Unknown
Hexadecanoic acid, trimethylsilyl ester	Aquatic toxicity
Pentadecane	Unknown
Octadecanoic acid	EDCs
Cinnamic acid-α-phenyl-trimethylsilyl ester	Aquatic toxicity
Cis,13-docosenoic acid	Unknown
9-[2,6-Diethylphenyl]2,8-dimethyl-9-*h*-purin-6-amine	Aquatic toxicity
2-Monopalmitin TMS ether	Unknown
1,2,-Diphenyl-s (t-butyl) acephenanthrylene	Data not available
Octacosane	EDCs
Nonacosane	Aquatic toxicity

EDC, Endocrine-disrupting chemical.

chlorophenol compounds are contributed from the pulping and bleaching process. Butler and Dal Pont (1992) reported that chlorine and its derivatives also form derived chlorophenolic compounds. These compounds are formed particularly due to breakdown of lignin in pulping process. The formation of these compounds depends mainly on the ratio of chlorine dioxide and chlorine, as well as on the types of woods used in the process. These chlorophenols may also act as precursors of highly toxic compounds such as dioxins and furans. The natural properties of chlorophenols make them lipophilic, and therefore tend to absorb on solids and accumulate in soils, sediments, and sludge. During the chemical processing of wood, the paper and pulp industry release high amounts of lignin, which persists in the environment even after the secondary treatment. Lignin is the product of a pulping process based on kraft in paper mills. This is mainly used as black liquor in the pulp mill for power generation, processing steam, and recovery. Craig et al. (1986) reported about the pulp and paper mill disposal site which highlighted that lignin and its derivatives not only give color but also cause an unpleasant smell. In effluent, when lignin exceeds the permissible limits, it inhibits the photosynthesis process by absorbing the solar light. Bio-stimulation and bioaugmentation processes reported by Chandra et al. (2009) are used to remove color and organic compounds at different environmental conditions. Next to cellulose, the lignin is the second most commonly used natural organic polymer in paper and pulp mills. Along with natural lignin, other derivatives such as chlorinated lignin are also used in pulping and bleaching processes (Ali and Sreekrishnan, 2001).

2.3.2 Colored and gaseous pollutants from the pulp paper industry

After secondary treatment, the discharged waste, typically having dark brown color, might depend on the raw material of industry, lignin, and its derivatives. During the pulping and bleaching step, coloring compounds exerts TSS, TDS, ions (K^+, Na^+, and Cl^-), and lignin, along with other organic polymer compounds and polymerized tannin acid, fatty acid, and plant derivatives. After the bleaching process, the dark brown color is converted to a bright color through pH adjustment in natural condition. The origin compounds, responsible for such color, are lignin, its derivatives, and tannins which are a well-known toxic substance to microorganisms and aquatic lives. In addition, the discharged waste also contains potential gaseous pollutants including dioxin, dioxin-like compounds, and furans in the environment. These pollutants are already listed in the category of highly hazardous organic pollutants. In particular, halogenated aromatic hydrocarbons such as dibenzofurans, polychlorinated dibenzodioxins, and biphenyls are of great concern nowadays. These compounds are lipophilic in nature, and have resistance for biological and chemical degradation (Vallejo et al., 2015). The common compounds used as a precursor in this industry are generated from the wood treatment, where it works as a bacterial and fungicidal agent (Vallejo et al., 2015). Due to low solubility, although the natural concentrations of dioxins and furans generally are low, it can enter into the water environment through effluent released from the sewage/municipal treatment plants. Among various wastewater treatment methods, solar light-based photocatalysis methods have been reported effective in the removal or degradation of dioxins and furans.

2.3.3 Fatty acids, resin acids, and extractives in wastewater

A typical paper pulp industry waste is also known for rich concentrations of various types of wood extracts which are observed during the paper making process. Further, the extractive type also depends on the wood species, various parts, and growth conditions. For example, trees grown in warm climates are able to produce appreciable amounts of fatty acid with less variation throughout the season. Among various process streams, secondary effluents and wasted sludge are the most dominating sources of wood extracts such as alcohols, terpenes, and low molecular weight carboxylic acids. These compounds can be extracted through organic solvents such as dichloromethane, methanol, acetone, diethyl ether, ethanol, etc. Furthermore, these extracted can also be removed/degraded or transformed/adsorbed by the activated sludge of bioreactors. Similarly, fats and waxes are defined as esters of carboxylic acids or fatty acid with glycerol or alcohols (Ali and Sreekrishnan, 2001). The number of dry wood fats was observed to be in the range of 0.3%–0.4%. With respect to the wood type, the heartwood is found to have greater content than sapwood. The extractives used in paper and pulp mill processes also contain fatty acid. To extract this fat and waxes from wood, organic solvents such as diethyl ether and acetone etc. are mainly used. So far, around 20 saturated/unsaturated fatty acids have been reported in the literature, which are isolated from softwoods. Some of these acids are removed in chemical pulping or saponification, and the esters and waxes are hydrolyzed. Further, long-chain fatty acids, which enter into wastewater treatment plants, inhibit the growth of bacterial communities (Ali and Sreekrishnan, 2001). These acids can be degraded anaerobically, but only in low concentrations at industrial level treatments. In addition, waste discharged after secondary treatment is rich in resin acids. Such acids are reported as being toxic for the aquatic system as well as for human health. These acids such as tricyclic and diterpenic carboxylic acids etc. are mainly released from tree bark and softwood trees, which are nonvolatile and hydrophobic in nature (McMartin et al., 2002). Moreover, the concentration of this resin may reach as high as hundreds of parts per million, however, double concentration can be expected in hardwood trees (Ali and Sreekrishnan, 2001). Literature review revealed that acids like abietic acid, neoabietic acid, dehydroabietic acid, pimaric acid, and isopimaric acid, etc. are generally released from paper and pulp mills. The further effect depends upon the pH and the solubility of these substances. The lower pH values promote higher toxicity than their dissociated counterparts.

2.3.4 Biocides and benzothiazoles in discharged waste

Various types of aerobic and anaerobic bacteria, fungi, and yeast are involved in the paper manufacturing process. Different types of biocide such as 2-(thiocyanomethylthio)-benzothiazole and 2-Mercaptobenzothiazole (MBT) are used in paper and pulp mills. Due to their high antifungal activity, these compounds are used to preserve wood from algal, fungal, and microbial growth, hence works as a replacement of traditional chlorophenols. Some of the examples of biocides, which work as oxidizing agents too, are chlorine dioxide, hydrogen peroxide, and thiocyanates, isothiazole, and cyclobutane. Further, these compounds may also be classified on the basis of mode of action of their chemical structure such as membrane-active biocides, cytotoxic agents, and genotoxicity agents. Furthermore, benzothiazoles are also used in the paper and pulp industry, along with algicides and fungicides. These compounds are also expected to enter in the environment through air or water means (Nawrocki et al., 2005). The derivatives of benzothiazoles are commonly used as a biocide for the prevention and protection in pulp paper industry (Mouchetant-Rostaing et al., 2000).

2.4 Phytotoxicity and genotoxicity of pulp paper industry waste

The pulp paper industry waste is the sixth largest pollution sources after the textile, steel, oil, cement, and leather industries (Ali and Sreekrishnan, 2001). The long-term exposure of these contaminated streams results in significant loss of human and animal health. Phytotoxicity of the pulp paper industry waste after secondary treatment is reported by several workers at different concentration of wastewater levels, that is, 25%, 50%, 75%, and 100%, respectively.

2.4.1 *Phytotoxicity in* Phaseolus mungo *and* Triticum aestivum

The phytotoxicity effects of pulp paper industry waste have been previously published by various researchers. The growth and production of crop plants were reported to be significantly affected by the release of pulp paper industry waste. A seed germination test is a common tool for toxicity assessment of industrial waste for environmental safety (OECD, 2003). In addition, the phytotoxicity in *Phaseolus mungo* and *Triticum aestivum* showed a maximum inhibition (above 90%) after exposure with pulp paper industry waste. When compared, *P. mungo* was found to be more affected than *T. aestivum*. Further, *P. mungo* plant having a germination index with more than 80% was observed to be severely affected by toxic compounds. In this chapter, evidence of phytotoxicity from secondary effluents of paper pulp mill is also discussed.

2.4.2 *Genotoxicity in* Allium cepa *plants*

The continued discharged pollution load from pulp paper industry waste is affecting the balance of the ecosystem and causes mutation in various plants and animals. In addition, higher plants are mainly considered as excellent genetic models to assess the effects of environmental pollutants discharged after secondary treatment. This approach is used to evaluate the possibilities of several genetic end points, ranging from point mutations to chromosome abbreviation in cells of various organs and/or tissues such as roots, leaves, and pollen, etc. (Grant, 1994). One of the examples is abnormalities in the chromosome of *Allium cepa*, shown at different stages of meiosis and varying time exposures and concentrations. The microscopic examination showed abnormalities of prophase, metaphase, anaphase, and telophase after the exposure of waste. Moreover, chromosomal fragments and bridges were also induced at anaphase, indicating mutagenic events in the cell. Moreover, the loss of the telomeric side of chromosomes resulted in ring chromosomes and can be attributed to the alteration in the spindle.

2.4.3 Mutagenicity, chromosome aberrations, and nuclear abnormalities

The pulp paper industry waste also contains various types of EDC nature compounds. Some of the EDC compounds identified in paper pulp sludge are tetradecanoic acid methyl ester, hexadecanoic acid, trimethylsilyl ester or palmitic acid TMS, and β-sitosterol trimethylsilyl ether, etc. Among these, β-sitosterol trimethylsilyl ether is also responsible for DNA and cell chromosome damage, and it is considered as a valuable biomarker for environmental toxicity of discharged waste. Other similar compounds include 2-methoxy phenol, 9,12-octadecadienoic acid (*Z,Z*)-2,3-dihydroxy propyl ester, which are also carcinogenic to both humans and animals. These compounds are listed as EDCs by USEPA. Besides these, the chromosome aberrations are demonstrated by a change in the genetic structure or chromosome number which can occur due to long-term exposure of industrial wastes. Several factors can cause the breakdown of DNA, inhibition of DNA synthesis, and replication of altered DNA, which consequently induce structural chromosomal abnormalities. Such tests can be performed through detailed knowledge of detected pollutants from pulp paper industry waste and assessment of toxicity parameters at a different stage of cell division shown by the potentially toxic element. Furthermore, due to the action of chemical agents used in the pulp paper industry, nuclear abnormalities are typically characterized by morphological alterations in interphase nuclei. Such alteration may be evidenced by lobulated nuclei, nuclei carrying nuclear buds, polynuclear cells, and mini cells in *A. cepa* test. Such abnormalities (lobulated nuclei, polynuclear cells, etc.) are not typically observed in F1 cells of *A. cepa* roots.

2.4.4 Mitotic index and micronucleus

One of the known parameters, used for the assessment of cytotoxicity of several agents, is mitotic index calculation. This method is based on the total number of dividing cells in a particular cell cycle. The change in growth and development of exposed organisms, resulting from the chemical action, may be observed through the lower mitotic index as compared to control. Contrary, a higher value of mitotic index represents an increase in the cell divisions. The higher values led to the disordered cell proliferation and tumor tissue formation. Therefore mitotic Index can be considered as an important indicator for monitoring the pollution potential of cytotoxic compounds. A decrease in the mitotic Index of *A. cepa* meristematic cells can be considered as a reliable tool in estimating the levels of cytotoxic agents. Several studies reported the utility of this test, and most of them showed satisfactory results. Moreover, many authors also considered micronucleus as the most effective and simplest endpoint to analyze the chemical-promoted mutagenic effect. This is due to the fact that in daughter cells micronucleus is easily observed as a similar structure to the main nucleus in the parental cells as a result of damage, not repaired or wrongly repaired, but in a reduced size. Thus the chromosome breaks, and losses arise from the development of some chromosome aberrations. Micronucleus may still derive from other processes such as polyploidization, in which they originate from the elimination of the main nucleus exceeding DNA in an attempt to restore normal ploidy conditions. Evaluation of micronucleus in *A. cepa* testing can be carried out in this species meristematic and F1 root cells. Micronucleus analysis is usually performed with the chromosome aberrations one in meristematic cells, which takes a longer time to perform. Both analyses, however, demonstrated sensitivity to mutagens detection in the environment. Micronucleus analysis also allows an investigation of the mechanisms of action of chemical agents in addition to evaluating mutagenic effects. The size of micronucleus is generally considered as an effective parameter in evaluating the clastogenic and aneugenic effects in *A. cepa*. This may be attributed to the presence of asymmetric karyotype which is directly linked with the size of chromosome. A large sized micronucleus indicates the aneugenic effect which probably results due to the loss of the chromosome. Contrary, small sized micronucleus indicates the clastogenic action which is expected to result from the chromosome breakage. Beside this, other advanced analysis techniques like chromosomal banding and in situ hybridization can be used for detailed investigations of such remediation strategies.

2.5 Phytoremediation of heavy metals from complex organometallic pollutants

Today, various industries are known for their pollution potential, especially through the discharge of metal-laden streams into water and soil environment. Like metals, some POPs are reported to be harmful to humans and animals, and have the tendency to accumulate in the food chain (Singh and Prasad, 2015). The phytoremediation mostly has been focused on hyperaccumulators plants from a contaminated site, which are capable to accumulate the heavy metals in various parts of the plant and tolerate the stress condition of residual organic pollutants. Hyperaccumulators plants accumulate more than a hundred times, in comparison to non-hyperaccumulator plants. In general, phytoremediation strategy is divided into five groups. The first one is phytostabilization in which removal of the pollutants takes place through bioavailability and immobilization process. The second one is phytoextraction techniques in which removal of pollutants at the contaminated site takes place by harvestable parts of the plant. Rhizofiltration is also related under phytoremediation, and in this technique, specific plants absorb the metals in their roots from fed waste streams. Photodegradation is another important technique for degrading the hazardous pollutants of a contaminated site with the help of microorganisms. Some hyperaccumulators plants, grown at the site of pulp paper industry waste site are *Alternanthera*, *P. australis*, *Brassica campestris*, *S. nigrum*, *P. dermatitis*, *C. sativa*, and *Ricinus*. These plants allow high accumulation of

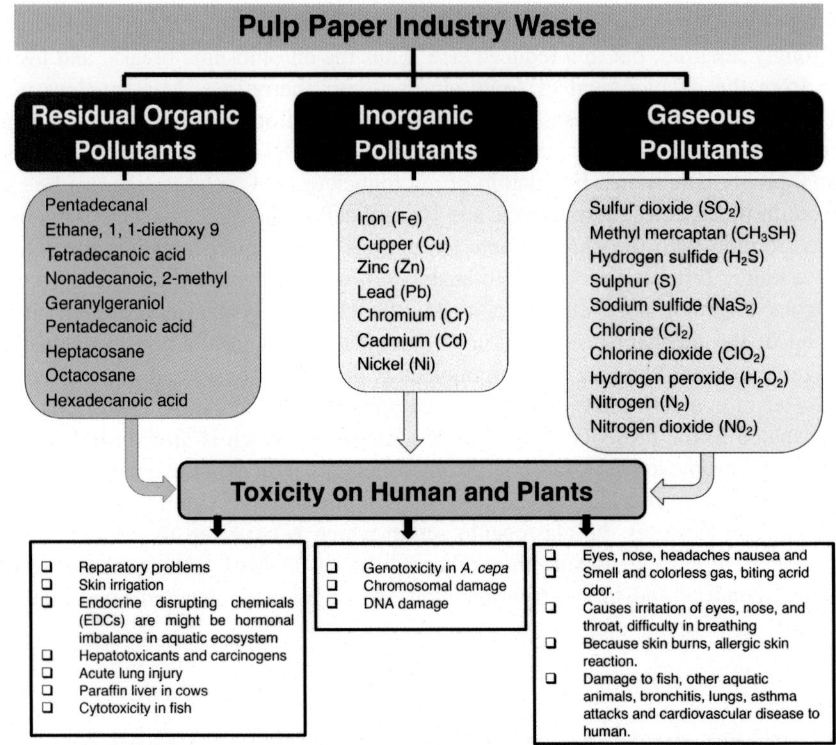

Figure 2.1 Degradation mechanism for different strategies of phytoremediation potential and pollutants for detoxification of residual organic pollutants from the pulp paper industry disposal site.

heavy metals and also tolerate the high concentration of hazardous pollutant-containing with chlorolignin. However, the management of these pollutants through bioremediation is still a challenge, as all the properties are still unknown. The constraints, which lie in the bioaccumulation of metals by plant species, include their strong binding tendency with other lignocellulosic waste and humic substances. The phytoextraction potential of some native plants for various metal and persistent organic copollutants, grown near the paper and pulp mill waste sites, has also been reported in the literature (Mazumdar and Das, 2015). Recently, more than 25 wetland species, grown near the paper and pulp mill site, were found to be effective in phytoremediation of Zn, Fe, Pb, and Mn. The potential of floating and marginal wetland plants, collected from the nearby sites, is presented in Fig. 2.1.

2.6 Strategy of phytoremediation

Phytoremediation strategies are used to improve the potential of industrial waste degradation methods. Developing countries, which generally lack economic assistance and incentives to implement remedial processes, could benefit from such cost-effective plant-based technologies.

2.6.1 Phytofiltration

Heavy metal pollution is a serious environmental threat that is affecting the modern world. Several methods for the removal of heavy metals from industrial waste sites have been documented in literature. These methods are mainly based on ion exchange, chemical, and microbiological precipitation (Janson et al., 1982; Moore and Kaplan, 1994). Phytofiltration might be a cost-effective approach for treating heavy metals contaminated disposal wastewater sites of pulp paper industry. Phytofiltration refers to the usage of aquatic and terrestrial plants for treatment of groundwater, surface water, and wastewater polluted with metals, metalloids, and radionuclides (Dushenkov et al., 1995). In this regard, plants like hyperaccumulating ferns (*Pteris vittata*, *Pteris cretica*), Indian mustard (*Brassica juncea*), sunflower (*Helianthus annuus*), and water hyacinth (*Eichornia crassipes*) have been reported to treat streams laden with As, Cu, Pb, Ni, Zn, Cr, Cd, Pu, U, and Hg. The possible mechanism for concentrating these elements in their roots or shoot tissues include precipitation, adsorption, or absorption (Dushenkov et al., 1995; Lytle et al., 1998; Riddle et al., 2002; Axtell et al., 2003; Huang et al., 2004).

2.6.2 Phytoaccumulation

The technique of remediation of organic pollutants and heavy metals from the waste of pulp paper industry by phytoaccumulation is cost-effective and a green technology. In this regard, some potential native wetland and grass plants have been reported to have the capability to remove soil and aqueous contaminants (Ye et al., 2004). Previous studies also highlighted the usage of wetland plant species for having tolerance to metals, accumulation, and translocation ability (Deng et al., 2006). The latest research also revealed high metal tolerance, appreciable biomass production, and metal uptake capacity of such plants. Specific parametric studies, based on wetlands, also revealed their potential against flooding, salinity, and toxic metals (As, Cd, Zn, etc.) (Chabbi et al., 2000; Rogers et al., 2008; Li et al., 2011; Wang et al., 2011; Yang et al., 2014).

2.6.3 Phytoextraction

The term phytoextraction refers to removal of soil contaminants through the application of pollutants accumulating plants. These plants basically favor the transfer and accumulation of metallic and organic pollutants from soil to above ground—shoots. In some cases, the grown roots on these pollutants can also be harvested as well (Kumar et al., 1995). In general, plant phytoextraction safety and effectiveness are determined by two main characteristics, that is, hyperaccumulating capability and biomass of plants (McGrath and Zhao, 2003). These plants have been developed to know how to handle a huge range of needed gas transfers, including carbon dioxide, water vapor, oxygen, ethylene, and some signaling molecules. Volatile organic compounds are also transferred through the plant, depending upon plant physiology and contaminant characteristics whenever entering the plant.

2.6.4 Phytostabilization

Phytostabilization process reflects the combined effects of sorption, precipitation, and complexation in plants. Through these mechanisms, plants allow remediation of heavy metals from soil. Thus plants reduce the bioavailability of pollutants in the environment and minimize the loss to soil and the environment. Consequently, environmental degradation caused by leaching of soil pollutants is also minimized by phytostabilization. The potential plants used for such purpose include *Sorghum* sp., *S. nigrum*, *Eucalyptus urophylla*, *Eucalyptus saligna*, *Vigna unguiculata*, etc. for the remediation of heavy metals (Cd, Cu Ni, Pb, and Zn) (Jadia and Fulekar, 2008).

2.6.5 Phytovolatilization

The use of plants is to volatilize pollutants from their aerial parts for removing the volatile compounds at the contaminated site of the pulp paper industry. Plants extract volatile pollutants (e.g., selenium, mercury) from the soil and volatilize them from the foliage. Native plants can interact with a different type of organic compounds, and thereby affect the fate and transport of many environmental pollutants.

2.6.6 Phytodegradation

In this process, along with plants, associated microorganisms assist the bioremediation of organic and inorganic pollutants. This approach may also be useful in degrading the pollutants from the waste disposal sites. In particular, rhizospheric microorganisms work in conjunction with plant roots in remediating the soils contaminated with organics/inorganics pollutants. The same mechanism was also evidenced in the process of air purification by the plants. Examples of such plants are *Datura innoxia* and *Lycopersicon peruvianum* containing peroxidase, laccase and nitrilase, *Blumea malcolm*, *Erythrina cristagalli*, and *Chlorella pyrenoidosa*.

2.6.7 Rhizofiltration

Rhizofiltration is based on the adsorption process, in which aqueous contaminants are adsorbed onto the plant roots. Further, absorption into the plant roots also favors this bioremediation. This remediation is facilitated by root zone (rhizosphere) surrounded by contaminated waters. Such treatment strategies are generally adopted in control environments where plants are grown in greenhouses in water. To develop implementable strategies for pulp paper industry secondary effluents, waste is brought into contact to acclimatize the plants with the environment. Later, these acclimatized plants are brought to the waste disposal sites where the roots adsorb/absorb the groundwater and contaminants as well. Once the roots are saturated with pollutants, the plants are harvested. The application of sunflowers to remove radioactive groundwater contaminants are already reported from Chernobyl, Ukraine. The commonly encountered heavy metals (Pb, Cd, Cu, Ni, Zn, and Cr) in contaminated waters can easily be extracted using rhizofiltration. The examples of such plants are sunflower, tobacco, spinach, rye, and Indian mustard. Besides these species, hyperaccumulators can also be used to reduce the concentration of heavy metals.

2.7 Heavy metals uptake by transport, translocation, and transformation

The presence of heavy metals in aqueous environment poses several health risks (Rengel and Zhang, 2003). Initially, root cells take up the metal ions, and transport them to the aerial parts of the plants. Nutrients are transported to cells through membrane channels or through membrane-bound proteins that bind the chemical to the cell (active transportation). Organic chemicals can also be absorbed into roots and taken up by plants, translocated, metabolized, or transpired. The engineered use of the green plants comprises of removing toxic pollutants such as pulp and paper mill wastes (lignin, chlorolignin, heavy metals, organic compounds, etc.). The basic mechanisms, involved in this process, are absorption, sequestration, degradation, and metabolism. Free-living microorganism and plants create the plant rhizosphere. Improved translocation of the absorbed metal to shoot a plant is an important feature of hyperaccumulating plant species. Transport, translocation, and transformation are important factors in the different parts of the plant's pollutant relationship. Plant growth and development depend on the water and soil sample acquisition and appropriation of heavy metal. The absorption of all types of nutrients is energized by proton pump H^+, and ATPase is ubiquitin in plant membrane, generating electrochemical potential. For these nutrients, the membrane transport mechanism depends on the electrical charge and the transport direction. Many plant transporters are found in the plasma membrane outer layer where they mediate cell uptake and efflux. These are the best-characterized group of transporters because they can express themselves for functional description in foreign cell types such as yeast or *Xenopus* oocytes. Root surfaces, which are specifically known for adsorption of elementary nutrients of soil/water, have a large surface area and high affinity for pollutants. In its infancy, plant transport systems for elemental nutrients and pollutants are immobile. Two associated subfamilies of zinc transporter (ZIP) proteins, involved in Fe(II) and Zn(II) uptake, are available in the preliminary sequence. Organic cheaters such as Ethylenediamine-tetra-acetic acid (EDTA) also promote the formation of auxin plant growth regulators, thus enhance the growth rate of plants. Hence, it plays a significant role in increasing metal uptake and translocation. The effect of toxic heavy metals causes DNA strands to break, DNA protein crosslink, DNA damage to oxidation, chromosomal aberrations, exchange of sister chromatids, the effect on apoptosis and modification of gene expression in plant metabolism and DNA control point. Heavy metal translocation from plant root to the above plant shoot surface depends on plant species, soil superiority, and heavy metals. Different metals have different mobility capacities in plants, for example, Zn and Cd are reported to be more mobile than Pb and C. Metal is bound mainly on the cell wall of the root during the transport, which results in accumulation of heavy metal in the root of the plant. A metal chelating agent such as organic acid, citric acid, malic acid, benzoic acid, lactic acid, amino acid, phenol, uric acid, and thiols facilitates the mobility of metals from root to shoot. The plant takes heavy metal from a different soil source into its root due to the high cation exchange capacity of xylem cell. After entering the root of the plant-heavy metal ion, it is possible to either store it in the root or translocate it to the shoot first. It may be deposited predominantly in the leaf vacuole. In particular, anion NO^{3-} can be stored in vacuoles where balancing the charge as a driver for cell expression and growth becomes the driving force. In the transport of the organic and inorganic molecules in a plant, mitochondrion and plastid membrane also play an important role. Some nutrient assimilation occurs in the plastid; inorganic ion transport is important in this membrane. An internal candidate gene has been identified, but very little is known about their mechanism or regulation of transport. Some family of plant genes has a key role to play in both carbon and nitrogen transport. Further, the rate of heavy metal uptake depends on plant species, heavy metal, and soil pH. A membrane protein belongs to a group of a different family of plant metal carriers and can act as tissue level. Mesophylls and epidermal cell or root to shoot are the most transportation of the heavy metals by native hyperaccumulators on contaminated site of pulp and paper mill waste. Heavy metal associate transporter from ATPase family was involved across the plasma membrane of the cytoplasm. Their carcinogenic effect in animal and humans are probably caused by their mutagenic ability. ATP-binding cassette (ABC) transporters participate in the translocation of the hormone auxins, a dominant developmental regulator, and the hormone abscisic acid, implicated in abiotic stress responses.

2.8 Toxicity of heavy metals on human health

Various heavy metals such as Fe, Ni, Cd, Cr, Mg, Cu, Zn, and Mn are well known for toxicity in plants and human beings. These elements improve the plant growth/ yield at low concentrations and benefit the entire food chain, but their high concentration is harmful to those that possess a serious threat to humans and living organism environment (Verkleij et al., 2009). In particular, these negative effects can be evidenced by decreased photosynthetic rate and chlorophyll content. The high concentration of metals is also observed to be responsible for cell membrane damage and destruction of other biomolecules (Ekmekçi et al., 2008). This can be attributed

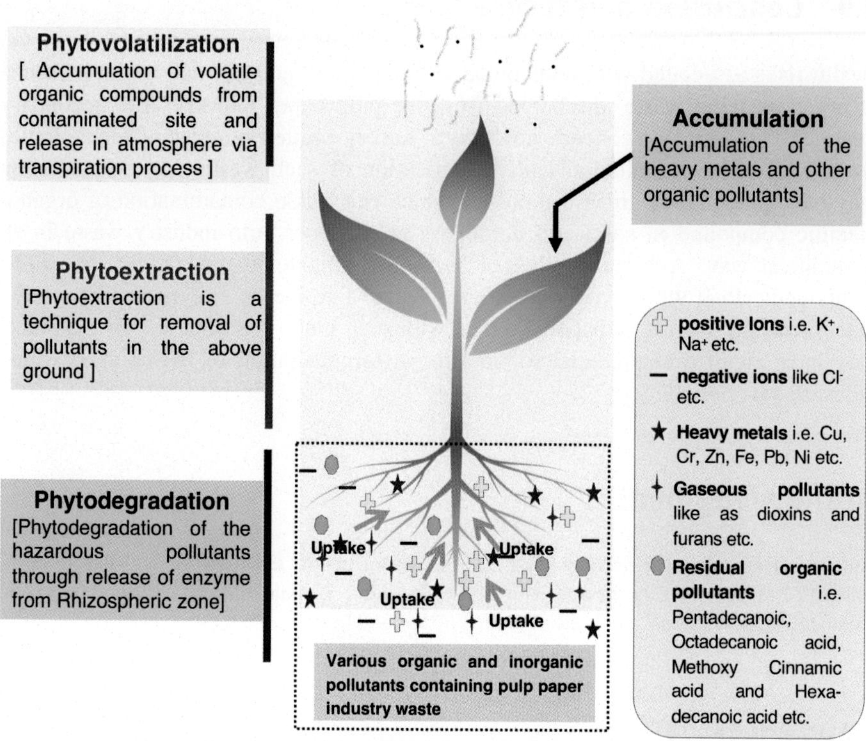

Figure 2.2 Classification of potential pollutants along with heavy metals and their toxicity.

to the increased production of Reactive oxygen species (ROS) in metal stressed plants. A number of health issues, associated with heavy metal toxicity to humans, depend on the type of metal, concentration level, and oxidation status (Fig. 2.2).

A brief summary of these health risks is provided below.

1. Ni inhalation can cause lung cancer. Some studies also reported cases of throat and stomach cancer (Khan et al., 2007).
2. Cr is reported to be responsible for rapid hair fall (Salem et al., 2000).
3. The presence of As is observed to be responsible for disturbances in cellular processes (Tripathi et al., 2007).
4. Pb toxicity is found to be responsible for cardiovascular disease, short-term memory loss, and coordination-related problems in children (Salem et al., 2000).
5. Toxicity of Cu was found to be linked with brain and kidney damage, and intestinal irritation (Salem et al., 2000).
6. Excess concentrations of Zn were observed to be responsible for dizziness and fatigue in humans (Hess and Schmid, 2002).
7. Cd, which is a well-known mutagenic and carcinogenic element; disturbs the body's metabolism of calcium. This problem is known as hypercalciuria, and it causes kidney failure and severe anemia (Awofolu, 2005).

2.9 Conclusion and future scope

Health risks associated with contaminated food, water, and air, due to the presence of organometallic waste, discharged from the pulp paper industry after secondary treatment, are a major concern nowadays. A very limited number of technologies are available for remediation and detoxification of such waste. Phytoremediation may have potential as green technology which can tackle contamination of organometallic compound in solid and liquid wastes of paper pulp industry waste in an economical way. A large number of various carcinogenic, mutagenic, and EDC compounds along with heavy metals are released into the environment, through the manufacturing process of paper. To deal with such kind of wastes, the native plants may have significant potential for in situ phytoremediation of heavy metals, and other organic pollutants.

Acknowledgment

Authors would like to acknowledge the Department of Environmental Microbiology, Babasaheb Bhim Rao Ambedkar University (BBAU), Lucknow, for providing necessary technical support.

References

Ali, M., Sreekrishnan, T.R., 2001. Aquatic toxicity from pulp and paper mill effluents: a review. Adv. Environ. Res. 5 (2), 175–196.

Awofolu, O.R., 2005. A survey of trace metals in vegetation, soil and lower animal along some selected major roads in the metropolitan city of Lagos. Environ. Monit. Assess. 105 (1–3), 431–447.

Axtell, N.R., Sternberg, S.P.K., Claussen, K., 2003. Lead and nickel removal using *Microspora* and *Lemna* minor. Bioresour. Technol. 89, 41–48.

Butler, E.C.V., Dal Pont, G., 1992. Liquid chromatography-electrochemistry procedure for the determination of chlorophenols compounds in pulp mill effluents and receiving waters. J. Chromatogr. A 609 (1–2), 113–123.

Chabbi, A., McKee, K.L., Mendelssohn, I.A., 2000. The fate of oxygen losses from *Typha domingensis* (Typhaceae) and *Cladium jamaicense* (Cyperaceae) and consequences for root metabolism. Am. J. Bot. 87, 1081–1090.

Chandra, R., Sharma, P., Yadav, S., Tripathi, S., 2018. Biodegradation of endocrine-disrupting chemicals and residual organic pollutants of pulp paper mill effluent by biostimulation. Front. Microbiol. 9.

Chandra, R., Yadav, S., Yadav, S., 2017. Phytoextraction potential of heavy metals by native wetland plants growing on chlorolignin containing sludge of pulp and paper industry. Ecol. Eng. 98, 134–145.

Chandra, R., Singh, R., 2012. Decolourisation and detoxification of rayon grade pulp paper mill effluent by mixed bacterial culture isolated from pulp paper mill effluent polluted site. Biochem. Eng. J. 61, 49–58.

Chandra, R., Raj, A., Yadav, S., Patel, D.K., 2009. Reduction of pollutants in pulp paper mill effluent treated by PCP-degrading bacterial strains. Environ. Monit. Assess. 155 (1–4), 1.

Craig, J., Hsu, P., Sastry, S., 1986. Adaptive control of mechanical manipulators. In: Proceedings. 1986 IEEE International Conference on Robotics and Automation, vol. 3. pp. 190–195.

Demirel, B.G., Altin, A., 2017. Production of sorbent from paper industry solid waste for oil spill cleanup. Mar. Pollut. Bull. 125, 341–349.

Deng, H., Ye, Z.H., Wong, M.H., 2004. Accumulation of lead, zinc, copper and cadmium by 12 wetland plant species thriving in metal-contaminated sites in China. Environ. Pollut. 132, 29–40.

Deng, H., Ye, Z.H., Wong, M.H., 2006. Lead and zinc accumulation and tolerance in populations of six wetland plants. Environ. Pollut. 141, 69–80.

Dushenkov, V., Nanda Kumar, P.B.A., Motto, H., Raskin, I., 1995. Phytofiltration: the use of plants to remove heavy metals from aqueous streams. Environ. Sci. Technol. 29, 1239–1245.

Ekmekçi, Y., Tanyolac, D., Ayhan, B., 2008. Effects of cadmium on antioxidant enzyme and photosynthetic activities in leaves of two maize cultivars. J. Plant Physiol. 165 (6), 600–611.

Grant, W.F., 1994. The present status of higher plant bioassays for the detection of environmental mutagens. Mutat. Res. 310 (2), 175–185.

Gupta, A.K., Sinha, S., 2006. Chemical fractionation and heavy metal accumulation in the plant of *Sesamum indicum* (L.) var. T55 is grown on soil amended with tannery sludge: selection of single extractants. Chemosphere 64 (1), 161–173.

Hess, R., Schmid, B., 2002. Zinc supplement overdose can have toxic effects. J. Pediatr. 24, 582–584.

Hewitt, L.M., Kovacs, T.G., Dubé, M.G., MacLatchy, D.L., Martel, P.H., McMaster, M.E., et al., 2008. Altered reproduction in fish exposed to pulp and paper mill effluents: roles of individual compounds and mill operating conditions. Environ. Toxicol. Chem. 27 (3), 682–697.

Huang, J.W., Poynton, C.Y., Kochian, L.V., Elless, M.P., 2004. Phytofiltration of arsenic from drinking water using arsenic-hyperaccumulating ferns. Environ. Sci. Technol. 38, 3412–3417.

Jadia, C.D., Fulekar, M.H., 2008. Phytoremediation: the application of vermicomposting to remove zinc, cadmium, copper, nickel, and lead by sunflower plant. Environ. Eng. Manag. J. 7 (5).

Janson, C.E., Kenson, R.E., Tucker, L.H., 1982. Treatment of heavy metals in wastewaters. What wastewater-treatment method is most cost-effective for electroplating and finishing operations? Here are the alternatives. Environ. Prog. Sustain. Energy 1, 212–216.

Khan, M.A., Ahmad, I., Rahman, I.U., 2007. Effect of environmental pollution on heavy metals content of *Withania somnifera*. J. Chin. Chem. Soc. 54 (2), 339–343.

Kumar, P.B.A.N., Dushenkov, V., Motto, H., Raskin, L., 1995. Phytoextraction: the use of plants to remove heavy metals from soils. Environ. Sci. Technol. 29 (1232), 1238.

Li, H., Ye, Z.H., Wei, Z.J., Wong, M.H., 2011. Root porosity and radial oxygen loss related to arsenic tolerance and uptake in wetland plants. Environ. Pollut. 159 (1), 30–37.

Lytle, C.M., Lytle, F.W., Yang, N., Qian, J.-H., Hansen, D., Zayed, A., et al., 1998. Reduction of Cr(VI) to Cr(III) by wetland plants: potential for in situ heavy metal detoxification. Environ. Pollut. 32, 3087–3093.

Mani, D., Kumar, C., 2014. Biotechnological advances in bioremediation of heavy metals contaminated ecosystems: an overview with special reference to phytoremediation. Int. J. Environ. Sci. Technol. 11 (3), 843–872.

Mazumdar, K., Das, S., 2015. Phytoremediation of Pb, Zn, Fe, and Mg with 25 wetland plant species from a paper mill contaminated site in North East India. Environ. Sci. Pollut. Res. 22 (1), 701–710.

McGrath, S.P., Zhao, F.J., 2003. Phytoextraction of metals and metalloids from contaminated soils. Curr. Opin. Biotechnol. 14, 277–282.

McMartin, D.W., Peru, K.M., Headley, J.V., Winkler, M., Gillies, J.A., 2002. Evaluation of liquid chromatography negative ion electrospray mass spectrometry for the determination of selected resin acids in river water. J. Chromatogr. A952, 289–293.

Moore, M.D., Kaplan, S., 1994. ASM News 60, 17–23.

Mouchetant-Rostaing, Y., Giard, M.H., Bentin, S., Aguera, P.E., Pernier, J., 2000. Neurophysiological correlates of face gender processing in humans. Eur. J. Neurosci. 12 (1), 303–310.

Nawrocki, S.T., Carew, J.S., Dunner, K., Boise, L.H., Chiao, P.J., Huang, P., et al., 2005. Bortezomib inhibits PKR-like endoplasmic reticulum (ER) kinase and induces apoptosis via ER stress in human pancreatic cancer cells. Cancer Res. 65 (24), 11510–11519.

OECD, 2003. OECD Guideline for Testing of Chemicals. Terrestrial Plant Tests: 208: Seedling Emergence and Seedling Growth Test. pp. 1–19.

Pokhrel, D., Viraraghavan, T., 2004. Treatment of pulp and paper mill wastewater—a review. Sci. Total Environ. 333 (1–3), 37–58.

Rengel, Z., Zhang, W.H., 2003. Role of dynamics of intracellular calcium in aluminum toxicity syndrome. New Phytol. 159, 295–314.

Reveille, V., Mansuy, L., Jarde, É., Garnier-Sillam, E., 2003. Characterisation of sewage sludge-derived organic matter: lipids and humic acids. Org. Geochem. 34 (4), 615–627.

Riddle, S.G., Tran, H.H., Dewitt, J.G., Andrews, J.C., 2002. Field, laboratory, and X-ray absorption spectroscopic studies of mercury accumulation by water hyacinths. Environ. Sci. Technol. 36 (9), 1965–1970.

Rogers, M.E., Colmer, T.D., Frost, K., Henry, D., Cornwall, D., Hulm, E., et al., 2008. Diversity in the genus *Melilotus* for tolerance to salinity and water logging. Plant Soil 304, 89–101.

Rosselli, W., Keller, C., Boschi, K., 2003. Phytoextraction capacity of trees growing on metal contaminated soil. Plant Soil 256 (2), 265–272.

Salem, H.M., Eweida, E.A., Farag, A., 2000. Heavy Metals in Drinking Water and Their Environmental Impact on Human Health. ICEHM2000. Cairo University, Egypt, pp. 542–556.

Sarwar, N., Malhi, S.S., Zia, M.H., Naeem, A., Bibi, S., Farid, G., 2010. Role of mineral nutrition in minimizing cadmium accumulation by plants. J. Sci. Food Agric. 90 (6), 925–937.

Shu, W.S., Ye, Z.H., Lan, C.Y., Zhang, Z.Q., Wong, M.H., 2002. Lead, zinc and copper accumulation and tolerance in populations of *Paspalum distichum* and *Cynodon dactylon*. Environ. Pollut. 120 (2), 445–453.

Singh, A., Prasad, S.M., 2015. Remediation of the heavy metal contaminated ecosystem: an overview of technology advancement. Int. J. Environ. Sci. Technol. 12 (1), 353–366.

Tiwari, S., Rai, P., Yadav, S.K., Gaur, R., 2013. A novel thermotolerant *Pediococcus acidilactici* B-25 strain for color, COD, and BOD reduction of distillery effluent for end-use applications. Environ. Sci. Pollut. Res. 20, 4046–4058.

Toczylowska-Maminska, R., 2017. Limits and perspectives of pulp and paper industry wastewater treatment – a review. Ren. Sustain. Energy Rev. 78, 764–772.

Tripathi, R.D., Srivastava, S., Mishra, S., Singh, N., Tuli, R., Gupta, D.K., et al., 2007. Arsenic hazards: strategies for tolerance and remediation by plants. Trends Biotechnol. 25 (4), 158–165.

Ugurlu, M., Gurses, A., Dogar, C., Yalcin, M., 2007. The removal of lignin and phenol from paper mill effluents by electrocoagulation. J. Environ. Manag. 87, 420–428.

US Environmental Protection Agency and Endocrine Disruptor Screening Program (USEPA and EDSP), 2012. Universe of Chemicals. The United States Environmental Protection Agency, Washington, DC.

Vallejo, M., Fernández-Castro, P., San Román, M.F., Ortiz, I., 2015. Assessment of PCDD/Fs formation in the Fenton oxidation of 2-chlorophenol: Influence of the iron dose applied. Chemosphere 137, 135–141.

Verkleij, J.A., Golan-Goldhirsh, A., Antosiewisz, D.M., Schwitzguébel, J.P., Schröder, P., 2009. Dualities in plant tolerance to pollutants and their uptake and translocation to the upper plant parts. Environ. Exp. Bot. 67 (1), 10–22.

Wang, M.Y., Chen, A.K., Wong, M.H., Qiu, R.L., Cheng, H., Ye, Z.H., 2011. Cadmium accumulation in and tolerance of rice (*Oryza sativa* L.) varieties with different rates of radial oxygen loss. Environ. Pollut. 159 (6), 1730–1736.

Yang, J.X., Tam, N.F.Y., Ye, Z.H., 2014. Root porosity, radial oxygen loss and iron plaque on roots of wetland plants in relation to zinc tolerance and accumulation. Plant Soil 374, 815–828.

Ye, Z.H., Yang, Z.Y., Chan, G.Y.S., Wong, M.H., 2001. Growth response of *Sesbania rostrata* and *S. cannabina* to sludge-amended lead/zinc mine tailings: greenhouse study. Environ. Int. 26 (5), 449–455.

Ye, Z.H., Wong, M.H., Lan, C.Y., 2004. Use of a wetland system for treating Pb/Zn mine effluent: a case study in southern China from 1984 to 2002. Wetland Ecosystems in Asia: Function and Management. Elsevier, pp. 413–434.

Transgenic plants in phytoremediation of organic pollutants

3

Santosh Kumar Mishra[1,], Priya Ranjan Kumar[1] and Ravi Kant Singh[2]*
[1]Department of Biotechnology, IMS Engineering College, Ghaziabad, India,
[2]Amity Institute of Biotechnology, Amity University Chhattisgarh, Raipur, India
*Corresponding author

3.1 Introduction

Rapid urbanization and industrialization influenced the increased use of organic compounds in the form of polycyclic aromatic hydrocarbons (PAHs), petroleum hydrocarbons, dyestuffs, halogenated hydrocarbons, insecticides, pesticides, explosives, and pharmaceutical products. It has resulted in serious environmental pollution and over the past few decades, considerable attention has been given by the scientific community (Meagher, 2000; Macek et al., 2000). Many nitro derivatives of aromatic compounds, aromatic hydrocarbons, polychlorinated biphenyls, PAHs, dioxins, and their derivatives are extremely toxic and mutagenic. These compounds are the major causes of cancer in humans, animals, and natural micro flora. The conventional methods of remediation like physical, chemical, and biological methods can clear contamination sites but have their own limitations like increased costs and lower efficiency. This is the main region for the development of new techniques for remediation (Lee and Huffman, 1989; Johnson et al., 1997). Phytoremediation, which uses plants to remove pollutants from soil, is one of the current emerging and very efficient treatment technologies. There are many modern tools and devices for analytical assessment that have been developed which provide clear and insightful analysis for the selection and optimization of remediation processes by various plant species (Macek et al., 2008; Salt et al., 1998). Hyper accumulators are required for the heavily polluted sites containing organic contaminants. Genetic engineering approaches can be used for the development of such tools. But, by altering the environmental and nutritional requirements of certain plants, such efficient accumulate can be made. Thus, phytoremediation offers a very efficient approach for the removal of contaminants from polluted soil (Susarla et al., 2002; Pilon-Smits, 2005). However, some other technologies like microbe-assisted phytoremediation, including rhizoremediation proved to be very effective for removing organic contaminants from impacted soils. It offers more efficiency when used with appropriate agronomic techniques (Zhuang et al., 2007; Frick et al., 1999).

When microorganisms are used for remediation of harmful organic pollutants this has also observed that both inoculation of microorganisms and use of nutrient are

Bioremediation of Pollutants. DOI: https://doi.org/10.1016/B978-0-12-819025-8.00003-X

essential for their maintenance over long periods (Goldstein et al., 1985). Microorganisms with the proven capabilities of biodegradation in laboratory may not show a similar result in the actual contaminated site (Eapen et al., 2007; Rugh, 2004). The potential capabilities of genetic engineering for strain improvement in order to make them more efficient in bioremediation has appeared in the last three decades (Sung et al., 2003; Doty et al., 2007; Singleton, 2007). There are two major problems associated with the introduction of transgenic microorganisms, that is, the policy related to the use of genetically modified microorganisms and the poor survival rate of developed engineered microbial strains which are being introduced into the contaminated soil. These issues focused the scientific communities to explore the phytoremediation abilities of plants for removing organic pollutants. Since plants are robust in growth, they are a renewable natural resource and can be used for in situ bioremediation practices (Suresh and Ravishankar, 2004; Parameswaran et al., 2007). Due to the availability of various environmental stresses, plants have developed very unique characteristics of metabolizing certain water soluble contaminants. Plants use natural filtering and a pumping system to uptake contaminants from soil, and with using energy from the sun, plants transport the contaminants to various tissues and cells for its metabolization (Abhilash et al., 2009; Dietz and Schnoor, 2001; Doty et al., 2007).

Use of plants in phytoremediation has almost complete public acceptance in comparison to microbes. Plants have the ability to survive high concentrations of contaminants as compared to most of the micro-organisms frequently used for the purpose of bioremediation. The result of phytoremediation is higher availability of organic carbon in the soil, which leads to more efficient microbial activity as well as also supporting pollutants degradation due to enhanced rhizospheric actions. Soil stabilization, carbon sequestration is a process that is useful for phytoremediation, and can also possibly be used for bio-fuel or fiber productions which are other benefits of phytoremediation. Significant interest is taken for plant-based remediation technologies for removal of a wide variety of contaminants present in soil. (Eapen et al., 2007; Hooker and Skeen, 1999; Dietz and Schnoor, 2001; Schnoor, 1997).

Various strategies are being used by plants to deal with environmental chemicals like phytovolatilization, phytodegradation, phytoextraction, and rhizodegradation (Fig. 3.1). In rhizosphere, xenobiotic compounds can be stabilized or degraded. These are adsorbed or accumulated in roots and from there transported to plants specially in the aerial part for further metabolism. At last, these components may be volatilized or degraded inside the plant tissue through a set of biochemical reaction. Mainly, conversion or enzymatic modification with conjugation and active sequestration are followed by plants for the detoxification of pollutants (Schnoor, 1997; Newman and Reynolds, 2004; Pilon-Smits, 2005).

Phytoextraction involves the removal and subsequent storage of contaminants by the plant, and this process plays an important role in the storage and exclusion of metallic contaminants. However these compounds cannot be metabolized. However, certain organic contaminants may also be treated by these methods that have an inherent resistance to degradation. Conversely, phytodegradation is a process in which plants metabolize the contaminants through a biochemical reaction in the presence of various enzymes released by the plants. This process may be useful in the degradation

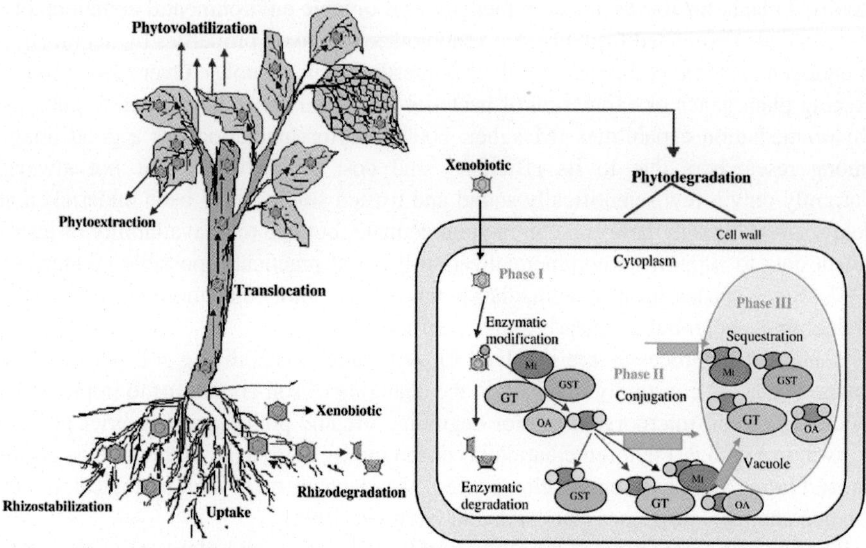

Figure 3.1 Typical mechanism of phytoremediation possessed by plants against xenobiotics. *Source*: Adopted with permission from Abhilash, P. C., Jamil, S., Singh, N. 2009. Transgenic plants for enhanced biodegradation and phytoremediation of organic xenobiotics. *Biotechnol. Adv. 27* (4), 474-488, Elsevier Inc

of herbicides and some other organic pollutants. Phytodegradation has strong similarities to degradation mechanism, those used by animals used for modification and degradation of drugs as well as several other toxins (Sanderman, 1994).

Detoxification of organic pollutants by plants is often slow and has several limitations, leading to the accumulation of toxic compounds in plants. Further these compounds are released into the environment and may be harmful for microorganisms present in the soil (Abou-Shanab et al., 2019). There are also concerns over the introduction of these contaminants into the food chain. Genetic engineering approaches are highly specific and decisive in designing the transgenic plants for the enhanced biodegradation as well as detoxification of environmental pollutants for sustainable development (Abou-Shanab et al., 2019). In the environmental remediation genetically modified organisms offer great potential, for effective environmental management including risks involved, constraints, and challenges.

3.2 Phytoremediation

Phytoremediation is considered as a better alternative compared to physical remediation methods which are environmentally destructive in nature. Plants have many biochemical and physiological properties and are considered as potential agents for remediation for various contaminants present in soil and water. In the recent few decades, significant progress has been made for identification of native or genetically modified plants for the enhanced remediation of organic environmental contaminants. Plants are also protected from organic xenobiotics with toxic properties by degradation of endogenous toxic compounds or their sequestration in vacuoles. Over expression of existing plant genes or expression of bacterial or animal genes is required to enhance phytoremediation capabilities (Meagher, 2000). Phytoremediation has a good image among researchers due to its efficiency and cost effectiveness, but not always. Currently only a few scientifically sound and trusted studies have been undertaken at field scale. Although there is a commercial demand, but due to unavailability of insufficient data to support its commercialization it is not practically possible (Watanabe, 1997). Since successful phytoremediation takes time, sometimes more than a decade, evaluating its potential at an early stage is difficult.

A number of processes are involved for removing or stabilizing soil pollutants in plants. Plants are constantly involved in the changing of soil conditions to improve the efficiency of soil microorganisms for degrading organic pollutants. The brief process is given in Fig. 3.2. Phytoremediation is carried out by a group of higher plants, bacteria, and fungi, and phytoremediation process also depends on several biological, physical, and chemical processes (Siciliano and Germida, 1998).

The specific interaction of a pollutant with soil, water, and plants are variable in nature and depend on the physicochemical properties of the organic contaminants, the physiological properties of the plant species, as well as the state of the contaminants to some extent. Organic compounds are translocated to other plant tissues and finally converted into volatilized form (Salt et al., 1998). Partial or complete degradation or transformation of these compounds is possible for conversion to relatively

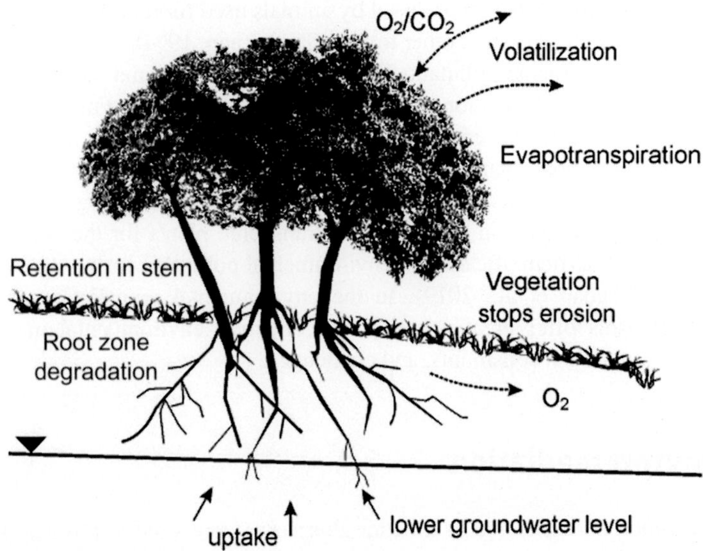

Figure 3.2 Relevant processes during phytoremediation.
Source: Adapted from Black, H., 1999. Phytoremediation: a growing field with some concerns. Scientist 13, 5–6

less phytotoxic compounds. These conjugates are present in bounded form in plant tissues. In all cases, the process of phytoremediation begins with contaminant transport to the plant. Xenobiotics, which are mostly organic in nature, appear to undergo a degree of transformation in plant cells before being excreted in vacuoles or bound to insoluble cellular structures. The xenobiotic conjugates can also be incorporated into biopolymers such as lignin. However in the aquatic/wetland plants they get excreted for storage outside the plant (Fig. 3.3)

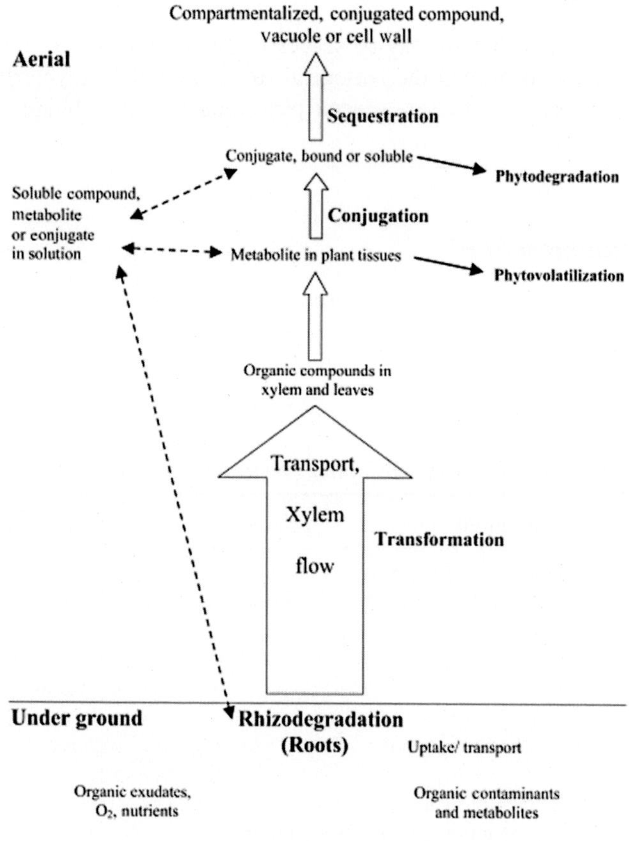

Figure 3.3 Phytoremediation mechanism of organic contaminants for metabolism of xenobiotics in plants. Transport mechanism in aquatic plants is representated by dashed lines showing possible while solid/hollow arrows representing contaminant transportation routes in plant.
Source: Adopted with permission from Singh, O. V., Jain, R. K. 2003. Phytoremediation of toxic aromatic pollutants from soil. Appl. Microbiol. Biotechnol. 63 (2), 128-135, Springer-Verlag

Plant roots released degradative enzymes to their surrounding environment which contribute to transformation of contaminants. These enzymes usually transform substances into products that can be easily absorbed by plant roots or rhizosphere microorganism for further metabolism (Gianfreda and Rao, 2004). The presence of these enzymes can significantly increase transformation of contaminants occurring in the rhizosphere, particularly xenobiotic organic compounds.

3.3 Some Selected Plants Used for Phytoremediation

In summary, phytoremediation may be successful by an influence of the vegetation on the physical, chemical, and the biological (roots, microbes, mycorrhiza) factors in soil. Based on these processes, several phytoremediation techniques have been developed (Table 3.1).

3.3.1 Phytoextraction

Phytoextraction techniques are mostly used for bioremediation of heavy metals. In these processes toxic pollutants are accumulated in plants. Harvested products, which concentrate on the pollutants, may be used or disposed through sequential activities.

Table 3.1 Some commonly used plants having potential for phytoremediation.

Plant species	Compound	Reference
Hybrid poplar	Trichloroethene, perchloroethylene	Schnabel et al. (1997)
Basket willow hybrid	Diesel oil	
Hybrid poplar	TNT	Thompson et al. (1998)
Basket willow hybrid	3,5-Dichlorophenol	Trapp et al. (2000)
Barley	Hexachlorobenzene, PCBs, pentachlorobenzene, trichlorobenzene	McCrady (1987)
Forage grasses	Chlorinated benzoic acids	Siciliano and Germida (1998)
Parrot feather	Tetrachloroethane (PCE), trichloroethane (TCE), TNT	Best et al. (1997)
Eurasian watermilfoil	TNT	Vanderford et al. (1997)
Waterweed	Pentachlorophenol, PCE, TCE	Roy and Hänninen (1994)

3.3.2 Rhizofiltration

In this process contaminants are absorbed by the plant through roots or any other specialized part of the plants. Heavy metals or lipophilic compounds can be extracted from water by using rhizofiltration technique.

3.3.3 Phytostabilization

Phytostabilization plays a significant role in the immobilization of organic contaminants present in the soil resulting in the prevention of soil erosion. During this process organic pollutants are converted into a non-soluble by redox reaction, especially in the root zone of the plants.

3.3.4 Rhizo and phytodegradation

Phytodegradation is the degradation of pollutants by plants whereas rhizodegradation involves the degradation of organic contaminants in the root zone with the involvement of microbial activity or by roots, or both may be involved. In the root zone, several processes enhance the degradation of some compounds such as am petroleum, PAH, BTEX, TNT, chlorinated solvents, and pesticides.

3.4 Transgenic plants in phytoremediation

Microorganisms and mammals are heterotrophic organisms that have an enzymatic system that completes the mineralization and degradation of organic pollutants and also can play complementary capabilities to plants for metabolizing various organic pollutants (Macek et al., 2000). With recent developments over the past few decades in genetic engineering, now it is possible to express bacterial and mammalian genes into the plant species for the development of transgenic plants having better and improved capabilities to metabolize organic pollutants (Abhilash and Singh, 2008).

3.5 Biochemistry of Phytoremediation

Recent research on the chlorinated aliphatic organic compounds reveals that degradation of these compounds and the consequences on human beings are mainly cellular toxicity carcinogenicity or mutagenic effects. The rate of metabolism of these compounds varies within organism to organism whether it may be of the same homologous series. The effect of carbon tetrachloride has been observed in some extent among other chlorinated metahnes (Dietz and Schnoor, 2001).

Chlorinated ethylenes involve the formation of epoxides, with polychlorinated ethenes (PCE) and tetrachloroethylene (PCE) during the dechlorination pathway and alternatively being conjugated to glutathione. Most of the researches on P450 reveals that this enzyme has a key role in the epoxide formation, whereas glutathione *S*-transferase (GST) catalyzes reactions with glutathione (Dietz and Schnoor, 2001). This has been observed that epoxideintermediate is highly transient in nature and difficult to detect. Therefore its role in the overall metabolism of TCE is still controversial and uncertain. TCE metabolism in mammals is given in Fig. 3.4. Phytotransformation has been studied mostly with pesticides in commercial crop plants. Since these compounds undergo a series of metabolic processes, therefore its convenient to detect metabolic activities related to these compounds. In the initial phase, functional groups such as −OH, $-NH_2$, or −SH are converted into a wide range of compounds by a series of reactions such as oxidation, reduction, or hydrolysis (Sanderman, 1994). For lipophilic compounds, oxygenation is also a typical reaction during the first phase with increased solubility of the compound formed (Abou et al., 2019). The typical catalytic process of phase I reaction is carried out by the enzyme P450. The second phase involves conjugation with D-glucose, glutathione, or amino acids, resulting in the formation of soluble and polar compounds (Carvalho, 2006). However some insoluble conjugates with cell wall components also form in plants

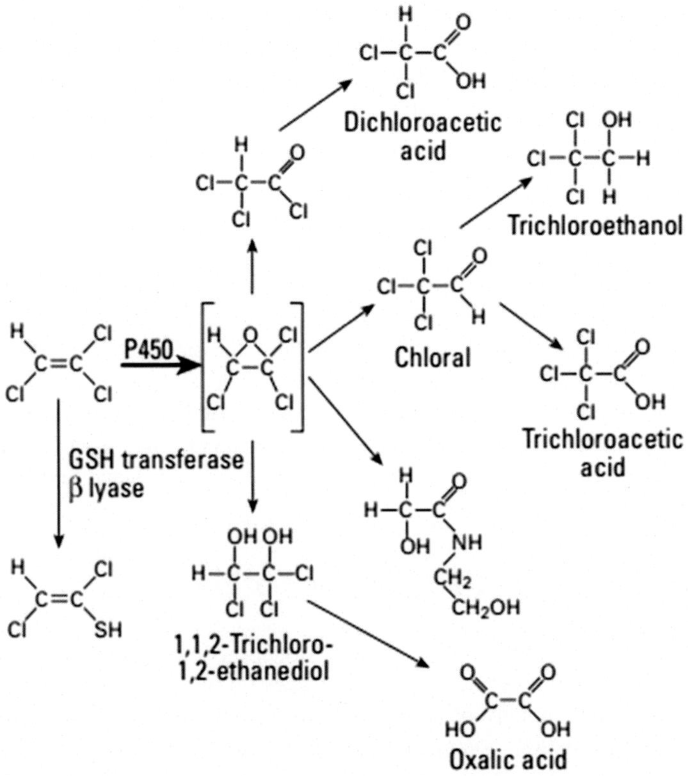

Figure 3.4 TCE metabolism in mammalian systems.
Source: Adopted with permission from Henschler, D., Hoos, W. R., Fetz, H., Dallmeier, E., Metzler, M., 1979. Reactions of trichloroethylene epoxide in aqueous systems. Biochem. pharmacol. *28*(4), 543-548.

during this process. These compounds formed through a series of nonselective reactions with the activity of free radicals used in lignin synthesis or by a more selective reaction involved in the biosynthesis of hemicelluloses (Hooker and Skeen, 1999). Insoluble conjugates are found to be bound residue and difficult in further characterization. Detoxification of herbicides in plants is attributed with glutathione and further catalyzed by GST (Pilon-Smits, 2005). Many herbicide safeners promote glutathione conjugation and detoxification by increasing levels of glutathione or enhanced activity of GST (Lima et al., 2005). *O*-Glucosyltransferases, *N*-glucosyltransferases, and malonyltransferases are other enzymes involved in phase II reactions (Sanderman, 1994). Compartmentation and storage are observed in the third phase of plant metabolism. Plants do not have a well-developed excretory mechanism in comparison to mammals *for* excreting unwanted compounds therefore; soluble metabolites are stored in the vacuole or in the cell wall. There are several other potential mechanisms for the uptake, and transformation of chlorinated aliphatic in a plant-soil system was observed by researchers and these mechanisms can be better understood looking at Fig. 3.5. Possible mechanisms throughout the plant include microbial transformation in the rhizosphere, uptake of the chemical and its metabolites into the roots, xylem transfer of the compounds to the leaves and finally excreted by stomata, uptake of chemicals from the air, phloem transfer, and also bound residue formation.

3.6 Role of transgenic plants in phytoremediation of organic pollutants

Recent advances in the area of phytoremediation, due to the development of genetically modified plants, are able to take up and degrade contaminants with better efficacy. Due to a better understanding of enzymatic activities and understanding of metabolism of xenobiotic chemicals, there was an increase in t phytoremediation capabilities of genetically engineered plants (Kawahigashi et al., 2005a,b,c). This type of technology has been used for several decades in agricultural practices in various countries. Due to the development of genetic engineering tools and techniques, several plant species are being developed to keep the phytoremediation potentialities of the plants. Transgenic tobacco plants containing human P4502 E1 gene are able to transform up to 640 times amount of TCE compared to the reference tobacco plant Such genetically engineered plats showed increased uptake and metabolism of ethylene dibromide and other halogen derivatives of hydrocarbon present in the groundwater. Transgenic tobacco plants expressing a microbial pentaerythritol tetranitrate reductase showed a higher tolerance to explosives such as trinitrotoluene and glycerol nitrate (Dixon et al., 1998a,b). Denitration of glycerol nitrate is more rapidly and completely transformed in the transgenic seedlings. IUn, which is most of the application sterile clone, could be used for phytoremediation application because there is no need for plant reproduction in such applications. The major applications of genetically engineered plants are specific enzymes that are involved in degradation of contaminants, which could be transferred to a

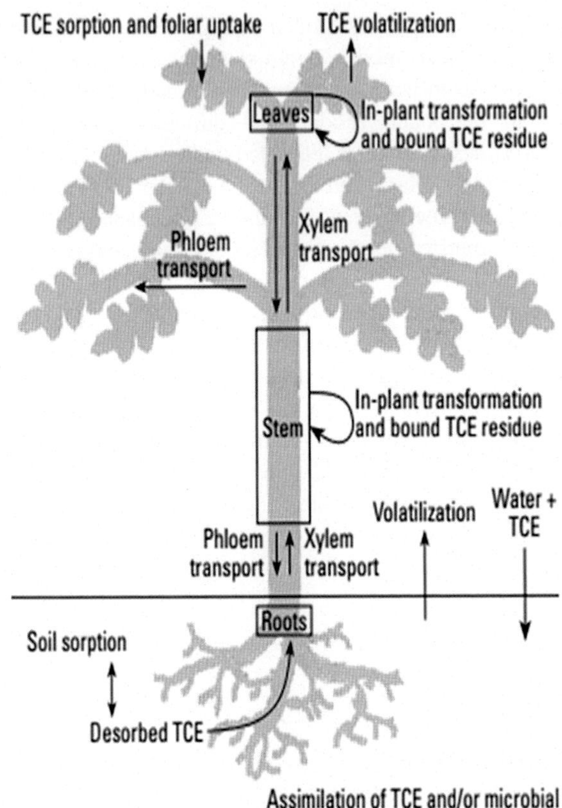

Figure 3.5 Potential uptake and transformation pathways of TCE in a plant—soil phytoremediation system.
Source: Adopted with permission from Schnabel, W. E., Dietz, A. C., Burken, J. G., Schnoor, J. L., Alvarez, P. J., 1997. Uptake and transformation of trichloroethylene by edible garden plants. Water Res. *31* (4), 816-824, Elsevier Inc.

particular plant species that are indigenous to an ecosystem or having other desired properties i.e. rapid growth, root structure, and high water uptake capability.

Initially the phytoremediation was limited to removal of inorganic pollutants present in the soil but in recent years this technology has proven its important role in the remediation of organic pollutants such as, aromatic hydrocarbons, explosives, and chlorinated solvents, etc. (Pilon-Smits, 2005). In the early development of commercially available transgenic plants, such plants were developed to reduce the amount of pesticides required to reduce the loss of crop yields. Transgenic plants in the early days were limited to remediation of heavy metal contaminants of soil, that is, Nicotiana tabaccum in which metallothionein gene of yeast is expressed for efficient tolerance to cadmium and Arabidopsis thaliana in which over expression of mercuric ion reductase gene results in higher and effective tolerance of mercury containing compounds present in soil and water (Misra and Gedama, 1989; Rugh et al., 1996). The first attempt was to develop genetically engineered plants for phytoremediation of organic pollutants for phytoremediation of explosives and halogenated organic compounds as contaminants. Now transgenic plants have a wide application in the phytoremediation of explosives, pesticides, organic solvents, and phenolics, etc. These plants contain either transgenes responsible for metabolizing the organic pollutants or having transgenes that are effective in increasing resistance of pollutants (Doty et al., 2000, Eapen et al., 2007; Macek et al., 2008; Mackova et al., 1998; Meagher, 2000).

3.7 Degradative pathways in plants

Metabolism of toxic organic compounds in humans, animals, and higher plants are involved basically in three main biochemical processes (1) transformation which is generally considered as phase I, (2) conjugation which is also known as phase II, and (3) compartmentalization which is also known as phase III. In t phase I different types of reactions involved are dealkylation, hydroxylation, epoxidation, peroxidation, sulphoxidation, or reduction by cytochrome P450s. In this step hydrophobic pollutants are converted to less hydrophobic metabolites. Reactions catalyzed by cytochrome P450 are vital steps that lead to detoxification, inactivation, and excretion of byproducts formed through a biochemical conversion. In phase II metabolites as well as organic pollutants of phase I directly conjugated with glutathione, amino acids, and sugars result in the formation of hydrophilic compounds. In the final phase (phase III) of the process conjugated metabolites are deposited in the vacuoles as well as to some extent in the cell walls. Recent findings in phase III metabolism reveal that phases can be categorized into two independent phases; one is confined to transport and storage of metabolites in the vacuole and the second one involves cell wall binding as well as involved excretory processes.

3.8 Role of cytochrome P450s in plants

Cytochome P450 is a group of enzymes that play a crucial role for reactions like oxidative and reductive as well as peroxidative metabolism of a diverse group of chemical compounds e.g. steroids fatty acids xenobiotics, therapeutics drugs, and environmental pollutants. Existence of these enzymes with microsomal carbon monoxide binding pigments was first reported in 1958. On the basis of absorption behavior at 450 nm with unique absorption peak, its heamoprotein nature was observed and finally known as cytohrome P450. These groups of enzymes are considered as one of the largest families of enzymes. These enzymes are present in all living organisms in different forms. However, the number of enzymes of various families varies among different organisms. Molecular activity and mechanism of metabolism of P450 for degradation of organic pollutants is limited. Some P450 species show metabolism of xenobiotics in the microsomes of the human liver (Inui and Ohkawa, 2005). It has been estimated that about 53 different CYP genes and nearly 24 pseudogenes are present in humans. In metabolic processes in plants, P450s oxidize carbon and nitrogen, which results in the formation of hydroxyl groups and subsequently the removal of alkyl group (Kawahigashi et al., 2008). Research data reveals that 11 human P450s using recombinant yeast expressing system can metabolize more than 27 herbicides as well as 4 insecticides (Inui and Ohkawa, 2005). To date most of the transgenic plants expressing the mammalian gene are limited for laboratory assessment. However these finding suggest that cytochrome P450 may play a key role for phytoremediation of organic pollutants present in the soil and groundwater.

3.9 Glutathione *S*-transferases and its role in transgenic plants for phytoremediation

Glutathione in conjugation with P450 helps in the detoxification of organic pollutants. GST is an enzyme that belongs to the multifunctional enzyme family found in animals, plants, and microorganisms. Major functions of these enzymes in plants are detoxification and excretion endogenous excretory substances (Wilce and Parker, 1994). In the metabolism of xenobiotics GSTs play a very important role. In normal conditions glutathione is present in reduced form and a small amount is also present in oxidized state (Dixon et al., 1998a,b). GST catalyzes the thiol of reduced glutathione in organic compounds with the help of nucleophilic addition reaction. The glutathione conjugates formed are more hydrophilic in nature and facilitate their exclusion.

All GSTs with plant origins having relative masses of approximately 50 kDa and made up of two subunits with equal relative masses give rise to homodimers or may be heterodimers encoded by a different gene. Each subunit behaves as a kinetically independent active site with distinct binding sites for glutathione with the substrate. Plants contain multiple homo as well as heterodimeric families of GST isoenzymes Dixon et al., 1998a,b). The presence of multiple numbers of isoenzymes facilitates differential and overlapping substrate selectivity resulting in the detoxification of a wide range of organic pollutants. The extent of detoxification or activation depends on the number as well as the amount of isoenzymes present in the plant tissue. Overexpression of these genes in appropriate plant species can enhance the metabolic activities of the plant resulting in the effective phytoremediation of organic pollutants (Abhilash et al., 2009).

3.10 Transgenic plants over expressing glutathione *S*-transferases for enhanced degradation of organic pollutants

Genetically modified poplar plant over expresses the bacterial gene γ ECS (glutabyl cysteine synthetase), a regulatory enzyme which limits the rate of reaction in the biological synthesis of GSH. The transformed plant shows an enhanced level of GSH and its precursor supports the antioxidative protection to the plant cells against various stresses caused by environmental pollutants and other climatic conditions (Noctor and Foyer, 1998). An increased level of glutathione is effective to increase the resistant against chloroacetanilide herbicide in transgenic plant expressing γ ECS. Transgenic tobacco plant over expressing GST I of maize shows enhanced phytoremediation of herbicide like chloroacetanilide. Isoenzyme GST I from maize shows improved catabolic activity against chloroacetanilide herbicide and is considered as involved in the detoxification of several other organic pollutants (Karavangeli et al., 2005). *Phragmites australis* plant showed glutathione dependent on the detoxification of organic pollutants such as acetyl salicylic acid, lamotrigin, etc.. Enhanced tolerance towards atrazine, metolachlor, and phenanthrene was observed by transgenic Brassica juncea plants over expressing γ-glutamyl-cysteine-synthetase and glutathione synthetase (Flocco et al., 2004). Researchers succeeded to introduce human P4502E1 and GST from fungus *Trcihoderma virens* in tobacco plant. The transgenic plant has shown enhanced degradation of anthracene and chloropyriphos (Dixit et al., 2008). It is expected that the transgenic expression of both human P450 and glutathione enzymes in plants will provide enhanced detoxification of organic xenobiotics and results in improved remediation process. The role of GSTs in xenobiotic detoxification and endogenous metabolism is given in Fig. 3.6.

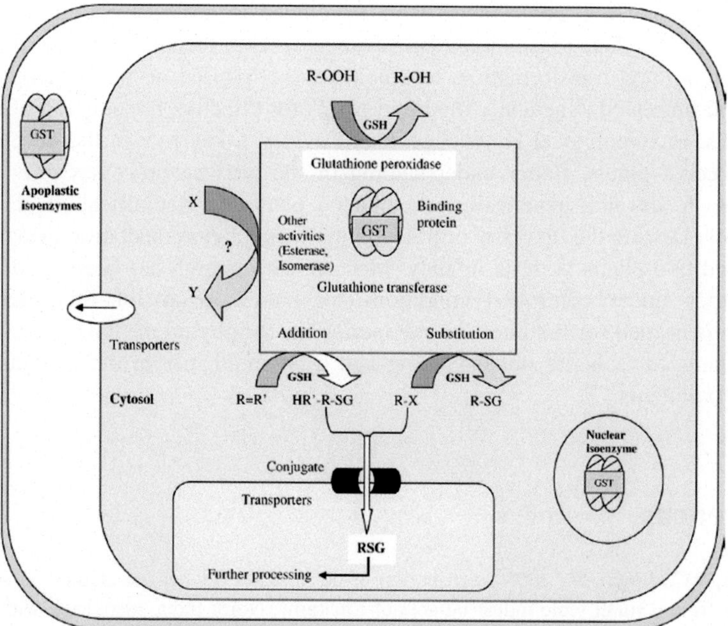

Figure 3.6 Schematic representation of the role of GSTs in xenobiotic detoxification and endogenous metabolism.
Source: Reproduced with permission from Abhilash, P. C., Jamil, S., Singh, N., 2009. Transgenic plants for enhanced biodegradation and phytoremediation of organic xenobiotics. *Biotechnol. Adv.* 27(4), 474-488, Elsevier Inc

It has been already been proven that detoxification reactions occur in phase II and the reactions are performed by GSTs, resulting in the conversion of less toxic less form because of conjugation. Most of the plant GSTs are supposed to be cytosolic in nature. However, there is experimental evidence that isoenzymes are present (Dixon et al., 1998a,b). The alternative activities of GSTs include enzymes like glutathione peroxidase, isomerase, esterase, and binding activities which play additional roles in further metabolism (Dixon et al., 1998a,b).

3.11 Conclusion

Plant association of plants and microbes results in combined action for the degradation, and transformation plays a key role in most of phytoremediation technologies for the degradation of organic pollutants present in soil, water, and sediments. Phytoremediation has been used to treat many classes of organic contaminants including petroleum products, chlorinated solvents, pesticides, and explosives, etc. Phytoremediation has several advantages over conventional technologies such as low cost, environmentally friendly, public acceptance, and versatile nature to degrade different organic pollutants. To improve the efficiency of phytoremediation, a variety of transgenic plants were developed in different countries over the last few decades. Due to technological advancements in the areas of gene identification, cloning, and transformation, techniques were focused on the development of novel and improved genetically modified plants for effective remediation of organic pollutants. Enzymological knowledge also provided a key role in the development of engineered plants. Better understanding of the enzyme processes provide new directions to develop genetically manipulated plants for detoxification of organic pollutants. Despite the diversity of potential options, phytoremediation using genetically modified plants is in its infancy. Most of the research has been conducted in laboratories under controlled conditions for short periods of time. Extensive research is needed for field condition experiments for phytoremediation using transgenic plants for a better understanding and a potential role in the degradation of organic pollutants.

References

Abhilash, P.C., Singh, N., 2008. Distribution of hexachlorocyclohexane isomers in soil samples from a small scale industrial area of Lucknow, North India, associated with lindane production. Chemosphere 73, 1011–1116.

Abhilash, P.C., Jamil, S., Singh, N., 2009. Transgenic plants for enhanced biodegradation and phytoremediation of organic xenobiotics. Biotechnol. Adv. 27 (4), 474–488.

Abou-Shanab, R.A., El-Sheekh, M.M., Sadowsky, M.J., 2019. Role of rhizobacteria in phytoremediation of metal-impacted sites. Emerging and Eco-Friendly Approaches for Waste Management. Springer, Singapore, pp. 299–328.

Best, E.P., Zappi, M.E., Fredrickson, H.L., Sprecher, S.L., Larson, S.L., Ochman, M., 1997. Screening of aquatic and wetland plant species for phytoremediation of explosives-contaminated groundwater from the Iowa Army Ammunition Plant. Ann NY Acad Sci 829, 179–194.

Black, H., 1999. Phytoremediation: a growing field with some concerns. Scientist 13, 5–6.

Carvalho, F.P., 2006. Agriculture, pesticides, food security and food safety. Environ. Sci. Policy 9, 685–692.

Dietz, A., Schnoor, J.L., 2001. Advances in phytoremediation. Environ. Health Perspect. 109, 163–171.

Dixit, P., Singh, S., Mukherjee, P.K., Eapen, S., 2008. Development of transgenic plants with cytochrome P450E1 gene and glutathione-*S*-transferase gene for degradation of organic pollutants. J. Biotechnol. 136, 692–695.

Dixon, D.P., Cole, D.J., Edwards, R., 1998a. Purification, regulation and cloning of a glutathione transferase (GST) from maize resembling the auxin-inducible type-Ill GSTs. Plant Mol Biol 36, 75–87.

Dixon, D.P., Cumminus, I., Cole, D.J., Edwards, R., 1998b. Glutathione-mediated detoxification system in plants. Curr. Opin. Plant Biol. 1, 258–266.

Doty, S.L., James, C.A., Moore, A.L., Vajzovic, A., Singleton, G.L., Ma, C., et al., 2007. Enhanced phytoremediation of volatile environmental pollutants with transgenic trees. P Natl Acad Sci 104 (43), 16816–16821.

Doty, S.L., Shang, T.Q., Wilson, A.M., Tangen, J., Westergreen, A.D., Newman, L.A., et al., 2000. Enhanced metabolism of halogenated hydrocarbons in transgenic plants containing mammalian cytochrome P450 2E1. P Natl Acad Sci 97 (12), 6287–6291.

Eapen, S., Singh, S., D'Souza, S.F., 2007. Advances in development of transgenic plants for remediation of xenobiotic pollutants. Biotechnol. Adv. 25, 442–451.

Flocco, C.G., Lindblom, S.D., Smits, E.A.H.P., 2004. Overexpression of enzymes involved in glutathione synthesis enhances tolerance to organic pollutants in *Brassica juncea*. Int. J. Phytoremediat. 6, 289–304.

Frick, C.M., Farrell, R.E., Germida, J.J., 1999. Assessment of phytoremediation as an in-situ technique for cleaning oil-contaminated sites. In: Petroleum Technology Alliance of Canada Report. Available from: <http://cluin.org/download/remed/phyassess.pdf>.

Gianfreda, L., Rao, M.A., 2004. Potential of extra cellular enzymes in remediation of polluted soils: a review. Enzyme Microbial Technol. 35, 339–354.

Goldstein, R.M., Mallory, L.M., Alexander, M., 1985. Reasons for possible failure of inoculation to enhance bioremediation. Appl. Environ. Microbiol. 50, 977–983.

Hughes, J.B., Shanks, J., Vanderford, M., Lauritzen, J., Bhadra, R., 1997. Transformation of TNT by aquatic plants and plant tissue cultures. J. Chromatogr 518, 361–364.

Hooker, B.S., Skeen, R.S., 1999. Transgenic phytoremediation blasts onto the scene. Nat. Biotechnol. 17, 428.

Inui, H., Ohkawa, H., 2005. Herbicide resistance plants with mammalian P450 monooxygenase genes. Pest Manage. Sci. 61, 286–291.

Johnson, A.C., Worall, F., White, C., Walker, A., Besien, T.J., Williams, R.J., 1997. The potential of incorporated organic matter to reduce pesticide leaching. Toxicol. Environ. Chem. 58, 47–61.

Karavangeli, M., Labrou, N.E., Clonis, Y.D., Tsaftaris, A., 2005. Development of transgenic tobacco plants overexpressing glutathione *S*-transferase I fro chloroacetanilide herbicides phytoremediation. Biomolec. Eng. 22, 121–128.

Kawahigashi, H., Hirose, S., Ohkawa, H., Ohkawa, Y., 2005a. Transgenic rice plants expressing human CYP2B6. J. Agric. Food Chem. 53, 9155–9160.

Kawahigashi, H., Hirose, S., Ohkawa, H., Ohkawa, Y., 2005b. Transgenic rice plants expressing human CYP1A1 remediate the triazine herbicides atrazine and simazine. J. Agric. Food Chem. 53, 8557–8564.

Kawahigashi, H., Hirose, S., Ozawa, K., Ido, Y., Kojima, M., Ohkawa, H., et al., 2005c. Analysis of substrate specificity of pig CYP2B22 and CYP2C49 towards herbicides by transgenic rice plants. Transg. Res. 14, 907–917.

Kawahigashi, H., Hirose, S., Ohkawa, H., Ohkawa, Y., 2008. Transgenic rice plants expressing human P450 genes involved in xenobiotic metabolism for phytoremediation. J Mol Microbiol Biotechnol 15, 212–219.

Lee, C.C., Huffman, G.L., 1989. Innovative thermal destruction technologies. Environ. Prog. 8, 190–199.

Lima, A.L.C., Farrington, J.W., Reddy, C.M., 2005. Combustion-derived polycyclic aromatic hydrocarbons in the environment: a review. Environ. Forens 6, 109–131.

Macek, T., Kotrba, P., Svatos, A., Novakova, M., Demnerova, K., Mackova, M., 2008. Novel roles for genetically modified plants in environmental protection. Trends Biotechnol. 26, 146–152.

Macek, T., Mackova, M., Kas, J., 2000. Exploitation of plants for the removal of organics in environmental remediation. Biotechnol. Adv. 18, 23–34.

Mackova, M., Macek, T., Kucerova, P., Burkhard, J., Trisk, J., Demnerova, K., 1998. Plant tissue cultures in model studies of transformation of polychlorinated biphenyls. Chem. Papers 52, 599–600.

McCrady, J.K., McFarlane, C., Lindstrom, F.T., 1987. The transport and affinity of substituted benzenes in soybean stems. J. Exp. Bot. 38 (11), 1875–1890.

Meagher, R.B., 2000. Phytoremediation of toxic elemental and organic pollutants. Curr. Opin. Plant Biol. 3 (2), 153–162.

Misra, S., Gedamu, L., 1989. Heavy metal tolerant transgenic Brassica napus L. and Nicotiana tabacum L. plants. Theor Appl Genet 78, 161–168.

Newman, L.A., Reynolds, C.M., 2004. Phytodegradation of organic compounds. Curr. Opin. Biotechnol. 15, 225–230.

Noctor, G., Foyer, C.H., 1998. Ascorbate and glutathione: keeping active oxygen under control. Annu. Rev. Plant Physiol. Plant Mol. Biol. 49, 249–279.

Parameswaran, A., Leitenmaier, B., Yang, M., Kroneck, P.M., Welte, W., Lutz, G., et al., 2007. Native Zn/Cd pumping P (1B) ATPase from natural overexpression in a hyperaccumulator plant. Biochem. Biophys. Res. Commun. 36, 51–56.

Pilon-Smits, E., 2005. Phytoremediation. Annu. Rev. Plant Biol. 56, 15–39.

Rugh, C.L., 2004. Genetically engineered phytoremediation: one man's trash is another man's transgene. Trends Biotechnol. 22, 496–498.

Rugh, C.L., Wilde, D., Stack, N.M., Thompson, D.M., Summer, A.O., Meagher, R.B., 1996. Mercuric ion reduction and resistance in transgenic Arabidopsis thaliana plants expressing a modified bacterial merA gene. Proc Natl Acad Sci USA 93, 3182–3187.

Roy, S., Hänninen, O., 1994. Pentachlorophenol: uptake/elimination kinetics and metabolism in an aquatic plant, Eichhornia crassipes. Environ. Toxicol. Chem. Int. J. 13 (5), 763–773.

Salt, D.E., Smith, R.D., Raskin, I., 1998. Phytoremediation. Annu. Rev. Plant Physiol. Plant Mol. Biol. 49, 643–668.

Sanderman, J.H., 1994. Higher plant metabolism of xenobiotics: the 'green liver' concept. Pharmacogenetics 4, 225–241.

Schnabel, W.E., Dietz, A.C., Burken, J.G., Schnoor, J.L., Alvarez, P.J., 1997. Uptake and transformation of trichloroethylene by edible garden plants. Water Res. 31 (4), 816–824.

Schnoor, J.L., 1997. Phytoremediation. Pittsburgh: National Environmental Technology Applications Center. Technology Evaluation Report. TE-97-01.

Siciliano, S.D., Germida, J.J., 1998. Mechanisms of phytoremediation: biochemical and ecological interactions between plants and bacteria. Environ. Rev. 6 (1), 65–79.

Singh, O.V., Jain, R.K., 2003. Phytoremediation of toxic aromatic pollutants from soil. Appl. Microbiol. Biotechnol. 63 (2), 128–135.

Singleton, G.L., 2007. Genetic Analysis of Transgenic Plant for Enhanced Phytoremediation. University of Washington, Seattle, WA.

Sung, K., Munster, C.L., Rhykerd, R., Drew, M.C., Corapcioglu, M.Y., 2003. The use of vegetation to remediate soil freshly contaminated by recalcitrant contaminants. Water Res. 37, 2408–2418.

Suresh, B., Ravishankar, G.A., 2004. Phytoremediation—a novel and promising approach for environmental clean-up. Crit. Rev. Biotechnol. 24, 97–124.

Susarla, S., Medina, V.F., McCutcheon, S.C., 2002. Phytoremediation: an ecological solution to organic chemical contamination. Ecol. Eng. 18, 647–658.

Thompson, P.L., Ramer, L.A., Schnoor, J.L., 1998. Uptake and transformation of TNT by hybrid poplar trees. Environ Sci Technol 32 (7), 975–980.

Trapp, S., Zambrano, K.C., Kusk, K.O., Karlson, U., 2000. A phytotoxicity test using transpiration of willows. Arch. Environ. Con. Tox 39 (2), 154–160.

Vanderford, M., Shanks, J.V., Hughes, J.B., 1997. Phytotransformation of trinitrotoluene (TNT) and distribution of metabolic products in Myriophyllum aquaticum. Biotechnol. Lett 19 (3), 277–280.

Watanabe, M.E., 1997. Phytoremediation on the brink of commercialisation. Environ. Sci. Technol. 31, 182–186.

Wilce, M.C.J., Parker, M.W., 1994. Structure and function of glutathione *S*-transferases. Biochem. Biophys. Acta 1205, 1–18.

Zhuang, X., Chen, J., Shim, H., Bai, Z., 2007. New advances in plant growth-promotingrhizobacteria for bioremediation. Environ. Int. 33, 406–413.

Further reading

Brenter, L.B., Mukherji, S.T., Merchie, K.M., Yoon, J.M., Schnoor, J.L., 2008. Expression of glutathione Stranferases in poplar trees (*Populus trichocarpa*) exposed to 2,4,6-trinitrotoluene (TNT). Chemosphere. 73, 657–662.

Greenberg, B.M., Hunag, X.D., Gurska, Y., Gerhardt, K.E., Wang, W., Lampi, M.A., et al., 2006. Successful field tests of a multi-process phytoremediation system for decontamination of persistent petroleum and organic contaminants. In: Proceedings of the Twenty-Ninth Artic and Marine Oil Spill Program (AMOP). Technical Seminar Vol. 1. Environment Canada; pp. 389–400.

Hayes, J.D., Wolf, C.R., 1990. Molecular mechanisms of drug resistence. Biochem. J. 272, 281–295.

Ishikawa, T., 1992. The ATP-dependent glutathione *S*-conjugate export pump. Trends Biochem. Sci. 17, 463–468.

Kawahigashi, H., Hirose, S., Ohkawa, H., Ohkawa, Y., 2006. Phytoremediation of herbicide atrazine and metolachlor by transgenic rice plants expressing human CYP1A1, CYP2B6 and CYP2C19. J. Agric. Food Chem. 54, 2985–2991.

Kawahigashi, H., 2009. Transgenic plants for phytoremediation of herbicides. Curr. Opin. Biotechnol. 20 (2), 225–230.

Kawahigash, H., Hirose, S., Inui, H., Ohkawa, H., Ohkawa, Y., 2003. Transgenic rice plants expressing human CYP1A1 exudes herbicide metabolites from their roots. Plant Sci. 165, 373–381.

Methods in Biotechnology, 2007. In: Willey, N. (Ed.), Phytoremediation Methods and Reviews, 23. Humana Press, Totowa, NJ.

Meyers, S.K., Deng, S.P., Basta, N.T., Clarkson, W.W., Wilber, G.G., 2007. Long-term explosive contamination in soil. Soil Sedim. Contam. 16, 61–77.

Noctor, G., Strohm, M., Jouanin, L., Kunert, K.J., Foyer, C.H., Rennenberg, H., 1996. Synthesis of glutathione in leaves of transgenic poplar overexpressing [gamma]-glutamylcysteine synthetase. Plant Physiol 112, 1071–1078.

Salvato, J.A., Nemerow, N.L., Agardy, F.J., 2003. Environmental Engineering. John Wiley & Sons, Inc., Hoboken, New Jersey, 1568 pp.

Progress, prospects, and challenges of genetic engineering in phytoremediation

4

*Shweta Jha**
Plant Functional Genomics Lab, Biotechnology Unit, Department of Botany (UGC-CAS), J.N.V. University, Jodhpur, India
*Corresponding author

4.1 Introduction

The rise in various industrial and agricultural anthropogenic activities resulted in a dramatic increase in the concentration of different pollutants in the soil, air, and water, causing environmental pollution. This represents a major global threat to the entire ecosystem, including deterioration of the quality of groundwater, soil properties, human and animal health. A wide range of remediation technologies has been employed to treat polluted sites, including physical, chemical, and biological methods. However, the conventional decontamination techniques like mechanical or chemical treatment and incineration are much more expensive and harmful for soil structure and its microbial fauna, as well as lac a sufficient environmental compatibility and public acceptability. Here, in situ biological remediation methods present a promising and more ecofriendly alternative approach, which consists of the use of microorganisms (bioremediation), and plants (phytoremediation) for efficient clean-up of soil, offering many benefits related to cost and environmental safety (Dhankher et al., 2011; Pandey and Souza-Alonso, 2019). Bioremediation using microbes has its own limitations like, the results of efficient microbial biodegradation ability under controlled lab conditions may not be replicated in actual field conditions. In addition, constant inoculation of microorganisms and application of nutrient media along with maintenance of optimal pH for proper microbial growth are essential over a long time, thereby increasing the maintenance cost of the site.

In this context, phytoremediation [*phyto* (Greek) = plant; *remedium* (Latin) = restoring balance], has been identified as an alternative method for in situ remediation of the contaminated site. It refers to the use of living plants to accumulate, detoxify, or modify harmful pollutants to nonhazardous compounds in the soil, air, or water. Certain hyperaccumulators naturally contain this ability to concentrate and metabolize toxic elements/compounds from the environment, like *Brassica* and *Allium* species, *Salix* spp. *Poplar* spp., *Pteris vitata*, *Helianthus annuus*, *Medicago sativa*, *Thlaspi caerulescens*, etc. (Koźmińska et al., 2018; Pandey and Bajpai, 2019). The common properties of a good hyperaccumulator are high biomass, fast

Bioremediation of Pollutants. DOI: https://doi.org/10.1016/B978-0-12-819025-8.00004-1

growth, and pollutant-tolerance. Phytoremediation has several advantages over other remediation methods, such as cost-efficacy, environmental safety, good public acceptance, robust growth, bioeconomy, renewability, etc. (Pandey and Bajpai, 2019; Pandey and Souza-Alonso, 2019). Furthermore, plants can tolerate much higher concentrations of toxic compounds as compared to microbes, and yield some additional benefits such as substantial reduction of disposition volume of the polluted material, prevention of soil erosion, carbon sequestration, biofuel production, and application to a wider range of neglected sites (Pandey and Souza-Alonso, 2019).

Plants can use different strategies for clean-up of contaminated/polluted sites: phytoextraction, phytostabilization/phytosequestration, phytovolatilization, phytodegradation, and rhizoremediation consisting of rhizofiltration and rhizodegradation (Dhankher et al., 2011; Pandey and Bajpai, 2019). *Phytoextraction* refers to uptake and removal of contaminates from the polluted substrate and subsequent translocation and accumulation in plant aboveground organs, like shoots (Suman et al., 2018). Phytoextracted pollutants cannot be degraded or metabolized but are excluded from the environment. This method is beneficial for removing inorganic pollutants like heavy metals from the soil. However, some organic pollutants may also be processed in the same way due to intrinsic resistance to decomposition. Phytoextraction requires harvesting of the plant tissue with accumulated pollutant. The metals can be recycled by incineration of plant material. If contaminants are adsorbed predominantly by plant root system and precipitated from polluted water, it is called *rhizofiltration.* In some cases the contaminants are sequestered or immobilized/stabilized in the rhizosphere by absorption or adsorption to the root surface, or some compounds are released by the plant roots into the soil or groundwater. This process is called *phytostabilization or phytosequestration.* Plants can also remove toxic volatile compounds through *phytovolatilization*, by uptaking the compounds by roots, followed by in-planta conversion into gaseous metabolites and releasing into the atmosphere by transpiration. This mechanism occurs for both volatile organic compounds (e.g., trichloroethylene or TCE) and heavy metals (e.g., Hg and Se). *Phytodegradation* is the process by which plants can transform or metabolize and degrade the pollutant after uptake from soil or groundwater, in a similar manner in which animals do for degradation of toxins (the "green liver" model). This method is mostly observed for remediation of organic compounds, like herbicides or petroleum hydrocarbons. If degradation takes place in the rhizosphere by microbes inhabiting the root zone of plants, it is called *rhizodegradation.* The combined rhizospheric processes occurring in the vicinity of plants roots, which promote remediation of toxic compounds are sometimes called *rhizoremediation* (Dhankher et al., 2011; Praveen et al., 2019).

However, remediation potential of plants is limited due to the slow rate of degradation, therefore a long time is required to remove toxic compounds from polluted sites, slow growth and low biomass, less efficient metabolism of toxic compounds, and difficult cultivation of hyperaccumulators inhabiting specific environment in other climatic conditions. In addition, potential accumulation of toxic metabolites inside plants and lack of proper disposal of plant tissues with contaminants might pose a risk of food chain contamination. All these problems can be minimized through phytomanagement of polluted sites (Pandey et al., 2015; Pandey and Bajpai, 2019).

4.2 Overview of biotechnological approaches to improve efficiency of phytoremediation

There are mainly two classes of pollutants: inorganic pollutants (e.g., heavy metals) and organic pollutants (herbicides, explosives, polycyclic aromatic hydrocarbon or PAH, halogenated compounds, etc.). The overall efficacy of phytoremediation of both kinds of pollutants can be increased by using genetic engineering approaches. Transgenic technology can introduce novel specific traits in high biomass plants that are not naturally present in them, through harnessing of bacterial or mammalian genes involved in metabolism, uptake, or translocation of various pollutants. These may encode for metal transporters/metal chelators, enzymes for degradation pathway, metal homeostasis, oxidative stress tolerance, or xenobiotic detoxification. The first transgenic plants were generated for developing heavy metal tolerance (Cd and Hg) in tobacco and *Arabidopsis*, expressing yeast metallothionein (MT) and mercuric ion reductase, respectively (Misra and Gedamu, 1989; Rugh et al., 1996). Similarly, the first genetically modified (GM) plant for clean-up of organic pollutants was developed for remediation of explosives and halogenated organic compounds (Doty et al., 2000; French et al., 1999). Afterwards, extensive research attempts were made to generate transgenic plants with enhanced phytoremediation capabilities using different genes for removal of heavy metals, chlorinated solvents, explosives, phenolics, etc., as reviewed extensively by many scientific groups (Dhankher et al., 2011; Koźmińska et al., 2018). In this chapter, we have presented a comprehensive review of current developments in genetic engineering of plants for enhanced phytoremediation potential, and discussed major challenges and future prospects of biotechnological approaches toward improving efficiency of plants for clean-up of the polluted environment.

4.2.1 Engineering for enhanced phytoremediation of inorganic pollutants

Environmental pollution by inorganic pollutants such as metals or metalloids (As, Hg, Se, Cd, Pb, Cu, Mn, Zn) has become a serious concern due to fast growth of industrialization/ and urbanization, thus posing a great threat to human and environmental health (Dhankher et al., 2011). Arsenic (As) is recorded as the most toxic carcinogenic metalloid. Edible crops grown widely in As-contaminated soil (like rice) are a major source of As in the human food supply. As-contamination of drinking water is widespread and has been reported by more than 70 countries, especially in Bangladesh and India, causing adverse health effects on humans (Dhankher et al., 2011). The most common inorganic species of As (i.e., oxoanionic form) consist of arsenate (AsV) and arsenite (AsIII), which are more poisonous and phytotoxic than its organic forms [monomethyl arsenate and dimethyl arsenate (DMA)]. AsV is predominantly found in aerobic soils, and has a close chemical analogy to inorganic phosphate (Pi), thus can penetrate into plant cells through plasma membrane (PM)-localized high affinity phosphate transporters family—PHT1. Being a phosphate analog, it can inhibit several key metabolic processes inside the cell by replacing inorganic Pi such as ATP synthesis, thereby causing exceptional toxicity (Dhankher et al., 2011). In addition, AsV may compete against important nutrients like phosphate due to its Pi analogy and impede their uptake from soil (Dhankher et al., 2011). Another species, AsIII, is mostly found in anaerobic submerged soil and can bind to Cys residues in proteins due to its high affinity to thiol group, which causes perturbation of structure and function of enzymes. It may lead to inhibition of important cellular metabolic reactions, such as ATP synthesis, oxidative phosphorylation, gluconeogenesis, etc. It can also bind to glutathione, resulting in oxidative stress due to excess level of reactive oxygen species (ROS).

Another highly toxic environmental pollutant is mercury (Hg) which has spread globally and poses a great threat to the health of living beings. Its level is continuously increasing in the environment due to natural (such as volcanic eruptions, fire, etc.) as well as man-made activities (fossil fuel burning, gold/silver mining, industrial activities, etc.). It exists in various forms in the environment: inorganic elemental metallic Hg(0) or ionic Hg(II), and organic forms (R-Hg) like methylmercury, dimethylmercury, and phenylmercury. In contrast to the As, organic forms of Hg, especially methylmercury is more toxic than its inorganic forms. This organic species of Hg can cause extensive damage in membrane-bound subcellular organelles like chloroplast, inhibiting photosynthesis and electron-transport chain. It can also get concentrated and biomagnified in the food chain/web, leading to significant health risks to humans. On the other hand, the ionic form Hg(II) can affect water and nutrient transport by damaging PM transporters and aquaporins. It is highly thiol-reactive and can negatively affect activity of enzymes vital for cellular functions. It may also strongly bind to soil organic components, thereby reducing their availability. While metallic Hg is relatively inert and the least toxic among all forms of Hg, it can easily be volatilized and released into atmosphere through transpiration.

Selenium (Se) is an essential micronutrient for several organisms, but shows toxicity at higher concentrations. Since it has a close chemical similarity to sulfur, plants can easily uptake its selenate (SeVI) or selenite (SeIV) forms by sulfate transporters. It can be accumulated in all plant organs, and incorporated into organic compounds by sulfate assimilation pathway or volatilized into the atmosphere. It can transform amino acids, forming selenocysteine (SeCys) and selenomethionine (SeMet), which can further be incorporated into important proteins and enzymes, thereby causing high toxicity. Some hyperaccumulator plants can tolerate high concentrations of Se up to 1% of dry weight. The key mechanism of Se tolerance consists of conversion of SeCys into relatively nontoxic elemental Se or methylation of SeCys or SeMet. The resulting methylated SeCys or SeMet can be further methylated to volatile dimethyl diselenide or dimethyl selenide (DMS), respectively (Dhankher et al., 2011).

Apart from these three important metallic pollutants, trace metals like cadmium (Cd), Manganese (Mn), cobalt (Co), nickel (Ni), and lead (Pb) are other dangerous heavy metal contaminants with high toxicity. Heavy metals cannot be completely metabolized or degraded by the action of microbes or plants, and can only be transported inside plants, and transformed into less toxic forms, or sometimes into volatile form (Suman et al., 2018). As discussed earlier, phytoremediation is a promising tool for removing heavy metals from contaminated sites, and the remediation potential of plants can be enhanced by genetic modifications of plants. A large number of studies have been published describing the use of phytoremediation for heavy metals, including both natural hyperaccumulator and GM plants (reviewed by Suman et al., 2018; Koźmińska et al., 2018; Kotrba et al., 2009; Mosa et al., 2016; Fasani et al., 2018; Krämer and Chardonnens, 2001). There are several pathways that can be genetically manipulated to enhance heavy metal tolerance in plants, including overexpression of genes responsible for metal uptake from soil, accumulation in roots, long-distance translocation to shoots, or sequestration in vacuoles (Suman et al., 2018). The major classes of these genes encode for metal binding ligands or chelator, metal transporters, and detoxifiers. In the following sections, we describe each of these classes in detail, which have employed so far for enhanced phytoremediation of heavy metals/metalloids. In addition, the recent advancements in "omics" approaches have identified some additional novel putative candidates imparting heavy metal tolerance (Mohammed et al., 2019; Visioli and Marmiroli, 2013).

4.2.1.1 Metal binding ligands/chelators

There are mainly three classes of metal binding ligands: phytochelatins (PCs), MTs, and low molecular weight (LMW) organic acids (LMWOA) and amino acids. The genes encoding enzymes for biosynthesis of these compounds can be exploited for genetic engineering of plants for enhanced detoxification of heavy metals and metalloids.

4.2.1.1.1 Phytochelatins

PCs constitute a large family of cysteine-rich thiol-active peptides enzymatically synthesized from its precursor glutathione (GSH). It has been established that GSH-dependent PC biosynthesis pathway is an important mechanism towards heavy metal tolerance in plants. Two key enzymes for PC biosynthesis are γ-glutamyl cysteine synthetase (γ-ECS) and PC synthase (PCS), catalyzing the first step in GSH biosynthesis and the final step in PC biosynthesis, respectively. It has been reported that intermediates of PC biosynthesis pathway act as carriers for long-distance translocation of thiol-active compounds and confer heavy metal tolerance in plants (Table 4.1). A modified bacterial γ-*ECS* gene, either alone or in combination with arsenate reductase (*arsC*) or glutathione synthase (GS), has been shown to enhance accumulation and tolerance for Cd, Cr, As, Pb, Zn, Hg, and Se in transgenic poplar,

Table 4.1 Comprehensive list of transgenic plants for enhanced phytoremediation of heavy metals/metalloids.

Gene transferred	Product	Donor organism	Target plant	Transgene effect	Reference(s)
Metal binding ligands/chelators					
ECS	γ-Glutamyl cysteine synthetase	*Escherichia coli*	*Populus tremula* × *Populus alba*	Enhanced Cd accumulation in roots	Koprivova et al. (2002)
ECS	γ-Glutamyl cysteine synthetase	*E. coli*	*P. tremula* × *P. alba*	Enhanced Cd tolerance and accumulation	He et al. (2015)
ECS	γ-Glutamyl cysteine synthetase	*E. coli*	*Brassica juncea*	Enhanced Se accumulation in leaves	Bañuelos et al. (2005)
γ-ECS + *arsC*	γ-Glutamyl cysteine synthetase and arsenate reductase	*E. coli*	*Arabidopsis thaliana*	Enhanced As tolerance and accumulation	Dhankher et al. (2002)
GS + *ECS*	Glutathione synthase + γ-glutamyl cysteine synthase	*E. coli*	*B. juncea*	Enhanced As, Cd, Pb, Zn, Cr tolerance	Reisinger et al. (2008)
gsh1	γ-Glutamyl cysteine synthetase	*E. coli*	*P. tremula* × *P. alba* (*cytosol*)	Affects the photosynthetic apparatus and improved tolerance for heavy metals	Ivanova et al. (2011)
GCS–GS operon	γ-Glutamyl cysteine synthetase and glutathione synthetase operon	*Streptococcus thermophilus*	*Beta vulgaris*	Enhanced Cd, Cu, Zn tolerance and accumulation	Liu et al. (2015)

AtPCS1	Phytochelatin synthase	*A. thaliana*	*Nicotiana tabacum*	Enhanced Cd tolerance and accumulation	Pomponi et al. (2006)
AtPCS1	Phytochelatin synthase	*A. thaliana*	*B. juncea*	Enhanced tolerance for As and Cd	Gasic and Korban (2007)
AtPCS1	Phytochelatin synthase	*A. thaliana*	*A. thaliana*	Enhanced As tolerance and Cd hypersensitivity	Li et al. (2004)
AtPCS1	Phytochelatin synthase	*A. thaliana*	*N. tabacum*	As and Cd accumulation and detoxification, root cyto-histological damages	Zanella et al. (2016)
AtPCS1	Phytochelatin synthase	*A. thaliana*	*A. thaliana (Chloroplast)*	No significant effect on Cd tolerance and accumulation, As sensitivity	Picault et al. (2006)
AtPCS1	Phytochelatin synthase	*A. thaliana*	*A. thaliana, N. tabacum*	Cd tolerance	Brunetti et al. (2011)
AtPCS1 and CePCS	Phytochelatin synthase	*A. thaliana, Caenorhabditis elegans*	*N. tabacum*	Cd hypersensitivity or enhanced Cd tolerance	Wojas et al. (2008)
CdPCS1	Phytochelatin synthase	*Ceratophyllum demersum*	*A. thaliana, tobacco*	Enhanced accumulation of As, Cd	Shukla et al. (2012, 2013)
TaPCS1	Phytochelatin synthase	*Triticum aestivum*	*Oryza sativa*	Enhanced Cd accumulation and sensitivity	Wang et al. (2012)
TaPCS1	Phytochelatin synthase	*T. aestivum*	*N. tabacum*	Enhanced Pb and Cd tolerance	Gisbert et al. (2003)

(*Continued*)

Table 4.1 (Continued)

Gene transferred	Product	Donor organism	Target plant	Transgene effect	Reference(s)
OsPCS1	Phytochelatin synthase	*O. sativa*	*O. sativa*	Reduction of As in grains	Hayashi et al. (2017)
OsPCS1	Phytochelatin synthase	*O. sativa*	*Rice T-DNA mutant and Tos17 insertion lines*	Increased sensitivity to As and Cd	Uraguchi et al. (2017)
OsPCS2	Phytochelatin synthase	*O. sativa*	*Saccharomyces cerevisiae*	Mitigation of cadmium and arsenic stresses	Das et al. (2017)
PCS1 + GSH1	Phytochelatin synthase and γ-glutamyl cysteine synthetase	*Allium sativum + S. cerevisiae*	*A. thaliana*	Enhanced tolerance and accumulation for As, Cd	Guo et al. (2008)
GS	Glutathione synthase	*E. coli*	*B. juncea*	Enhanced Se accumulation in leaves	Bañuelos et al. (2005)
gshII	Glutathione synthetase	*E. coli*	*B. juncea*	Enhanced Cd tolerance and accumulation	Liang Zhu et al. (1999)
CUP 1	Yeast metallothionein	*S. cerevisiae*	*N. tabacum*	Enhanced Cu uptake	Thomas et al. (2003)
MT	Metallothionein	*Silene vulgaris*	*N. tabacum*	Enhanced Cd accumulation	Gorinova et al. (2007)
MTA	Metallothionein A	*Pisum sativum*	*A. thaliana*	Enhanced Cu^{2+} accumulation in roots	Evans et al. (1992)
MTA1	Metallothionein A1	*P. sativum*	*P. alba*	Enhanced tolerance for Cu, Zn	Balestrazzi et al. (2009)

MTA1	Metallothionein A1	*P. sativum*	*P. alba*	Enhanced accumulation for Cu, Zn	Turchi et al. (2012)
MT1	Metallothionein 1	*Cajanus cajan*	*A. thaliana*	Enhanced accumulation and tolerance for Cu, Cd	Sekhar et al. (2011)
MT1	Metallothionein 1	*Elsholtzia haichowensis*	*N. tabacum*	Enhanced tolerance and accumulation for Cu	Xia et al. (2012)
MT4	Metallothionein 4	*A. thaliana*	*A. thaliana*	Enhanced Cu accumulation and Cu and Zn tolerance	Rodríguez-Llorente et al. (2010)
MT2	Metallothionein 2	*B. juncea*	*A. thaliana*, *E. coli*	Enhanced Cu and Cd tolerance	Zhigang et al. (2006)
MT2a	Metallothionein 2a	*Iris lactea* var. chinensis	*A. thaliana*	Enhanced Cd tolerance	Gu et al. (2014)
MT2b	Metallothionein 2b	*I. lactea* var. chinensis	*A. thaliana*	Enhanced accumulation and tolerance for Cu	Gu et al. (2015)
MTII	Metallothionein II	*S. cerevisiae*	*N. tabacum*	Enhanced Cd and Zn accumulation	Daghan et al. (2013)
MT1 and MT2	Metallothioneins 1,2	*Brassica campestris*	*A. thaliana*	Enhanced Cu accumulation and Cu, Cd tolerance	Lv et al. (2013)
AtNAS1	Nicotianamine synthase	*A. thaliana*	*N. tabacum*	Enhanced Fe accumulation and Ni tolerance	Douchkov et al. (2005)
HvNAS1	Nicotianamine synthase	*Hordeum vulgare*	*A. thaliana*, *N. tabacum*	Enhanced accumulation and tolerance for Cd, Ni, Cu, Mn, Zn, Fe	Kim et al. (2005)
NcNAS1	Nicotianamine synthase	*Noccaea caerulescens*	*A. thaliana*	Enhanced accumulation and tolerance for Ni	Pianelli et al. (2005)

(*Continued*)

Table 4.1 (Continued)

Gene transferred	Product	Donor organism	Target plant	Transgene effect	Reference(s)
Metal transporters					
HMA4	Heavy metal transporting ATPase4	*A. thaliana*	*A. thaliana*	Enhanced translocation and tolerance for Zn, Cd	Verret et al. (2004)
HMA4 + MT2	Heavy metal transporting ATPase4 and metallothionein 2	*A. thaliana*	*N. tabacum*	Enhanced translocation of Cd and Zn and Cd tolerance	Grispen et al. (2011)
HMA3	Heavy metal transporting ATPase3	*Sedum alfredii*	*S. alfredii*	Enhanced tolerance and hyperaccumulation of Cd	Zhang et al. (2016)
CAX2 and CAX4	Calcium exchangers, divalent cation/ proton antiporters	*A. thaliana*	*N. tabacum*	Enhanced tolerance for Cd, Mn, Zn	Korenkov et al. (2007)
CAX2	Calcium exchanger 2/ vacuolar Ca^{2+}/H^{+} antiporter	*A. thaliana*	*N. tabacum*	Enhanced Mn tolerance	Hirschi et al. (2000)
CAXcd	Mutated form of CAX1	*A. thaliana*	*Petunia x hybrida*	Enhanced accumulation and tolerance for Cd	Wu et al. (2011)
CzcC, CzcB and CzcA	CzcCBA efflux system, cation–proton antiporter	*Pseudomonas putida*	*N. tabacum*	Reduced accumulation and Cd uptake	Nesler et al. (2017)
OsACA6	P-type 2B Ca(2 +) ATPase	*O. sativa*	*N. tabacum*	Enhanced tolerance for Cd	Shukla et al. (2014)

MRP7	Vacuolar transporter	*A. thaliana*	*N. tabacum*	Enhanced Cd tolerance and root accumulation	Wojas et al. (2009)
VP1	Vacuolar H^+ pyrophoshatase	*T. aestivum*	*N. tabacum*	Enhanced Cd accumulation and tolerance	Khoudi et al. (2012)
PvACR3	As(III) vacuolar antiporter	*Pteris vittata*	*A. thaliana*	Enhanced tolerance and accumulation of As	Chen et al. (2013)
YCF1	Vacuolar transporter	*S. cerevisiae*	*A. thaliana*	Enhanced accumulation and tolerance for Cd, Pb	Song et al. (2003)
YCF1	Vacuolar transporter	*S. cerevisiae*	*P. alba* × *P. tremula*	Enhanced accumulation and tolerance for Cd	Shim et al. (2013)
PHT1 and PHT7 + *YCF1*	Phosphate transporter and vacuolar transporter	*A. thaliana* + *S. cerevisiae*	*N. tabacum*	Enhanced accumulation and tolerance for As	LeBlanc et al. (2013)
AtABCC1 and AtABCC2	ABCC-type transporters	*A. thaliana*	*S. cerevisiae*, *A. thaliana*	Enhanced Cd, Hg tolerance	Park et al. (2012)
OsABCC1	C-type ABC transporter	*O. sativa*	*O. sativa*	Detoxification and reduction of As in rice grains	Song et al. (2014)
AtABCC3	ABC-type transporter	*A. thaliana*	*A. thaliana*	Phytochelatin-mediated Cd tolerance	Brunetti et al. (2015)
AtABCC1 and AtABCC2	ABCC-type phytochelatin transporters	*A. thaliana*	*Yeast*, *A. thaliana*	Increased As tolerance	Song et al. (2010a)
ScYCF1, *OsABCC1*, *γ-ECS*	Yeast cadmium factor, ABC transporter, γ-glutamylcysteine synthetase	*S. cerevisiae*, *O. sativa*, *E. coli*	*O. sativa*	Lower grain As	Deng et al. (2018)

(*Continued*)

Table 4.1 (Continued)

Gene transferred	Product	Donor organism	Target plant	Transgene effect	Reference(s)
OsPTR7 (OsNPF8.1) (KO/KD)	Peptide transporter	*O. sativa*	*O. sativa*	Reduction in dimethyl arsenate accumulation in grains	Tang et al. (2017)
ATM3	Mitochondrial transporter	*A. thaliana*	*A. thaliana*	Enhanced shoot accumulation of Cd and Cd, Pb tolerance	Kim et al. (2006)
ZntA	Zinc transporter	*E. coli*	*A. thaliana*	Enhanced tolerance to Pb and Cd but decreased accumulation	Lee et al. (2003)
tzn1	Zinc transporter	*Neurospora crassa*	*N. tabacum*	Enhanced Zn accumulation without cotransport of Cd	Dixit et al. (2010)
NiCoT	Co–Ni permease/ transporter	*Rhodopseudomonas palustris*	*N. tabacum*	Increased acquisition of Co and Ni	Nair et al. (2012)
rhlA and rhlB	Metal transporters	*Pseudomonas aeruginosa*	*N. tabacum*	Increased Al tolerance	Brichkova et al. (2007)
tcu-1	Copper transporter	*N. crassa*	*N. tabacum*	Enhanced uptake of Cu	Singh et al. (2011)

Detoxifiers					
merA	Mercuric ion reductase	*E. coli*	*A. thaliana*	Enhanced volatilization and tolerance for Hg	Rugh et al. (1996)
merA	Mercuric ion reductase	*E. coli*	*Liriodendron tulipifera*	Enhanced volatilization and tolerance for Hg	Rugh et al. (1998)
merA	Mercuric ion reductase	*E. coli*	*Populus deltoides*	Enhanced volatilization and tolerance for Hg	Che et al. (2003)
merA	Mercuric ion reductase	*E. coli*	*N. tabacum*	Enhanced volatilization and tolerance for Hg	He et al. (2001), Heaton et al. (2005), and Haque et al. (2010)
mer A	Mercuric ion reductase	*E. coli*	*O. sativa*	Enhanced volatilization and tolerance for Hg	Heaton et al. (2003)
merA	Mercuric ion reductase	*E. coli*	*Liquidambar styraciflua* × *Liquidambar formosana*	Enhanced volatilization and tolerance for Hg	Dai et al. (2009)
merB	Organomercurial lyase	*E. coli*	*A. thaliana*	Enhanced tolerance for monomethyl mercuric chloride and phenyl mercuric acetate	Bizily et al. (1999)
merA + *merB*	Mercuric ion reductase and organomercurial lyase	*E. coli*	*A. thaliana*	Enhanced tolerance and degradation of organic Hg	Bizily et al. (2000, 2003)
merA + *merB*	Mercuric ion reductase and organomercurial lyase	*E. coli*	*N. tabacum* (*chloroplast*)	Enhanced tolerance for phenyl mercuric acetate	Ruiz et al. (2003)

(*Continued*)

Table 4.1 (Continued)

Gene transferred	Product	Donor organism	Target plant	Transgene effect	Reference(s)
merA + *merB*	Mercuric ion reductase and organomercurial lyase	*E. coli*	*P. deltoides*	Enhanced tolerance for phenyl mercuric acetate and Hg degradation	Lyyra et al. (2007)
merA + *merB*	Mercuric ion reductase and organomercurial lyase	*E. coli*	*N. tabacum (chloroplast)*	Enhanced uptake, translocation and volatilization of phenylmercuric acetate	Hussein et al. (2007)
MerC	Mercury transporter	*Acidithiobacillus ferrooxidans*	*A. thaliana; N. tabacum*	Enhanced Hg accumulation and sensitivity	Sasaki et al. (2006)
MerC	Mercury transporter	*Pseudomonas K-62*	*A. thaliana (root epidermis)*	Efficient mercury accumulation in shoots	Uraguchi et al. (2019)
merT gene, polyphosphate kinase gene (ppk)	Mercury transporter mercury chelator, polyphosphate (polyP)	*merT—Pseudomonas K-62, ppk—Klebsiella aerogenes*	*N. tabacum*	Accelerated uptake and enhanced accumulation of Hg	Nagata et al. (2009)
merB,ppk, merT	Organomercurial lyase (MerB), polyphosphate kinase and mercury transporter	*merB—E. coli, merT—Pseudomonas K-62, ppk—K. aerogenes*	*N. tabacum*	Enhanced tolerance for methylmercury	Nagata et al. (2010)

merP	Mercury binding protein	*Bacillus megaterium*	*A. thaliana*	Enhanced tolerance and accumulation of Hg, Cd and Pb	Hsieh et al. (2009)
APS	ATP sulfurylase	*A. thaliana*	*B. juncea*	Enhanced selenate uptake, reduction to organic Se forms, and Se tolerance	Pilon-Smits et al. (1999)
APS	ATP sulfurylase	*A. thaliana*	*B. juncea*	Enhanced Se accumulation in leaves	Bañuelos et al. (2005)
APS1	ATP sulfurylase	*A. thaliana*	*B. juncea*	Enhanced shoot accumulation of Cd, Cu, Cr, Mo, V, W	Wangeline et al. (2004)
SL	Selenocysteine lyase	*Mus musculus*	*A. thaliana*	Enhanced Se tolerance and accumulation	Pilon et al. (2003)
SMT	Selenocysteine methyltransferase	*Astragalus bisulcatus*	*A. thaliana*, *B. juncea*	Enhanced Se accumulation, tolerance and volatilization	LeDuc et al. (2004)
APS1 + SMT	ATP sulfurylase and selenocysteine methyltransferase	*A. thaliana and A. bisulcatus*	*B. juncea*	Enhanced tolerance and accumulation of Se	LeDuc et al. (2006)
SL + SMT	Selenocysteine lyase and selenocysteine methyltransferase	*Mus musculus and A. bisulcatus*	*B. juncea*	Enhanced Se accumulation	Bañuelos et al. (2007)
CGS	Cystathionine-*c*-synthase	*A. thaliana*	*B. juncea*	Enhanced Se volatilization	Van Huysen et al. (2003)
arsC	Arsenate reductase	*E. coli*	*N. tabacum and A. thaliana*	Enhanced accumulation and tolerance for Cd	Dhankher et al. (2003)

(*Continued*)

Table 4.1 (Continued)

Gene transferred	Product	Donor organism	Target plant	Transgene effect	Reference(s)
AtACR2	Arsenic reductase 2	*A. thaliana*	*N. tabacum*	Enhanced tolerance to As	Nahar et al. (2017)
polyP/ppk	Polyphosphate /polyphosphate kinase	*K. aerogenes*	*N. tabacum*	Enhanced accumulation and remediation of Hg	Nagata et al. (2006a,b)
RCS1	Cysteine synthase	*O. sativa*	*N. tabacum*	Enhanced tolerance for Cd	Harada et al. (2001)
CS	Cysteine synthase	*Spinacia oleracea*	*N. tabacum*	Enhanced accumulation of Cd and tolerance for Cd, Se, Ni	Kawashima et al. (2004)
Signaling and regulatory proteins					
ZAT6	ZAT transcription factor	*A. thaliana*	*A. thaliana*	Increased Cd tolerance and accumulation	Chen et al. (2016)
WRKY12	WRKY transcription factor	*A. thaliana*	*A. thaliana*	Negative regulation of Cd accumulation and tolerance	Han et al. (2019)
WRKY28 (loss of function)	WRKY transcription factor	*O. sativa*	*O. sativa*	Reduced phosphate and arsenate accumulation, inhibition of root elongation and lower grain yield	Wang et al. (2018)

CBP4	Plasma membrane calmodulin-binding protein	*N. tabacum*	*N. tabacum*	Enhanced Ni tolerance and Pb hypersensitivity	Arazi et al. (1999)
CBP4 (truncated version) CBP4ΔC, CNGC1	Plasma membrane channel calmodulin-binding protein, cyclic nucleotide-gated ion channel-1 related protein	*N. tabacum, A. thaliana*	*N. tabacum*	Enhanced tolerance for Pb	Sunkar et al. (2000)
DEP1	G protein	*O. sativa*	*A. thaliana*	Enhanced accumulation and tolerance for Cd	Kunihiro et al. (2013)
CAT3	Catalase	*B. juncea*	*N. tabacum*	Enhanced tolerance for Cd	Guan et al. (2009)
APX1 (knockout)	Cytosolic ascorbate peroxidase	*A. thaliana*	*A. thaliana*	Enhanced tolerance for Pb	Jiang et al. (2017)
APX1 (loss of function)	Cytosolic ascorbate peroxidase	*A. thaliana*	*A. thaliana*	Enhanced tolerance for Se	Jiang et al. (2016)
GST	Glutathione transferase	*Trichoderma virens (fungus)*	*N. tabacum*	Enhanced Cd tolerance without enhancing its accumulation	Dixit et al. (2011a)
GST + CYP2E1	Cytochrome P450 and glutathione-*S*-transferase	*Homo sapiens*	*Medicago sativa*	Enhanced tolerance for heavy metals and organic pollutants	Zhang and Liu (2011)
accd	1-ACC deaminase	*Enterobacter cloacae*	*Solanum lycopersicum*	Enhanced accumulation and tolerance for heavy metals	Grichko et al. (2000)
accd	ACC deaminase	*E. cloacae*	*Brassica napus*	Enhanced accumulation and tolerance for As	Nie et al. (2002)

(*Continued*)

Table 4.1 (Continued)

Gene transferred	Product	Donor organism	Target plant	Transgene effect	Reference(s)
accd	ACC deaminase	*P. putida*	*Brassica napus*	Enhanced tolerance for Ni	Stearns et al. (2005)
iaaM + accd	Tryptophan monooxygenase and ACC deaminase	*Agrobacterium tumefaciens + P. putida*	*Petunia x hybrida; N. tabacum*	Increased accumulation and tolerance for Co, Cu	Zhang et al. (2008)
Other proteins					
ACBP1	Acyl-CoA-binding protein	*A. thaliana*	*A. thaliana*	Enhanced accumulation and tolerance for Pb	Xiao et al. (2008)
Hb1	Class 1 hemoglobin	*N. tabacum*	*N. tabacum*	Enhanced metal tolerance by reducing Cd uptake and accumulation	Lee and Hwang (2015)
MAN3/XCD1	Endo-β-mannase	*A. thaliana*	*A. thaliana*	Enhanced accumulation and tolerance for Cd	Chen et al. (2015)
Lhcb2	Chlorophyll a/b-binding protein	*S. alfredii*	*N. tabacum*	Enhanced tolerance for Cd	Zhang et al. (2011a)
PSE1	Pb-sensitive1	*A. thaliana*	*A. thaliana*	Enhanced tolerance for Pb	Fan et al. (2016)
FC1	Ferrochelatase-1	*A. thaliana*	*A. thaliana*	Enhanced tolerance for Cd	Song et al. (2017)
WaarsM	Arsenic methyltransferase	*Westerdykella aurantiaca (soil fungus)*	*O. sativa, S. cerevisiae*	Reduced As accumulation in grains	Verma et al. (2018, 2019)

SAT	Acetyltransferase	*Noccaea goesingense*	*A. thaliana*	Enhanced tolerance for Ni	Freeman et al. (2004)
MTP1	Metal tolerance protein	*N. goesingense*	*A. thaliana*	Enhanced accumulation and tolerance for Zn	Gustin et al. (2009)
MTP1	Metal tolerance protein	*O. sativa*	*N. tabacum*	Enhanced accumulation and tolerance for Cd	Das et al. (2016)

ABC, ATP-binding cassette; *ACC*, aminocyclopropane.

Indian mustard, tobacco, and *Arabidopsis* (Koprivova et al., 2002; Bañuelos et al., 2005; He et al., 2015; Dhankher et al., 2002; Reisinger et al., 2008; Li et al., 2006). Similarly, overexpression of bacterial *gsh1* encoding γ-glutamyl cysteine synthetase in the cytosol of poplar hybrid tree (*Populus tremula* × *Populus alba*) improved the performance of transgenic plants on heavy metal-contaminated soil under field conditions (Ivanova et al., 2011). Liu et al. (2015) observed that ectopic expression of *Streptococcus thermophilus* γ-glutamyl cysteine synthetase–GS (GCS–GS) operon in transgenic sugar beets exhibited enhanced tolerance for Cd, Zn, and Cu (Table 4.1).

Several studies have proven PCS gene for PC biosynthesis as a promising candidate for improving heavy metal tolerance in transgenic plants. Some important examples are presented in Table 4.1. In the majority of these studies, PCS from *Arabidopsis thaliana (AtPCS1)* has been widely exploited, and reported to enhance accumulation and detoxification for Cd and/or As in transgenic tobacco, Indian mustard, and *Arabidopsis* (Pomponi et al., 2006; Gasic and Korban, 2007; Li et al., 2004; Zanella et al., 2016; Brunetti et al., 2011). Li et al. (2004) reported that constitutive expression of *AtPCS1* under strong actin promoter resulted in 20–100 times more biomass accumulation in transgenic *Arabidopsis* plants on 250–300 μM arsenate, leading to enhanced As tolerance, but hypersensitivity to Cd, as compared to WT plants. The heterologous expression of *AtPCS1* in tobacco resulted in a twofold increase in Cd^{2+} accumulation in both roots and shoots, but did not affect long-distance root-to-shoot transport of Cd^{2+} (Pomponi et al., 2006). Transgenic plants of Indian mustard overexpressing *AtPCS1* exhibited significantly higher concentrations of PCs and thiols than wild-type (WT) plants, and subsequent tolerance to Cd and As, but did not increase the accumulation potential of As in shoopdelts of transgenic plants (Gasic and Korban, 2007). In a recent study, ectopic expression of *AtPCS1* in transgenic tobacco plants led to a significant increase in PCs concentration, accumulation in roots, and detoxification of As and Cd, but did not prevent cellular damages to lateral roots, increase in cell vacuolization, thinning and crushing of endodermal and stelar cells and cellular and nuclear hypertrophy (Zanella et al., 2016).

Chloroplasts or plastids are the site for accumulation of PC biosynthesis precursor—glutathione (GSH). Cytosolic overexpression of *AtPCS1* in transgenic *Arabidopsis* plants induced hypersensitivity to Cd, but tolerance to As. In contrast, chloroplast targeting of this gene using RbcS transit peptide resulted in marked sensitivity to As, and no significant effect on Cd tolerance and accumulation in transgenic plants as compared to WT plants (Picault et al., 2006). The species-specific differences for PC-dependent heavy metal tolerance were demonstrated by Brunetti et al. (2011). He showed that *Arabidopsis* seedlings overexpressing *AtPCS1* (*AtPCSox* lines) were less tolerant to Cd at relatively lower Cd concentrations, but more tolerant than WT at higher Cd concentrations. In contrast, transgenic tobacco plants expressing the same gene exhibited Cd hypersensitivity at higher Cd concentrations, whereas more Cd tolerance than WT at relatively lower Cd concentrations. This difference might be due to dissimilar levels of endogenous PCs and GSH in diverse plant species. The PCS genes from various other organisms have also been used for enhancing the ability of plants for remediation of heavy metals or metalloids. Wojas et al. (2008) reported contrasting Cd tolerance in transgenic tobacco plants overexpressing different PCS genes from *A. thaliana* and *Caenorhabditis elegans. AtPCS1* transgenic lines were Cd-hypersensitive showing no significant difference in Cd accumulation than WT, whereas plants expressing *CePCS* exhibited enhanced tolerance and accumulation for Cd. This disparity might be due to species-specific variations in PCS enzyme produced by different transgenes. The ectopic expression of *CdPCS1* from an aquatic submerged rootless macrophyte *Ceratophyllum demersum* (a heavy metal hyperaccumulator) in transgenic *Arabidopsis* and tobacco led to a several-fold increase in levels of PCs and its precursor nonprotein thiols, which subsequently resulted in enhanced accumulation and tolerance for heavy metal(loid)s Cd and As without any adverse effect on plant growth (Shukla et al., 2012, 2013). Similarly, transgenic tobacco seedlings overexpressing wheat PCS1 (*TaPCS1*) gene showed enhanced tolerance for Cd and Pb, and could be able to grow and accumulate almost double concentrations of Pb in heavily contaminated soils, as compared to WT plants (Gisbert et al., 2003). In contrast, the ectopic expression of *TaPCS1* in transgenic rice exhibited a substantial increase in Cd accumulation in shoots, but hypersensitivity to Cd, which might be due to altered composition of nonprotein thiols and depletion of GSH pool, leading to oxidative stress in transgenic plants (Wang et al., 2012). Recently, the rice PCS *OsPCS1* was also reported to play a crucial role in heavy metal tolerance. Transgenic plants overexpressing this gene exhibited a significant reduction in As in rice grains, as compared to WT (Hayashi et al., 2017). In a separate study, *OsPCS1* mutants showed hypersensitivity to As and Cd, but contrasting accumulation pattern for As and Cd in grains of the T-DNA and Tos17 insertion lines of rice, depicting diverse PC-dependent pathways for different heavy metals (Uraguchi et al., 2017). Another rice PC gene *OsPCS2* contains two alternatively spliced forms—*OsPCS2a*, and *OsPCS2b* containing the premature termination codon. In contrast to *OsPCS2b*, the transgenic yeast cells harboring *OsPCS2a* showed an increase in accumulation and tolerance for Cd and As (Das et al., 2017). The gene stacking or pyramiding have a great potential for enhancing phytoremediation capabilities of plants. The coexpression of PCS and γ-glutamylcysteine synthetase from garlic and baker's yeast (*AsPCS1* and *ScGSH1*) in *Arabidopsis* resulted in higher accumulation and more tolerance for As and Cd, as compared to single-gene transformants, due to increased synthesis of total PCs in dual-gene transgenic lines (Guo et al., 2008).

4.2.1.1.2 Metallothioneins

MTs are a direct product of gene expression consisting of Cys-rich proteins. They are classified into four groups on the basis of structure of their Cys-motifs. The MTs are able to chelate heavy metal ions via these Cys-rich motifs, thereby playing crucial roles in metal homeostasis or detoxification (Koźmińska et al., 2018). Several studies have conducted so far for overexpression of MT genes for enhanced phytoremediation potential of plants for heavy metals (Kotrba et al., 2009; Koźmińska et al., 2018). Some important examples are presented in Table 4.1. In earlier studies, pea MT gene (*PsMTA1*) was shown to enhance Cu2 + accumulation in roots of transgenic *Arabidopsis* plants (Evans et al., 1992). The same gene conferred tolerance for Cu and Zn, and protection against DNA damage when overexpressed in white poplar tree (Balestrazzi et al., 2009; Turchi et al., 2012). Similarly, the overexpression of type-2 MT from Indian mustard (*BjMT2*) in *Escherichia coli* and transgenic *Arabidopsis* plants resulted in enhanced tolerance for Cu^{2+} and Cd^{2+} exposure than control (Zhigang et al., 2006). Transgenic lines showed enhanced shoot growth and chlorophyll content but inhibited root elongation under control conditions, whereas identical root growth was under 50−100 μM Cu^{2+} to WT, depicting differential regulation of shoot and root development by overexpression of MT2 (Zhigang et al., 2006). The overexpression of MT gene from *Silene vulgaris* (*SvMT*) improved phytoaccumulation of Cd in roots and leaves of transgenic tobacco (Gorinova et al., 2007). Physiological and biochemical analysis of transgenic seedlings revealed tissue level-dependent visual Cd toxicity symptoms and water balance disturbance. The overexpression of MT in chloroplast was shown to increase Hg accumulation and tolerance in transgenic plants (Ruiz et al., 2011). In two separate studies, yeast MTs (*CUP1 or ScMTII*) was shown to enhance Cu uptake and seven times more Cu accumulation, or Cd/Zn accumulation in transgenic tobacco plants (Thomas et al., 2003; Daghan et al., 2013).

Heavy metal tolerance in plants might be associated with maintenance of cellular homeostasis by modulation of ROS scavenging in plant cells. Xia et al. (2012) observed that the heterologous expression of MT1 from *Elsholtzia haichowensis* (*EhMT1*) increased accumulation and tolerance for Cu, while reducing ROS production. The heterologous expression of the type-2 MTs from extreme halophytes *Iris lactea* (*IlMT2a and IlMT2b*) and *Salicornia brachiata* (*SbMT2*) resulted in improved tolerance for heavy metals in transgenic *Arabidopsis* and tobacco, suggesting some similarity in stress-responsive mechanisms for heavy metal and salinity stresses (Gu et al., 2014, 2015; Chaturvedi et al., 2014).

4.2.1.1.3 Low molecular weight organic and amino acid

The genes encoding LMWOA (e.g., citric/oxalic acid) and nonproteinogenic amino acids (e.g., nicotianamine) can also be exploited to improve heavy metal tolerance in plants (Table 4.1). The nicotianamine synthase (NAS1) is the primary enzyme for nicotianamine biosynthesis in plants, and is a potential candidate for chelation of heavy metal ions. It catalyzes trimerization of *S*-adenosyl methionine to nicotianamine. The overexpression of *AtNAS1* gene in transgenic tobacco resulted in enhanced tolerance for Ni and Fe accumulation (Douchkov et al., 2005). Similarly, elevated levels of nicotianamine by overexpressing *HvNAS1* or *NcNAS1* from *Hordeum vulgare* or *Noccaea caerulescens*, respectively in transgenic *Arabidopsis* or tobacco conferred more accumulation and tolerance to heavy metals, especially Ni (Kim et al., 2005; Pianelli et al., 2005). The transgenic lines were able to grow on soils having high contamination of Ni, which might be due to vacuolar sequestration of Ni−NA complex or translocation to shoots (Kim et al., 2005). LMWOA, such as citric acid and oxalic acid also have the ability to chelate heavy metal ions. The genes encoding metabolism (e.g., malate dehydrogenase, citrate synthase, phosphoenolpyruvate carboxylase) or transport (e.g., citrate transporters, MATE family transporters) of these compounds have been exploited for genetic engineering of plants for improved heavy metal tolerance (Koźmińska et al., 2018).

Besides these three major metal binding ligands, the bacterial polyphosphate (polyP) or polyphosphate kinase (ppk) were shown to decrease the cytotoxicity of Hg^{2+}, via metal chelation and conversion into less toxic Hg-polyP/ppk complex. The heterologous expression of *polyP* or *ppk* genes from *Klebsiella aerogenes* resulted in enhanced scavenging and significantly more accumulation of mercury in transgenic tobacco on semisolid medium containing toxic levels of Hg^{2+} and on simulated soils, as compared to WT control plants (Nagata et al., 2006a,b). The transgenic plants showed normal growth, flowering and seed setting similar to non-transgenic plants, without any morphological abnormality.

4.2.1.2 Metal transporters

Heavy metal transporters belong to large gene families in plants, and are localized to both PM and tonoplast. These transporters can be exploited to enhance phytoremediation potential of transgenic plants via increasing metal uptake and accumulation as well as sequestration of metals in vacuolar lumen, or to inhibit long-distance transport to shoots. Some important examples of these transporters are presented in Table 4.1 that have been used for plant genetic engineering, and include heavy metal transporting ATPase (HMA)3/4, cation exchanger calcium exchanger (CAX) 2/4, YCF1, phosphate transporters PHT1/7, MRP7, vacuolar H^{+}-phosphatase VP1, ZNT, ABCC-type transporters, ACR3, etc.

In plants, toxic metal ions are detoxified in a two-step process: formation of a metal–PC complex by chelation with PCs, and sequestration of this complex into vacuoles with the help of ABCC-type transporters. ATP-binding cassette (ABC) family is a large gene family encoding vacuolar transporters, and exploited largely for heavy metal phytoremediation. Two ABCC-type transporters, AtABCC1 and AtABCC2 play an important role in arsenic decontamination. Song et al. (2010a) reported that As tolerance in *Arabidopsis* is mediated by these two ABCC-type PC transporters, whereas Park et al. (2012) showed that overexpression of *AtABCC1* and *AtABCC2* can increase Cd(II) accumulation in *Arabidopsis*, resulting in enhanced tolerance to cadmium and also mercury. A rice C-type ABC transporter, *OsABCC1* was found to reduce arsenic accumulation in the grain via limiting As transport to grains by vacuolar sequestration of the metal in the companion cells of the node phloem (Song et al., 2014). Very recently Deng et al. (2018) observed that increasing As vacuolar sequestration inhibited its translocation and accumulation into the grains of transgenic rice cooverexpressing two different vacuolar sequestration genes in a tissue-specific manner, *ScYCF1* (yeast cadmium factor) in the cortical cells of roots and *OsABCC1* (rice ABC transporter) in phloem cells of internode, along with a bacterial γ-glutamylcysteine synthetase. The transgenic rice plants showed up to 70% reduction in As accumulation in the grains of brown rice due to a significant decrease in As translocation from roots to shoots and internodes to grain. The ABCC3 is a cadmium-regulated transporter that sequestered PC–Cd complex in vacuoles. The overexpression of *AtABCC3* in *Arabidopsis* resulted in PC-mediated detoxification and enhanced tolerance for Cd, whereas *Arabidopsis* mutant (abcc3) showed high sensitivity to Cd (Brunetti et al., 2015). The vacuolar transporters yeast cadmium factor 1 (*YCF1*) has been shown to improve Pb(II) and Cd(II) accumulation and tolerance in transgenic *Arabidopsis* and poplar trees (Song et al., 2003; Shim et al., 2013), whereas coexpression of *YCF1* with phosphate transporters *AtPHT1* and *AtPHT7* resulted in enhanced tolerance for As in transgenic tobacco plants (LeBlanc et al., 2013). Recently, a vacuolar phosphate transporter 1 (VPT1) was shown to affect As tolerance by regulation of phosphate homeostasis in transgenic *Arabidopsis* (Luan et al., 2018). Whereas a mutant of the phosphate transporter PHT1;1 exhibited increase in accumulation of As in *Arabidopsis* (Catarecha et al., 2007).

Cadmium can be directly imported into vacuoles by divalent cation Ca^{2+}/H^{+} antiporters or CAXs. The overexpression of *AtCAX2* in transgenic tobacco resulted in enhanced accumulation of Cd^{2+} and Mn^{2+} in isolated root tonoplast vesicles and significantly improved tolerance to high levels of Mn^{2+} (Hirschi et al., 2000). In a similar study, coexpression of *AtCAX2* and *AtCAX4* in tobacco exhibited less growth inhibition and higher tolerance to three metals, that is, Cd, Zn, and Mn due to the increase in selective ion transport into root vacuoles and impact on root/shoot Cd partitioning (Korenkov et al., 2007). CAXs are energized via proton pumps by establishing pH gradient across membrane. Ectopic expression of wheat proton pump/H^{+}-pyrophosphatase (*TaVP1*) resulted in increased Cd tolerance and higher accumulation in the biomass of transgenic tobacco plants as compared to WT plants (Khoudi et al., 2012). HMA are also potential candidates for root-to-shoot translocation of heavy metals in plants. Ectopic expression of *SaHMA3* resulted in enhanced accumulation and tolerance to Cd in hyperaccumulator species *Sedum alfredii* (Zhang et al., 2016). Whereas the combined expression of *AtHMA4* and MT *AtMT2b* in tobacco increased root-to-shoot translocation of Zn and Cd and Cd tolerance (Grispen et al., 2011). The members of P-type ATPases family also act as the metal transporter, and are important for ion homeostasis inside cell. The ectopic expression of *OsACA6*, a P-type 2B Ca(2 +) ATPase led to enhanced Cd tolerance in tobacco, via modulation of ROS scavenging pathway enzymes. Transgenic plants showed larger biomass, long roots, and higher chlorophyll content than untransformed control plants under Cd stress conditions (Shukla et al., 2014). In an important study by Chen et al. (2013), constitutive expression of the key As(III) antiporter *ACR3* from hyperaccumulator fern species *Pteris vittata* was shown to provide great As tolerance in transgenic *Arabidopsis*, with accumulation of 7.5-fold more As in aboveground tissues in comparison to WT plants. The seeds of transgenic plants were able to germinate at lethal concentrations of As, that is, 80 μM As(III) or 1200 μM arsenate [As(V)].

Biotechnological approaches can also be used to reduce the toxicity of edible crops grown in heavy metal polluted soils, by lowering the metal accumulation in plants. In this context, simultaneous expression of all functional components of the bacterial CzcCBA efflux system (*PpCzcC*, *PpCzcB*, and *PpCzcA*) caused a significant decrease in Cd uptake and accumulation in transgenic tobacco plants, as compared to single transgenes alone (Nesler et al., 2017). In a recent study, Tang et al. (2017) demonstrated the involvement of rice peptide transporter *OsPTR7* in long-distance transport of DMA. Knockout or knockdown mutants of this gene accumulated undetectable amounts of As in the rice grains, as compared to the WT plants. A bacterial Co–Ni transporter/permease from *Rhodopseudomonas palustris* specifically translocated Co and Ni, and resulted in two to five times increase in accumulation of both heavy metals in transgenic tobacco, as compared to WT non-transgenic plants (Nair et al., 2012). Zinc is an essential micronutrient and its deficiency is very common in humans, which adversely affects their health. Thus biofortification of Zn is necessary in edible crops, while excluding its toxic analog Cd. In this regard, overexpression of a fungal Zn transporter gene *tzn1* caused a significant increase in Zn accumulation in transgenic tobacco plants, without cotransporting Cd (Dixit et al., 2010). Apart from the above examples, many other metal transporters like MRP7, ACR3, rhlA, and rhlB, tcu-1 were also shown to play an important role in phytoremediation of heavy metals, as shown in Table 4.1.

4.2.1.3 Detoxifiers

Certain gene products from bacteria, plants, and animals are able to directly detoxify the heavy metal compounds such as Hg and Se, and confer tolerance for these metals (Dhankher et al., 2011; Kotrba et al., 2009). As discussed in Section 4.2.1, the organic form of Hg (like methylmercury and phenylmercury) is more toxic than its ionic or elemental form. Sequestration and removal of Hg from contaminated sites can be promoted by converting organomercurial and inorganic mercuric ions into much lesser toxic, nonreactive, volatile, elemental mercury. Only some bacteria have a natural ability for this metabolic conversion, by expressing genes organized in the *mer operon*. The mer opero constitutes the genes involved in mercury regulation (*merR*), transport (*merT*, *merP*, *merC*, *merF*), and electrochemical reduction (*merA*, *merB*). The *merB* encodes for a 24 kDa enzyme organomercury lyase, which catalyzes the protonolytic conversion of organic forms of mercury to a more reactive and less toxic ionic Hg(II), by cleavage of the R-Hg bond (R = methyl or phenyl group). This ionic Hg(II) is further electrochemically reduced to elemental Hg (0) by the action of mercuric ion reductase, encoded by *merA*. Metallic Hg(0) is less toxic than ionic or organic Hg, relatively inert, less soluble, and volatile. It can rapidly evaporate into the atmosphere at nontoxic levels (Dhankher et al., 2011). The genes for bacterial mercury detoxification pathway have been functionally characterized in plants (as shown in Table 4.1). The work was initiated by Rugh et al. (1996) by overexpressing a modified *merA* gene (*merApe9*) from *E. coli* in *Arabidopsis*. Transgenic plants exhibited seed germination, growth, flowering, and seed setting on medium containing 5–20 ppm of $HgCl_2$, the concentration lethal to WT plants. The transgenic *Arabidopsis* seedlings were also able to evaporate volatile Hg(0), as compared to untransformed control plants (Rugh et al., 1996). The similar results for Hg detoxification and volatilization were obtained in subsequent studies showing ectopic expression of *merA* in transgenic poplar trees, tobacco, rice, and hybrid sweetgum (Rugh et al., 1998; Che et al., 2003; Dai et al., 2009; He et al., 2001; Heaton et al., 2003; Haque et al., 2010). Heaton et al. (2005) observed that *merA*-engineered transgenic tobacco plants exhibited good growth and seed germination on semisolid media containing $HgCl_2$ concentrations toxic to WT control plants, and retained higher Hg(0) evapotranspiration rate on hydroponic media, as compared to untransformed plants. Vertical transport of Hg was also proved by reciprocal grafting of transgenic and WT plants (Heaton et al., 2005). In a separate study, the heterologous expression of a modified bacterial *merB* in *Arabidopsis* conferred tolerance to higher concentrations of organomercurials (i.e., monomethyl mercuric chloride and phenyl mercuric acetate) in transgenic plants due to conversion into less toxic ionic Hg(II) (Bizily et al., 1999).

Since MerA can only transform ionic Hg(II) to Hg(0), and cannot detoxify organomercurials, the gene stacking of both merA and merB is required to efficiently transform methylmercury to elemental Hg(0). This approach has been used by several groups for more efficient mercury phytoremediation, as shown in Table 4.1 (Bizily et al., 2000; Lyyra et al., 2007). In those studies, a high level of Hg tolerance was not achieved via nuclear expression of *merA* and *merB*, since organomercurials mainly affect subcellular organelles like chloroplast and endoplasmic reticulum. Therefore chloroplast targeting of Hg detoxification enzymes might be a good strategy for their optimum activity and subsequent higher Hg tolerance. Ruiz et al. (2003) was the first to use chloroplast genetic engineering for Hg detoxification by stable integration of the native bacterial operon containing the *merA* and *merB* genes into the chloroplast genome. The simultaneous expression of both genes in chloroplast conferred increased tolerance to phenylmercuric acetate (PMA), up to 400 μM concentrations. Transgenic tobacco plants expressing both *merA* and *merB* in their chloroplast exhibited enhanced tolerance for PMA and $HgCl_2$, increased uptake and high accumulation of both the organic and inorganic forms of Hg (up to 2000 μg/g in root tissues and 100-fold more shoot accumulation than WT control plants), and efficient conversion of Hg(II) into Hg(0) along with its rapid volatilization within a week (Hussein et al., 2007).

Another gene *merC* encodes a mercury transporter in bacteria. Very recently, the cell-type specific expression of *merC* driven by a root epidermal cell-specific promoter resulted in significantly enhanced accumulation of Hg in shoots of transgenic *Arabidopsis* (Uraguchi et al., 2019), which could be beneficial for improved phytoremediation of Hg. The *merC* was expressed as fusion protein with a plant SNARE (SYP121) for protein trafficking to PM. Another mercury transporter MerT from *Pseudomonas K-62* was simultaneously expressed with polyphosphate kinase (ppk) from *K. aerogenes* in transgenic tobacco. The double transgenic plants exhibited accelerated uptake and more accumulation of mercury as compared to single ppk-transgenic plants, thus reducing the time for purification (Nagata et al., 2009). The same group adopted another gene pyramiding approach by integrating a bacterial *merB* gene into *ppk/merT*-transgenic tobacco. The callus of *ppk/merT/merB*-transgenic plants exhibited enhanced tolerance for organic methylmercury, and more mercury accumulation than single or double transgenic and WT control plants (Nagata et al., 2010). MerB converted methylmercury to ionic Hg(II), which is subsequently chelated by polyP forming a less toxic Hg-polyP complex, thus preventing the release of volatile elemental Hg(0) in the environment and generating potentially recyclable mercury-rich plant residues (Nagata et al., 2010).

The biotechnological approaches for selenium (Se) phytoremediation involve enhanced Se uptake, assimilation, accumulation, and volatilization. Since Se has close chemical similarity to sulfur (S), it can be metabolized through S-assimilation pathway. Some hyperaccumulators species like *Astragalus bisulcatus* and *Stanleya pinnata* can accumulate organic forms of Se (e.g., methyl-SeCys) up to 1% of their dry weight, having anticarcinogenic properties and without any toxic effect. Several genes have been exploited for improving Se tolerance in plants, encoding enzymes involved in selenate to selenite reduction, methylation of selenocysteine, and transformation of selenocysteine to volatile DMS or elemental Se. Genetically engineered plants exhibited up to nine- and threefold increase in Se accumulation and volatilization, respectively under laboratory and field conditions (Pilon-Smits and LeDuc, 2009). The rate-limiting step for Se assimilation in plants is the reduction of selenate to selenite, catalyzed by the enzyme ATP sulfurylase (APS). In a pioneer study by Pilon-Smits et al. (1999), the constitutive expression of *AtAPS1* gene encoding a chloroplastic form of APS led to enhanced uptake and reduction of selenate to organic Se (selenomethionine), 1.5- to 3-fold higher Se accumulation, and enhanced Se tolerance in both young and mature transgenic plants of Indian mustard, as compared to untransformed control plants. Field trials of transgenic Indian mustard plants expressing the same gene exhibited significantly improved phytoremediation of Se-polluted soil (Bañuelos et al., 2005). The transgenic *Arabidopsis* plants expressing a mouse selenocysteine lyase (SL) showed enhanced accumulation of selenium and Se tolerance (Pilon et al., 2003), whereas overexpression of another enzyme selenocysteine methyltransferase (SMT) from the Se hyperaccumulator *A. bisulcatus* resulted in enhanced Se accumulation and tolerance in transgenic plants of *Arabidopsis* and Indian mustard (LeDuc et al., 2004). The coexpression of double transgenes for APS1 and SMT, or SL and SMT resulted in more improved potential for Se phytoremediation, as compared to single transgenic or WT control plants (LeDuc et al., 2006; Bañuelos et al., 2007). Cystathionine-*c*-synthase (CGS) has also been reported to play an important role in Se volatilization, as it catalyzes the first step for the transformation of selenocysteine to volatile DMS. The transgenic plants of Indian mustard expressing *AtCGS* exhibited two- to threefold higher Se volatilization rate and enhanced tolerance to selenite as compared to WT control plants (Van Huysen et al., 2003). Some other enzymes also have a detoxifying capacity for heavy metals. The arsenic reductase2 (*AtACR2*) showed enhanced As tolerance of up to 200 μM of arsenate and more accumulation (up to 2400 μg/g dry weight) in roots of transgenic tobacco plants than WT control plants (Nahar et al., 2017).

4.2.1.4 Signaling and regulatory proteins

Various abiotic stresses, including heavy metal stress, cause an increase in the cellular ROS levels leading to oxidative stress. To scavenge the excess ROS and maintain cellular homeostasis, the antioxidant system including antioxidative enzymes get induced under such stress conditions. Genetic manipulation of these radical scavenging enzymes may result in improved phytoremediation potential for heavy metals, although there are no such consensuses. It has been reported that the enzymes of antioxidant machinery can convert ionic form of heavy metals to a less toxic elemental form (Koźmińska et al., 2018). Enhanced Cd tolerance was observed in transgenic tobacco expressing catalase (CAT) enzyme from Indian mustard (Guan et al., 2009). Transgenic plants exhibited longer roots and good growth in the presence of 100 μM Cd stress, while untransformed control plants showed chlorosis and death at this toxic concentration. Similarly, glutathione transferase from *Trichoderma virens* (*TvGST*) was shown to improve tolerance for Cd without affecting its accumulation in transgenic tobacco (Dixit et al., 2011a). The transgenic plants exhibited less lipid peroxidation, and increased activity of various antioxidative enzymes such as CAT, peroxidases (APX, GPX), superoxide dismutase, etc. The simultaneous coexpression of GST and human CYP2E1 in transgenic alfalfa resulted in greatly enhanced tolerance for mixed complex contaminants harboring both heavy metal (Cd) and organic pollutant (TCE), as compared to single transgenes or WT control plants (Zhang and Liu, 2011). In a recent contradictory report, a loss-of-function APX1 mutant of *Arabidopsis* exhibited increased tolerance for Pb and Se, mediated by increased activity of ABC transporters, PCs biosynthetic enzymes or GSH1 enzyme leading to higher glutathione (GSH) levels (Jiang et al., 2016, 2017). The overexpression of γ-tocopherolmethyltransferase gene (*γ-TMT*) in chloroplast has been shown to increase resistance to metal-induced oxidative stresses via reducing the ROS levels (Jin and Daniell, 2014).

Besides these, there are some recent studies that have shown an important role of transcription factors (TFs) and DNA repair enzymes for enhanced heavy metal phytoremediation. The overexpression of these TFs subsequently leads to up regulation of antioxidative enzymes, metal transporters and MTs, resulting in enhanced tolerance for metals. WRKY TFs belong to a large gene family and regulate growth and developmental processes as well as abiotic stress responses in plants. The overexpression of *AtWRKY12* was shown to negatively regulate Cd tolerance in *Arabidopsis*, via down regulation of PC biosynthesis genes in a GSH-dependent manner, whereas the loss-of-function mutants of *AtWRKY12* exhibited enhanced tolerance for Cd (Han et al., 2019). In another study, the expression of *OsWRKY28* was shown to be induced by arsenate and other oxidative stresses, thus it was functionally characterized for its role in heavy metal stress (Wang et al., 2018). It was found to positively regulate arsenate and phosphate accumulation, root system architecture, and fertility in rice in a jasmonic acid-dependent manner. The loss-of-function mutant of *OsWRKY28* exhibited a significant reduction in arsenate/phosphate accumulation in shoots, total root length, number of lateral roots and tillers, reduced fertility and grain yield, as compared to WT control plants (Wang et al., 2018). The overexpression of drought-responsive element binding (DREB) family TF, *StDREB* was also shown to improve Cd tolerance in transgenic potato plants (Charfeddine et al., 2017). The overexpression of zinc finger TF ZAT6 resulted in enhanced Cd tolerance in transgenic *Arabidopsis* via induction of GSH-dependent PC biosynthesis pathway; whereas reduced Cd tolerance was observed in the ZAT6 loss-of-function mutants (Chen et al., 2016). In a separate study, the overexpression of a DNA repairing gene *MtTdp2α* encoding tyrosyl-DNA phosphodiesterase-2 alleviated the Cu-mediated stress in transgenic *Medicago truncatula*, leading to enhanced tolerance for Cu (Faè et al., 2014).

Some hormone biosynthesis or signaling genes have also been reported to be involved in heavy metal phytoremediation (Table 4.1). Stearns et al. (2005) reported the role of bacterial *accd* gene encoding 1-aminocyclopropane-1-carboxylic acid deaminase (ACC deaminase, an ethylene biosynthesis gene) for Ni tolerance. The constitutive or root-specific expression of this gene in transgenic canola plants resulted in significant increase in Ni accumulation and tolerance, when grown in Ni-contaminated soils, as compared to untransformed control plants (Stearns et al., 2005). Similarly, the coexpression of bacterial *iaaM* gene encoding tryptophan monooxygenase (involved in auxin biosynthesis) under the xylem-specific promoter, along with constitutively expressing *accd* gene in transgenic petunia and tobacco showed enhanced accumulation and tolerance for Cu^{2+} and Co^{2+}. The T_1 transgenic plants harboring both genes exhibited more biomass, longer roots, faster growth rate, more accumulation of heavy metals, and the ability to tolerate up to 150 mg/L $CuSO_4$, as compared to single *iaaM* transgenic or untransformed control plants (Zhang et al., 2008). They were also able to grow on mixed contaminate soils containing both inorganic and organic pollutants.

4.2.1.5 Other important genes for enhancing heavy metal phytoremediation potential

Apart from the above mentioned genes, some other genes involved in various cellular processes may act as potential candidates to enhance phytoremediation potential of plants for heavy metals. For example, the overexpression of *Arabidopsis MAN3/XCD1* encoding endo-β-mannase enzyme conferred an increase in accumulation and tolerance for Cd, while loss-of-function mutant of this gene showed inverse phenotype (Chen et al., 2015). This mannose-mediated metal tolerance is related to GSH-dependent PC biosynthetic pathway. The overexpression of *SaLhcb2* encoding chlorophyll a/b-binding protein from a Zn/Cd hyperaccumulator species *S. alfredii* resulted in 6%–35% higher Cd accumulation in shoots and higher biomass in transgenic tobacco plants as compared to WT control plants (Zhang et al., 2011a). Interestingly, transgenic plants showed a significant increase in shoot and root biomass (14%–57%) under unstressed conditions (Zhang et al., 2011a). The *Arabidopsis PSE1* gene encoding Pb-sensitive1 protein was found to modulate Pb tolerance via induction of expression of GSH-dependent PC biosynthesis genes, and the ABC transporter PDR12/ABCG40. The overexpression of *AtPSE1* in *Arabidopsis* resulted in enhanced Pb accumulation and tolerance, while loss-of-function mutant of PSE1 exhibited Pb sensitivity (Fan et al., 2016).

Ferrochelatase-1 (FC1) is the terminal enzyme of heme biosynthesis in plants and act as a positive regulator of Cd stress tolerance. The constitutive overexpression of *AtFC1* led to enhanced Cd tolerance and accumulation of Cd and nonprotein thiol compounds in *Arabidopsis*. The transgenic plants exhibited a significant increase in biomass, primary root length, chlorophyll content. And a reduction in oxidative stress under Cd exposure, as compared to untransformed control plants (Song et al., 2017). Whereas *fc1* mutant showed Cd sensitivity, and transcriptome analysis of this loss-of-function mutant revealed differential expression of genes related to metal transporters, oxidative stress response, GSH/PC biosynthesis, and metal detoxification (Song et al., 2017). In a separate study, the ectopic expression of a fungal *WaarsM* gene encoding arsenic methyltransferase in transgenic rice led to enhanced tolerance for As, while decreasing As accumulation up to 50%–52% in shoots, roots and grains, when grown in As-contaminated soil. The transgenic plants exhibited increased methylation of toxic inorganic As species and As volatilization (Verma et al., 2018). Another study revealed that overexpression of *OsMTP1* gene encoding a novel metal tolerance protein resulted in a significant decrease in Cd-induced toxic effects and enhanced Cd tolerance in tobacco. Transgenic plants exhibited 2.8-fold higher biomass, enhanced vacuolar thiol content and vacuolar sequestration and hyperaccumulation of Cd rather than untransformed control plants (Das et al., 2016).

4.2.2 Genetic engineering for phytoremediation of organic pollutants

Organic pollutants have become a major source of environmental pollution due to increasing urbanization and industrialization. These have higher environmental stability and bioaccumulation ability, thereby causing more harmful long-lasting toxic effects (Rylott et al., 2015). Unlike inorganic pollutants, organic pollutants can be completely degraded and detoxified by the organisms, if required enzymatic machinery is present inside the cells. Although plants have the natural ability for partial degradation of some xenobiotics and present an attractive platform for phytoremediation, the process is limited by its speed and efficiency, as well as inefficient uptake due to less solubility of organic compounds (Van Aken and Doty, 2010; Eapen et al., 2007). Whereas microbes are known to possess efficient catabolic pathways to completely degrade and mineralize those toxic organic xenobiotics. The xenobiotics are detoxified in a three-step pathway in plants: activation/transformation of the foreign organic compound by addition of some functional group (amino, hydroxyl, or sulfhydryl group) by activity of plant enzymes such as cytochrome P450s, conjugation with other moieties like glutathione by glutathione *S*-transferases (GSTs) or glucose by glycosyltransferases, and finally sequestration of the less toxic conjugated-complex to vacuoles or apoplast by activity of transporters like ABC transporters (Dhankher et al., 2011). Transgenic technology can be used for increasing the efficiency of phytoremediation and complete removal of organic pollutants from soil or water. The restoration potential of plants for xenobiotic detoxification and degradation can be enhanced by overexpression of plant genes encoding enzymes for any of the above steps, or transferring foreign genes involved in the metabolism of organic compounds from bacteria, fungi, or mammals (recently reviewed by Hussain et al., 2018; Rylott and Bruce, 2009;

Kawahigashi, 2009). Organic pollutants can be classified into five major groups: explosives (such as royal demolition explosive or RDX, 2,4,6-trinitrotoluene or TNT), herbicides, pesticides, and insecticides (atrazine, alachlor, metolachlor, 2,4-D, chlorpyrifos or CPS), halogenated aliphatic compounds or solvents (TCE), polycyclic aromatic hydrocarbons or PAH (anthracene, naphthalene, pyrene, petroleum products like benzene, toluene, ethylbenzene, and xylene) and polychlorinated biphenyls or PCBs (2,3,4-trichlorobiphenyl). In the following sections, various approaches for plant genetic manipulation for enhanced phytoremediation of different groups of organic pollutants are discussed in detail, and presented in Table 4.2.

4.2.2.1 Explosives

The live military fire training ranges are severely contaminated worldwide by nitroaromatic explosives primarily due to their extensive manufacturing, use, and detonation. The main examples of these explosives are TNT and hexahydro-1,3,5-trinitro-1,3,5-triazine or RDX. These are highly persistent in the environment, recalcitrant to degradation, and toxic to all living beings. RDX damages the central nervous system and TNT may cause liver damage and anemia, and is highly phytotoxic, rendering it difficult to remove by phytoremediation. RDX is less toxic, but has high water-solubility and leaches in groundwater easily, causing contamination of drinking water (Rylott and Bruce, 2009). It is listed as a carcinogenic priority pollutant by the US Environmental Protection Agency (USEPA).

Although different physical/chemical/biological approaches exist for removing explosives from contaminated sites such as incineration, composting, and bioremediation using microbes, these are expensive, less efficient, and not adequate for a high magnitude of explosive pollution. Phytoremediation approaches are also not so effective due to phytotoxicity of explosives and low degradation ability of plants (Rylott and Bruce, 2009). Therefore genetic engineering approaches have been used for enhancing potential of plants for efficient and speedy clean-up of explosives from soil and groundwater (Rylott et al., 2015). Some examples are shown in Table 4.2. A bacterial *nfsI* gene from *Enterobacter cloacae* encoding nitroreductase (NR) enzyme has been shown to enhance detoxification of TNT, via catalyzing the reduction of TNT to 4-hydroxylamino-2,6-dinitrotoluene, and amino-dinitrotoluene derivatives, followed by enhanced conjugation to macromolecules in transgenic plants. Transgenic tobacco plants overexpressing this gene were able to remove 0.5 mM of TNT from aqueous medium, the concentration toxic to WT control plants (Hannink et al., 2001, 2007). To alleviate the pollen-mediated transfer of foreign genes to wild populations, the bacterial NR-*nfsI* was expressed in tobacco plastids (Zhang et al., 2017a). The transplastomic plants showed higher biomass production and regeneration, efficient removal of TNT from solid or liquid media and enhanced tolerance for TNT rather than untransformed control plants. Similarly, the overexpression of a NR from *E. coli* (*nfsA/ntr*) resulted in increased NR activity and seven to eight times more uptake of TNT in transgenic *Arabidopsis*, as compared to untransformed control plants, leading to enhanced ability of transgenic plants to tolerate and detoxify TNT (Kurumata et al., 2005). A

Table 4.2 Comprehensive list of transgenic plants for enhanced phytoremediation of organic xenobiotics.

Gene transferred	Product	Source (donor) organism	Target (recipient) plant	Transgene effect	Reference(s)
Explosives					
onr	Pentaerythritol tetranitrate reductase	*Enterobacter cloacae*	*Nicotiana tabacum*	Increased denitration and biodegradation of TNT and GTN	French et al. (1999)
nfsI	Nitroreductase	*E. cloacae*	*N. tabacum*	Phyto-detoxification of TNT	Hannink et al. (2001, 2007)
nfsA/ntr	Nitroreductase	*Escherichia coli*	*Arabidopsis thaliana*	Enhanced uptake and detoxification of TNT	Kurumata et al. (2005)
nfsI	Nitroreductase	*E. cloacae*	*N. tabacum (plastids)*	Enhanced detoxification of TNT	Zhang et al. (2017a)
pnrA	Nitroreductase	*Pseudomonas putida*	*Populus tremula* × *tremuloides var. Etropole*	Enhanced tolerance for TNT	van Dillewijn et al. (2008)
opr	Oxophytodienoate reductases	*A. thaliana*	*A. thaliana*	Enhanced detoxification of TNT	Beynon et al. (2009)
743B4, 73C1	UGTs	*A. thaliana*	*A. thaliana*	Enhanced detoxification of TNT	Gandia-Herrero et al. (2008)

GST U24 and U25	Glutathione transferases U24 and U25	*A. thaliana*	*A. thaliana*	Enhanced detoxification of TNT and environmental pollutants	Gunning et al. (2014)
xplA	Cytochrome P450	*Rhodococcus rhodrochrous*	*A. thaliana*	Phytoremediation of RDX	Rylott et al. (2006)
xplA, xplB	Cytochrome P450	*R. rhodrochrous*	*A. thaliana*	Enhanced detoxification of RDX	Jackson et al. (2007)
xplA, xplB, nfsI	Cytochrome P450 and associated reductase, nitroreductase	*R. rhodrochrous strain 11Y, E. cloacae*	*A. thaliana*	Tolerance for RDX and TNT	Rylott et al. (2011)
xplA, xplB, nfsI (NR)	Fused flavodoxin–cytochrome P450 and partnering flavodoxin reductase, nitroreductase	*Rhodococcus rhodochrous strain 11Y, E. cloacae*	*Panicum virgatum, Agrostis stolonifera*	RDX degradation and TNT detoxification	Zhang et al. (2017b)
xplA, xplB, nfsI	Fused flavodoxin–cytochrome P450 and partnering flavodoxin reductase, nitroreductase	*R. rhodochrous strain 11Y, E. cloacae*	*Pascopyrum smithii*	RDX degradation and TNT detoxification	Zhang et al. (2019)
Herbicides					
CYP1A1 and NADPH-CYP P450	Fused enzyme containing cytochrome P450 1A1 and NADPH–cytochrome P450 oxidoreductase	*Rattus rattus, S. cerevisiae*	*N. tabacum*	Resistance to the herbicide CTU	Shiota et al. (1994)

(*Continued*)

Table 4.2 (Continued)

Gene transferred	Product	Source (donor) organism	Target (recipient) plant	Transgene effect	Reference(s)
CYP71A10	Cytochrome P450 monoxygenase	*Glycine max*	*N. tabacum, A. thaliana*	Tolerance to phenylurea herbicide	Siminszky et al. (1999)
CYP1A1 and CYP1A2	Fused enzyme containing cytochrome P450 1A1 and cytochrome P450 1A2	*Homo sapiens, S. cerevisiae*	*N. tabacum*	Higher resistance to herbicide CTU	Shiota et al. (2000)
CYP76B1	Cytochrome P450 Monooxygenase 76B1	*Helianthus tuberosus*	*N. tabacum, A. thaliana*	Increased herbicide metabolism and tolerance for linuron, isoproturon or CTU	Didierjean et al. (2002)
CYP1A1 and CYP1A2	Cytochrome P450 monooxygenases	*H. sapiens*	*N. tabacum (cell culture)*	Biotransformation of atrazine	Bode et al. (2004)
CYP1A1 or CYP1A2	Cytochrome P450 monooxygenases	*H. sapiens*	*N. tabacum (cell cultures)*	Biotransformation of metamitron	Bode et al. (2006)
CYP2B6	Cytochrome P450 monooxygenases	*H. sapiens*	*Oryza sativa L. cv. Nipponbare*	Enhanced degradation of metolachlor	Kawahigashi et al. (2005a)
CYP1A1	Cytochrome P450 monoxygenase	*H. sapiens*	(*O. sativa cv. Nipponbare*)	Enhanced remediation of triazine herbicides atrazine and simazine	Kawahigashi et al. (2005b)

CYP2B22, CYP2C49	Cytochrome P450 monoxygenase	*Sus scrofa*	*O. sativa*	Enhanced herbicide tolerance	Kawahigashi et al. (2005c)
CYP1A1, CYP2B6, CYP2C19	Cytochrome P450 monoxygenase	*H. sapiens*	*O. sativa cv. Nipponbare*	Enhanced degradation of atrazine and metolachlor	Kawahigashi et al. (2006, 2007, 2008)
CYP1A1, CYP2B6, CYP2C9 and CYP2C19	Cytochrome P450 monoxygenase	*H. sapiens*	*Solanum tuberosum, O. sativa*	Enhanced degradation of sulfonylurea and other herbicides, insecticides and industrial chemicals	Inui and Ohkawa (2005)
CYP1A2	Cytochrome P450 monooxygenases	*H. sapiens*	*A. thaliana*	Resistance to the phenylurea herbicide CTU	Kebeish et al. (2014)
p450-1a2	P450 isozymes CYP1A2	*H. sapiens*	*A. thaliana*	Phytoremediation of the organic xenobiotic simazine	Azab et al. (2016)
CYP1A2	Cytochrome P450 monoxygenase	*H. sapiens*	*A. thaliana and E. coli*	Enhanced degradation of linuron	Azab et al. (2018)
gstI-6His	6His-tagged GST I, glutathione *S*-transferases	*Zea mays*	*N. tabacum*	Enhanced detoxification of herbicides—alachlor and chloroacetanilide	Karavangeli et al. (2005)
CYP450E1 and GST	Cytochrome P450 monoxygenase	*H. sapiens*	*N. tabacum*	Increased tolerance and detoxification of chlorpyriphos and anthracene	Dixit et al. (2008)

(*Continued*)

Table 4.2 (Continued)

Gene transferred	Product	Source (donor) organism	Target (recipient) plant	Transgene effect	Reference(s)
GSTU4	Tau class GST isoenzyme, glutathione transferases, GPOX	*G. max*	*N. tabacum*	Enhanced tolerance for diphenyl ether and chloroacetanilide herbicide	Benekos et al. (2010)
atzA	Atrazine chlorohydrolase	Bacteria	*N. tabacum*, *A. thaliana*, *Medicago sativa*	Enhanced degradation of atrazine	Wang et al. (2005)
Protox	Protoporphyrinogen IX oxidase	*Bacillus subtilis*	*O. sativa*	Increased detoxification of diphenyl ether and oxyflufen herbicide	Jung et al. (2008)
γ-ECS	γ-Glutamycysteine synthetase	*Populus trichocarpa*	*P. trichocarpa*	Enhanced remediation of acetochlor and metolachlor	Gullner et al. (2001)
γ-ECS, GS	γ-Glutamycysteine synthetase; glutathione synthetase	*Brassica juncea*	*B. juncea*	Increased resistance for metolachlor, phenanthrene, atrazine and 1-chloro-2, 4-dinitrobenzene	Flocco et al. (2004)

Halogenated aliphatic compounds					
CYP2E1	Cytochrome P450 2E1	*H. sapiens*	*N. tabacum cv. Xanthii*	Enhanced detoxification of low molecular weight volatile organic compounds	James et al. (2008)
CYP2E1	Cytochrome P450 2E1	*H. sapiens*	*P. tremula* × *Populus alba*	Enhanced detoxification of halogenated hydrocarbons	Doty et al. (2007)
CYP2E1	Cytochrome P450 (2E1)	*Oryctolagus cuniculus*	*A. thaliana*, *Sesbania grandiflora*	Enhanced degradation of TCE DDT	Mouhamad et al. (2012)
dhlA, *dhlB*	Haloalkane dehalogenase, haloacid dehalogenase	*Xanthobacter autotrophicus GJ10*	*N. tabacum*	Degradation of DCE	Mena-Benitez et al. (2008)
PAH					
GST	Glutathione-*S*-transferase	*Trichoderma virens*	*N. tabacum*	Enhanced degradation of anthracene	Dixit et al. (2011b)
NahAa, *NahAb*	Flavoprotin reductase, Ferredoxin	*P. putida G7 Naphthalene dioxygenase system*	*O. sativa*, *A. thaliana*	Phytoremediation of phenanthrene	Peng et al. (2014a)
nidA, *nidB*. *NahAa and NahAb*	(Hybrid bacterial dioxygenase complex), large and small subunits of naphthalene dioxygenase, flavoprotein reductase and ferredoxin	*Mycobacterium vanbaalenii*, *P. putida*	*A. thaliana*	Increased resistance for 2–4 rings polycyclic aromatic hydrocarbons	Peng et al. (2014b)

(*Continued*)

Table 4.2 (Continued)

Gene transferred	Product	Source (donor) organism	Target (recipient) plant	Transgene effect	Reference(s)
ADI1	Ferredoxin-like protein	*O. sativa*	*A. thaliana*	Increased tolerance to naphthalene	Fu et al. (2016)
PCB					
bphc	2,3-Dihydroxybiphenyl-1,2-dioxygenase	*Pandoraea pnomenusa*	*N. tabacum*	Increased detoxification of PCBs	Chrastilova et al. (2007)
bphC	2,3-Dihydroxybiphenyl-1,2-dioxygenase	*P. pnomenusa*	*N. tabacum*	Increased detoxification of PCBs	Novakova et al. (2010)
bphC/His	2,3-Dihydroxybiphenyl-1,2-dioxygenase	*P. pnomenusa B-356*	*N. tabacum*	Increased detoxification of PCB mixture—delor 103 and congener 28	Viktorovtá et al. (2014)
bphC.B	2,3-Dihydroxybiphenyl-1,2-dioxygenase	Soil metagenomic library	*M. sativa*	Increased detoxification of PCB and 2,4-dichlorophenol	Wang et al. (2015)
ZFP	Zinc finger protein	*Cucurbita pepo*	*N. tabacum*	Accumulation of hydrophobic contaminants—PCBs, dioxins and dioxin-like compounds	Inui et al. (2015)

Other organic pollutants					
tpxI	Basic peroxidase	*Solanum lycopersicum*	*S. lycopersicum (hairy roots)*	Increased remediation of phenol	Oller et al. (2005)
tpx1 and tpx2	Basic peroxidases	*Solanum lycopersicum*	*N. tabacum (hairy roots)*	Increased remediation of phenol	Alderete et al. (2009)
tpx1 and tpx2	Basic peroxidases	*S. lycopersicum*	*N. tabacum*	Phytoremediation of 2,4-dichlorophenol	Talano et al. (2012)
MnP	Mn-peroxidase	*Coriolus versicolor*	*N. tabacum*	Increased remediation of PCP	Iimura et al. (2002)
UGT72B1	Glucosyltransferase	*P. trichocarpa*	*A. thaliana*	Tolerance for TCP	Su et al. (2012)
LACI	Root-specific secretory laccase	*Gossypium arboretum*	*A. thaliana*	Enhanced ex-planta resistance to phenolic allelochemicals and TCP	Wang et al. (2004)
LAC	Fungal laccase	*Coriolus versicolar*	*N. tabacum*	Increased remediation of bisphenol-A and PCP	Sonoki et al. (2005)
POXA1b, POXC	Fungal laccase	*Pleurotus ostreatus*	*N. tabacum, microalgae*	Reduced phenol content	Chiaiese et al. (2011, 2012)
ophc2	Organophosphorus hydrolase	*Pseudomonas pseudoalcaligenes*	*N. tabacum*	Increased remediation of methyl parathion	Wang et al. (2008)

(*Continued*)

Table 4.2 (Continued)

Gene transferred	Product	Source (donor) organism	Target (recipient) plant	Transgene effect	Reference(s)
ADH2	Glutathione-dependent formaldehyde dehydrogenase	*A. thaliana*	*A. thaliana*	Enhanced formaldehyde detoxification	Achkor et al. (2003)
faldh	Formaldehyde dehydrogenase	*Brevibacillus brevis*	*N. tabacum*	Enhanced formaldehyde tolerance and detoxification	Nian et al. (2013)
rmpA, *rmpB*, *hps and phi*	RuMP pathway, HPS and PHI	*Mycobacterium gastri MB19*	*A. thaliana and N. tabacum (chloroplast)*	Enhanced assimilation of formaldehyde	Chen et al. (2010)
hps and phi	HPS/PHI fusion enzyme, HPS and PHI	*M. gastri MB19*	*N. tabacum*	Enhanced formaldehyde purification	Sawada et al. (2007)
hps and phi	HPS/PHI fusion enzyme, HPS and PHI	*M. gastri MB19*	*Pelargonium sp. Frensham (chloroplasts)*	Enhanced assimilation and phytoremediation of formaldehyde	Song et al. (2010b)
das/dak	DAS and DAK, XuMP pathway	Methylotrophic yeasts	*Geranium (Pelargonium sp. Frensham), tobacco*	Purification of formaldehyde	Zhou et al. (2015) and Xiao et al. (2012)

CYP2E1	Cytochrome P450 2E1	*H. sapiens*	*Petunia hybrida*	Efficient removal of benzene, toluene and formaldehyde	Zhang et al. (2011b)
tmr	Triphenylmethane reductase	*Citrobacter* sp.	*A. thaliana*	Increased remediation of Triphenylmethane dyes crystal violet and malachite green	Fu et al. (2013)

CTU, Chlortoluron; *DAK*, dihydroxyacetone kinase; *DAS*, dihydroxyacetone synthase; *DCE*, 1,2-dichloroethane; *DDT*, dichlorodiphenyltrichloroethane; *GPOX*, glutathione-dependent peroxidase; *GTN*, glycerol trinitrate; *HPS*, 3-hexulose-6-phosphate synthase; *PCB*, polychlorinated biphenyl; *PCP*, pentachlorophenol; *PHI*, 6-phosphate-3-hexuloisomerase; *RDX*, royal demolition explosive; *RuMP*, ribulose monophosphate; *TCE*, trichloroethylene; *TCP*, trichlorophenol; *TNT*, 2,4,6-trinitrotoluene; *UGTs*, uridine diphosphate glycosyltransferases; *XuMP*, xylulose monophosphate.

different NR (*pnrA*) from *Pseudomonas putida* was shown to confer improved uptake and tolerance for higher amounts of TNT from contaminated soil and water when ectopically expressed in transgenic aspen (hybrid). Transgenic plants were able to tolerate up to 57 mg/L and 1000 mg/kg TNT in the liquid media and soil respectively, the levels otherwise toxic to WT control plants. After rapid uptake/adsorption, most of the TNT remained in the roots, with very less transport to shoots (van Dillewijn et al., 2008). Apart from NR, some bifunctional *O*- and *C*-glucosyltransferases were also reported to detoxify TNT in plants. The overexpression of *Arabidopsis* uridine diphosphate glycosyltransferases resulted in enhanced root growth and decontamination of TNT in transgenic *Arabidopsis*, as compared to WT control plants (Gandia-Herrero et al., 2008). Whereas some glutathione transferases from *Arabidopsis* (U24 and U25) were reported to confer enhanced detoxification for a range of environmental pollutants and TNT (Gunning et al., 2014). It has been shown that the bacterial *xplA* and *xplB* genes from *Rhodococcus rhodochrous* strain 11Y1, encoding cytochrome P450 system may cause denitration and ring cleavage in the nitroaromatic explosives, thus catalyzing the degradation of RDX in the absence of oxygen (Rylott et al., 2015). *Arabidopsis* plants were genetically engineered to express bacterial *xplA* gene encoding cytochrome P450 fused to a flavodoxin redox partner. Transgenic plants exhibited greater biomass, and efficiently removed and detoxified RDX from hydroponic solution, as well as RDX-contaminated soil, without any phytotoxic effect, as compared to untransformed control plants (Rylott et al., 2006). Later on Jackson et al. (2007) reported enhanced explosive-degrading potential of transgenic *Arabidopsis* plants expressing fused flavodoxin–cytochrome P450 *XplA* and partnering flavodoxin reductase *XplB*. In a further study, the bacterial cytochrome P450 enzymes and TNT-detoxifying NR (*xplA*, *xplB*, *nfsI*) were coexpressed in *Arabidopsis* for phytoremediation of both explosives RDX and cocontaminating TNT simultaneously (Rylott et al., 2011). These transgenic plants were able to grow on heavily contaminated soil with both explosives and more efficiently removed RDX and TNT as compared to plants harboring single transgene (*xplA*) or control plants; thus suitable for in situ phytoremediation of explosives. In addition, a very low level of RDX was found in aerial shoots and leaves, reducing the possibility for bioaccumulation (Rylott et al., 2011). The same multiple bacterial genes (*xplA*, *xplB*, and an NR-*nfsI*) were stacked and ectopically expressed in two grass species (switchgrass and creeping bentgrass), which are agronomically important and well-suited for uptake and removal of explosives from root zone leachates (Zhang et al., 2017b). Transgenic plants were able to detoxify both RDX and TNT, and able to significantly remove explosives from liquid media, with less retention in aerial tissues, as compared to untransformed control plants (Zhang et al., 2017b). Recently, the same multiple transgene pyramiding approach using *xplA*, *xplB*, and *nfsI* was exploited first time in a field-applicable perennial grass species (Western wheatgrass), which is able to grow in the polluted environments and used for the phytoremediation and re-vegetation of military training ranges (Zhang et al., 2019). Transgenic plants showed significant removal and degradation of RDX from liquid media and less retention of RDX in aerial tissues, along with increased detoxification of TNT than untransformed control plants (Zhang et al., 2019).

4.2.2.2 Herbicides, pesticides, and insecticides

Herbicides, pesticides, and insecticides present a persistent problem by adversely affecting the health of the surrounding environment and living beings (including chronic abnormalities in humans), despite their important use in agriculture. According to USEPA, the exposure to insecticides like CPS may cause nervous system and respiratory disorders, birth defects and male sterility in humans [US Environmental Protection Agency (USEPA), 2002]. This compound also shows phytotoxicity symptoms, like delayed seed germination, abnormalities in cell division, and fruit formation (Lee et al., 2012).

Various physical and chemicals methods like land filling and incineration are being used for removing these compounds from contaminated sites, but these are expensive and not efficient. Plants can metabolize these pollutants, but activity of metabolic enzymes is often very low. Therefore genetic manipulation of plants using various genes involved in metabolism of these chemical compounds (isolated mainly from mammals) presents a promising approach to improve phytoremediation potential (Kawahigashi, 2009). Some related examples are shown in Table 4.2. In plants, herbicides can be metabolized predominantly by *N*-dealkylation, in addition to hydrolytic dehalogenation and GSH conjugation, which results in almost nontoxic metabolites (Kawahigashi, 2009). Cytochrome P450 monooxygenases (CYPs) appear to be the key enzymes for catalyzing the rapid oxidative dealkylation of herbicides. These are induced by various xenobiotics and play a major role in the oxidative primary metabolism of exogenous lipophilic compounds such as drugs/pharmaceuticals in mammals. It has been reported that only 11 isoforms of human CYP enzymes can metabolize >90% of drugs in the liver, mostly belonging to CYP1A, CYP12A/2B/2C/2D/2E and CYP13A (Bode et al., 2004). Due to their broad substrate specificity, these are recognized as potential candidates for oxidative metabolism of various xenobiotics including herbicides. Human CYP1A1 and CYP1A2 were reported as major cytochrome P450 species for the oxidative metabolism of the phenylurea herbicides chlortoluron, linuron (LIN), and triazine herbicide atrazine. Earlier studies have reported that ectopic expression of a rat cytochrome P4501A1 (CYP1A1)/human cytochromes P4501A1 or P4501A2 and yeast NADPH–cytochrome P450 oxidoreductase fusion enzyme conferred resistance to the herbicide chlortoluron in transgenic tobacco (Shiota et al., 1994, 2000). Transgenic plants expressing CYP1A2 fusion enzyme were able to metabolize chlortoluron more efficiently via *N*-demethylation and ring-methyl hydroxylation, and were more tolerant to this herbicide than the plants expressing CYP1A1 fusion enzyme (Shiota et al., 2000). The constitutive expression of CYP76B1, a cytochrome P450 enzyme from Jerusalem artichoke resulted in increased metabolism and tolerance for herbicides in transgenic tobacco and *Arabidopsis*. Transgenic plants exhibited a 10- and 20-fold increase intolerance towards isoproturon or chlortoluron and LIN herbicides, respectively without showing any adverse phenotypic symptom (Didierjean et al., 2002). In another study by Bode et al. (2004), human P450 enzymes CYP1A1 and CYP1A2 were shown to catalyze *N*-dealkylation of atrazine herbicide, leading to biotransformation and enhanced metabolism of atrazine, when expressed in tobacco cell culture. The same plant host−transgene system catalyzed deamination and biotransformation of metamitron via aromatic hydroxylation of the herbicide (Bode et al., 2006). Transgenic Nipponbare rice plants expressing human cytochrome P450 enzyme CYP2B6 exhibited enhanced phytoremediation of chloroacetanilide herbicides—alachlor and metolachlor, and were able to remove more amount of herbicides from soil or water, as compared to untransformed control plants under greenhouse conditions (Kawahigashi et al., 2005a). On the other hand, overexpression of human P450 CYP1A1 conferred tolerance up to 8.8 and 50 μM of atrazine and simazine, respectively in transgenic rice due to more rapid metabolism and rate of removal of the herbicides from the soil compared to WT control plants (Kawahigashi et al., 2005b). The coexpression of three human cytochrome P450 monooxygenases—CYP1A1, CYP2B6, and CYP2C19 was shown to enhance tolerance for a broad range of chloroacetanilide and triazine herbicides—atrazine, metolachlor, and norflurazon and their mixture in transgenic rice (Kawahigashi et al., 2006). Transgenic plants removed the various herbicides from plant tissue, growth media, as well as from soil more rapidly than untransformed control plants. In another study, coexpression of human CYP1A1, CYP2B6, CYP2C9, and CYP2C19 in transgenic potato and rice conferred increased tolerance for different classes of herbicides, and certain insecticides and industrial chemicals like sulfonylurea (Inui and Ohkawa, 2005). Similarly, transgenic *Arabidopsis* plants expressing the mammalian cytochrome P450 enzyme CYP1A2 exhibited marked increase in tolerance to a phenylurea herbicide chlortoluron with reduced herbicide accumulation in-planta (Kebeish et al., 2014). The metabolic pathway for a triazine herbicide—simazine was engineered by genetic manipulation of *Arabidopsis* plants using human P450 isozymes CYP1A2 (p450-1a2). Transgenic plants showed enhanced tolerance up to 250 μmol of this herbicide supplemented in culture medium or foliar spray (the concentration toxic to control) without any harmful effect on rosette diameter and primary root length compared to WT control plants (Azab et al., 2016). Recently, Azab et al. (2018) showed that ectopic expression of human P450 enzyme CYP1A2 increased tolerance and detoxification of the phenylurea herbicide LIN in transgenic *Arabidopsis*. Transgenic plants were able to grow at a high concentration of LIN supplemented in culture medium or in foliar spray, and exhibited improved detoxification of the herbicide in their leaves without any phenotype, as compared to untransformed control plants (Azab et al., 2018).

GSTs are enzymes encoded by a multigene family, and are known to be involved in various cellular processes along with response to abiotic stresses. These are also reported to detoxify or biotransform various herbicides (Karavangeli et al., 2005; Benekos et al., 2010). The His-tagged GST I isozyme from maize significantly enhanced metabolism and detoxification of a chloroacetanilide herbicide alachlor, when overexpressed in transgenic tobacco (Karavangeli et al., 2005). On the other hand, the overexpression of a tau class GST isoenzyme GSTU4 from soybean that acts as glutathione-dependent peroxidase resulted in enhanced detoxification and tolerance for diphenyl ether herbicides (fluorodifen or oxyfluorfen) and chloroacetanilides herbicide (alachlor) in transgenic tobacco (Benekos et al., 2010). Transgenic plants were able to tolerate up to 200 μM fluorodifen or oxyfluorfen and 7.5 mg/L alachlor compared to untransformed control plants, while no effect was observed for metolachlor (Benekos et al., 2010). The - γ-glutamyl cysteine synthetase (γ-ECS) enzyme was also shown to be involved in herbicide detoxification in addition to heavy metal tolerance. Overexpression of γ-ECS in the cytosol or chloroplast of poplar species resulted in increased tolerance and GSH-dependent detoxification for chloroacetanilide herbicides—acetochlor and metolachlor (Gullner et al., 2001). Whereas coexpression of *Bjγ-ECS* and *BjGS* encoding glutathione synthetase conferred enhanced tolerance to a broad spectrum of organic pollutants—atrazine, 1-chloro-2,4-dinitrobenzene, phenanthrene, and metolachlor in Indian mustard plants (Flocco et al., 2004). In another study by Wang et al. (2005), a bacterial *atzA* gene encoding the atrazine chlorohydrolase, the first enzyme for atrazine metabolic pathway was overexpressed in tobacco, *Arabidopsis*, and alfalfa, and confers increased resistance for a wide range of atrazine concentrations.

CPS [*O*,*O*-diethyl*O*-(3,5,6-trichloro-2-pyridinyl)-phosphorothioate] is the most commonly used organophosphorus insecticides posing a great risk to the environment and human health. Poplar and willow plants are natural plant species for its phytoremediation that can uptake and metabolize the insecticide in plant tissues (Lee et al., 2012), however the rate of detoxification is very low and inefficient. Dixit et al. (2008) reported the efficient phytoremediation and enhanced degradation of CPS by coexpressing human CYP450E1 and GST in transgenic tobacco. Transgenic plants also showed resistance for other organic pollutants such as anthracene.

4.2.2.3 Halogenated aliphatic compounds

Halogenated aliphatic compounds such as TCE solvent and 1,2-dichloroethane (1,2-DCA) are the most common widespread environmental pollutants of soil, groundwater, and air, and can cause cancer and damage of liver and CNS in human beings. Conventional remediation techniques for its removal include air sparging, soil extraction, chemical/oxidants treatment, and bioremediation using anaerobic bacteria, but have limitations due to difficulties in establishment and highly expensive management (Van Aken and Doty, 2010). Phytoremediation of TCE using poplar trees is an effective alternative approach because of their rapid growth, high water uptake, extensive deep root system, and easy propagation by vegetative means. However, this approach is also limited due to slow metabolism of TCE in plants and poor activity of cytochrome P450 2E1 (CYP2E1) enzyme, which is the key enzyme for the rate-limiting first step of TCE degradation pathway (i.e., the activation step) in plants and mammals. Therefore plant genetic engineering using CYP2E1 may present a powerful strategy to enhance the phytoremediation potential of plants (Van Aken and Doty, 2010; Hussain et al., 2018). This hypothesis was proved by enhanced metabolism of halogenated hydrocarbons including TCE through overexpression of mammalian CYP2E1 in transgenic hybrid poplar and transgenic tobacco (Doty et al., 2007; James et al., 2008). The transgenic tobacco plants exhibited increased removal rate of LMW volatile organic compounds such as TCE, ethylene dibromide, vinyl chloride, carbon tetrachloride, benzene, toluene, chloroform, and bromodichloromethane under hydroponic culture, than untransformed or vector transformed control plants (James et al., 2008). Kang et al. (2010) reported CYP2E1-induced differential gene expression in response to TCE in a transgenic hybrid poplar using poplar genome microarrays. Transgenic plants overexpressing mammalian CYP2E1 exhibited a great number of differentially expressed genes, mainly cytochrome p450s, GSTs, glucosyltransferases, and ABC transporters, suggesting the triggering of TCE degradation pathway by the activity of CYP2E1 in transgenic poplar (Kang et al., 2010). In another study, ectopic expression of rabbit CYP2E1 in transgenic *Arabidopsis* and *Sesbania* has been shown to enhance phytoremediation of TCE and dichlorodiphenyl trichloroethane from contaminated water (Mouhamad et al., 2012).

A key component of organic pollutants is haloalkanes which include 1,2-DCA, and is listed as a priority pollutant and a carcinogen by the USEPA. It is highly stable and water-soluble, recalcitrant to biodegradation, produces toxic intermediates, leaches rapidly in groundwater and can bioaccumulate, posing a great threat to the whole ecosystem. Plants lack the metabolic pathway for degradation of this compound, however, two bacterial enzymes, haloalkane dehalogenase (DhlA) and haloacid dehalogenase (DhlB) from *Xanthobacter autotrophicus* strain GJ10, are able to dehalogenate a range of halogenated aliphatic hydrocarbons, including 1,2-DCA. Mena-Benitez et al. (2008) reported engineering of a catabolic pathway for 1,2-DCA degradation by coexpression of bacterial *dhlA* and *dhlB* genes in transgenic tobacco. Double transgenic plants were able to survive at high concentrations of 1,2-DCA toxic to control or single transgenic plants, expressing only *dhlA* (Mena-Benitez et al., 2008).

4.2.2.4 Polyaromatic hydrocarbons

Polycyclic aromatic hydrocarbons (PAH) are highly toxic, recalcitrant carcinogenic organic compounds produced by incomplete combustion of fossil fuels, and listed as priority pollutants (USEPA). It can be categorized into two groups: LMW PAHs having two to three benzene rings (e.g., naphthalene, acenaphthylene, acenaphthene, fluorene), and high molecular weight (HMW) PAHs with four or more benzene rings. Phytoremediation is a cheaper and effective environmentally friendly technique for clean-up of PAH-polluted sites, however, its success depends on the type of plant species. Recently, Sivaram et al. (2018) compared the C3 and C4 carbon fixation pathways for remediation of PAH-contaminated soils, and reported the significantly better removal of HMW PAHs from the contaminated soil by C4 plant species (maize, Sudan grass, and vetiver) compared to the C3 plants (cowpea, sunflower, and wallaby grass). The remediation potential of plant species for degradation or removal of PAHs from soil or water can be greatly enhanced by heterologous expression of foreign genes. As discussed earlier, metabolism of organic compounds takes place in three steps in plants, and GST mediates the second step of xenobiotic conjugation. Plant GST enzymes cannot efficiently degrade recalcitrant PAHs. A fungal GST from *T. virens* has been explored for its efficiency for remediation of a recalcitrant PAH-anthracene by overexpression in tobacco (Dixit et al., 2011b). Transgenic plants showed significantly enhanced tolerance and degradation of anthracene to naphthalene derivatives, compared to WT control plants.

Naphthalene, a LMW PAH, is highly toxic to plants and animals, causing inhibition of growth, chlorosis, and wilting in plants and cancer in humans. Due to its high phytotoxicity, phytoremediation approach for its clean-up is not so viable. Some genes encoding dioxygenase system have been identified from bacteria for efficient degradation of this recalcitrant organic pollutant. Ferredoxin (Fd) is important for biological electron transfer chain in the PAH degradation process. The complex naphthalene dioxygenase system of *P. putida* strain G7 containing *NahAa* and *NahAb* encoding flavoprotein reductase and ferredoxin of the electron-transport chain was overexpressed in rice and *Arabidopsis* (Peng et al., 2014a). The transgenic plants exhibited enhanced uptake, tolerance, and degradation of high concentration of phenanthrene, as compared to untransformed control plants. Later on, a hybrid bacterial dioxygenase complex containing *nidA*, *nidB* from *Mycobacterium vanbaalenii* PYR-1 and *NahAa* and *NahAb* from *P. putida* G7, encoding large and small subunits of naphthalene dioxygenase, flavoprotein reductase and ferredoxin, was overexpressed in *Arabidopsis* (Peng et al., 2014b). Transgenic plants showed enhanced tolerance for LMW PAHs, through faster degradation of PAH by naphthalene metabolic dioxygenase pathway. In another study, Fu et al. (2016) reported enhanced tolerance and degradation efficiency of naphthalene in transgenic *Arabidopsis* plants overexpressing the photosynthetic-type ferredoxin-like protein (ADI1) from rice.

4.2.2.5 Polychlorinated biphenyls

PCBs are highly stable, persistent, lipophilic aromatic compounds having two benzene rings substituted by chlorine. These are listed as priority pollutants (USEPA), that can cause chronic toxicity and damage of endocrine system in humans, and bioaccumulate in the food chain (Novakova et al., 2010). Some microbes possess ability to degrade PCBs under aerobic or nonaerobic conditions. The bacterial 2,3-dihydroxybiphenyl (2,3-DHB)−1,2-dioxygenase enzyme catalyzes critical steps of PCB degradation pathway and cleavage of the biphenyl ring, thus is a potential candidate for engineering plants for enhanced phytoremediation of PCBs. The ectopic expression of *bphc* gene from *Pandoraea pnomenusa* encoding 2,3-DHB resulted in improved metabolism and degradation of PCBs in transgenic plants, either by rhizoremediation or phytodegradation (Sylvestre et al., 2009; Chrastilova et al., 2007; Novakova et al., 2010). Transgenic tobacco plants expressing the same gene produced large biomass, and exhibited greater tolerance for commercial PCB mixture delor 103 under greenhouse conditions, and higher removal of congener 28 from contaminated soil, as compared to untransformed control plants (Viktorovtá et al., 2014). In a separate study, Wang et al. (2015) reported significantly enhanced phytoremediation of mixed contaminates of PCBs and 2,4-dichlorophenol (2,4-DCP) by overexpression of *BphC.B* from soil metagenomic library in transgenic Alfalfa plants. Some members of Cucurbitaceae family are a good accumulator of persistent hydrophobic organic pollutants. A *CpZFP* gene encoding a zinc finger protein from *Cucurbita pepo* has been overexpressed in a noncucurbit plant tobacco (Inui et al., 2015). Transgenic plants exhibited greater accumulation of highly hydrophobic PCBs, dioxins and dioxin-like compounds in shoots when cultivated in the PCB-contaminated soil, compared to nontransgenic control plants.

4.2.2.6 Other organic pollutants

Phenolic compounds present in effluents of industries are hazardous to human health and the environment. Basic peroxidase isozymes present in hairy roots of plants are involved in oxidation and removal of phenol. The overexpression of tomato *tpxI* gene alone or in combination with *tpx2* encoding basic peroxidases in transgenic hairy roots of tomato and tobacco resulted in significantly higher removal efficiency for phenol than untransformed control plants (Oller et al., 2005; Alderete et al., 2009). Talano et al. (2012) showed that double transgenic plants of tobacco harboring both TPX1 and TPX2 exhibited higher tolerance to 2,4-DCP at an early stage of development and better removal efficiencies (98%) at a toxic (25 mg/L) concentrations of the pollutant, compared to single TPX1 transgenic or control plants. A recombinant Mn-peroxidase gene (*MnP*) from *Coriolus versicolor* exhibited a twofold higher rate of removal of pentachlorophenol (PCP) by transgenic tobacco than control plants, when supplemented with 250 μM concentration of the pollutant, without any adverse effect on phenotype (Iimura et al., 2002).

Some enzymes secreted by plant roots in rhizosphere are capable to perform ex-planta phytoremediation (rhizoremediation) of organic pollutants, and exploited for genetic engineering plants for enhanced remediation potential. One such important example is secretory laccase, which can catalyze oxidation of a broad range of phenolic compounds. The overexpression of a root-specific secretory laccase from cotton (*LACI*) in transgenic *Arabidopsis* has been shown to enhance tolerance towards various recalcitrant phenolic allelochemicals and transformation of 2,4,6-trichlorophenol (Wang et al., 2004). Whereas transgenic tobacco plants expressing a fungal laccase from *Coriolus versicolor* exhibited improved efficiency for removal of bisphenol-A and PCP, up to 20 μmol/g dry weight, as compared to WT control plants (Sonoki et al., 2005). In studies by Chiaiese et al. (2011, 2012), two secretory fungal laccase *POXA1b* and *POXC* from oyster mushroom were able to reduce phenol content in olive oil mill wastewater, when overexpressed in transgenic tobacco or microalgae, through secretion of the recombinant protein in the root exudates and the culture medium, respectively.

Formaldehyde (HCHO) is an important toxic gaseous or aqueous air pollutant, representing the risk of cancer and other syndromes in humans. In plants, glutathione-dependent formaldehyde dehydrogenase (FALDH) enzyme plays an important role for HCHO detoxification. The overexpression of *Arabidopsis ADH2* encoding FALDH in transgenic *Arabidopsis* plants resulted in enhanced uptake (up to 25% increase) and detoxification of high concentrations of exogenous HCHO (Achkor et al., 2003). Similarly, ectopic expression of a bacteria *faldh* from *Brevibacillus brevis* in tobacco conferred higher tolerance to exogenous HCHO. Transgenic plants were able to take up and oxidize higher concentrations of gaseous/aqueous HCHO to nontoxic formate and other metabolites, as compared to nontransgenic control plants (Nian et al., 2013). Some methylotrophic bacteria are capable to fix and assimilate HCHO directly. The 3-hexulose-6-phosphate synthase (HPS) and 6-phosphate-3-hexuloisomerase (PHI) are two important enzymes of the HCHO assimilation pathway in methylotrophs like *Mycobacterium gastri* strain MB19. Overexpression of the HPS/PHI fusion protein in transgenic tobacco resulted in up to 20% increased efficiency for HCHO-removal rather than control plants (Sawada et al., 2007). Whereas heterologous expression of the fusion enzyme in the chloroplasts of an ornamental plant—geranium conferred enhanced uptake, assimilation, tolerance, and purification of HCHO from the polluted air, compared to nontransgenic control plants (Song et al., 2010b). Furthermore, Chen et al. (2010) reported the ribulose monophosphate (RuMP) pathway as another bacterial formaldehyde-fixing pathway that can bypass to the Calvin-Benson cycle in plants. The coexpression of *M. gastri rmpA* and *rmpB* genes of RuMP pathway along with *hps/phi* fusion system in the chloroplast of *Arabidopsis* and tobacco exhibited improved tolerance and removal of gaseous HCHO due to its fixation as a sugar phosphate (Chen et al., 2010). In addition, dihydroxyacetone synthase (DAS) and dihydroxyacetone kinase (DAK), the key enzymes for xylulose monophosphate (XuMP) pathway, are reported to be involved in formaldehyde-assimilation pathway. The overexpression of *das/dak* fusion genes from methylotrophic yeasts in transgenic geranium and tobacco chloroplast resulted in enhanced purification of high levels of gaseous HCHO in the polluted environment, compared to WT control plants (Zhou et al., 2015; Xiao et al., 2012).

Another important group of recalcitrant environmental pollutants are triphenylmethane dyes, which are spread globally and widely utilized for staining purposes in textile, medical, and food processing industries. It represents a great threat to human health and are potential carcinogen and mutagenic agents. The genes from microbial degradation pathway can be exploited for enhanced phytoremediation of these dyes. In this respect, the ectopic expression of a bacterial *tmr* gene encoding Triphenylmethane reductase from *Citrobacter* sp. yielded increased tolerance for crystal violet (CV) and malachite green dyes in transgenic *Arabidopsis*. The transgenic plants exhibited significantly improved efficiency for removal and transformation of toxic CV to nontoxic leucocrystal violet, compared to untransformed control plants (Fu et al., 2013).

4.3 Major concerns and future perspectives

In the past decade, there has been a huge development in the field of genetic engineering to raise transgenic plants for enhanced in situ remediation of toxic contaminants from soil, water, and air for sustainable protection of environment and human health. Genetic engineering of a single or multiple gene(s) in plants is proving to be a useful tool in achieving significant clean-up of inorganic and organic pollutants from the contaminated sites in an ecofriendly and efficient manner. This chapter summarizes the recent progress in genetic engineering of plants to develop enhanced tolerance, accumulation, and detoxification of heavy metals/metalloids and persistent xenobiotics for environmental clean-up.

Although genetic manipulation of plants has immense potential for enhancing phytoremediation capability, yet there are some concerns regarding the use of this powerful technology, such as techno-economic and ecological impact. Those concerns should be carefully addressed and strategies should be developed to mitigate the associated risks. Most of the studies have been performed in model plant species such as *Arabidopsis* and tobacco under controlled laboratory or greenhouse conditions, which provided a clear idea about the function of important genes for decontamination of pollutants. However, results from these studies cannot be extrapolated to field-applicable plant species. There is a critical requirement to examine the function of promising genes in other nonmodel economically important plants, suited for phytoremediation (e.g., the species having large biomass, deep root system, and efficient metabolism). Selection of appropriate plant species, field testing, and environmental risk assessment should be considered while developing transgenic plants for improved potential for degradation of recalcitrant contaminants. Validation and actual field trials of these transgenic plants in real contaminated sites are required, as in natural environment various pollutants can be found as a mixture, instead of a single factor tested under lab conditions. Proper disposal of contaminant-enriched plant biomass and economical aspects of this process is also a major concern.

Despite the great potential of genetic engineering for phytoremediation, there have been ethical and public concerns related to the testing of GM plants in field conditions and their commercial release, due to potential risks of transgene spread to wild relatives by cross-pollination, and increased invasiveness of highly tolerant engineered plants resulting in loss of diversity. The likelihood of an adverse effect on related soil microbial flora and fauna should also be taken into consideration. One possible solution to overcome this problem could be growing these plants in isolated contaminated sites (e.g., industrial area) instead of open agricultural fields. The early harvesting of transgenic plants before onset of flowering and the use of male-sterile plants could be effective approaches to reduce the risk of interbreeding and uncontrolled spread of pollen or seeds. The risk of transgene escape could also be reduced by employing novel techniques of cisgenesis and intragenesis using gene(s) from sexually compatible plant species for genetic manipulation of plants. In addition, barnase-barstar system acting as poison/antidote and genetic use restriction technologies could be employed for reducing the dispersion of transgenes in the environment (Kuvshinov et al., 2001; Millwood et al., 2016). Chloroplast or plastid engineering presents a good choice to restrict the pollen-mediated transgene flow because of maternal inheritance of chloroplastic DNA. In addition, the entire bacterial operon can be transferred to transgenic plants for high level expression of components of xenobiotic-degradation pathway. The use of antibiotic resistance genes as selection marker for transformation is often criticized due to the possibility of horizontal gene transfer from plants to microbes. This situation can be avoided by deletion of the antibiotic resistance gene from plant genome using bacteriophage *cre-lox* or yeast FLP-FRT recombination system (Song et al., 2008; Wang et al., 2011), or use of some nonantibiotic marker systems, such as screening on specific growth medium, or visual detection of transformants. Although stringent regulatory guidelines exist for the assessment of risks associated with GM plants (such as transgene flow, loss of biodiversity, allergenicity, toxicity, and environmental risks), the process to get regulatory approval is very much cumbersome involving high cost and time. These laws and regulatory frameworks should be relaxed and reevaluated regularly for easy implementation of this promising technology for phytoremediation of contaminated fields.

The main issue with using transgenic technology for developing tolerance for a wide variety of inorganic and organic pollutants is the absence of a single uniform defense or degradation mechanism, thereby causing differential responses among various plant species, and even in different tissues and organs of the same plant. Therefore the need arises for analyzing the cell and tissue-specific expression of transgenes. The effect of various isoforms of the same gene (e.g., antioxidative enzymes) should also be taken into account while selecting candidate genes for genetic modification of plants. Apart from that, constitutive overexpression of some genes may negatively affect plant growth and development under unstressed conditions, which can be overcome by the use of stress-inducible promoter for transgene expression. In the last few years, two strategies for plant genetic manipulation have been routinely employed for remediation of various environmental pollutants, consisting of either native plant pathway-engineering for xenobiotic uptake, metabolism, and degradation, or introduction of entirely new microbial pathways in plants. However, the complete knowledge of these plant processes and responses toward pollutant-stress is still lacking, which is a major procedural constrain for designing novel biotechnological approaches for efficient engineering of plants toward improved phytoremediation. Therefore, intensive research efforts are needed to elucidate the molecular mechanism of cross-talk among various stress responses and plant enzymatic detoxification processes for identification of major genetic determinants for phytoremediation; which has become possible by recent advancement of tools and techniques of functional genomics and other "omics" technologies, like next generation sequencing for whole genome and transcriptome, proteomics, and metabolomics (Kotrba et al., 2009; Mohammed et al., 2019; Visioli and Marmiroli, 2013). With the advent of newly emerging technologies, such as gene silencing or in situ genome editing by CRISPR/Cas9, it is now possible to determine the gene function accurately, and their precise engineering for increased phytoremediation ability.

Although some reports already exist for engineering of complete plant metabolic pathway for xenobiotic degradation through multiple-gene expression instead of single-gene transformation, more research is needed in this direction for efficient removal of hazardous chemicals from polluted sites using gene stacking/pyramiding approaches. The role of novel regulators for plant stress responses like micro RNA and long noncoding RNA also remain to be explored in future studies.

Apart from conventional plant biotechnological approaches, some other approaches need a substantial amount of consideration for enhancement of phytoremediation efficiency, such as transgene expression in hairy root system for increased uptake of pollutants, stimulating synergistic interactions among transgenic plants and genetically modified microbes for enhanced rhizoremediation, identification of novel genes from promising hyperaccumulator algal species for improving xenobiotic-chelating ability and engineering of plant cell wall proteins, etc. Future investigations in this field would open up new possibilities to overcome major concerns associated with GM plants and provide an excellent opportunity to generate precisely engineered plants towards increased phytoremediation potential.

Acknowledgments

The author thankfully acknowledges Department of Science and Technology, Government of India for DST-SERB Young Scientist grant (SB/YS/LS-39/2014); University Grants Commission, Government of India for UGC-BSR Start-up research grant (F.30-50/2014/BSR); UGC-SAPII CAS program and DST-FIST program in Department of Botany, J.N.V. University, Jodhpur.

The author declares no financial or commercial conflict of interest.

References

Achkor, H., Díaz, M., Fernández, M.R., Biosca, J.A., Parés, X., Martínez, M.C., 2003. Enhanced formaldehyde detoxification by overexpression of glutathione-dependent formaldehyde dehydrogenase from *Arabidopsis*. Plant Physiol. 132 (4), 2248–2255.

Alderete, L.G.S., Talano, M.A., Ibannez, S.G., Purro, S., Agostini, E., 2009. Establishment of transgenic tobacco hairy roots expressing basic peroxidases and its application for phenol removal. J. Biotechnol. 39, 273–279.

Arazi, T., Sunkar, R., Kaplan, B., Fromm, H., 1999. A tobacco plasma membrane calmodulin-binding transporter confers Ni^{2+} tolerance and Pb^{2+} hypersensitivity in transgenic plants. Plant J. 20 (2), 171–182.

Azab, E., Hegazy, A.K., El-Sharnouby, M.E., Abd. Elsalam, H.E., 2016. Phytoremediation of the organic Xenobiotic simazine by p450-1a2 transgenic *Arabidopsis thaliana* plants. Int. J. Phytoremed. 18 (7), 738–746.

Azab, E., Kebeish, R., Hegazy, A.K., 2018. Expression of the human gene CYP1A2 enhances tolerance and detoxification of the phenylurea herbicide linuron in *Arabidopsis thaliana* plants and *Escherichia coli*. Environ. Pollut. 238, 281–290.

Balestrazzi, A., Botti, S., Zelasco, S., Biondi, S., Franchin, C., Calligari, P., 2009. Expression of the *PsMTA1* gene in white poplar engineered with the MAT system is associated with heavy metal tolerance and protection against 8-hydroxy-2′-deoxyguanosine mediated-DNA damage. Plant Cell Rep. 28 (8), 1179–1192.

Bañuelos, G., Terry, N., LeDuc, D.L., Pilon-Smits, E.A.H., Mackey, B., 2005. Field trial of transgenic Indian mustard plants shows enhanced phytoremediation of selenium-contaminated sediment. Environ. Sci. Technol. 39, 1771–1777.

Bañuelos, G., Leduc, D.L., Pilon-Smits, E.A.H., Terry, N., 2007. Transgenic Indian mustard overexpressing selenocysteine lyase or selenocysteine methyltransferase exhibit enhanced potential for selenium phytoremediation under field conditions. Environ. Sci. Technol. 41 (2), 599–605.

Benekos, K., Kissoudis, C., Nianiou-Obeidat, I., Labrou, N., Madesis, P., Kalamaki, M., et al., 2010. Overexpression of a specific soybean *GmGSTU4* isoenzyme improves diphenyl ether and chloroacetanilide herbicide tolerance of transgenic tobacco plants. J. Biotechnol. 150 (1), 195–201.

Beynon, E.R., Symons, Z.C., Jackson, R.G., Lorenz, A., Rylott, E.L., Bruce, N.C., 2009. The role of oxophytodienoate reductases in the detoxification of the explosive 2,4,6-trinitrotoluene by *Arabidopsis*. Plant Physiol. 151, 253–261.

Bizily, S.P., Rugh, C.L., Summers, A.O., Meagher, R.B., 1999. Phytoremediation of methyl mercury pollution: *merB* expression in *Arabidopsis thaliana* confers resistance to organomercurials. Proc. Natl. Acad. Sci. U.S.A 96, 6808–6813.

Bizily, S.P., Rugh, C.L., Meagher, R.B., 2000. Phytodetoxification of hazardous organomercurials by genetically engineered plants. Nat. Biotechnol. 18, 213–217.

Bizily, S.P., Kim, T., Kandasamy, M.K., Meagher, R.B., 2003. Subcellular targeting of methylmercury lyase enhances its specific activity for organic mercury detoxification in plants. Plant Physiol. 131, 463–471.

Bode, M., Stöbe, P., Thiede, B., Schuphan, I., Schmidt, B., 2004. Biotransformation of atrazine in transgenic tobacco cell culture expressing human P450. Pest. Manag. Sci. 60 (1), 49–58.

Bode, M., Haas, M., Faymonville, T., Thiede, B., Schuphan, I., Schmidt, B., 2006. Biotransformation of metamitron by human p450 expressed in transgenic tobacco cell cultures. J. Environ. Sci. Health B 41 (3), 201–222.

Brichkova, G.G., Shishlova, A.M., Maneshina, T.V., Kartel, N.A., 2007. Tolerance to aluminum in genetically modified tobacco plants. Cytol. Genet. 41, 151–155.

Brunetti, P., Zanella, L., Proia, A., De Paolis, A., Falasca, G., Altamura, M.M., et al., 2011. Cadmium tolerance and phytochelatin content of *Arabidopsis* seedlings over-expressing the phytochelatin synthase gene *AtPCS1*. J. Exp. Bot. 62 (15), 5509–5519.

Brunetti, P., Zanella, L., De Paolis, A., Di Litta, D., Cecchetti, V., Falasca, G., et al., 2015. Cadmium-inducible expression of the ABC-type transporter *AtABCC3* increases phytochelatin-mediated cadmium tolerance in *Arabidopsis*. J. Exp. Bot. 66 (13), 3815–3829.

Catarecha, P., Segura, M.D., Franco-Zorrilla, J.M., García-Ponce, B., Lanza, M., Solano, R., et al., 2007. A mutant of the *Arabidopsis* phosphate transporter PHT1;1 displays enhanced arsenic accumulation. Plant Cell 19, 1123–1133.

Charfeddine, M., Charfeddine, S., Bouaziz, D., Messaoud, R.B., Bouzid, R.G., 2017. The effect of cadmium on transgenic potato (*Solanum tuberosum*) plants overexpressing the *StDREB* transcription factors. Plant Cell Tissue Organ. Cult. 128 (3), 521–541.

Chaturvedi, A.K., Patel, M.K., Mishra, A., Tiwari, V., Jha, B., 2014. The *SbMT-2* gene from a halophyte confers abiotic stress tolerance and modulates ROS scavenging in transgenic tobacco. PLoS One 9 (10), e111379.

Che, D., Meagher, R.B., Heaton, A.C., Lima, A., Rugh, C.L., Merkle, S.A., 2003. Expression of mercuric ion reductase in Eastern cottonwood (*Populus deltoides*) confers mercuric ion reduction and resistance. Plant Biotechnol. J. 1, 311–319.

Chen, L.M., Yurimoto, H., Li, K.Z., Orita, I., Akita, M., Kato, N., et al., 2010. Assimilation of formaldehyde in transgenic plants due to the introduction of the bacterial ribulose monophosphate pathway genes. Biosci. Biotechnol. Biochem. 74 (3), 627–635.

Chen, Y., Xu, W., Shen, H., Yan, H., Xu, W., He, Z., et al., 2013. Engineering arsenic tolerance and hyperaccumulation in plants for phytoremediation by a *PvACR3* transgenic approach. Environ. Sci. Technol. 47, 9355–9362.

Chen, J., Yang, L., Gu, J., Bai, X., Ren, Y., Fan, T., et al., 2015. *MAN3* gene regulates cadmium tolerance through the glutathione-dependent pathway in *Arabidopsis thaliana*. N. Phytol. 205 (2), 570–582.

Chen, J., Yang, L., Yan, X., Liu, Y., Wang, R., Fan, T., et al., 2016. Zinc-finger transcription factor ZAT6 positively regulates cadmium tolerance through the glutathione-dependent pathway in *Arabidopsis*. Plant Physiol. 171 (1), 707–719.

Chiaiese, P., Palomba, F., Tatino, F., Lanzillo, C., Pinto, G., Pollio, A., et al., 2011. Engineered tobacco and microalgae secreting the fungal laccase *POXA1b* reduce phenol content in olive oil mill wastewater. Enzyme Microb. Technol. 49, 540–546.

Chiaiese, P., Palomba, F., Galante, C., Esposito, S., De Biasi, M.G., Filippone, E., 2012. Transgenic tobacco plants expressing a fungal laccase are able to reduce phenol content from olive mill wastewaters. Int. J. Phytorem. 14, 835–844.

Chrastilova, Z., Mackova, M., Novakova, M., Macek, T., Szekeres, M., 2007. Transgenic plants for effective phytoremediation of persistent toxic organic pollutants present in the environment. J. Biotechnol. 131S, S38.

Daghan, H., Arslan, M., Uygur, V., Koleli, N., 2013. Transformation of tobacco with *ScMTII* gene-enhanced cadmium and zinc accumulation. Clean Soil Air Water 41, 503–509.

Dai, J., Balish, R., Meagher, R.B., Merkle, S.A., 2009. Development of transgenic hybrid sweetgum (*Liquidambar styraciflua* × *L. formosana*) expressing γ-glutamylcysteine synthetase or mercuric reductase for phytoremediation of mercury pollution. N. For. 38 (1), 35–52.

Das, N., Bhattacharya, S., Maiti, M.K., 2016. Enhanced cadmium accumulation and tolerance in transgenic tobacco overexpressing rice metal tolerance protein gene *OsMTP1* is promising for phytoremediation. Plant Physiol. Biochem. 105, 297–309.

Das, N., Bhattacharya, S., Bhattacharyya, S., Maiti, M.K., 2017. Identification of alternatively spliced transcripts of rice phytochelatin synthase 2 gene *OsPCS2* involved in mitigation of cadmium and arsenic stresses. Plant Mol. Biol. 94 (1–2), 167–183.

Deng, F., Yamaji, N., Ma, J.F., Lee, S.K., Jeon, J.S., Martinoia, E., et al., 2018. Engineering rice with lower grain arsenic. Plant Biotechnol. J. 16 (10), 1691–1699.

Dhankher, O.P., Li, Y., Rosen, B.P., Shi, J., Salt, D., Senecoff, J.F., et al., 2002. Engineering tolerance and hyperaccumulation of arsenic in plants by combining arsenate reductase and γ-glutamylcysteine synthetase expression. Nat. Biotechnol. 20, 1140–1145.

Dhankher, O.P., Shasti, N.A., Rosen, B.P., Fuhrmann, M., Meagher, R.B., 2003. Increased cadmium tolerance and accumulation by plants expressing bacterial arsenate reductase. N. Phytol. 159, 431–441.

Dhankher, O.P., Doty, S.L., Meagher, R.B., Pilon-Smits, E.A.H., 2011. Biotechnological approaches for phytoremediation. In: Altman, A., Hasegawa, P.M. (Eds.), Plant Biotechnology and Agriculture. Academic Press, Oxford, pp. 309–328.

Didierjean, L., Gondet, L., Perkin, R., Lau, S.M., Schaller, H., O'Keefe, et al., 2002. Engineering herbicide metabolism in tobacco and *Arabidopsis* with CYP76B1, a cytochrome P450 enzyme from Jerusalem artichoke. Plant Physiol. 130, 179–189.

Dixit, P., Singh, S., Mukherjee, P.K., Eapen, S., 2008. Development of transgenic plants with cytochrome P450E1 gene and glutathione-*S*-transferase gene for degradation of organic pollutants. J. Biotechnol. 136S, S692–S693.

Dixit, P., Singh, S., Vancheeswaran, R., Patnala, K., Eapen, S., 2010. Expression of a *Neurospora crassa* zinc transporter gene in transgenic *Nicotiana tabacum* enhances plant zinc accumulation without co-transport of cadmium. Plant Cell Environ. 33 (10), 1697–1707.

Dixit, P., Mukherjee, P.K., Ramachandran, V., Eapen, S., 2011a. Glutathione transferase from *Trichoderma virens* enhances cadmium tolerance without enhancing its accumulation in transgenic *Nicotiana tabacum*. PLoS One 6 (1), e16360.

Dixit, P., Mukherjee, P.K., Sherkhane, P.D., Kale, S.P., Eapen, S., 2011b. Enhanced tolerance and remediation of anthracene by transgenic tobacco plants expressing a fungal glutathione transferase gene. J. Hazard. Mater. 192, 270–276.

Doty, S.L., Shang, Q.T., Wilson, A.M., Moore, A.L., Newman, L.A., Strand, S.E., et al., 2000. Enhanced metabolism of halogenated hydrocarbons in transgenic plants contain mammalian P450 2E1. Proc. Natl. Acad. Sci. U.S.A. 97, 6287–6291.

Doty, S.L., Shang, Q.T., Wilson, A.M., Moore, A.L., Newman, L.A., Strand, S.E., et al., 2007. Enhanced metabolism of halogenated hydrocarbons in transgenic plants contain mammalian P450 2E1. Proc. Natl. Acad. Sci. U.S.A. 97, 6287–6291.

Douchkov, D., Gryczka, C., Stephan, U.W., Hell, R., Bäumlein, H., 2005. Ectopic expression of nicotianamine synthase genes results in improved iron accumulation and increased nickel tolerance in transgenic tobacco. Plant Cell Environ. 28 (3), 365–374.

Eapen, S., Singh, S., D'Souza, S.F., 2007. Advances in development of transgenic plants for remediation of xenobiotic pollutants. Biotechnol. Adv. 25 (5), 442–451.

Evans, K.M., Gatehouse, J.A., Lindsay, W.P., Shi, J., Tommey, A.M., Robinson, N.J., 1992. Expression of the pea metallothionein-like gene *PsMTA* in *Escherichia coli* and *Arabidopsis thaliana* and analysis of trace metal ion accumulation: implications for *PsMTA* function. Plant Mol. Biol. 20, 1019–1028.

Faè, M., Balestrazzi, A., Confalonieri, M., Donà, M., Macovei, A., Valassi, A., et al., 2014. Copper-mediated genotoxic stress is attenuated by the overexpression of the DNA repair gene *MtTdp2*α (tyrosyl-DNA phosphodiesterase 2) in *Medicago truncatula* plants. Plant Cell Rep. 33 (7), 1071–1080.

Fan, T., Yang, L., Wu, X., Ni, J., Jiang, H., Zhang, Q., et al., 2016. The *PSE1* gene modulates lead tolerance in *Arabidopsis*. J. Exp. Bot. 67 (15), 4685–4695.

Fasani, E., Manara, A., Martini, F., Furini, A., DalCorso, G., 2018. The potential of genetic engineering of plants for the remediation of soils contaminated with heavy metals. Plant Cell Environ. 41 (5), 1201–1232.

Flocco, C.G., Lindblom, S.D., Smits, E.A.H.P., 2004. Overexpression of enzymes involved in glutathione synthesis enhances tolerance to organic pollutants in *Brassica juncea*. Int. J. Phytoremed. 6, 289–304.

Freeman, J.L., Persans, M.W., Nieman, K., Albrecht, C., Peer, W., Pickering, I.J., et al., 2004. Increased glutathione biosynthesis plays a role in nickel tolerance in *Thlaspi* nickel hyperaccumulators. Plant Cell 16 (8), 2176–2191.

French, C.J., Rosser, S.J., Davies, G.J., Nicklin, S., Bruce, N.C., 1999. Biodegradation of explosives by transgenic plants expressing pentaerythritol tetranitrate reductase. Nat. Biotechnol. 17, 491–494.

Fu, X.Y., Zhao, W., Xiong, A.S., Tian, Y.S., Zhu, B., Peng, R.H., et al., 2013. Phytoremediation of triphenylmethane dyes by overexpressing a *Citrobacter* sp. triphenylmethane reductase in transgenic *Arabidopsis*. Appl. Microbiol. Biotechnol. 97, 1799–1806.

Fu, X.Y., Zhu, B., Han, H.J., Zhao, W., Tian, Y.S., Peng, R.H., et al., 2016. Enhancement of naphthalene tolerance in transgenic *Arabidopsis* plants overexpressing the ferredoxin-like protein (ADI1) from rice. Plant Cell Rep. 35 (1), 17–26.

Gandia-Herrero, F., Lorenz, A., Larson, T., Graham, I.A., Bowles, D.J., Rylott, E.L., et al., 2008. Detoxification of the explosive 2,4,6-trinitrotoluene in *Arabidopsis*: discovery of bifunctional *O*- and *C*-glucosyltransferases. Plant J. 56, 963–974.

Gasic, K., Korban, S.S., 2007. Transgenic Indian mustard (*Brassica juncea*) plants expressing an *Arabidopsis* phytochelatin synthase (*AtPCS1*) exhibit enhanced As and Cd tolerance. Plant Mol. Biol. 64, 361–369.

Gisbert, C., Ros, R., DeHaro, A., Walker, D.J., PilarBernal, M., Serrano, R., et al., 2003. A plant genetically modified that accumulates Pb is especially promising for phytoremediation. Biochem. Biophys. Res. Commun. 303, 440–445.

Gorinova, N., Nedkovska, M., Todorovska, E., Simova-Stoilova, L., Stoyanova, Z., Georgieva, K., et al., 2007. Improved phytoaccumulation of cadmium by genetically modified tobacco plants (*Nicotiana tabacum* L.). Physiological and biochemical response of the transformants to cadmium toxicity. Environ. Pollut. 145 (1), 161–170.

Grichko, V.P., Filby, B., Glick, B.R., 2000. Increased ability of transgenic plants expressing the bacterial enzyme ACC deaminase to accumulate Cd, Co, Cu, Ni, Pb, and Zn. J. Biotechnol. 81, 45–53.

Grispen, V.M., Hakvoort, H.W., Bliek, T., Verkleij, J.A., Schat, H., 2011. Combined expression of the *Arabidopsis* metallothionein MT2b and the heavy metal transporting ATPase HMA4 enhances cadmium tolerance and the root to shoot translocation of cadmium and zinc in tobacco. Environ. Exp. Bot. 72 (1), 71–76.

Gu, C.S., Liu, L.Q., Zhao, Y.H., Deng, Y.M., Zhu, X.D., Huang, S.Z., 2014. Overexpression of *Iris lactea* var. chinensis metallothionein *llMT2a* enhances cadmium tolerance in *Arabidopsis thaliana*. Ecotoxicol. Environ. Saf. 105, 22–28.

Gu, C.S., Liu, L.Q., Deng, Y.M., Zhu, X.D., Huang, S.Z., Lu, X.Q., 2015. The heterologous expression of the *Iris lactea* var. chinensis type 2 metallothionein *IlMT2b* gene enhances copper tolerance in *Arabidopsis thaliana*. Bull. Environ. Contam. Toxicol. 94 (2), 247–253.

Guan, Z., Chai, T., Zhang, Y., Xu, J., Wei, W., 2009. Enhancement of Cd tolerance in transgenic tobacco plants overexpressing a Cd-induced catalase cDNA. Chemosphere 76 (5), 623–630.

Gullner, G., Komives, T., Rennenberg, H., 2001. Enhanced tolerance of transgenic poplar plants overexpressing gamma-glutamylcysteine synthetase towards chloroacetanilide herbicides. J. Exp. Bot. 52, 971–979.

Gunning, V., Tzafestas, K., Sparrow, H., Johnston, E.J., Brentnall, A.S., Potts, J.R., et al., 2014. *Arabidopsis* glutathione transferases U24 and U25 exhibit a range of detoxification activities with the environmental pollutant and explosive, 2,4,6-trinitrotoluene. Plant Physiol. 165, 854–865.

Guo, J., Dai, X., Xu, W., Ma, M., 2008. Overexpressing *GSH1* and *AsPCS1* simultaneously increases the tolerance and accumulation of cadmium and arsenic in *Arabidopsis thaliana*. Chemosphere 72 (7), 1020–1026.

Gustin, J.L., Loureiro, M.E., Kim, D., Na, G., Tikhonova, M., Salt, D.E., 2009. MTP1-dependent Zn sequestration into shoot vacuoles suggests dual roles in Zn tolerance and accumulation in Zn-hyperaccumulating plants. Plant J. 57, 1116–1127.

Han, Y., Fan, T., Zhu, X., Wu, X., Ouyang, J., Jiang, L., et al., 2019. *WRKY12* represses *GSH1* expression to negatively regulate cadmium tolerance in *Arabidopsis*. Plant Mol. Biol. 99 (1–2), 149–159.

Hannink, N.K., Rosser, S.J., French, C.E., Basran, A., Murray, J.A., Nicklin, S., et al., 2001. Phytodetoxification of TNT by transgenic plants expressing a bacterial nitroreductase. Nat. Biotechnol. 19, 1168–1172.

Hannink, N.K., Subramanian, M., Rosser, S.J., Basran, A., Murray, J.A.H., Shanks, J.V., et al., 2007. Enhanced transformation of TNT by tobacco plants expressing a bacterial nitroreductase. Int. J. Phytoremed. 9, 385–401.

Haque, S., Zeyaullah, M., Nabi, G., Srivastava, P.S., Ali, A., 2010. Transgenic tobacco plant expressing environmental *E. coli merA* gene for enhanced volatilization of ionic mercury. Microbiol. Biotechnol. 20 (5), 917–924.

Harada, E., Choi, Y.E., Tsuchisaka, A., Obata, H., Sano, H., 2001. Transgenic tobacco plants expressing a rice cysteine synthase gene are tolerant to toxic levels of cadmium. J. Plant Physiol. 158 (5), 655–661.

Hayashi, S., Kuramata, M., Abe, T., Takagi, H., Ozawa, K., Ishikawa, S., 2017. Phytochelatin synthase *OsPCS1* plays a crucial role in reducing arsenic levels in rice grains. Plant J. 91 (5), 840–848.

He, Y.K., Sun, J.G., Feng, X.Z., Czakó, M., Márton, L., 2001. Differential mercury volatilization by tobacco organs expressing a modified bacterial *merA* gene. Cell Res. 11, 231–236.

He, J., Li, H., Ma, C., Zhang, Y., Polle, A., Rennenberg, H., et al., 2015. Overexpression of bacterial γ-glutamylcysteine synthetase mediates changes in cadmium influx, allocation and detoxification in poplar. N. Phytol. 205, 240–254.

Heaton, A.C.P., Rugh, C.L., Kim, T., Meagher, R.B., 2003. Toward detoxifying mercury-polluted aquatic sediments with rice genetically engineered for mercury resistance. Environ. Toxicol. Chem. 22 (12), 2940–2947.

Heaton, A.C.P., Rugh, C.L., Wang, N.J., Meagher, R.B., 2005. Physiological Responses of Transgenic *merA*-tobacco (*Nicotiana tabacum*) to foliar and root mercury exposure. Water Air Soil Pollut. 161, 137–155.

Hirschi, K.D., Korenkov, V.D., Wilganowski, N.L., Wagner, G.J., 2000. Expression of *Arabidopsis* CAX2 in tobacco. Altered metal accumulation and increased manganese tolerance. Plant Physiol. 124, 125–133.

Hsieh, J.L., Chen, C.Y., Chiu, M.H., Chein, M.F., Chang, J.S., Endo, G., et al., 2009. Expressing a bacterial mercuric ion binding protein in plant for phytoremediation of heavy metals. J. Hazard. Mater. 161, 920–925.

Hussain, I., Aleti, G., Naidu, R., Puschenreiter, M., Mahmood, Q., Rahman, M.M., et al., 2018. Microbe and plant assisted-remediation of organic xenobiotics and its enhancement by genetically modified organisms and recombinant technology: a review. Sci. Total Environ. 628–629, 1582–1599.

Hussein, S.H., Ruiz, O.N., Terry, N., Daniell, H., 2007. Phytoremediation of mercury and organomercurials in chloroplast transgenic plants: enhanced root uptake, translocation to shoots and volatilization. Environ. Sci. Technol. 41, 8439–8446.

Iimura, Y., Ikeda, S., Sonoki, T., Hayakawa, T., Kajita, S., Kimbara, K., et al., 2002. Expression of a gene for Mn-peroxidase from *Coriolus versicolor* in transgenic tobacco generates potential tool for phytoremediation. Appl. Microbiol. Biotechnol. 59, 246–251.

Inui, H., Ohkawa, H., 2005. Herbicide resistance in transgenic plants with mammalian P450 monooxygenase genes. Pest Manage. Sci. 61 (3), 286–291.

Inui, H., Hirota, M., Goto, J., Yoshihara, R., Kodama, N., Matsui, T., et al., 2015. Zinc finger protein genes from *Cucurbita pepo* are promising tools for conferring non-Cucurbitaceae plants with ability to accumulate persistent organic pollutants. Chemosphere 123, 48–54.

Ivanova, L.A., Ronzhina, D.A., Ivanov, L.A., Stroukova, L.V., Peuke, A.D., Rennenberg, H., 2011. Over-expression of *gsh1* in the cytosol affects the photosynthetic apparatus and improves the performance of transgenic poplars on heavy metal-contaminated soil. Plant Biol. (Stuttg.) 13 (4), 649–659.

Jackson, E.G., Rylott, E.L., Fournier, D., Hawari, J., Bruce, N.C., 2007. Exploring the biochemical properties and remediation applications of the unusual explosive-degrading P450 system XplA/B. Proc. Natl. Acad. Sci. U.S.A. 104, 16822–16827.

James, C.A., Xin, G., Doty, S.L., Strand, S.E., 2008. Degradation of low molecular weight volatile organic compounds by plants genetically modified with mammalian cytochrome P450 2E1. Environ. Sci. Technol. 42 (1), 289–293.

Jiang, L., Chen, Z., Gao, Q., Ci, L., Cao, S., Han, Y., et al., 2016. Loss-of-function mutations in the *APX1* gene result in enhanced selenium tolerance in *Arabidopsis thaliana*. Plant Cell Environ. 39 (10), 2133–2144.

Jiang, L., Wang, W., Chen, Z., Gao, Q., Xu, Q., Cao, H., 2017. A role for *APX1* gene in lead tolerance in *Arabidopsis thaliana*. Plant Sci. 256, 94–102.

Jin, S., Daniell, H., 2014. Expression of gamma-tocopherol methyltransferase in chloroplasts results in massive proliferation of the inner envelope membrane and decreases susceptibility to salt and metal-induced oxidative stresses by reducing reactive oxygen species. Plant Biotechnol. J. 12, 1274–1285.

Jung, S., lee, H.J., Lee, Y., Kang, K., Kim, Y.S., Grimm, B., et al., 2008. Toxic tetrapyrrole accumulation in protoporphyrinogrn IX oxidase overexpressing transgenic rice plants. Plant Mol. Biol. 67, 535–546.

Kang, J.W., Wilkerson, H.-W., Farin, F.M., Bammler, T.K., Beyer, R.P., Strand, S.E., et al., 2010. Mammalian cytochrome CYP2E1 triggered differential gene regulation in response to trichloroethylene (TCE) in a transgenic poplar. Funct. Integr. Genomics 10 (3), 417–424.

Karavangeli, M., Labrou, N.E., Clonis, Y.D., Tsaftaris, A., 2005. Development of transgenic tobacco plants overexpressing glutathione *S*-transferase I for chloroacetanilide herbicides phytoremediation. Biomol. Eng. 22, 121–128.

Kawahigashi, H., 2009. Transgenic plants for phytoremediation of herbicides. Curr. Opin. Biotechnol. 20 (2), 225–230.

Kawahigashi, H., Hirose, S., Ohkawa, H., Ohkawa, Y., 2005a. Phytoremediation of metolachlor by transgenic rice plants expressing human CYP2B6. J. Agric. Food Chem. 53 (23), 9155–9160.

Kawahigashi, H., Hirose, S., Ohkawa, H., Ohkawa, Y., 2005b. Transgenic rice plants expressing human CYP1A1 remediate the triazine herbicides atrazine and simazine. J. Agric. Food Chem. 53 (22), 8557–8564.

Kawahigashi, H., Hirose, S., Ozawa, K., Ido, Y., Kojima, M., Ohkawa, H., et al., 2005c. Analysis of substrate specificity of pig CYP2B22 and CYP2C49 towards herbicides by transgenic rice plants. Transgenic Res. 14, 907–917.

Kawahigashi, H., Hirose, S., Ohkawa, H., Ohkawa, Y., 2006. Phytoremediation of the herbicides atrazine and metolachlor by transgenic rice plants expressing human CYP1A1, CYP2B6, and CYP2C19. J. Agric. Food Chem. 54 (8), 2985–2991.

Kawahigashi, H., Hirose, S., Ohkawa, H., Ohkawa, Y., 2007. Herbicide resistance of transgenic rice plants expressing human CYP1A1. Biotechnol. Adv. 25, 75–85.

Kawahigashi, H., Hirose, S., Ohkawa, H., Ohkawa, Y., 2008. Transgenic rice plants expressing human P450 genes involved in xenobiotic metabolism for phytoremediation. J. Mol. Microbiol. Biotechnol. 15, 212–219.

Kawashima, C.G., Noji, M., Nakamura, M., Ogra, Y., Suzuki, K.T., Saito, K., 2004. Heavy metal tolerance of transgenic tobacco plants over-expressing cysteine synthase. Biotechnol. Lett. 26 (2), 153–157.

Kebeish, R., Azab, E., Peterhaensel, C., El-Basheer, R., 2014. Engineering the metabolism of the phenylurea herbicide chlortoluron in genetically modified *Arabidopsis thaliana* plants expressing the mammalian cytochrome P450 enzyme CYP1A2. Environ. Sci. Pollut. Res. Int. 21 (13), 8224–8232.

Khoudi, H., Maatar, Y., Gouiaa, S., Masmoudi, K., 2012. Transgenic tobacco plants expressing ectopically wheat H^+-pyrophosphatase (H^+-PPase) gene *TaVP1* show enhanced accumulation and tolerance to cadmium. J. Plant Physiol. 169 (1), 98–103.

Kim, S., Takahashi, M., Higuchi, K., Tsunoda, K., Nakanishi, H., Yoshimura, E., et al., 2005. Increased nicotianamine biosynthesis confers enhanced tolerance of high levels of metals, in particular nickel, to plants. Plant Cell Physiol. 46 (11), 1809–1818.

Kim, D.Y., Bovet, L., Kushnir, S., Noh, E.W., Martinoia, E., Lee, Y., 2006. *AtATM3* is involved in heavy metal resistance in *Arabidopsis*. Plant Physiol. 140 (3), 922–932.

Koprivova, A., Kopriva, S., Jäger, D., Will, B., Jouanin, L., Rennenberg, H., 2002. Evaluation of transgenic poplars over-expressing enzymes of glutathione synthesis for phytoremediation of cadmium. Plant Biol. 4, 664–670.

Korenkov, V., Hirschi, K., Crutchfield, J.D., Wagner, G.J., 2007. Enhancing tonoplast Cd/H antiport activity increases Cd, Zn, and Mn tolerance, and impacts root/shoot Cd partitioning in *Nicotiana tabacum* L. Planta 226, 1379–1387.

Kotrba, P., Najmanova, J., Macek, T., Ruml, T., Mackova, M., 2009. Genetically modified plants in phytoremediation of heavy metal and metalloid soil and sediment pollution. Biotechnol. Adv. 27 (6), 799–810.

Koźmińska, A., Wiszniewska, A., Hanus-Fajerska, E., Muszyńska, E., 2018. Recent strategies of increasing metal tolerance and phytoremediation potential using genetic transformation of plants. Plant Biotechnol. Rep. 12, 1–14.

Krämer, U., Chardonnens, A.N., 2001. The use of transgenic plants in the bioremediation of soils contaminated with trace elements. Appl. Microbiol. Biotechnol. 55 (6), 661–672.

Kunihiro, S., Saito, T., Matsuda, T., Inoue, M., Kuramata, M., Taguchi-Shiobara, F., et al., 2013. Rice *DEP1*, encoding a highly cysteine-rich G protein γ subunit, confers cadmium tolerance on yeast cells and plants. J. Exp. Bot. 64 (14), 4517–4527.

Kurumata, M., Takahashi, M., Sakamotoa, A., Ramos, J.L., Nepovim, A., Vanek, T., et al., 2005. Tolerance to, and uptake and degradation of 2,4,6-trinitrotoluene (TNT) are enhanced by the expression of a bacterial nitroreductase gene in *Arabidopsis thaliana*. Z. Naturforsch. C 60, 272–278.

Kuvshinov, V., Koivu, K., Kanerva, A., Pehu, E., 2001. Molecular control of transgene escape from genetically modified plants. Plant Sci. 160 (3), 517–522.

LeBlanc, M.S., McKinney, E.C., Meagher, R.B., Smith, A.P., 2013. Hijacking membrane transporters for arsenic phytoextraction. J. Biotechnol. 163 (1), 1–9.

LeDuc, D.L., Tarun, A.S., Montes-Bayon, M., Meija, J., Malit, M.F., Wu, C.P., et al., 2004. Overexpression of selenocysteine methyltransferase in *Arabidopsis* and Indian mustard increases selenium tolerance and accumulation. Plant Physiol. 135 (1), 377–383.

LeDuc, D.L., AbdelSamie, M., Móntes-Bayon, M., Wu, C.P., Reisinger, S.J., Terry, N., 2006. Overexpressing both ATP sulfurylase and selenocysteine methyltransferase enhances selenium phytoremediation traits in Indian mustard. Environ. Pollut. 144 (1), 70–76.

Lee, B.R., Hwang, S., 2015. Over-expression of *NtHb1* encoding a non-symbiotic class 1 hemoglobin of tobacco enhances a tolerance to cadmium by decreasing NO (nitric oxide) and Cd levels in *Nicotiana tabacum*. Environ. Exp. Bot. 113, 18–27.

Lee, J., Bae, H., Jeong, J., Lee, J.Y., Yang, Y.Y., Hwang, I., et al., 2003. Functional expression of a bacterial heavy metal transporter in *Arabidopsis* enhances resistance to and decreases uptake of heavy metals. Plant Physiol. 133, 589–596.

Lee, K.Y., Strand, S.E., Doty, S.L., 2012. Phytoremediation of chlorpyrifos by *Populus* and *Salix*. Int. J. Phytoremed. 14 (1), 48–61.

Li, Y., Dhankher, O.P., Carreira, L., Lee, D., Chen, A., Schroeder, J.I., et al., 2004. Overexpression of phytochelatin synthase in *Arabidopsis* leads to enhanced arsenic tolerance and cadmium hypersensitivity. Plant Cell Physiol. 45, 1787–1797.

Li, Y., Dankher, O.P., Carreira, L., Smith, A.P., Meagher, R.B., 2006. The shoot-specific expression of γ-glutamylcysteine synthetase directs the long-distance transport of thiol-peptides to roots conferring tolerance to mercury and arsenic. Plant Physiol. 141, 288–298.

Liang Zhu, Y., Pilon-Smits, E.A.H., Jouanin, L., Terry, N., 1999. Overexpression of glutathione synthetase in Indian mustard enhances cadmium accumulation and tolerance. Plant Physiol. 119, 73–80.

Liu, D., An, Z., Mao, Z., Ma, L., Lu, Z., 2015. Enhanced heavy metal tolerance and accumulation by transgenic sugar beets expressing *Streptococcus thermophilus* STGCS-GS in the presence of Cd, Zn and Cu alone or in combination. PLoS One 10 (6), e0128824.

Luan, M., Liu, J., Liu, Y., Han, X., Wenzhi, G.S., Luan, L.S., 2018. Vacuolar phosphate transporter 1 (VPT1) affects arsenate tolerance by regulating phosphate homeostasis in *Arabidopsis*. Plant Cell Physiol. 59 (7), 1345–1352.

Lv, Y., Deng, X., Quan, L., Xia, Y., Shen, Z., 2013. Metallothioneins *BcMT1* and *BcMT2* from *Brassica campestris* enhance tolerance to cadmium and copper and decrease production of reactive oxygen species in *Arabidopsis thaliana*. Plant Soil. 367 (1–2), 507–519.

Lyyra, S., Meagher, R.B., Kim, T., Heaton, A., Montello, P., Balish, R.S., et al., 2007. Coupling two mercury resistance genes in Eastern cottonwood enhances the processing of organomercury. Plant Biotechnol. J. 5, 254–262.

Mena-Benitez, G.L., Gandia-Herrero, F., Graham, S., Larson, T.R., McQueen-Mason, S.J., French, C.E., et al., 2008. Engineering a catabolic pathway in plants for the degradation of 1,2-dichloroethane. Plant Physiol. 147, 1192–1198.

Millwood, R.J., Moon, H.S., Poovaiah, C.R., Muthukumar, B., Rice, J.H., Abercrombie, J.M., et al., 2016. Engineered selective plant male sterility through pollen-specific expression of the *EcoRI* restriction endonuclease. Plant Biotechnol. J. 14, 1281–1290.

Misra, S., Gedamu, L., 1989. Heavy metal tolerant transgenic *Brassica napus* L. and *Nicotiana tabacum* L. plants. Theor. Appl. Genet. 78, 161–168.

Mohammed, J.A., Arlene, A.B., Asthana, A.M., Qureshi, I., 2019. Proteomics of cadmium tolerance in plants. In: Uzzaman, M.H., Vara Prasad, M.N., Nahar, K. (Eds.), Cadmium Tolerance in Plants, Agronomic, Molecular, Signaling, and Omic Approaches. Academic Press, Elsevier, pp. 143–175.

Mosa, K.A., Saadoun, I., Kumar, K., Helmy, M., Dhankher, O.P., 2016. Potential biotechnological strategies for the clean-up of heavy metals and metalloids. Front. Plant Sci. 7, 303.

Mouhamad, R., Ghanem, I., AlOrfi, M., Ibrahim, K., Ali, N., Al-Daoude, A., 2012. Phytoremediation of trichloroethylene and dichlorodiphenyl trichloroethane-polluted water using transgenic *Sesbania grandiflora* and *Arabidopsis thaliana* plants harboring rabbit cytochrome p450 2E1. Int. J. Phytoremed. 14, 656–668.

Nagata, T., Ishikawa, C., Kiyono, M., Pan-Hou, H., 2006a. Accumulation of mercury in transgenic tobacco expressing bacterial polyphosphate. Biol. Pharma. Bullet 29, 2350–2353.

Nagata, T., Kiyono, M., Pan-Hou, H., 2006b. Engineering expression of bacterial polyphosphate kinase in tobacco for mercury remediation. Appl. Microbiol. Biotechnol. 72, 777–782.

Nagata, T., Nakamura, A., Akizawa, T., Pan-Hou, H., 2009. Genetic engineering of transgenic tobacco for enhanced uptake and bioaccumulation of mercury. Biol. Pharm. Bull. 32 (9), 1491–1495.

Nagata, T., Morita, H., Akizawa, T., Pan-Hou, H., 2010. Development of a transgenic tobacco plant for phytoremediation of methylmercury pollution. Appl. Microbiol. Biotechnol. 87 (2), 781–786.

Nahar, N., Rahman, A., Nawani, N.N., Ghosh, S., Mandal, A., 2017. Phytoremediation of arsenic from the contaminated soil using transgenic tobacco plants expressing *ACR2* gene of *Arabidopsis thaliana*. J. Plant Physiol. 218, 121–126.

Nair, S., Joshi-Saha, A., Singh, S., Ramachandran, V., Singh, S., Thorat, V., et al., 2012. Evaluation of transgenic tobacco plants expressing a bacterial Co-Ni transporter for acquisition of cobalt. J. Biotechnol. 161 (4), 422–428.

Nesler, A., DalCorso, G., Fasani, E., Manara, A., Di Sansebastiano, G.P., Argese, E., et al., 2017. Functional components of the bacterial *CzcCBA* efflux system reduce cadmium uptake and accumulation in transgenic tobacco plants. Nat. Biotechnol. 35, 54–61.

Nian, H., Meng, Q., Zhang, W., Chen, L., 2013. Overexpression of the formaldehyde dehydrogenase gene from *Brevibacillus brevis* to enhance formaldehyde tolerance and detoxification of tobacco. Appl. Biochem. Biotechnol. 169 (1), 170–180.

Nie, L., Shah, S., Rashid, A., Burd, G.I., Dixon, D.G., Glick, B.R., 2002. Phytoremediation of arsenate contaminated soil by transgenic canola and the plant growth-promoting bacterium *Enterobacter cloacae* CAL2. Plant Physiol. Biochem. 40, 355–361.

Novakova, M., Mackova, M., Antosova, Z., Viktorova, J., Szekeresm, M., Demnerova, K., et al., 2010. Cloning the bacterial *bphC* gene into *Nicotiana tabacum* to improve the efficiency of phytoremediation of polychlorinated biphenyls. Bioeng. Bugs 1, 419–423.

Oller, A.L.W., Agostini, E., Talano, M.A., Capozucca, C., Milrad, S.R., Tigier, H.A., et al., 2005. Overexpression of a basic peroxidase in transgenic tomato (*Lycopersicon esculentum* Mill. cv. Pera) hairy roots increases phytoremediation of phenol. Plant Sci. 169, 1102–1111.

Pandey, V.C., Bajpai, O., 2019. Phytoremediation: from theory towards practice. In: Pandey, V.C., Bauddh, K. (Eds.), Phytomanagement of Polluted Sites. Elsevier, Amsterdam, pp. 1–49. , <https://doi.org/10.1016/B978-0-12-813912-7.00001-6>.

Pandey, V.C., Souza-Alonso, P., 2019. Market opportunities in sustainable phytoremediation. In: Pandey, V.C., Bauddh, K. (Eds.), Phytomanagement of Polluted Sites. Elsevier, Amsterdam, pp. 51–82. , <https://doi.org/10.1016/B978-0-12-813912-7.00002-8>.

Pandey, V.C., Pandey, D.N., Singh, N., 2015. Sustainable phytoremediation based on naturally colonizing and economically valuable plants. J. Clean. Prod. 86, 37–39.

Park, J., Song, W.Y., Ko, D., Eom, Y., Hansen, T.H., Schiller, M., et al., 2012. The phytochelatin transporters *AtABCC1* and *AtABCC2* mediate tolerance to cadmium and mercury. Plant J. 69 (2), 278–288.

Peng, R.H., Fu, X.Y., Zhao, W., Tian, Y.S., Zhu, B., Han, H.J., et al., 2014a. Phytoremediation of phenanthrene by transgenic plants transformed with a naphthalene dioxygenase system from *Pseudomonas*. Environ. Sci. Technol. 48, 12824–12832.

Peng, R., Fu, X., Tian, Y., Zhao, W., Zhu, B., Xu, J., et al., 2014b. Metabolic engineering of *Arabidopsis* for remediation of different polycyclic aromatic hydrocarbons using a hybrid bacterial dioxygenase complex. Metabol. Eng. 26c, 100–110.

Pianelli, K., Mari, S., Marquès, L., Lebrun, M., Czernic, P., 2005. Nicotianamine over-accumulation confers resistance to nickel in *Arabidopsis thaliana*. Transgenic Res. 14 (5), 739–748.

Picault, N., Cazalé, A.C., Beyly, A., Cuiné, S., Carrier, P., Luu, D.T., et al., 2006. Chloroplast targeting of phytochelatin synthase in *Arabidopsis*: effects on heavy metal tolerance and accumulation. Biochimie 88 (11), 1743–1750.

Pilon-Smits, E.A.H., LeDuc, D.L., 2009. Phytoremediation of selenium using transgenic plants. Curr. Opin. Biotechnol. 20 (2), 207–212.

Pilon-Smits, E.A.H., Hwang, S., MelLytle, C., Zhu, Y., Tai, J., Bravo, R., et al., 1999. Overexpression of ATP sulfurylase in Indian mustard leads to increased selenate uptake, reduction, and tolerance. Plant Physiol. 119, 123–132.

Pilon, M., Owen, J.D., Garifullina, G.F., Kurihara, T., Mihara, H., Esaki, N., et al., 2003. Enhanced selenium tolerance and accumulation in transgenic *Arabidopsis* expressing a mouse selenocysteine lyase. Plant Physiol. 131 (3), 1250–1257.

Pomponi, M., Censi, V., Di Girolamo, V., De Paolis, A., di Toppi, L.S., Aromolo, R., et al., 2006. Overexpression of *Arabidopsis* phytochelatin synthase in tobacco plants enhances Cd^{2+} tolerance and accumulation but not translocation to the shoot. Planta 223 (2), 180–190.

Praveen, A., Pandey, V.C., Marwa, N., Singh, D.P., 2019. Rhizoremediation of polluted sites: harnessing plant-microbe interactions. In: Pandey, V.C., Bauddh, K. (Eds.), Phytomanagement of Polluted Sites. Elsevier, Amsterdam, pp. 389–407. , <https://doi.org/10.1016/B978-0-12-813912-7.00015-6>.

Reisinger, S., Schiavon, M., Terry, N., Pilon-Smits, E.A.H., 2008. Heavy metal tolerance and accumulation in Indian mustard (*Brassica juncea* L.) expressing bacterial γ-glutamylcysteine synthetase or glutathione synthetase. Int. J. Phytoremed. 10, 1–15.

Rodríguez-Llorente, I.D., Pérez-Palacios, P., Doukkali, B., Caviedes, M.A., Pajuelo, E., 2010. Expression of the seed-specific metallothionein mt4a in plant vegetative tissues increases Cu and Zn tolerance. Plant Sci. 178 (3), 327–332.

Rugh, C.L., Wilde, H.D., Stack, N.M., Thompson, D.M., Summers, A.O., Meagher, R.B., 1996. Mercuric ion reduction and resistance in transgenic *Arabidopsis thaliana* plants expressing a modified bacterial *merA* gene. Proc. Natl. Acad. Sci. U.S.A. 93, 3182–3187.

Rugh, C.L., Senecoff, J.F., Meagher, R.B., Merkle, S.A., 1998. Development of transgenic yellow poplar for mercury phytoremediation. Nat. Biotechnol. 16, 925–928.

Ruiz, O.N., Hussein, H.S., Terry, N., Daniell, H., 2003. Phytoremediation of organomercurial compounds via chloroplast genetic engineering. Plant Physiol. 132, 1344–1352.

Ruiz, O.N., Alvarez, D., Torres, C., Roman, L., Daniell, H., 2011. Metallothionein expression in chloroplasts enhances mercury accumulation and phytoremediation capability. Plant Biotechnol. J. 9, 609–617.

Rylott, E.L., Bruce, N.C., 2009. Plants disarm soil: engineering plants for the phytoremediation of explosives. Trends Biotechnol. 27 (2), 73–81.

Rylott, E.L., Jackson, R.G., Edwards, J., Womack, G.L., Seth-Smith, H.M., Rathbone, D.A., et al., 2006. An explosive degrading cytochrome P450 activity and its targeted application for the phytoremediation of RDX. Nat. Biotechnol. 24, 216–219.

Rylott, E.L., Budarina, M.V., Barker, A., Lorenz, A., Strand, S.E., Bruce, N.C., 2011. Engineering plants for the phytoremediation of RDX in the presence of the co-contaminating explosive TNT. N. Phytol. 192, 405–413.

Rylott, E.L., Johnston, E.J., Bruce, N.C., 2015. Harnessing microbial gene pools to remediate persistent organic pollutants using genetically modified plants—a viable technology? J. Exp. Bot. 66 (21), 6519–6533.

Sasaki, Y., Hayakawa, T., Inoue, C., Miyazaki, A., Silver, S., Kusano, T., 2006. Generation of mercury-hyperaccumulating plants through transgenic expression of the bacterial mercury membrane transport protein *MerC*. Transgenic Res. 15, 615–625.

Sawada, A., Oyabu, T., Chen, L.M., Li, K.Z., Hirai, N., Yurimoto, H., et al., 2007. Purification capability of tobacco transformed with enzymes from a methylotrophic bacterium for formaldehyde. Int. J. Phytoremed. 9 (6), 487–496.

Sekhar, K., Priyanka, B., Reddy, V.D., Rao, K.V., 2011. Metallothionein 1 (*CcMT1*) of pigeonpea (*Cajanus cajan*, L.) confers enhanced tolerance to copper and cadmium in *Escherichia coli* and *Arabidopsis thaliana*. Environ. Exp. Bot. 72 (2), 131–139.

Shim, D., Kim, S., Choi, Y.I., Song, W.Y., Park, J., Youk, E.S., et al., 2013. Transgenic poplar trees expressing yeast cadmium factor 1 exhibit the characteristics necessary for the phytoremediation of mine tailing soil. Chemosphere 90 (4), 1478–1486.

Shiota, N., Nagasawa, A., Sakaki, T., Yabusaki, Y., Ohkawa, H., 1994. Herbicide-resistant tobacco plants expressing the fused enzyme between rat cytochrome P4501A1 (CYP1A1) and yeast NADPH-cytochrome P450 oxidoreductase. Plant Physiol. 106 (1), 17–23.

Shiota, N., Kodama, S., Inui, H., Ohkawa, H., 2000. Expression of human cytochromes P450 1A1 and P450 1A2 as fused enzymes with yeast NADPH-cytochrome P450 oxidoreductase in transgenic tobacco plants. Biosci. Biotechnol. Biochem. 64 (10), 2025–2033.

Shukla, D., Kesari, R., Mishra, S., Dwivedi, S., Tripathi, R.D., Nath, P., et al., 2012. Expression of phytochelatin synthase from aquatic macrophyte *Ceratophyllum demersum* L. enhances cadmium and arsenic accumulation in tobacco. Plant Cell Rep. 31 (9), 1687–1699.

Shukla, D., Kesari, R., Tiwari, M., Dwivedi, S., Tripathi, R.D., Nath, P., et al., 2013. Expression of *Ceratophyllum demersum* phytochelatin synthase, *CdPCS1*, in *Escherichia coli* and *Arabidopsis* enhances heavy metal(loid)s accumulation. Protoplasma 250, 1263–1272.

Shukla, D., Huda, K.M., Banu, M.S., Gill, S.S., Tuteja, R., Tuteja, N., 2014. *OsACA6*, a P-type 2B Ca(2 +) ATPase functions in cadmium stress tolerance in tobacco by reducing the oxidative stress load. Planta. 240 (4), 809–824.

Siminszky, B., Corbin, F.T., Ward, E.R., Fleischmann, T.J., Dewey, R.E., 1999. Expression of a soybean cytochrome P450 monooxygenase cDNA in yeast and tobacco enhances the metabolism of phenyl urea herbicides. Proc. Natl. Acad. Sci. U.S.A. 96, 1750–1755.

Singh, S., Korripally, P., Vancheeswaran, R., Eapen, S., 2011. Transgenic *Nicotiana tabacum* plants expressing a fungal copper transporter gene show enhanced acquisition of copper. Plant Cell Rep. 30 (10), 1929–1938.

Sivaram, A.K., Logeshwaran, P., Subash Chandra bose, S.R., Lockington, R., Naidu, R., Megharaj, M., 2018. Comparison of plants with C3 and C4 carbon fixation pathways for remediation of polycyclic aromatic hydrocarbon contaminated soils. Sci. Rep. 8, 2100.

Song, W., Sohn, E.J., Martinoia, E., Lee, Y.J., Yang, Y., Jasinski, M., et al., 2003. Engineering tolerance and accumulation of lead and cadmium in transgenic plants. Nat. Biotechnol. 21, 914–919.

Song, H.-Y., Ren, X.-S., Si, J., Li, C.-Q., Song, M., 2008. Construction of marker-free GFP transgenic tobacco by Cre/lox site-specific recombination system. Agricult. Sci. China 7 (9), 1061–1070.

Song, W.Y., Park, J., Mendoza-Cózatl, D.G., Suter-Grotemeyer, M., Shim, D., Hörtensteiner, S., et al., 2010a. Arsenic tolerance in *Arabidopsis* is mediated by two ABCC-type phytochelatin transporters. Proc. Natl. Acad. Sci. U.S.A. 7 (49), 21187–21192.

Song, Z., Orita, I., Yin, F., Yurimoto, H., Kato, N., Sakai, Y., et al., 2010b. Overexpression of an HPS/PHI fusion enzyme from *Mycobacterium gastri* in chloroplasts of geranium enhances its ability to assimilate and phytoremediate formaldehyde. Biotechnol. Lett. 32 (10), 1541–1548.

Song, W.Y., Yamaki, T., Yamaji, N., Ko, D., Jung, K.H., Fujii-Kashino, M., et al., 2014. A rice ABC transporter, OsABCC1, reduces arsenic accumulation in the grain. Proc. Natl. Acad. Sci. U.S.A. 111 (44), 15699–15704.

Song, J., Feng, S.J., Chen, J., Zhao, W.T., Yang, Z.M., 2017. A cadmium stress-responsive gene *AtFC1* confers plant tolerance to cadmium toxicity. BMC Plant Biol. 17 (1), 187.

Sonoki, T., Kajita, S., Ikeda, S., Uesugi, M., Tatsumi, K., Katayama, Y., et al., 2005. Transgenic tobacco expressing fungal laccase promotes the detoxification of environmental pollutants. Appl. Microbiol. Biotechnol. 67, 138–142.

Stearns, J.C., Shah, S., Greenberg, B.M., Dixon, D.G., Glick, B.R., 2005. Tolerance of transgenic canola expressing 1-aminocyclopropane-1-carboxylic acid deaminase to growth inhibition by nickel. Plant Physiol. Biochem. 43, 701–708.

Su, Z.H., Xu, Z.S., Peng, R.H., Tian, Y.S., Zhao, W., Han, H.J., et al., 2012. Phytoremediation of trichlorophenol by Phase II metabolism in transgenic *Arabidopsis* overexpressing a *Populus* glucosyltransferase. Environ. Sci. Technol. 46, 4016–4024.

Suman, J., Uhlik, O., Viktorova, J., Macek, T., 2018. Phytoextraction of heavy metals: a promising tool for clean-up of polluted environment? Front. Plant Sci. 9, 1476.

Sunkar, R., Kaplan, B., Bouché, N., Arazi, T., Dolev, D., Talke, I.N., et al., 2000. Expression of a truncated tobacco *NtCBP4* channel in transgenic plants and disruption of the homologous *Arabidopsis CNGC1* gene confer Pb^{2+} tolerance. Plant J. 24, 533–542.

Sylvestre, M., Macek, T., Mackova, M., 2009. Transgenic plants to improve rhizoremediation of polychlorinated biphenyls (PCBs). Curr. Opin. Biotechnol. 20 (2), 242–247.

Talano, M.A., Busso, D.C., Paisio, C.E., González, P.S., Purro, S.A., Medina, M.I., et al., 2012. Phytoremediation of 2,4-dichlorophenol using wild type and transgenic tobacco plants. Environ. Sci. Pollut. Res. Int. 19 (6), 2202–2211.

Tang, Z., Chen, Y., Chen, F., Ji, Y., Zhao, F.J., 2017. A putative peptide transporter in rice, is involved in dimethylarsenate accumulation in rice grain. Plant Cell Physiol. 58 (5), 904–913.

Thomas, J.C., Davies, E.C., Malick, F.K., Endreszl, C., Williams, C.R., Abbas, M., et al., 2003. Yeast metallothionein in transgenic tobacco promotes copper uptake from contaminated soils. Biotechnol. Prog. 19 (2), 273–280.

Turchi, A., Tamantini, I., Camussi, A.M., Racchi, M.L., 2012. Expression of a metallothionein A1 gene of *Pisum sativum* in white poplar enhances tolerance and accumulation of zinc and copper. Plant Sci. 183, 50–56.

Uraguchi, S., Tanaka, N., Hofmann, C., Abiko, K., Ohkama-Ohtsu, N., Weber, M., et al., 2017. Phytochelatin synthase has contrasting effects on cadmium and arsenic accumulation in rice grains. Plant Cell Physiol. 58 (10), 1730–1742.

Uraguchi, S., Sone, Y., Kamezawa, M., Tanabe, M., Hirakawa, M., Nakamura, R., et al., 2019. Ectopic expression of a bacterial mercury transporter *MerC* in root epidermis for efficient mercury accumulation in shoots of *Arabidopsis* plants. Sci. Rep. 9 (1), 4347.

US Environmental Protection Agency (USEPA), 2002. Interim reregistration eligibility decision for chlorpyrifos. In: Prevention, Pesticides and Toxic Substances (7508C), EPA 738-R-01-007.

Van Aken, B., Doty, S.L., 2010. Transgenic plants and associated bacteria for phytoremediation of chlorinated compounds. Biotechnol. Genet. Eng. Rev. 26, 43–64.

van Dillewijn, P., Couselo, J., Corredoira, E., Delgado, A., Wittich, R., Ballester, A., et al., 2008. Bioremediation of 2,4,6-trinitrotoluene by bacterial nitroreductase expressing transgenic aspen. Environ. Sci. Technol. 42, 7405–7410.

Van Huysen, T., Abdel-Ghany, S., Hale, K.L., LeDuc, D., Terry, N., Pilon-Smits, E.A.H., 2003. Overexpression of cystathionine-*c*-synthase enhances selenium volatilization in *Brassica juncea*. Planta 218, 71–78.

Verma, S., Verma, P.K., Meher, A.K., Bansiwal, A.K., Tripathi, R.D., Chakrabarty, D., 2018. A novel fungal arsenic methyltransferase, *WaarsM* reduces grain arsenic accumulation in transgenic rice (*Oryza sativa* L.). J. Hazard. Mater. 344, 626–634.

Verma, S., Verma, P.K., Chakrabarty, D., 2019. Arsenic bio-volatilization by engineered yeast promotes rice growth and reduces arsenic accumulation in grains. Int. J. Environ. Res. 1–11. Available from: https://doi.org/10.1007/s41742-019-00188-7.

Verret, F., Gravot, A., Auroy, P., Leonhardt, N., David, P., Nussaume, L., et al., 2004. Overexpression of *AtHMA4* enhances root-to-shoot translocation of zinc and cadmium and plant metal tolerance. FEBS Lett. 576 (3), 306–312.

Viktorovtá, J., Novakova, M., Trbolova, L., Vrchotova, B., Lovecka, P., Mackova, M., et al., 2014. Characterization of transgenic tobacco plants containing bacterial *bphC* gene and study of their phytoremediation ability. Int. J. Phytoremed. 16 (7–12), 937–946.

Visioli, G., Marmiroli, N., 2013. The proteomics of heavy metal hyperaccumulation by plants. J. Proteomics. 79, 133–145.

Wang, G.D., Li, Q.J., Luo, B., Chen, X.Y., 2004. Ex planta phytoremediation of trichlorophenol and phenolic allelochemicals via an engineered secretory laccase. Nat. Biotechnol. 22 (7), 893–897.

Wang, L., Samac, D.A., Shapir, N., Wackett, L.P., Vance, C.P., Olszewski, N.E., et al., 2005. Biodegradation of atrazine in transgenic plants expressing a modified bacterial atrazine chlorohydrolase (*atzA*) gene. Plant Biotechnol. J. 3, 475–486.

Wang, X.X., Wu, N.F., Guo, J., Chu, X.Y., Tian, J., Yao, B., et al., 2008. Phytodegradation of organophosphorus compounds by transgenic plants expressing a bacterial organophosphorus hydrolase. Biochem. Biophys. Res. Commun. 365, 453–458.

Wang, Y., Yau, Y.-Y., Perkins-Balding, D., Thomson, J.G., 2011. Recombinase technology: applications and possibilities. Plant Cell Rep. 30 (3), 267–285.

Wang, F., Wang, Z., Zhu, C., 2012. Heteroexpression of the wheat phytochelatin synthase gene (*TaPCS1*) in rice enhances cadmium sensitivity. Acta Biochim. Biophys. Sin. (Shanghai) 44 (10), 886–893.

Wang, Y., Ren, H., Pan, H., Liu, J., Zhang, L., 2015. Enhanced tolerance and remediation to mixed contaminates of PCBs and 2,4-DCP by transgenic alfalfa plants expressing the 2,3-dihydroxybiphenyl-1,2-dioxygenase. J. Hazard. Mater. 286, 269–275.

Wang, P., Xu, X., Tang, Z., Zhang, W., Huang, X.Y., Zhao, F.J., 2018. *OsWRKY28* regulates phosphate and arsenate accumulation, root system architecture and fertility in rice. Front. Plant Sci. 9, 1330.

Wangeline, A.L., Burkhead, J.L., Hale, K.L., Lindblom, S.D., Terry, N., Pilon, M., et al., 2004. Overexpression of ATP sulfurylase in Indian mustard. J. Environ. Qual. 33 (1), 54–60.

Wojas, S., Clemens, S., Hennig, J., Sklodowska, A., Kopera, E., Schat, H., et al., 2008. Overexpression of phytochelatin synthase in tobacco: distinctive effects of *AtPCS1* and *CePCS* genes on plant response to cadmium. J. Exp. Bot. 59 (8), 2205–2219.

Wojas, S., Hennig, J., Plaza, S., Geisler, M., Siemianowski, O., Skłodowska, A., et al., 2009. Ectopic expression of *Arabidopsis* ABC transporter MRP7 modifies cadmium root-to-shoot transport and accumulation. Environ. Pollut. 157, 2781–2789.

Wu, Q., Shigaki, T., Williams, K.A., Han, J.S., Kim, C.K., Hirschi, K.D., et al., 2011. Expression of an *Arabidopsis* Ca^{2+}/H^{+} antiporter CAX1 variant in petunia enhances cadmium tolerance and accumulation. J. Plant Physiol. 168, 167–173.

Xiao, S., Gao, W., Chen, Q.F., Ramalingam, S., Chye, M.L., 2008. Overexpression of membrane-associated acyl-CoA-binding protein ACBP1 enhances lead tolerance in *Arabidopsis*. Plant J. 54 (1), 141–151.

Xia, Y., Qi, Y., Yuan, Y., Wang, G., Cui, J., Chen, Y., et al., 2012. Overexpression of *Elsholtzia haichowensis* metallothionein 1 (*EhMT1*) in tobacco plants enhances copper tolerance and accumulation in root cytoplasm and decreases hydrogen peroxide production. J. Hazard. Mater. 233, 65–71.

Xiao, S.-Q., Sun, Z., Wang, S.-S., Zhang, J., Li, K.-Z., Chen, L.-M., 2012. Overexpression of dihydroxyacetate synthase and dihydroxyacetone kinase in chloroplasts install a novel photosynthetic HCHO-assimilation pathway in transgenic tobacco using modified Gateway entry vector. Acta Physiol. Plant. 34 (5), 1975–1985.

Zanella, L., Fattorini, L., Brunetti, P., Roccotiello, E., Cornara, L., D'Angeli, S., et al., 2016. Overexpression of *AtPCS1* in tobacco increases arsenic and arsenic plus cadmium accumulation and detoxification. Planta. 243 (3), 605–622.

Zhang, Y., Liu, J., 2011. Transgenic alfalfa plants co-expressing glutathione *S*-transferase (GST) and human CYP2E1 show enhanced resistance to mixed contaminates of heavy metals and organic pollutants. J. Hazard. Mater. 189, 357–362.

Zhang, Y., Zhao, L., Wang, Y., Yang, B., Chen, S., 2008. Enhancement of heavy metal accumulation by tissue specific co-expression of *iaaM* and *ACC deaminase* genes in plants. Chemosphere 72, 564–571.

Zhang, M., Senoura, T., Yang, X., Chao, Y., Nishizawa, N.K., 2011a. *Lhcb2* gene expression analysis in two ecotypes of *Sedum alfredii* subjected to Zn/Cd treatments with functional analysis of *SaLhcb2* isolated from a Zn/Cd hyperaccumulator. Biotechnol. Lett. 33 (9), 1865–1871.

Zhang, D., Xiang, T., Li, P., Bao, L., 2011b. Transgenic plants of *Petunia hybrida* harboring the CYP2E1 gene efficiently remove benzene and toluene pollutants and improve resistance to formaldehyde. Genet. Mol. Biol. 34 (4), 634–639.

Zhang, J., Zhang, M., Shohag, M.J.I., Tian, S., Song, H., Feng, Y., et al., 2016. Enhanced expression of *SaHMA3* plays critical roles in Cd hyperaccumulation and hypertolerance in Cd hyperaccumulator *Sedum alfredii* Hance. Planta 243 (3), 577–589.

Zhang, L., Rylott, E.L., Bruce, N.C., Strand, S.E., 2017a. Phyto detoxification of TNT by transplastomic tobacco (*Nicotiana tabacum*) expressing a bacterial nitroreductase. Plant Mol. Biol. 95 (1–2), 99–109.

Zhang, L., Routsong, R., Nguyen, Q., Rylott, E.L., Bruce, N.C., Strand, S.E., 2017b. Expression in grasses of multiple transgenes for degradation of munitions compounds on live-fire training ranges. Plant Biotechnol. J. 15, 624–633.

Zhang, L., Rylott, E.L., Bruce, N.C., Strand, S.E., 2019. Genetic modification of western wheatgrass (*Pascopyrum smithii*) for the phytoremediation of RDX and TNT. Planta 249, 1007–1015.

Zhigang, A., Cuijie, L., Yuangang, Z., Yejie, D., Wachter, A., Gromes, R., et al., 2006. Expression of *BjMT2*, a metallothionein 2 from *Brassica juncea*, increases copper and cadmium tolerance in *Escherichia coli* and *Arabidopsis thaliana*, but inhibits root elongation in *Arabidopsis thaliana* seedlings. J. Exp. Bot. 57 (14), 3575–3582.

Zhou, S., Xiao, S., Xuan, X., Sun, Z., Li, K., Chen, L., 2015. Simultaneous functions of the installed DAS/DAK formaldehyde-assimilation pathway and the original formaldehyde metabolic pathways enhance the ability of transgenic geranium to purify gaseous formaldehyde polluted environment. Plant Physiol. Biochem. 89, 53–63.

Recent advances in phytoremediation using genome engineering CRISPR–Cas9 technology

5

Pallavi Saxena[1], Nitin Kumar Singh[2], Harish[1], Amit Kumar Singh[3], Siddhartha Pandey[4], Arti Thanki[5] and Tara Chand Yadav[6,]*
[1]Department of Botany, Mohanlal Sukhadia University, Udaipur, India, [2]Department of Environment Science and Engineering, Marwadi University, Rajkot, India, [3]Department of Biochemistry, University of Allahabad, Allahabad, India, [4]Department of Civil Engineering, Chalapathi Institute of Technology, Guntur, India, [5]Department of Environment Science and Engineering, Marwadi University, Rajkot, India, [6]Department of Biotechnology, Indian Institute of Technology, Roorkee, India
*Corresponding author

5.1 Introduction

Technological advances made by humans have been phenomenal ever since man adopted technology. The development pace has jumped by leaps and bounds, since the industrial revolution in England during the 17th century, but it also created a Frankenstein called pollution which continues to engulf the Earth and its environment. Over the years, pollution has been continuously rising with major forms being in the air, water, and soil, which has had many health and well-being implications for humans. Among others industrial activities, human activities such as mining, burning of fossil fuels, industrial waste, and massive usage of agrochemicals are the main stream sources of contamination nowadays (Wuana and Okieimen, 2011). The scientific community had been at odds in finding a solution for mitigating its ill effects, without hampering the development trajectory. The term phytoremediation refers to the cleaning act of pollutants with the help of plants. Basically, it is a technique to restore soil fertility and to cleanse the soil. Today more than 500 plant species are reported to have appreciable phytoremediation potential (Krämer, 2010). Besides, it also helps in mitigation of soil erosion along with for the maintenance of biodiversity, sequestration of carbon, and increased biomass production. Furthermore, it also helps in restoration of natural flora and fauna resulting in formation of surmising landscapes. Phytoremediation is an ecofriendly, solar energy driven, and cost-effective approach in comparison to the conventional methods,

Bioremediation of Pollutants. DOI: https://doi.org/10.1016/B978-0-12-819025-8.00005-3

which involve intense labor, cost, and chemicals (Kotrba et al., 2009). In this approach, plants either stabilize the contaminants in their roots or accumulate them in tissues of various plant parts (Dhankher et al., 2012). The way to achieve these goals relies mainly on identification of effective phytoremediators, since phytoremediation is limited by environmental extremes and the nature of pollutants. To date, several phytoremediators have been identified that are conducive to the environment in which they are being used and are suitable for their bioremediation application in the soil (Abhilash et al., 2009; Basharat et al., 2018). In spite of these advantages, phytoremediation also has its own limitations. Most of the phytoremedial plants possess a very slow growth rate which consequently results in less biomass. In order to meet the remediation challenge, biomass production of phytoremediation plants is not sufficient (Kärenlampi et al., 2000). Moreover, they are very selective in terms of metal and rigid to grow only in specified environmental conditions. Therefore it is not convenient to move the technique forward in a raw manner to meet the upcoming challenges. Plants used for phytoremediation need to be more tolerant toward elevated metal levels, and should possess an intensive root system potent enough to immobilize significant amounts of pollutants (Gratão et al., 2005). An advance approach to develop efficacy of the phytoremediation plants involves omics which include metabolomics, proteomics, genomics, and transcriptomics. Application of genetic engineering techniques in bioremediation assists in developing high biomass content, an intensive root system, and hyperaccumulating plants which can be grown in minimal specified environmental conditions (Mosa et al., 2016). Characterization and development of genomic regions that correlate with the accumulation and survival with varying tolerance are a major concern among researchers. Hence, genetic manipulation of the potent plants to develop into efficient pollutant removal can be achieved using combinational omics approaches (Fig. 5.1) (Shriram et al., 2016).

In general, genome editing enables researchers to alter the gene expression regulation at specified sites, and expedite new insights into the plant functional genomics (Wolt et al., 2016). The recent advancement in this technology has certainly achieved a breakthrough for effectively tackling it by way of clustered regularly interspaced palindromic repeats (CRISPR)-aided phytoremediation (Kaur et al., 2019). CRISPR is a revolutionary genome-engineering tool that enables the enhancement of targeted traits in plants. It has emerged as one of the promising tools in functional genomics (Perez-Pinera et al., 2013). CRISPR collectively with Cas proteins are known as CRISPR−Cas system (Bortesi and Fischer, 2015). This technique provides a pragmatic approach to develop advance phenotypes; however, it works on the plant genome effortlessly. Further, it is found to be more inclined toward monocotyledons crop plants because of their high GC content (Miao et al., 2013). Some of the most difficult plant modifications are reported to be done easily through this tool, which were reportedly very difficult to modify earlier (Bolotin et al., 2005). CRISPR technology helps in altering DNA which can be used for changing the gene sequencing of plants for ameliorating the ill effects of heavy metals and other pollutants present in the soil. This approach also strengthens the nutrient uptake and enhances the bioavailability of metals for plants through plant−microbe interactions (Abhilash et al., 2012). In other words,

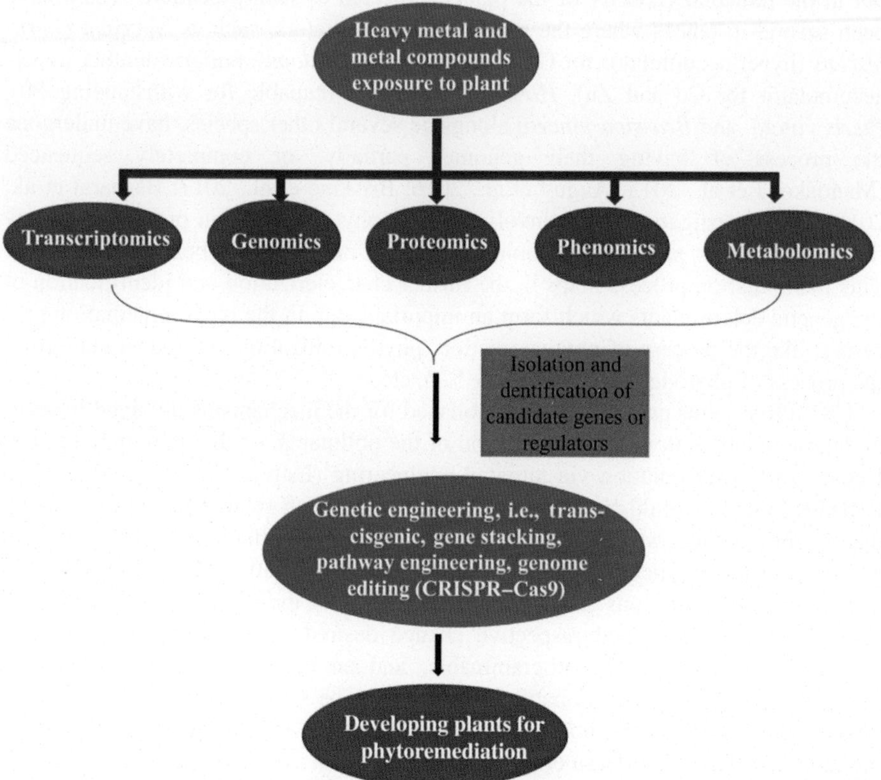

Figure 5.1 Omics approaches for improving phytoremediation.

CRISPR-based phytoremediation can effectively be used for purifying the soil profile, removing toxic elements form the rhizosphere, and making it reusable for agricultural purposes as well as other beneficial projects like agroforestry and social forestry (Thijs et al., 2016). For effective phytoremediation, identifying key genes involved in the signaling network of plant–microbe interactions with appropriate understanding of gene functions is required. Target genes could be modified using a gene editing technique such as CRISPR. Thus enhancing the phytoremediation properties of the plant which can be used in various environments subjected to its extremities (Basu et al., 2018).

5.2 Recent advances in genetic engineering using CRISPR–Cas9 for phytoremediation

Indeed, CRISPR systems are used for the purpose of phytoremediation of plants, so that genome editing may be done on them; but it is also a field that holds considerable potential yet to be explored, wherein a considerable increment can be carried out in the remedial capacity of the plants subjected to such procedure. There have been several instances where the model phytoremediators, such as *Noccaea caerulescens* (hyper-accumulator for Cd, Zn, and Ni), *Arabidopsis halleri* (another hyper-accumulator for Cd and Zn), *Hirschfeldia incana* (capable for withstanding Pb), *Pteris vittata*, and *Brassica juncea*, alongside several other species, have undergone the process of having their genomes partially, or completely sequenced (Mandáková et al., 2015; Auguy et al., 2016; Briskine et al., 2017; Basharat et al., 2018). The investigation phase involved in the phytoremediation process is greatly facilitated by this process of genomic sequence of the plants being manipulated. This manipulation process helps in the further characterization and identification of the genetic determinants which form an important step in the phytoremediation processes, like the process of phytoextraction, phytostabilization, phytodegradation, or the process of phytodesalination (Table 5.1), etc.

CRISPR systems can indeed be established for the mechanisms involved in accumulation, volatilization, and degradation of the pollutants, so that they may be subjected to phytoremediation via targeted engineering (Estrela and Cate, 2016). The CRISPR systems would be better off if the information related to the sequencing data of these plants, which have undergone genetic manipulation, could be utilized for this targeted engineering for phytoremediation. Application of CRISPR allows a specific set of instructions straight into the genome of the plant, wherein the set of instructions can be related respective change desired to be observed. Since this technology is significantly programmable, and an advanced method of genetic manipulation allows programmable instructions to be sent in a candid mode, much unlike what the older methods, such as ZFNs and TALENs allowed and others mentioned in Table 5.2 (Basu et al., 2018).

The CRISPR technology as it is available now had a significant impact in widening the area in which the technology can be exercised, or to say, the variety of plants that can be subjected to genome editing (Seth, 2016). A variety of factors have played a role in this, especially the availability of the sequences for the plant genomes, which was further aided by the presence of software tools and the other bioinformatics-based approaches. The constraint-based modeling system is used to explain the plant physiology using mathematical modeling (Zhu et al., 2016). Similar tools are flux balance analysis for metabolic pathway assessment (Cheung et al., 2015), and omics tool for genome integration assessment (Yoshida et al., 2015; do Amaral and Souza, 2017; Flexas and Gago, 2018), etc. The availability of the Cas9 versions, which were codon optimized for dicots as well as monocots, also helped in allowing the genome manipulation being conducted on a wide variety of plants. The process of phytoremediation generally revolves its focus around the genes expressed by way of CRISPR-mediation, so as to have the metal ligands (e.g., metallothioneins and phytochelatins) increased in synthesis, and also have an increase in the synthesis of the metal transport proteins (e.g., from the CDF, HMA, and ZIP families) along with plant growth hormones (CKs and GAs), as well as the root exudates (siderophores, especially) (Kim et al., 2005; Prasad and Aranda, 2018). Phytoremediation has a tendency to be advantageously affected by several genetic transfers, as shown by several studies that have been conducted since 2000

Table 5.1 Different mechanism reported in plants.

Phytoremediation	Mechanism	Some of the reported plants	References
Phytostablization	It involves adsorption or accumulation of contaminants in the plant tissues in that way the contaminants present in the soil or water get restricted within the plant tissue only. By using this method the root system of the plant also helps in entrapping the contaminants and preventing its migration to other sites.	• *Agrostis castellana* • *Phyla nodiflora* • *Gentiana pennelliana* • *Lolium italicum* • *Festuca arundinacea* • *Anthyllis vulneraria*	• Numan et al. (2018) • Kumar and Verma (2018) • Pastor et al. (2015) • Yoon et al. (2006)
Phytoextraction	It involves adsorption of the contaminants which takes place in the root system but gets stored into the aerial parts of the plant. Usually those parts are burnt and metals are recycled from the ashes.	• *Youngia erythrocarpa* • *Thlaspi caerulescens* • *Brachiaria decumbens* • *Rhizobium metallidurans* • *Alyssum vulneraria*	
Phytodegradation	It consists of either enzymes produced or present within the plant which can degrade.	• *Mariophyllum aquaticum* • *Elodea canadensis* • *Spirodela oligorrhiza*	
Phytovolatilization	Plants absorb the pollutants then convert them into a less toxic form then further release them in low volatilize concentrations into the environment.	• *Brassica juncea* • *Arabidopsis thaliana*	
Phytofiltration	The plants are then planted on the site of contaminated groundwater where the roots take up the water and contaminants.	• *Manihot sculenta Cranz* • *B. juncea Czern.* • *Berkheya coddii* • *Eichhornia crassipes*	

(*Continued*)

Table 5.1 (Continued)

Phytoremediation	Mechanism	Some of the reported plants	References
		• *Hydrocotyle umbellata* L. • *Lemna minor* L. • *Micranthemum umbrosum* • *Callitriche stagnalis* S. • *Potamogeton natans* L.	

Table 5.2 Types of clustered regularly interspaced palindromic repeats–Cas9 systems last two are not used commonly for genetic engineering.

Types	Target	Host	Endonuclease protein
I	DNA	Bacteria and Archaea	Cas3
II	DNA	Bacteria	HNH and RuvC nuclease domain of Cas9
III	DNA and RNA	Archaea and in some bacteria	Cas10
IV		Bacteria	
V		Bacteria and (*Candidatus Methanomethylophilus alvus*)	Cpf1

involving transfer of specific plants, as well as bacterial genes, which were transferred to the targeted plants, and demonstrated beneficial effects for the processes of phytoremediation (Pandey, and Singh, 2019). One such example can be taken of the *Arabidopsis* and the tobacco plants which are enhanced by the NAS1 gene; the plant tends to show significantly increased tolerance against metals like Cd, Cu, Fe, Ni, and Zn, while the uptake of metals like Mn and Ni was also observed. When the metallothioneins encoding-genes (such as MTA1, MT1, and MT2) happened to be overexpressed, the plants of tobacco and *Arabidopsis* exhibited an increase in the capacity of enduring and accumulating the metals such as Cd, Cu, and Zn (Xia et al., 2012; Sebastian et al., 2019). When the metallothionein gene MT2b is expressed, the *H. incana*, is known to exhibit an increase in its capacity for toleration and thereafter, accumulation of Pb. The *B. juncea* plants showed an increased tolerance for Se when the genes APS and SMT were transferred to it. These two genes are responsible for the synthesis of the ATP sulfurylase, as well as selenocysteine methyltransferase; respectively (LeDuc et al., 2006; Lv et al., 2013).

CRISPR systems can be used to enhance these genes to significantly newer and higher levels, and all these genes that can be so enhanced, as well as several others have been examined recently by Fasani et al. (2018). They reviewed all these genes with the potential of enhancement using the CRISPR technology, in a paper concerning reclamation of the soils polluted with metals, by plants that have been modified transgenically. Their work in the paper, which is fairly comprehensive, includes a detailed indication of the specific genes that were transferred, as well as their specific sources, along with the effects that were observed to come about in the target plant (Table 5.3).

The effects varied from increment in the tolerance to the toxic levels of metal presence, an enhancement in the capacity to uptake metal, and also ventured as far as to result in the hyper-accumulation of metals in some cases (Huang et al., 2019). These effects can be a useful boost to the phytoremediation. However, without fail, it must also be noted that often, when the accumulation of a certain metal in the target plant is allowed to happen by way of introducing any specific gene, it can cause a development of hypersensitivity to that particular element in the plant in some cases, invariably with the chance of causing plant decay. It was noted that when the plasma membrane protein (NtCBP4) happened to become overexpressed in tobacco plants, it resulted in an increased capacity for accumulation of Pb, but simultaneously, it also caused an enhancement of the sensitivity of the plant to Pb. On a similar note, when MerC gene was expressed in the *Arabidopsis* and tobacco plants, it caused an increase in the accumulation of the Hg metal, but also left the plant hypersensitive to Hg (Fasani et al., 2018). This approach also covers several other organic pollutants as well, which include polycyclic aromatic hydrocarbons, as well as the polychlorinated biphenyls (Banerjee and Roychoudhury, 2019). Explosives such as the hexahydro-1,3,5-trinitro-1,3,5-triazine (RDX), or the 2,4,6-trinitro-toluene (TNT) are also within the scope of this approach. Genes involved in the process of detoxification of the organic xenobiotics existing in plants have been reported by several studies (Jaiswal et al., 2019). The data that is provided by these studies can be further utilized as the seed material for the trials conducted in the future using CRISPR-based enhancement of the systems in the plant enzymes, which are responsible for the removal, and thereafter detoxification of the organic contaminants. These approaches could be employed on the contaminants like polychlorinated biphenyl and polycyclic aromatic hydrocarbons also. Enzyme systems of plants which are responsible for removing pollutants should be targeted further for CRISPR-mediated advancement (Pandey and Singh, 2019). Recently, genes expressing tolerance against naphthalene and phenanthrene were developed by editing the genes in rice and *Arabidopsis* responsible for the naphthalene dioxygenase production (Peng et al., 2014). Tolerance increased significantly to Polychlorinated biphenyls (PCBs) and 2,4-dichlorophenol after altering the bph gene expression within the alfalfa plants (Wang et al., 2015). Another study performed using *Arabidopsis* plant reported efficiently cleaned up RDX contaminants by using cytochrome P450-reductase complex. Transfer of XplA and XplB genes which belong to the

Table 5.3 Recent gene editing tools and techniques used to improve the phytoremediation efficacy are mentioned below.

Editing tool		References
CRISPR–Cas9	RNA-guided DNA endonuclease that is targeted to specific sequences in the genome	Ebbs et al. (1997), Feng et al. (2015), Tang et al. (2017), Habibi et al. (2017), Agnihotri, and Seth (2019), Kaur et al. (2019), Pandey and Singh (2019)
CRMAGE	CRISPR–Cas9 combines with the Lambda (l) and later engineered with the MAGE technique	Ronda et al. (2016), Mukherjee (2017)
MuGENT	Integration of genomes with mutants consisting of high efficiency	Agnihotri and Seth (2019)
RGENs	Lambda (l) mediated homologous recombination coupled to CRISPR–Cas9	
TALENs	DNA-binding domain targeted by nonspecific exonuclease for modifications	Basharat et al. (2018)
ZFNs	DNA-binding proteins that facilitate genome editing at targeted sites by commencing double-strand breaks in the DNA	Jagtap et al. (2016), Noman et al. (2016)

CRISPR, Clustered regularly interspaced palindromic repeats; *ZFNs*, zinc-finger nucleases.

bacterial genome were reported to be implemented while altering the cytochrome genetic make-up (Rylott et al., 2015). CRISPR also acknowledged to design more growth promoting rhizobacteria. The plant–microbe interaction or microbiome also plays a vital role in imparting tolerance to the plant to withstand adverse conditions (Mosa et al., 2016) by editing the plant interaction regarding rhizobacteria, phytohormones levels, and nitrogen fixation capacity. Phosphate ion solubility and other direct and indirect mechanisms could be controlled, which altogether upgrade the plant for reclamation of contaminants (Boivin et al., 2016; Chinnaswamy et al., 2018; Thode et al., 2018). However, even though CRISPR–Cas9 is a magnificent tool for genome editing results vary in plants. So many parameters like target sites within plants, delivery systems, and plants themselves with different genetic makeup play a vital role in determining the success response rate. Although it has manifested excellent potential in enhancing the remediation process within the plants still, all these results are still confined to the lab spaces for now. Success ratio in the field is still needed to be documented. The framework of the forthcoming methodologies which could be a futuristic approach for phytoremediation improvement may integrate the immediate transfection of Cas9 alongside gRNAs into the plant protoplasts, T-DNA-conveyed gRNA–Cas9, and plant retrieval from single-cells (Mikami et al., 2015). T-DNA-delivered gRNA-Cas9 also has been performed but activity is observed in somatic tissues due to induction restrictions. CRISPR–Cas9 holds significant potential in using gRNA-guided gene expression alteration without cloning. Moreover, the binding ability of cas9 protein with the transcription factors helps us to control the transcription level activity by subsequently regulating the transcription factor (Miglani, 2017), this modulation can work over 1000-fold range with precision (Piatek et al., 2015; Lowder et al., 2015). Recently, metal transporter gene "OsNramp5" was reported to be controlled by CRISPR in rice plant. It has opened up a new direction for using CRISPR in metal transportation regulation (Peng et al., 2017; Tang et al., 2017). Undoubtedly, CRISPR–Cas9 has facilitated an effortlessly genome of even such plants that were capable, but due to the genome complexity they were not considered to be used for phytoremediation like maize and poplar. But recently the complex genome of maize due to high ploidy was reported to be manipulated by the CRISPR–Cas9 system (Agarwal et al., 2018). As maize possesses the pronounced potential for metal accumulation and is advantageous due to being a fast-growing crop with significant biomass yield, it now could be utilized further for remediation of multiple contaminants. Likewise, another crop plant poplar was one of the recurrent choices for phytoremediation because it has an intensive root system which penetrates deep in the soil and allows the substantial accumulation of pollutants (Baldantoni et al., 2014).

5.3 Future perspective

Over the last two decades, genome editing has been an area of interest to develop advance phenotypes to meet global demands. It was dealt with right from the nucleotide alterations to the mega-base deletion. Combinational omics approaches play a vital role in revolutionizing the genetic engineering tools. Availability of CRISPR–Cas9 system has a great prospective to facilitate ambitious assignments in plant biology. CRISPR–Cas9 tools will support the growing functional genomics and systems biological sciences data to be subjugated very comprehensively by accelerating discovery of genes and interrelated traits improvement among plant species (Table 5.4).

Most of the CRISPR–Cas9 based studies have been performed by mammalian experimentation. Deceptively it was assessed that the findings can be implemented in other model organisms as well but yet it is imperative to assess the findings based on plant systems. Forthcoming investigations for improving the CRISPR system will require optimization of sgRNA scaffold, which possesses a magnificent binding affinity for Cas9. More complex genome plants like wheat, sugarcane, etc. should be explored further by using this tool. Producing more competent Cas

Table 5.4 Identified genes to improve the efficacy for phytoremediation.

Gene	Role	References
Toluene-*o*-monooxygenase	Play vital role in removing trichloroethylene	Mahendra and Alvarez-Cohen (2006), Aburto-Medina et al. (2017)
Toluene 4-monooxygenases	NDMA removal	Scott et al. (2008), Li et al. (2013), Bouhajja et al. (2017), Yamaguchi et al. (2018)
Polyphosphate kinase	Uranium removal	Ibañez et al. (2015), Daghan (2019), Kaur et al. (2019)
MerR	Resistance against Hg(II)	Checcucci et al. (2017), Wei et al. (2018), Goyal (2019), Singh et al. (2019)
mtL	Lignin and poly-aromatic hydrocarbons degradation	Peng et al. (2017), Khan et al. (2018), Yasin et al. (2018)
DBT monooxygenase	Eliminates sulfur	Pandey et al. (2018), Borah (2018)
bph Operon	Polychlorinated biphenyls and biphenyls degradation	Jiang et al. (2018), Kour et al. (2019)
Organophosphorus hydrolase	Organophosphorus deduction	Scott et al. (2017), Verma and Shukla (2016)
lux	Specific aromatic compounds like naphthalene which helps in bioluminescence	Ariani et al. (2015), Stoláriková-Vaculíková et al. (2015), Romè et al. (2016), Kumar and Prasad (2019)
Cytochrome $P450_{CAM}$	Hexane and 3-methylpentane oxidation	Saleem (2016), Azab et al. (2016), Legault et al. (2017), Daudzai et al. (2018), Zhang et al. (2018)
Ortho-dechlorination gene	Chlorobenzoic acids degradation	Chakraborty and Das (2016)
Chlorobenzoate dehalogenase	2,4-Dinitrotoluene	Agulló et al. (2019), Kumar and Pannu (2018), Tiwari et al. (2017)
Pnp Operon	Degraded paraoxon	Ye et al. (2015), Gupta and Kumar (2017)
XylR	Atrazine reduction	Wu et al. (2015), Zhao and Huang (2018), Chauhan and Mathur (2018)
ArsB/C	Arsenate removal	Mergeay et al. (2003)
Vgb	Improved growth	Khleifat et al. (2006), Aburto-Medina et al. (2017), Jaiswal et al. (2019)

NDMA, Nonaromatic nitrosodimethylamine; dibenzothiophene (*DBT*) sulfone monooxygenase.

protein by exploring the diverse bacterial community may open up new frontiers for genetic engineering. Screening of transgenic plants in real ecological conditions needs to be assessed further. And for that, assessment development of new combinational omics approaches need to be continuous. The scope of off-target mutations and variances in cleavage competence need to be evaluated more precisely. Another conspicuous task ahead is absence of high throughput screening approaches to recognize transgenic plants with edited gene events, along with other technological developments advancement of CRISPR system to deal with a huge data set of diverse and complex plant genome which need to be advanced.

5.4 Conclusion

Omics-based tools like genomics, proteomics, metagenomics, transcriptomics, along with computational tools developed for analysis of statistics produced from these techniques offer substantial evidence for understanding of intricate behavior of microbes which play a vital role in bioremediation. CRISPR–Cas9, ZFNs, and TALEN are a few gene editing tools which aim to retrieve function of an explicit gene of microbe with specific enzymes required for bioremediation. Even though substantial improvement has been made in the advancement of diverse in silico databanks, software, and computational prototypes for investigation of microbial processes, key obstacles arise in examining output results via appropriate, user-friendly, and easy bioinformatics tools to mark significant inferences. To obtain a substantial breakthrough in the bioremediation process to develop a sustainable and clean environment, these methods can be mapped and integrated from theoretical to experimental methods for conformity.

References

Abhilash, P.C., Jamil, S., Singh, N., 2009. Transgenic plants for enhanced biodegradation and phytoremediation of organic xenobiotics. Biotechnol. Adv. 27 (4), 474–488.

Abhilash, P.C., Powell, J.R., Singh, H.B., Singh, B.K., 2012. Plant–microbe interactions: novel applications for exploitation in multipurpose remediation technologies. Trends Biotechnol. 30 (8), 416–420.

Aburto-Medina, A., Taha, M., Shahsavari, E., Ball, A.S., 2017. Degradation of the dinitrotoluene isomers 2,4- and 2,6-DNT. Appraising the Role of Microorganisms. In Enhancing Cleanup of Environmental Pollutants. Springer, Cham, pp. 5–20.

Agarwal, A., Yadava, P., Kumar, K., Singh, I., Kaul, T., Pattanayak, A., et al., 2018. Insights into maize genome editing via CRISPR/Cas9. Physiol. Mol. Biol. Plants 24 (2), 175–183.

Agnihotri, A., Seth, C.S., 2019. Transgenic brassicaceae: a promising approach for phytoremediation of heavy metals. Transgenic Plant Technology for Remediation of Toxic Metals and Metalloids. Academic Press, pp. 239–255.

Agulló, L., Pieper, D.H., Seeger, M., 2019. Genetics and biochemistry of biphenyl and PCB biodegradation. In: Aerobic Utilization of Hydrocarbons, Oils, and Lipids, Springer, Cham, pp. 595–622.

Ariani, A., Di Baccio, D., Romeo, S., Lombardi, L., Andreucci, A., Lux, A., et al., 2015. RNA sequencing of Populus x canadensis roots identifies key molecular mechanisms underlying physiological adaption to excess zinc. PLoS One 10 (2), e0117571.

Auguy, F., Fahr, M., Moulin, P., El Mzibri, M., Smouni, A., Filali-Maltouf, A., et al., 2016. Transcriptome changes in *Hirschfeldia incana* in response to lead exposure. Front. Plant. Sci. 6, 1231.

Azab, E., Hegazy, A.K., El-Sharnouby, M.E., Abd Elsalam, H.E., 2016. Phytoremediation of the organic xenobiotic simazine by p450-1a2 transgenic *Arabidopsis thaliana* plants. Int. J. Phytoremediat. 18 (7), 738–746.

Baldantoni, D., Cicatelli, A., Bellino, A., Castiglione, S., 2014. Different behaviours in phytoremediation capacity of two heavy metal tolerant poplar clones in relation to iron and other trace elements. J. Environ. Econ. Manag. 146, 94–99.

Banerjee, A., Roychoudhury, A., 2019. Genetic engineering in plants for enhancing arsenic tolerance. Transgenic Plant Technology for Remediation of Toxic Metals and Metalloids. Academic Press, pp. 463–475.

Basharat, Z., Novo, L., Yasmin, A., 2018. Genome editing weds CRISPR: what is in it for phytoremediation? Plants 7 (3), 51.

Basu, S., Rabara, R.C., Negi, S., Shukla, P., 2018. Engineering PGPMOs through gene editing and systems biology: a solution for phytoremediation? Trends Biotechnol. 36 (5), 499–510.

Boivin, S., Fonouni-Farde, C., Frugier, F., 2016. How auxin and cytokinin phytohormones modulate root microbe interactions. Front. Plant Sci. 7, 1240.

Bolotin, A., Quinquis, B., Sorokin, A., Ehrlich, S.D., 2005. Clustered regularly interspaced short palindrome repeats (CRISPRs) have spacers of extrachromosomal origin. Microbiology 151 (8), 2551–2561.

Borah, D., 2018. Microbial bioremediation of petroleum hydrocarbon: an overview. Microbial Action on Hydrocarbons. Springer, Singapore, pp. 321–341.

Bortesi, L., Fischer, R., 2015. The CRISPR/Cas9 system for plant genome editing and beyond. Biotechnol. Adv. 33 (1), 41–52.

Bouhajja, E., McGuire, M., Liles, M.R., Bataille, G., Agathos, S.N., George, I.F., 2017. Identification of novel toluene monooxygenase genes in a hydrocarbon-polluted sediment using sequence- and function-based screening of metagenomic libraries. Appl. Microbiol. Biotechnol. 101 (2), 797–808.

Briskine, R.V., Paape, T., Shimizu-Inatsugi, R., Nishiyama, T., Akama, S., Sese, J., et al., 2017. Genome assembly and annotation of *Arabidopsis halleri*, a model for heavy metal hyperaccumulation and evolutionary ecology. Mol. Ecol. Resour. 17 (5), 1025–1036.

Chakraborty, J., Das, S., 2016. Molecular perspectives and recent advances in microbial remediation of persistent organic pollutants. Environ. Sci. Pollut. R. 23 (17), 16883–16903.

Chauhan, P., Mathur, J., 2018. Potential of *Helianthus annuus* for phytoremediation of multiple pollutants in the environment: a review. Int. J. Biol. Sci. 4 (3), 5–16.

Checcucci, A., Bazzicalupo, M., Mengoni, A., 2017. Exploiting nitrogen-fixing rhizobial symbionts genetic resources for improving phytoremediation of contaminated soils. Enhancing Cleanup of Environmental Pollutants. Springer, Cham, pp. 275–288.

Cheung, C.M., Ratcliffe, R.G., Sweetlove, L.J., 2015. A method of accounting for enzyme costs in flux balance analysis reveals alternative pathways and metabolite stores in an illuminated *Arabidopsis* leaf. Plant Physiol. 169 (3), 1671–1682.

Chinnaswamy, A., Coba de la Peña, T., Stoll, A., de la Peña Rojo, D., Bravo, J., Rincón, A., et al., 2018. A nodule endophytic *Bacillus megaterium* strain isolated from *Medicago polymorpha* enhances growth, promotes nodulation by Ensifer medicae and alleviates salt stress in alfalfa plants. Ann. Appl. Biol. 172 (3), 295–308.

Daghan, H., 2019. Transgenic tobacco for phytoremediation of metals and metalloids. Transgenic Plant Technology for Remediation of Toxic Metals and Metalloids. Academic Press, pp. 279–297.

Daudzai, Z., Treesubsuntorn, C., Thiravetyan, P., 2018. Inoculated *Clitoria ternatea* with *Bacillus cereus* ERBP for enhancing gaseous ethylbenzene phytoremediation: plant metabolites and expression of ethylbenzene degradation genes. Ecotoxicol. Environ. Saf. 164, 50–60.

Dhankher, O.P., Pilon-Smits, E.A., Meagher, R.B., Doty, S., 2012. Biotechnological approaches for phytoremediation. Plant Biotechnology and Agriculture. Academic Press, pp. 309–328.

do Amaral, M.N., Souza, G.M., 2017. The challenge to translate OMICS data to whole plant physiology: the context matters, Front. Plant Sci., 8. p. 2146.

Ebbs, S.D., Lasat, M.M., Brady, D.J., Cornish, J., Gordon, R., Kochian, L.V., 1997. Phytoextraction of cadmium and zinc from a contaminated soil. J. Environ. Qual. 26 (5), 1424–1430.

Estrela, R., Cate, J.H.D., 2016. Energy biotechnology in the CRISPR-Cas9 era. Curr. Opin. Biotechnol. 38, 79–84.

Fasani, E., Manara, A., Martini, F., Furini, A., DalCorso, G., 2018. The potential of genetic engineering of plants for the remediation of soils contaminated with heavy metals. Plant Cell Environ. 41 (5), 1201–1232.

Feng, J., Jester, B.W., Tinberg, C.E., Mandell, D.J., Antunes, M.S., Chari, R., et al., 2015. A general strategy to construct small molecule biosensors in eukaryotes. Elife 4, e10606.

Flexas, J., Gago, J., 2018. A role for ecophysiology in the 'omics' era. Plant J. 96 (2), 251–259.

Goyal, D., 2019. Investigations Into Endophytes in *Pimenta dioica* L Merr.

Gratão, P.L., Prasad, M.N.V., Cardoso, P.F., Lea, P.J., Azevedo, R.A., 2005. Phytoremediation: green technology for the cleanup of toxic metals in the environment. Braz. J. Plant Physiol. 17 (1), 53–64.

Gupta, P., Kumar, V., 2017. Value added phytoremediation of metal stressed soils using phosphate solubilizing microbial consortium. World J. Microbiol. Biotechnol. 33 (1), 9.

Habibi, P., Grossi-de-Sá, M.F., Makhzoum, A., Malik, S., da Silva, A.L.L., Hefferon, K., et al., 2017. Bioengineering hairy roots: phytoremediation, secondary metabolism, molecular pharming, plant-plant interactions and biofuels. Sustainable Agriculture Reviews. Springer, Cham, pp. 213–251.

Huang, Y., Xi, Y., Gan, L., Johnson, D., Wu, Y., Ren, D., et al., 2019. Effects of lead and cadmium on photosynthesis in *Amaranthus spinosus* and assessment of phytoremediation potential. Int. J. Phytoremediat. 21, 1–9.

Ibañez, S.G., Paisio, C.E., Oller, A.L.W., Talano, M.A., González, P.S., Medina, M.I., et al., 2015. Overview and new insights of genetically engineered plants for improving phytoremediation. Phytoremediation. Springer, Cham, pp. 99–113.

Jagtap, U.B., Bapat, V.A., Saladin, G., Chudzińska, E., Krzesłowska, M., Pawlaczyk, E.M., et al., 2016. Role of microbes and plants in phytoremediation: potential of genetic engineering. In: Global Challenges, Social Aspects and Environmental Benefits. p. 89.

Jaiswal, S., Singh, D.K., Shukla, P., 2019. Gene editing and systems biology tools for pesticide bioremediation: a review. Front. Microbiol. 10.

Jiang, L., Luo, C., Zhang, D., Song, M., Sun, Y., Zhang, G., 2018. Biphenyl-metabolizing microbial community and a functional operon revealed in E-waste-contaminated soil. Environ. Sci. Technol. 52 (15), 8558–8567.

Kärenlampi, S., Schat, H., Vangronsveld, J., Verkleij, J.A.C., van der Lelie, D., Mergeay, M., et al., 2000. Genetic engineering in the improvement of plants for phytoremediation of metal polluted soils. Environ. Pollut. 107 (2), 225–231.

Kaur, R., Yadav, P., Kohli, S.K., Kumar, V., Bakshi, P., Mir, B.A., et al., 2019. Emerging trends and tools in transgenic plant technology for phytoremediation of toxic metals and metalloids. Transgenic Plant Technology for Remediation of Toxic Metals and Metalloids. Academic Press, pp. 63–88.

Khan, W.U., Yasin, N.A., Ahmad, S.R., Ali, A., Ahmad, A., Akram, W., et al., 2018. Role of *Burkholderia cepacia* CS8 in Cd-stress alleviation and phytoremediation by *Catharanthus roseus*. Int. J. Phytoremediat. 20 (6), 581–592.

Khleifat, K.M., Abboud, M.M., Al-Mustafa, A.H., 2006. Effect of Vitreoscilla hemoglobin gene (vgb) and metabolic inhibitors on cadmium uptake by the heterologous host *Enterobacter aerogenes*. Process Biochem. 41 (4), 930–934.

Kim, S., Takahashi, M., Higuchi, K., Tsunoda, K., Nakanishi, H., Yoshimura, E., et al., 2005. Increased nicotianamine biosynthesis confers enhanced tolerance of high levels of metals, in particular nickel, to plants. Plant Cell Physiol. 46 (11), 1809–1818.

Kotrba, P., Najmanova, J., Macek, T., Ruml, T., Mackova, M., 2009. Genetically modified plants in phytoremediation of heavy metal and metalloid soil and sediment pollution. Biotechnol. Adv. 27 (6), 799–810.

Kour, D., Rana, K.L., Kumar, R., Yadav, N., Rastegari, A.A., Yadav, A.N., et al., 2019. Gene manipulation and regulation of catabolic genes for biodegradation of biphenyl compounds. New and Future Developments in Microbial Biotechnology and Bioengineering. Elsevier, pp. 1–23.

Krämer, U., 2010. Metal hyperaccumulation in plants. Annu. Rev. Plant Biol. 61, 517–534.

Kumar, A., Prasad, M.N.V., 2019. Plant genetic engineering approach for the Pb and Zn remediation: defense reactions and detoxification mechanisms. Transgenic Plant Technology for Remediation of Toxic Metals and Metalloids. Academic Press, pp. 359–380.

Kumar, A., Verma, J.P., 2018. Does plant—microbe interaction confer stress tolerance in plants: a review? Microbiol. Res. 207, 41–52.

Kumar, D., Pannu, R., 2018. Perspectives of lindane (γ-hexachlorocyclohexane) biodegradation from the environment: a review. Bioresour. Bioprocess. 5 (1), 29.

LeDuc, D.L., AbdelSamie, M., Móntes-Bayon, M., Wu, C.P., Reisinger, S.J., Terry, N., 2006. Overexpressing both ATP sulfurylase and selenocysteine methyltransferase enhances selenium phytoremediation traits in Indian mustard. Environ. Pollut. 144 (1), 70–76.

Legault, E.K., James, C.A., Stewart, K., Muiznieks, I., Doty, S.L., Strand, S.E., 2017. A field trial of TCE phytoremediation by genetically modified poplars expressing cytochrome P450 2E1. Environ. Sci. Technol. 51 (11), 6090–6099.

Li, M., Mathieu, J., Liu, Y., Van Orden, E.T., Yang, Y., Fiorenza, S., et al., 2013. The abundance of tetrahydrofuran/dioxane monooxygenase genes (thmA/dxmA) and 1,4-dioxane degradation activity are significantly correlated at various impacted aquifers. Environ. Sci. Tech. Let. 1 (1), 122–127.

Lowder, L.G., Zhang, D., Baltes, N.J., Paul, J.W., Tang, X., Zheng, X., et al., 2015. A CRISPR/Cas9 toolbox for multiplexed plant genome editing and transcriptional regulation. Plant Physiol. 169 (2), 971–985.

Lv, Y., Deng, X., Quan, L., Xia, Y., Shen, Z., 2013. Metallothioneins BcMT1 and BcMT2 from *Brassica campestris* enhance tolerance to cadmium and copper and decrease production of reactive oxygen species in *Arabidopsis thaliana*. Plant Soil 367 (1–2), 507–519.

Mahendra, S., Alvarez-Cohen, L., 2006. Kinetics of 1,4-dioxane biodegradation by monooxygenase-expressing bacteria. Environ. Sci. Technol. 40 (17), 5435–5442.

Mandáková, T., Singh, V., Krämer, U., Lysak, M.A., 2015. Genome structure of the heavy metal hyperaccumulator *Noccaea caerulescens* and its stability on metalliferous and nonmetalliferous soils. Plant Physiol. 169 (1), 674–689.

Mergeay, M., Monchy, S., Vallaeys, T., Auquier, V., Benotmane, A., Bertin, P., et al., 2003. *Ralstonia metallidurans*, a bacterium specifically adapted to toxic metals: towards a catalogue of metal-responsive genes. FEMS Microbiol. Rev. 27 (2–3), 385–410.

Miao, J., Guo, D., Zhang, J., Huang, Q., Qin, G., Zhang, X., et al., 2013. Targeted mutagenesis in rice using CRISPR-Cas system. Cell Res. 23 (10), 1233.

Miglani, G.S., 2017. Genome editing in crop improvement: present scenario and future prospects. J. Crop Improv. 31 (4), 453–559.

Mikami, M., Toki, S., Endo, M., 2015. Comparison of CRISPR/Cas9 expression constructs for efficient targeted mutagenesis in rice. Plant Mol. Biol. 88 (6), 561–572.

Mosa, K.A., Saadoun, I., Kumar, K., Helmy, M., Dhankher, O.P., 2016. Potential biotechnological strategies for the cleanup of heavy metals and metalloids. Front. Plant Sci. 7, 303.

Mukherjee, D., 2017. Microorganisms: role for crop production and its interface with soil agroecosystem. Plant-Microbe Interactions in Agro-Ecological Perspectives. Springer, Singapore, pp. 333–359.

Noman, A., Aqeel, M., He, S., 2016. CRISPR-Cas9: tool for qualitative and quantitative plant genome editing. Front. Plant Sci. 7, 1740.

Numan, M., Bashir, S., Khan, Y., Mumtaz, R., Shinwari, Z.K., Khan, A.L., et al., 2018. Plant growth promoting bacteria as an alternative strategy for salt tolerance in plants: a review. Microbiol. Res. 209, 21–32.

Pandey, V.C., Singh, V., 2019. Exploring the potential and opportunities of current tools for removal of hazardous materials from environments. Phytomanagement of Polluted Sites. Elsevier, pp. 501–516.

Pandey, A., Tripathi, P.H., Pandey, S.C., Pathak, V.M., Nailwal, T.K., 2018. Removal of toxic pollutants from soil using microbial biotechnology. Microbial Biotechnology in Environmental Monitoring and Cleanup. IGI Global, pp. 86–105.

Pastor, J., GutiÉrrez-ginÉs, M.J., HernÁndez, A.J., 2015. Heavy-metal phytostabilizing potential of *Agrostis castellana* Boiss. & Reuter. Int. J. Phytoremediat. 17 (10), 988–998.

Peng, R.H., Fu, X.Y., Zhao, W., Tian, Y.S., Zhu, B., Han, H.J., et al., 2014. Phytoremediation of phenanthrene by transgenic plants transformed with a naphthalene dioxygenase system from *Pseudomonas*. Environ. Sci. Technol. 48 (21), 12824–12832.

Peng, J.S., Wang, Y.J., Ding, G., Ma, H.L., Zhang, Y.J., Gong, J.M., 2017. A pivotal role of cell wall in cadmium accumulation in the *Crassulaceae* hyperaccumulator *Sedum plumbizincicola*. Mol. Plant 10 (5), 771–774.

Perez-Pinera, P., Kocak, D.D., Vockley, C.M., Adler, A.F., Kabadi, A.M., Polstein, L.R., et al., 2013. RNA-guided gene activation by CRISPR-Cas9–based transcription factors. Nat. Methods 10 (10), 973.

Piatek, A., Ali, Z., Baazim, H., Li, L., Abulfaraj, A., Al-Shareef, S., et al., 2015. RNA-guided transcriptional regulation in planta via synthetic dC as9-based transcription factors. Plant Biotechnol. J. 13 (4), 578–589.

Prasad, R., Aranda, E. (Eds.), 2018. Approaches in Bioremediation: The New Era of Environmental Microbiology and Nanobiotechnology. Springer.

Romè, C., Huang, X.Y., Danku, J., Salt, D.E., Sebastiani, L., 2016. Expression of specific genes involved in Cd uptake, translocation, vacuolar compartmentalisation and recycling in *Populus alba* Villafranca clone. J. Plant Physiol. 202, 83–91.

Ronda, C., Pedersen, L.E., Sommer, M.O., Nielsen, A.T., 2016. CRMAGE: CRISPR optimized mage recombineering. Sci. Rep. 6, 19452.

Rylott, E.L., Johnston, E.J., Bruce, N.C., 2015. Harnessing microbial gene pools to remediate persistent organic pollutants using genetically modified plants—a viable technology? J. Exp. Bot. 66 (21), 6519–6533.

Saleem, H., 2016. Plant-bacteria partnership: phytoremediation of hydrocarbons contaminated soil and expression of catabolic genes. J. Environ. Sci. 1 (1), 19.

Scott, C., Pandey, G., Hartley, C.J., Jackson, C.J., Cheesman, M.J., Taylor, M.C., et al., 2008. The enzymatic basis for pesticide bioremediation. Indian J. Microbiol. 48 (1), 65.

Scott, C., Oakeshott, J., Russell, R., French, N., Kotsonis, S., Liu, K., Commonwealth Scientific and Industrial Research Organization (CSIRO), 2017. Enzymes for Degrading Organophosphates. U.S. Patent 9,796,990.

Sebastian, A., Shukla, P., Nangia, A.K., Prasad, M.N.V., 2019. Transgenics in phytoremediation of metals and metalloids: from laboratory to field. Transgenic Plant Technology for Remediation of Toxic Metals and Metalloids. Academic Press, pp. 3–22.

Seth, K., 2016. Current status of potential applications of repurposed Cas9 for structural and functional genomics of plants. Biochem. Biophys. Res. Commun. 480 (4), 499–507.

Shriram, V., Kumar, V., Devarumath, R.M., Khare, T.S., Wani, S.H., 2016. MicroRNAs as potential targets for abiotic stress tolerance in plants. Front. Plant Sci. 7, 817.

Singh, D.K., Lingaswamy, B., Koduru, T.N., Nagu, P.P., Jogadhenu, P.S.S., 2019. A putative merR family transcription factor Slr0701 regulates mercury inducible expression of MerA in the cyanobacterium *Synechocystis* sp. PCC6803. MicrobiologyOpen 8, e838.

Stoláriková-Vaculíková, M., Romeo, S., Minnocci, A., Luxová, M., Vaculík, M., Lux, A., et al., 2015. Anatomical, biochemical and morphological responses of poplar *Populus deltoides* clone Lux to Zn excess. Environ. Exp. Bot. 109, 235–243.

Tang, L., Mao, B., Li, Y., Lv, Q., Zhang, L., Chen, C., et al., 2017. Knockout of OsNramp5 using the CRISPR/Cas9 system produces low Cd-accumulating indica rice without compromising yield. Sci. Rep. 7 (1), 14438.

Thijs, S., Sillen, W., Rineau, F., Weyens, N., Vangronsveld, J., 2016. Towards an enhanced understanding of plant–microbiome interactions to improve phytoremediation: engineering the metaorganism. Front. Microbiol. 7, 341.

Thode, S.K., Rojek, E., Kozlowski, M., Ahmad, R., Haugen, P., 2018. Distribution of siderophore gene systems on a *Vibrionaceae* phylogeny: database searches, phylogenetic analyses and evolutionary perspectives. PLoS One 13 (2), e0191860.

Tiwari, S., Tripathi, A., Gaur, R., 2017. Bioremediation of plant refuges and xenobiotics. Principles and Applications of Environmental Biotechnology for a Sustainable Future. Springer, Singapore, pp. 85–142.

Verma, A., Shukla, P.K., 2016. A prospective study on emerging role of phytoremediation by endophytic microorganisms. Toxicity and Waste Management Using Bioremediation. IGI Global, pp. 236–265.

Wang, Y., Ren, H., Pan, H., Liu, J., Zhang, L., 2015. Enhanced tolerance and remediation to mixed contaminates of PCBs and 2,4-DCP by transgenic alfalfa plants expressing the 2,3-dihydroxybiphenyl-1,2-dioxygenase. J. Hazard. Mater. 286, 269–275.

Wei, Q., Yan, J., Chen, Y., Zhang, L., Wu, X., Shang, S., et al., 2018. Cell surface display of MerR on *Saccharomyces* cerevisiae for biosorption of mercury. Mol. Biotechnol. 60 (1), 12–20.

Wolt, J.D., Wang, K., Yang, B., 2016. The regulatory status of genome-edited crops. Plant Biotechnol. J. 14 (2), 510–518.

Wu, Z., Zhao, X., Sun, X., Tan, Q., Tang, Y., Nie, Z., et al., 2015. Xylem transport and gene expression play decisive roles in cadmium accumulation in shoots of two oilseed rape cultivars (*Brassica napus*). Chemosphere 119, 1217–1223.

Wuana, R.A., Okieimen, F.E., 2011. Heavy metals in contaminated soils: a review of sources, chemistry, risks and best available strategies for remediation. ISRN Ecol. 2011.

Xia, Y., Qi, Y., Yuan, Y., Wang, G., Cui, J., Chen, Y., et al., 2012. Overexpression of *Elsholtzia haichowensis* metallothionein 1 (EhMT1) in tobacco plants enhances copper tolerance and accumulation in root cytoplasm and decreases hydrogen peroxide production. J. Hazard. Mater. 233, 65–71.

Yamaguchi, T., Nakamura, S., Hatamoto, M., Tamura, E., Tanikawa, D., Kawakami, S., et al., 2018. A novel approach for toluene gas treatment using a downflow hanging sponge reactor. Appl. Microbiol. Biotechnol. 102 (13), 5625–5634.

Yasin, N.A., Khan, W.U., Ahmad, S.R., Ali, A., Ahmed, S., Ahmad, A., 2018. Effect of *Bacillus* fortis 162 on growth, oxidative stress tolerance and phytoremediation potential of catharanthus roseus under Chromium stress. Int. J. Agric. Biol. 20 (7), 1513–1522.

Ye, D., Li, T., Liu, D., Zhang, X., Zheng, Z., 2015. P accumulation and physiological responses to different high P regimes in *Polygonum* hydropiper for understanding a P-phytoremediation strategy. Sci. Rep. 5, 17835.

Yoon, J., Cao, X., Zhou, Q., Ma, L.Q., 2006. Accumulation of Pb, Cu, and Zn in native plants growing on a contaminated Florida site. Sci. Total Environ. 368 (2–3), 456–464.

Yoshida, T., Mogami, J., Yamaguchi-Shinozaki, K., 2015. Omics approaches toward defining the comprehensive abscisic acid signaling network in plants. Plant Cell Physiol. 56 (6), 1043–1052.

Zhang, L., Rylott, E.L., Bruce, N.C., Strand, S.E., 2018. Genetic modification of western wheatgrass (*Pascopyrum smithii*) for the phytoremediation of RDX and TNT. Planta 249, 1–9.

Zhao, F.J., Huang, X.Y., 2018. Cadmium phytoremediation: call rice CAL1. Mol. Plant 11 (5), 640–642.

Zhu, X.G., Lynch, J.P., LeBauer, D.S., Millar, A.J., Stitt, M., Long, S.P., 2016. Plants in silico: why, why now and what?—an integrative platform for plant systems biology research. Plant Cell Environ. 39 (5), 1049–1057.

Part II

Microbial Remediation

Endophytes—the hidden world for agriculture, ecosystem, and environmental sustainability

6

Shubhi Srivastava[1], *Madhubanti Chaudhuri*[1] *and*
Vimal Chandra Pandey[2,*]
[1]Microbiology Laboratory, Department of Botany, University of Calcutta, Kolkata, India,
[2]Department of Environmental Science, Babasaheb Bhimrao Ambedkar University, Lucknow, India
*Corresponding author

6.1 Introduction

Endophytes are a group of bacteria that reside in plant tissues. They colonize into the internal tissues of the plants and make a relationship that's either symbiotic, trophobiotic, mutualistic, and commensalistic. Endophytes can also act as a plant growth promoter and act as a biocontrol agents. Endophytes can also provide some natural products that can be used in medicines, industries, and agricultures (Ryan et al., 2008). Endophytes exist in grasses, rainforest plants, and palms. Endophytes are able to colonize into the internal tissues and improve the quality of crops which makes them very important for agriculture. Endophytes can also improve the physiology of plants. Several endophytes are also reported to inhibit pathogen growth in plants (Sturz et al., 1997). Several researchers revealed the effective combined role of plants and microbes to remediate the contaminants from soil, water, and air (Vymazal and Březinová, 2015; Ávila et al., 2013; Verlicchi and Zambello, 2014). Plants play a major role in pollutant removal by directing or by enhancing the degradation ability with the help of rhizospheric microbes (Li et al., 2014).

Endophytes are known to have genes that assist plant growth promotion and degradation ability to cope with various environmental stresses (Ho et al., 2013; Visioli et al., 2015; Babu et al., 2013). There are a few studies that revealed the use of endophytic bacteria in wetlands treating sewage effluent (Ijaz et al., 2015) or textile effluent (Shehzadi et al., 2014). Srivastava et al. (2016) found that integrated use of *Pteris vittata* with the isolated endophyte *Bacillus pumilus* can enhance plant growth and uptake of arsenic, and have potential application for the remediation of arsenic contaminated sites. Visioli et al. (2015) described the combined endophytic inoculants enhance nickel phytoextraction from serpentine soil in the hyperaccumulator *Noccaea caerulescens.*

Endophytic microorganisms have been observed to produce novel bioactive metabolites with unique structures and diverse chemical nature such as alkaloids,

Bioremediation of Pollutants. DOI: https://doi.org/10.1016/B978-0-12-819025-8.00006-5

flavonoids, phenolic acids, sterols, terpenoids, etc. (Tan and Zou, 2001). People began to investigate microbes as a source for bioactive natural products after Pasteur discovered the process of fermentation by living cells. The antibiotic era started with Fleming. Since then, scientists have been engaged in the discovery and application of microbial metabolites for utility in medicine, agriculture, and industry. In this chapter we are going to discuss the beneficial traits of endophytic bacteria in agriculture, ecosystem, and environment.

6.2 Plant growth-promoting endophytes

Rhizobacteria are well known for their plant growth-promoting abilities but in the recent era endophytes can also work as biocontrol and plant growth promoters through the improved cycling of nutrients and minerals such as nitrogen, phosphate, and other nutrients. Endophytic bacteria have the same mechanisms as rhizobacteria. They have phosphate solubilization activity, 1-aminocyclopropane-1-carboxylic acid deaminase activity, indole acetic acid production, and siderophore production. Endophytes also supply the nutrients and vitamins to host plants. There are so many another benefits of endophytes such as osmotic adjustment, stomatal regulation, nitrogen fixation, and providing minerals and nutrients to host plants. In the recent era of development we can use these endophytes in plant regeneration and phytoremediation (Compant et al., 2005a,b). Endophytes are also able to prevent some harmful effects of pathogenic microbes and damages caused by insects and nematodes (Kerry, 2000; Ping and Boland, 2004; Berg and Hallmann, 2006).

The mechanisms of plant growth-promoting bacteria to promote plant growth are well described by certain mechanisms (Glick, 2012). Plant growth-promoting bacteria facilitates the growth by directly and indirectly. Direct promotion occurs by two methods, first is acquisition of nutrients including phosphorus, nitrogen, and iron from environmental resources. The second method is regulation of phytohormones like auxin, cytokinin, and ethylene. Indirect promotion of plant include production of certain metabolites like antibiotics, cell wall degrading enzymes, decreasing plant ethylene level, and synthesis of volatile compounds which inhibits the pathogens. According to Ryan et al. (2008) endophytes will be used in a wide range for plant growth promoting rhizobacteria (PGPR) applications, antibiotic production, and use in plants as biocontrol agents for various pathogenic agents (Arunachalam and Gayathri, 2010). Parallelly they will be used in phytoremediation of environmental contaminants. There are more examples of plant growth-promoting endophytes described in Table 6.1.

6.3 Natural products from endophytes

The requirement of new and useful compounds to provide sustenance to human civilization is ever increasing. Plants have served as sources of several bioactive

Table 6.1 Some recent plant growth-promoting endophytes isolated from plants.

Isolated from	Plant growth promoting endophytes	Year
Triticum aestivum L	*Aspergillus, Fusarium, Penicillium, Alternaria, Cladosporium, Talaromyces, Trichoderma*	Ripa et al. (2019)
Teucrium polium	*Bacillus, Penicillium*	El-Din Hassan (2017)
Malus × domestica Borkh.	*Curtobacterium, Pantoea, Pseudomonas*	Miliute et al. (2016)
Rice	*Azospirillum* sp. B510	Kaneko et al. (2010)
Rice, maize, wheat	*Azospirillum lipoferum* 4B	Wisniewski-Dyé et al. (2011)
Rice	*Burkholderia* spp. KJ006	Kwak et al. (2012)
Sweet potato	*Klebsiella* sp. Sal 1, *Enterobacter* sp.	Dhungana et al. (2018)
Solanum tuberosum L.	*Burkholderia, Azospirillum, Ideonella, Pseudacidovorax*, and *Bradyrhizobium*	Pageni et al. (2014)
Sugarcane	*Klebsiella, Enterobacter*, and *Pantoea*	Rodrigues et al. (2016)
Glycine max L.	*Enterobacter, Acinetobacter, Pseudomonas, Ochrobactrum, Bacillus*	Zhao et al. (2018)
Pteris vittata L.	*Bacillus pumilus*	Srivastava et al. (2016)

compounds for the welfare of mankind for centuries. However, drug resistance in bacteria, appearance of life-threatening viruses, and inadequacy of producing enough food in certain areas to support the population are a few problems that have come up in the last few decades. Environmental degradation and loss of biodiversity add to the problem. This led to the exploitation of microbes to produce novel natural products (Strobel and Daisy, 2003). Endophytes, or the microbial population residing in the internal tissues of plants are currently viewed as an outstanding source of bioactive natural products because of the presence of numerous endophytic microorganisms occupying literally millions of unique biological niches growing in so many unusual environments. If endophytes can produce the same rare and important bioactive compounds as their host plants, this would not only reduce the need to harvest slow-growing and rare plants, but will also help to preserve the world's ever-decreasing biodiversity.

Considering that only a handful of endophytes have been studied from the various species of plants growing on Earth, several research groups are working on evaluation and elucidation of the potential of these microorganisms for production of important bioactive metabolites (Pimentel et al., 2011). As expected, endophytes are already reported to produce a number of bioactive metabolites which can serve as excellent sources of drugs for treatment of various diseases and with potential applications in agriculture, medicine, food, and other industries (Gouda et al., 2016). The role of endophytes in the production of bioactive compounds and the microbial biotransformation process has been regarded as a novel alternative method to obtain such compounds for human use. Biotransformation is a useful method not only for production of novel compounds, but also for enhancement of production of a desired compound. For this reason, biotransformation using microbial cultures has received increasing attention as a method for the conversion of lipids, alkaloids, and terpenes carrying out stereo-specific and stereo-selective reactions for the production of novel bioactive metabolites (Pimentel et al., 2011). Fig. 6.1 represents several natural products from endophytes.

Antimicrobial compounds—Discovery of antimicrobial metabolites from endophytes is a solution to overcome the increasing drug resistance by plant and human pathogens and the insufficient number of effective antibiotics against diverse bacterial and fungal species (Pimentel et al., 2011). Natural products from endophytic microbes have been observed to inhibit a wide variety of disease-causing agents including phytopathogens, bacteria, and fungi affecting humans and animals

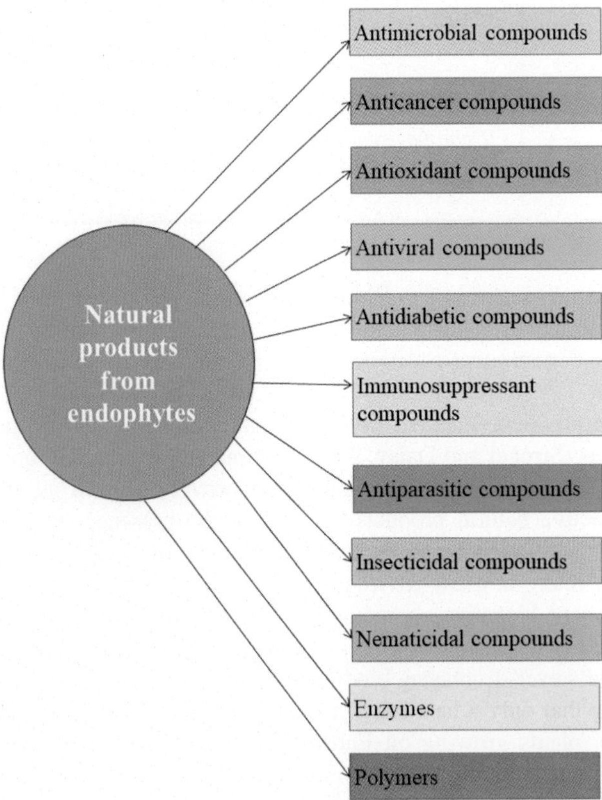

Figure 6.1 The several natural products from endophytes.

Table 6.2 Antimicrobial compounds produced by endophytes.

Antimicrobial compound	Endophyte	Host	Reference
Antifungal pestaloside, pestalopyrone, hydroxypestalopyrone	*Pestaliopsis microspora*	*Torreya taxifolia*	Lee et al. (1995)
Ecomycin active against *Candida albicans* and *Cryptococcus neoformans*	*Pseudomonas viridiflava*	Grass	Miller et al. (1998)
Cryptocandin active against *C. albicans*, *Trichophyton* spp., *Sclerotinia sclerotiorum*, *Botrytis cinerea*	*Cryptosporiopsis quercina*	*Tripterigeum wilfordii*	Strobel et al. (1999)
Antibacterial and antifungal colletotric acid	*Colletotrichum gleosporioides*	*Artemisia mongolica*	Zou et al. (2000)
Munumbicin active against *Bacillus anthracis*, MDR *Mycobacterium tuberculosis*	*Streptomyces* sp. NRRL 30562	*Kennedia nigriscans*	Castillo et al. (2002)
Kakadumycin, a broad spectrum antibiotic	*Streptomyces* sp. NRRL 30566	*Grevillea pteridifolia*	Castillo et al. (2003)
Coronamycin active against pythaceous fungi, *Cryotococcus neoformans*	*Streptomyces* sp.	*Monstera* sp.	Ezra et al. (2004)
Fumigaclavine C, fumitremorgin C, helvolic acid, physcion active against *C. albicans*	*Aspergillus fumigatus* CY018	*Cynodon dactylon*	Liu et al. (2004)
Griseofulvin active against human and veterinary myotic diseases	*Xylaria* p. F0010	*Abies holophylla*	Park et al. (2005)
7-Amino-4-methyl coumarin active against *Staphylococcus aureus*, *Escherichia coli*, *Salmonella typhi*, *Salmonella enteriditis*, *Vibrio parahaemolyticus*, *Yersinia*, *Shigella*, *C. albicans*, *Aspergillus niger*	*Xylaria* p. YX-28	*Ginkgo biloba*	Liu et al. (2008)
(3,1′-didehydro-3[2″(3‴,3‴-dimethyl-prop-2-enyl)-3″-indolylmethylene]-6-methyl pipera-zine-2,5-dione)	*Penicillium chrysogenum*	Mangrove plant *Porteresia coarctata*	Devi et al. (2012)
Munumbicin	*Phomopsis* sp.	*Allamanda cathartica*	Nithya and Muthumary (2011)
Clethramycin against *C. neoformans*	*Streptomycin hygroscopicus*	Medicinal plants	Golinska et al. (2015)
Saadamycin against *Campylobacter jejuni*	*C. quercina*	*T. wilfordii*	Dutta et al. (2014)

(Strobel and Daisy, 2003). Some known antimicrobial compounds are discussed in Table 6.2.

Anticancer compounds—There has been an increasing demand for anticancer drugs that allow earlier and more precise treatment. The anticancer properties of several secondary metabolites from endophytes have been investigated recently. Some examples of the potential of endophytes to produce anticancer agents are cited below.

The diterpenoid Taxol (also known as Paclitaxel) has generated more attention and interest than any other new drug since its discovery due to its unique mode of action compared to other anticancer agents. Taxol ($C_{47}H_{51}NO_{14}$) was first isolated from the bark of yew trees such as *Taxus brevifolia*, *Taxus baccata* (Wani et al., 1971). But these trees are rare and slow-growing. This fails to meet the market demand of this drug and other sources for production of Taxol have been investigated. The isolation of Taxol producing fungus *Taxomyces andreanae* endophytic to *T. brevifolia* has provided an alternative approach to obtain a cheaper and more available product via microbial fermentation (Stierle et al., 1993). Other genera of fungal endophytes associated or not to yew are *Taxodium distichum* (Li et al., 1996), *Wollemia nobilis* (Strobel et al., 1997), *Phyllosticta spinarum* (Kumaran et al., 2008), *Bartalinia robillardoides* (Gangadevi and Muthumary, 2008), *Pestaliopsis terminaliae* (Gangadevi and Muthumary, 2008), *Botrydiplodia theobromae* (Pandi et al., 2010).

A few other important anticancer compounds isolated from endophytic microbes include camptothecin from *Fusarium solani* endophytic to *Camptothecin acuminate* (Kusari et al., 2009), vinblastine and vincristine from *Fusarium oxysporum* endophytic to *Catharanthus roseus* (Kumar et al., 2013), podophyllotoxin from *Phialocephala fortinii* endophytic to *Podophyllum peltatum* (Eyberger et al., 2006), torreyanic acid from *Pestaliopsis microspora* endophytic to *Torreya taxifolia* (Lee et al., 1996).

Antioxidant compounds—Though medicinal plants, vegetables and fruits are common sources of natural antioxidants, and metabolites from endophytes can be potent sources of the same. Pestacin and isopestacin were obtained from *P. microspora* endophytic to *Terminalia morobensis* (Harper et al., 2003). The bacteria *Paenibacillus polymyxa* endophytic to the root of *Stemona japonica*, a traditional Chinese medicinal plant is reported to produce exopolysaccharide demonstrating strong scavenging activities on superoxide and hydroxyl radicals (Liu et al., 2009). More recently, extracts of *Lactobacillus* sp. endophytic to *Adhatoda beddomei* (Swarnalatha et al., 2015) and rhizome endophytes of *Curcuma longa* (Sulistiyani et al., 2016) are reported to have antioxidant activities.

Antiviral compounds—A fascinating use of metabolites obtained from endophytes is the inhibition of viruses. *Pestalotiopsis theae* endophytic to an unidentified Chinese tree produced pestalotheol (Liu et al., 2008) having antiviral property and inhibition of HIV-1 replication by secondary metabolites from endophytic fungi of desert plants (Wellensiek et al., 2013) have been reported. From the literature survey, it is evident that the search for antiviral compounds from endophytes is in its infancy.

Antidiabetic compounds—A nonpeptidal insulin-mimetic compound was isolated from *Pseudomassaria* sp. of African rainforest (Zhang et al., 1999) and endophytic actinobacteria from medicinal plants were observed to be α-glucosidase inhibitor (Pujiyanto et al., 2012). α-amylase inhibitor secreting *Streptomyces* sp. was isolated from well-known antidiabetic medicinal plant *Leucas ciliata* and *Rauwolfia densiflora* (Akshatha et al., 2013).

Immunosuppressant compounds—*Fusarium subglutinans* endophytic to *Taxus wilfordii* has been reported to produce immunosuppressive subglutinol A and B (Lee et al., 1995). Also, sydoxanthone A and B and 13-*O*-acetylsydowinin B were isolated from endophytic fungus *Aspergillus sydowii* endophytic to *Scapania ciliate* (Song et al., 2013).

Antiparasitic compounds—Antitrypanosomal, antileshmanial, and antimalarial compounds are obtained from endophytes. Zin et al. (2011) isolated antitrypanosomal compound from endophytic *Streptomycete*. Kakadumycin A having antimalarial activity has been isolated from *Streptomycete* endophytic to *Grevillea pteridifolia* (Castillo et al., 2003). Citrinin produced by *Penicillium janthinellum* endophytic to *Melia azedarach* has antileishmanial activity (Marinho et al., 2005).

Insecticidal compounds—Several endophytes are reported to have insecticidal properties. Nodulisporic acid, first isolated from *Nodulosporium* sp., endophytic to *Bontia daphnoides*, has potent insecticidal activities against blowfly larva (Demain, 2000). Endophytic fungus from wintergreen (*Gaultheria procumbens*) produces benzofuran derivatives which are toxic to spruce budworm (Findlay et al., 1997). Naphthalene is an important product produced by the fungus *Muscodor vitigenus* endophytic to *Paullinia paullinoides* (Daisy et al., 2002). Endophyte research continues for the discovery of safe alternatives for synthetic insecticides.

Nematicidal compounds—Endophytic bacteria of different crops in Tamil Nadu, India, have biocontrol potential against the root-knot nematode *Meloidogyne incognita* (Vetrivelkalai et al., 2010). Endophytic bacteria isolated from *Euphorbia pulcherrima*, *Pyrethrum cinerariifolium*, and *Heracleum candicans* displayed activity against *Caenorhabditis elegans* and *Bursaphelenchus xylophilus* (Zheng et al., 2008). Metabolites active against *B. xylophilus* and *Panagrellus redivivus* are produced by the fungus *Geotrichum* sp. endophytic to neem (*Azadirachta indica*) (Li et al., 2007).

Enzymes—Endophytes have been studied for their potential to produce enzymes having high biotechnological interest such as in processing of foods, manufacturing of detergents, textiles, leather industry, pharmaceutical products, and molecular biology. Several enzymes such as amylase, lipase, cellulase, laccase, protease, chitinase, xylanase have been detected from endophytic microorganisms. Some endophytes producing enzymes are discussed in Table 6.3

Polymers—Endophytic microorganisms have been described as producers of polymers such as exopolysaccharides (EPS) and polyhydroxyalkanoates. EPS production has been reported from endophytic fungi *Piper hispidum* (Orlandelli et al., 2016), diazotrophic bacteria *Burkholderia brasiliensis* endophytic to rice roots (Mattos et al., 2001) or more recently by *Bacillus cereus* endophytic to *Ricinus communis* (Mukherjee et al., 2017). Bioplastics are receiving increasing commercial attention and the most widely produced bioplastic is polyhydroxy butyrate (PHB). PHB production has been observed in *Gluconacetobacter diazotrophicus* endophytic to sugarcane (Cardoso et al., 2014), *B. cereus* endophytic to *Helianthus annuus* (Das et al., 2016).

Table 6.3 Enzymes obtained from endophytes.

Endophyte	Host	Enzymes	Reference
Colletotrichum, Fusarium, Phomopsis	*Brucea javanica*	Amylase, cellulase, ligninase, pectinase, xylanase	Choi et al. (2005)
Bacillus, Pseudomonas, Corynebacterium, Staphylococcus, Actinomyces	*Jacaranda decurrens*	Amylase, lipase, protease	Carrim et al. (2006)
Bjerkandera sp., *Mycelia sterilia*	*Drimys winteri, Prumnopitys andina*	Lignocellulolytic activities	Oses et al. (2006)
Paenibacillus polymyxa	*Stemona japonica*	Fibrinolytic enzymes	Lu et al. (2007)
Colletotrichum sp.	Thai medicinal plants	Asparaginase	Theantana et al. (2009)
Curvularia spp., *Colletotrichum gleosporoides, Penicillium citrinum, Phyllosticta* sp., *Cladosporium cladosporoides, Lasiodiplodia theobromae*	*Adhatod vasica, Coleus aromatics, Costas igneus, Lawsonia inermis*	Amylase, cellulase, laccase, lipase, protease	Amirita et al. (2012)
Streptomyces griseofulvus	Sweet pea	Chitinase	Tang-um and Niamsup (2012)
Fungal isolates	Medicinal plants of Western Ghat	Amylase, cellulase, pectinase, asparaginase	Uzma et al. (2016)
Alternaria sp., *Colletotrichum* sp.	*Azadirachta indica*	Laccase, tyrosinase, asparinase, xylanase	Taware et al. (2017)

6.4 Endophyte-assisted phytoremediation

Different remediation methods are mostly applicable to more contaminated sites like soil and ground water, but these methods are more expensive. This high cost of remediation increases the large number of contaminated sites. Therefore, natural remediation techniques have been developed to provide more environmentally friendly and cost effective techniques of contaminated sites. In these natural remediation techniques one cost effective technique is phytoremediation. Phytoremediation has the natural ability to extract chemicals from soil, water, and air with the help of sunlight. Phytoremediation was conducted in different forms like phytoextraction, rhizofiltration, absorption, phytostabilization, phytovolatilization. Phytoremediation has several advantages and cost effective than traditional remediation technologies (Srivastava et al., 2016). Other advantages of phytoremediation are carbon sequestration and production of bioenergy applications.

Phytoremediation has some constraints because of phytotoxicity, slow growth, and limited uptake. To overcome this problem plants associated with microbes are used to clean up the contaminated sites. Plant roots are also helpful in the movement of microbes. The combination of plants and microbes has become a more popular, cost effective, and environmentally friendly technology. Recently, endophytic bacteria are gaining attention for their role in phytoremediation process. Endophytic bacteria or fungi reside inside the plant tissue. These have the highest population in root tissues then stem and decrease in leaves (Ryan et al., 2008). There are so many plants reported for the asymptomatic relation with endophytic microbes. Enterobacteriaceae is the most common genera of endophytic species. There are some other groups that also reported on endophytic microbes like *Pseudomonadaceae* and *Burkholderiaceae*. Endophytic microbes have almost the same functions like rhizospheric microbes they help in plant growth promotion, plant nutrient uptake, and inhibits plant diseases caused by pathogens (Sturz et al., 1997, 2000). They enhance the tolerance to environmental stresses. Diazotrophic endophytes have the capability to fix the nitrogen of the host plant (Azevedo et al., 2000), Although there are several advantages and disadvantages, but careful selection of degrading endophytes and native plants can avoid some risks for plants.

Preston et al. (2001) describe the gene expression technology by providing genes required by bacteria to survive and colonize into the plant. Several endophytic bacteria are known to their published genome sequences like *Pseudomonas putida* W619, *Serratia proteamaculans* 568, *Azospirillum* spp. B510, *Acetobacter diazotrophicus* Pal5, *Enterobacter* spp. 638, *Stenotrophomonas maltophilia* R551-3, *Azoarcus* spp. BH72, and *Klebsiella pneumoniae* 342. Srivastava et al. (2016) clearly described the role of endophyte *B. pumilus* isolated from fern *P. vittata* which was capable to decontaminate the metalloid arsenic from soil by enhancing the phytoextraction in *P. vittata.* Li et al. (2016) described the role of endophyte enterobacter sp. K3-2 which increased the growth and root Cu accumulation of *Sorghum sudanense*. Babu et al. (2013) revealed the enhancement of heavy metal phytoremediation by *Alnus firma* with endophytic *Bacillus thuringiensis* GDB-1.

Although these results are not guaranteed for similar results in fields, therefore more research is needed in field conditions with field plants that are able to give good results in phytoremediation. This technology gives great opportunities in production of biomass and bioenergy crops with phytoremediation of contaminated soil and water.

6.5 Conclusion

In this chapter, we study that endophytes can be bacteria, fungi, or actinomycetes and that they have the same properties of plant growth promotion and bioremediation. They have the capability to produce so many natural products which can be useful in pharmaceutical and agricultural industries. Endophytes open some more doors for researchers to produce new compounds that can be useful in so many industries. They have a great importance for plants, humans, and the environment. Researchers can explore endophytes for a newer role to the humans and environment.

References

Akshatha, V.J., Nalini, M.S., D'Souza, C., Prakash, H.S., 2013. Streptomycete endophyte from anti-diabetic medicinal plants of the Western ghats inhibit alpha-amylase and promote glucose uptake. Lett. Appl. Microbiol. 58, 433–439.

Amirita, A., Sindhu, P., Swetha, J., Vasanthi, N.S., Kannan, K.P., 2012. Enumeration of endophytic fungi from medicinal plants and screening of extracellular enzymes. World J. Sci. Technol. 2, 13–19.

Arunachalam, C., Gayathri, P., 2010. Studies on bioprospecting of endophytic bacteria from the medicinal plant of *Andrographis paniculata* for their antimicrobial activity and antibiotic susceptibility pattern. Int. J. Curr. Pharm. Res. 2 (4), 63–68.

Ávila, C., Reyes, C., Bayona, J.M., García, J., 2013. Emerging organic contaminant removal depending on primary treatment and operational strategy in horizontal subsurface flow constructed wetlands: influence of redox. Water Res. 47, 315–325.

Azevedo, J.L., Maccheroni, J.W., Pereira, O.J., Araujo, L.W., 2000. Endophytic microorganisms: a review on insect control and recent advance on tropical plants. J. Biotechnol. 3 (1), 40–65.

Babu, A.G., Kim, J.D., Oh, B.T., 2013. Enhancement of heavy metal phytoremediation by *Alnus firma* with endophytic *Bacillus thuringiensis* GDB-1. J. Hazard. Mater. 25, 477–483. Available from: https://doi.org/10.1016/j.jhazmat.2013.02.014.

Berg, G., Hallmann, J., 2006. Control of plant pathogenic fungiwith bacterial endophytes. In: Schulz, B.J.E., Boyle, C.J.C., Sieber, T.N. (Eds.), Microbial Root Endophytes. Springer-Verlag, Berlin, pp. 53–69.

Cardoso, A.M., da Silva, C.V.F., Silva, A.S.P., Padua, V.L.M., 2014. Polyhydroybutyrate production by a sugarcane growth promoter bacterium. BMC Proc. 8, 246.

Carrim, A.J.I., Barbola, E.C., Viera, J.D.G., 2006. Enzymatic activity of endophytic bacterial isolates of *Jacaranda decurrens* Cham. (Carobinha-do-campo). Braz. Arch. Biol. Technol. 49, 353–359.

Castillo, U.F., Strobel, G.A., Ford, E.J., Hess, W.M., Porter, H., Jensen, J.B., et al., 2002. Munumbicins, wide-spectrum antibiotics produced by *Streptomyces*NRRL 30562, endophytic on *Kennedia nigriscans*. Microbiology 148, 2675–2685.

Castillo, U.F., Harper, J.K., Strobel, G.A., Sears, J., Alesi, K., 2003. Kakadumycins, novel antibiotics from *Streptomyces* sp. NRRL 30566, an endophyte of *Grevillea pteridifolia*. FEMS Microbiol. Lett. 224, 183–190.

Choi, Y.W., Hodgkiss, I.J., Hyde, K.D., 2005. Enzyme production by endophytes of *Brucea javanica*. J. Agric. Technol. 1, 55–66.

Compant, S., Duffy, B., Nowak, J., Cl, C., Barka, E.A., 2005a. Use of plant growth-promoting bacteria for biocontrol of plant diseases: principles, mechanisms of action, and future prospects. Appl. Environ. Microbiol. 71, 4951–4959.

Compant, S., Reiter, B., Sessitsch, A., Nowak, J., Clément, C., Barka, E.A., 2005b. Endophytic colonization of *Vitis vinifera* L. by a plant growth-promoting bacterium, *Burkholderia* sp. strain PsJN. Appl. Environ. Microbiol. 71, 1685–1693.

Daisy, B.H., Strobel, G.A., Castillo, U., Ezra, D., Sears, J., Weaver, D., et al., 2002. Naphthalene, an insect repellent, is produced by *Muscodor vitigenus*, a novel endophytic fungus. Microbiology 148, 3737–3741.

Das, R., Dey, A., Pal, A., Paul, A.K., 2016. Influence of growth conditions on production of poly (3-hydroxybutyrate) by *Bacillus cereus* HAL 03 endophytic to *Helianthus annus* L. J. Appl. Biol. Biotechnol. 4, 75–84.

Demain, A.L., 2000. Microbial natural products: a past within a future. In: Wrigley, S.K., Hayes, M.A., Thomas, R., Chystal, E.J., Nicholson, N. (Eds.), Biodiversity: New Heads for Pharmaceutical and Agrochemical Industries. The Royal Society of Chemistry, Cambridge, United Kingdom, pp. 3–16.

Devi, P., Rodrigues, C., Naik, C.G., D'Souza, L., 2012. Isolation and characterization of antibacterial compound from a mangrove endophytic fungus *Penicillium chrysogenum* MTCC 5108. Indian. J. Microbiol. 52, 617–623.

Dhungana, S.A., Adachi, F., Hayashi, S., Puri, R.R., Itoh, K., 2018. Plant growth promoting effects of Nepalese sweet potato endophytes. Horticulture 4 (4), 53. Available from: https://doi.org/10.3390/horticulturae4040053.

Dutta, D., Puzari, K.C., Gogoi, R., Dutta, P., 2014. Endophytes: exploitation as a tool in plant protection. Braz. Arch. Biol. Technol. 57, 621–629.

El-Din Hassan, S., 2017. Plant growth-promoting activities for bacterial and fungal endophytes isolated from medicinal plant of *Teucrium polium* L. J. Adv. Res. 8 (6), 687–695.

Eyberger, A., Dondapati, R., Porter, J.R., 2006. Endophyte fungal isolates from *Podophyllum peltatum* produce podophyllotoxin. J. Nat. Prod. 69, 1121–1124.

Ezra, D., Castillo, U.F., Strobel, G.A., Hess, W.M., Porter, H., Jensen, J.B., 2004. Coronamycins, peptide antibiotics produced by a verticillate *Streptomyces* sp. (MSU-2110) endophytic on *Monstera* sp. Microbiology 150, 785–793.

Findlay, J.A., Buthelezi, S., Guoqiang, L., Seveck, M., 1997. Insect toxins from an endophytic fungus from Wintergreen. J. Nat. Prod. 60, 1214–1215.

Gangadevi, V., Muthumary, J., 2008. Taxol, an anticancer drug produced by an endophytic fungus *Bartalinia robillardoides* Tassi, isolated from a medicinal plant, *Aegle marmelos* Correa ex Roxb. World J. Microbiol. Biotechnol. 24, 717–724.

Glick, B.R., 2012. Plant growth-promoting bacteria: mechanisms and applications. Scientifica 2012, Article ID 963401.

Golinska, P., Wypij, M., Agarkar, G., Rathod, D., Dahm, H., Rai, M., 2015. Endophytic actinobacteria of medicinal plants: diversity and bioactivity. Antonie van. Leeuwenhoek 108, 267–289.

Gouda, S., Das, G., Sen, S.K., Shin, H., Patra, J.K., 2016. Endophytes: a treasure house of bioactive compounds of medicinal importance. Front. Microbiol. 7, 1–8.

Harper, J.K., Arif, A.M., Ford, E.J., 2003. Pestacin: a 1,3-dihydro isobenzofuran from *Pestalotiopsis microspora* possessing antioxidant and antimycotic activities. Tetrahedron 59, 2471–2476.

Ho, Y.N., Hsieh, J.L., Huang, C.C., 2013. Construction of a plant-microbe phytoremediation system: combination of vetiver grass with a functional endophytic bacterium, achromobacter xylosoxidans F3B, for aromatic pollutants removal. Bioresour. Technol. 145, 43–47. Available from: https://doi.org/10.1016/j.biortech.2013.02.051.

Ijaz, A., Shabir, G., Khan, Q.M., Afzal, M., 2015. Enhanced remediation of sewage effluent by endophyte-assisted floating treatment wetlands. Ecol. Eng. 84, 58–66. Available from: https://doi.org/10.1016/j.ecoleng.2015.07.025.

Kaneko, T., Minamisawa, K., Isawa, T., Nakatsukasa, H., Mitsui, H., Kawaharada, Y., et al., 2010. Complete genomic structure of the cultivated rice endophyte azospirillum sp. B510. DNA Res. 17, 37–50. Available from: https://doi.org/10.1093/dnares/dsp026.

Kerry, B.R., 2000. Rhizosphere interactions and the exploitation of microbial agents for the biological control of plant-parasitic nematodes. Ann. Rev. Phytopath 38, 423–441.

Kumar, A., Patil, D., Rajamohan, P.R., Ahmad, A., 2013. Isolation, purification and characterization of vinblastin and vincristine from endophytic fungus *Fusarium oxysporum* isolated from *Catharanthus roseus*. PLoS One 8.

Kumaran, R.S., Muthumary, J., Hur, B.K., 2008. Production of taxol from *Phyllosticta spinarum*, an endophytic fungus of *Cupressus* sp. Eng. Life Sci. 8, 438–446.

Kusari, S., Lamshöft, M., Spiteller, M., 2009. *Aspergillus fumigates* Fresenius, an endophytic fungus from *Juniperus communis* L. Horstmann as a novel source of the anticancer prodrug deoxypodophyllotoxin. J. Appl. Microbiol. 107, 1019–1030.

Kwak, M.-J., Song, J.Y., Kim, S.-Y., Jeong, H., Kang, S.G., Kim, B.K., et al., 2012. Complete genome sequence of the endophytic bacterium *Burkholderia* sp. strain KJ006. J. Bacteriol. 194 (16), 4432–4433. Available from: https://doi.org/10.1128/JB.00821-12.

Lee, J., Lobkovsky, E., Pilam, N.B., Strobel, G.A., Clardy, J., 1995. Subglutinols A and B: immunosuppressive compounds from the endophytic fungus *Fusarium subglutinans*. J. Org. Chem. 60, 7076–7077.

Lee, J.C., Strobel, G.A., Lobkovsky, E., Clardy, J.C., 1996. Torreyanic acid: a selectively cytotoxic quinone dimer from the endophytic fungus *Pestaliopsis microspora*. J. Org. Chem. 61, 3232–3233.

Li, J.-Y., Strobel, G., Sidhu, R., Hess, W.M., Ford, E.J., 1996. Endophytic taxol-producing fungi from bald cypress, *Taxodium distichum*. Microbiology 142, 2223–2226.

Li, G.H., Zhang, K.Q., Xu, J.P., Dong, J.Y., Liu, Y.J., 2007. Nematicidal substances from fungi. Recent. Pat. Biotechnol. 1, 212–233.

Li, Y., Zhu, G., Ng, W.J., Tan, S.K., 2014. A review on removing pharmaceutical contaminants from wastewater by constructed wetlands: design, performance and mechanism. Sci. Total. Environ. 46, 908–932. Available from: https://doi.org/10.1016/j.scitotenv.2013.09.018.

Li, Y., Wang, Q., Wang, L., He, L.-Y., Sheng, X.-F., 2016. Increased growth and root Cu accumulation of *Sorghum sudanense* by endophytic enterobacter sp. K3-2: implications for *Sorghum sudanense* biomass production and phytostabilization. Ecotoxicol. Environ. Saf. 124, 163–168. Available from: https://doi.org/10.1016/j.ecoenv.2015.10.012.

Liu, J.Y., Song, Y.C., Zhang, Z., Wang, L., Guo, Z.J., Zou, W.X., et al., 2004. *Aspergillus fumigatus* CY018, an endophytic fungus in *Cynodon dactylon* as a versatile producer of new and bioactive metabolites. J. Biotechnol. 114, 279–287.

Liu, X., Dong, M., Chen, X., Jieng, M., Lv, X., Zhou, J., 2008. Antimicrobial activity of an endophytic *Xylaria* sp. YX-28 and identification of its antimicrobial compound 7-amino-4-methylcoumarin. Appl. Microbiol. Biotechnol. 78, 241−247.

Liu, L., Luo, J., Ye, H., Sun, Y., Lu, Z., Zeng, X., 2009. Production, characterization and antioxidant activities in vitro of exopolysaccharides from endophytic bacterium *Paenibacillus polymyxa* EJS-3. Carbohydr. Polym. 78, 275−281.

Lu, F., Sun, L., Lu, Z., Bie, X., Fang, Y., Liu, S., 2007. Isolation and identification of an endophytic strain EJS-3 producing novel fibrinolytic enzymes. Curr. Microbiol. 54, 435−439.

Marinho, A.M.R., Rodrigues-Filho, E., Moitinho, M.D.L.R., Santos, L.S., 2005. Biologically active polyketides produced by *Penicillium janthinellum* isolated as an endophytic fungus from fruits of *Melia azedarach*. J. Braz. Chem. Soc. 16, 280−283.

Mattos, K.A., Jones, C., Heise, N., Previato, J.O., Mendonca-Previato, L., 2001. Structure of an acidic exopolysaccharide produced by the diazotrophic endophytic bacterium *Burkholderia brasilinesis*. Eur. J. Biochem. 268, 3174−3179.

Miliute, I., Buzaite, O., Gelvonauskiene, D., Sasnauskas, A., Stanys, V., Baniulis, D., 2016. Plant growth promoting and antagonistic properties of endophytic bacteria isolated from domestic appleISSN 1392−3196/e-ISSN 2335−8947 Zemdirbyste-Agriculture 103 (1), 77−82.

Miller, R.V., Miller, C.M., Garton-Kinney, D., Redgrave, B., Sears, J., Condron, M., et al., 1998. Ecomycins, unique antimycotics from *Pseudomonas viridiflava*. J. Appl. Microbiol. 84, 937−944.

Mukherjee, A., Das, R., Sharma, A., Pal, A., Paul, A.K., 2017. Production and partial characterization of extracellular polysaccharide from endophytic *Bacillus cereus* RCR 08. Biosci. Biotechnol. Res. Commun. 10, 612−622.

Nithya, K., Muthumary, J., 2011. Bioactive metabolite produced by *Phomopsis* sp., an endophytic fungus in *Allamanda cathartica* Linn. Recent. Res. Sci. Technol. 3, 44−48.

Orlandelli, R.C., Vasconcelos, A.F.D., Azvedo, J.L., da Silva, M.L.C., Pamphile, J.A., 2016. Screening of endophytic sources of exopolysaccharides: preliminary characterization of crude exopolysaccharide produced by submerged culture of *Diaporthe* sp. JF6698 under different cultivation time. Biochim. Open 2, 33−40.

Oses, R., Valenzuela, S., Freer, J., Baeza, J., Rodriguez, J., 2006. Evaluation of fungal endophytes for lignocellulolytic enzyme production and wood biodegradation. Int. Biodeterior. Biodegrad. 57, 129−135.

Pageni, B.B., Lupwayi, N.Z., Akter, Z., Larney, F.J., Kawchuk, L.M., Gan, Y.T., 2014. Plant growth-promoting and phytopathogen-antagonistic properties of bacterial endophytes from potato (*Solanum tuberosum* L.) cropping systems. Can. J. Plant Sci. 94 (5), 835−844. Available from: https://doi.org/10.4141/cjps2013-356.

Pandi, M., Manikandan, R., Muthumary, J., 2010. Anticancer activity of fungal taxol derived from *Botryodiplodia theobromae* Pat., an endophytic fungus, against 7, 12 dimethyl benz(*a*)anthracene (DMBA)-induced mammary gland carcinogenesis in Sprague Dawley rats. Biomed. Pharm. 64, 48−53.

Park, J.H., Choi, G.J., Lee, H.B., 2005. Griseofulvin from *Xylaria* sp. strain F0010, an endophytic fungus of *Abies holophylla* and its antifungal activity against plant pathogenic fungi. J. Microbiol. Biotechnol. 15, 112−117.

Pimentel, M.R., Molina, G., Dionisio, A.P., Marostica Junior, M.R., Pastore, G.M., 2011. The use of endophytes to obtain bioactive compounds and their application in biotransformation process. Biotechnol. Res. Int. 2011, 1−11. Available from: https://doi.org/10.4061/2011/576286.

Ping, L., Boland, W., 2004. Signals from the underground: bacterial volatiles promote growth in Arabidopsis. Trends Plant. Sci. 9, 263–266.

Preston, G.M., Bertrand, N., Rainey, P.B., 2001. Type III secretion in plant growth-promoting *Pseudomonas fluorescens* SBW25. Mol. Microbiol. 41 (5), 999–1014.

Pujiyanto, S., Lestari, Y., Suwanto, A., Budiarti, S., Darusma, L.K., 2012. Alphaglucosidase inhibitor activity and characterization of endophytic Actinomycetes isolated from some Indonesian diabetic medicinal plants. Int. J. Pharm. Pharm. Sci. 4, 327–333.

Ripa, F.A., Cao, W.-d, Tong, S., Sun, J.-g, 2019. Assessment of plant growth promoting and abiotic stress tolerance properties of wheat endophytic fungi. BioMed. Res. Inter. . Available from: https://doi.org/10.1155/2019/6105865Article ID 6105865, 12 pages.

Rodrigues, A.A., Forzani, M.V., Soares, R., de, S.,S.T., Sibov, G., Vieira, J.D., 2016. Isolation and selection of plant growth-promoting bacteria associated with sugarcane. Pesq. Agropec. Trop., Goiânia 46 (2), 149–158.

Ryan, R.P., Germaine, K., Franks, A., Ryan, D.J., Dowling, D.N., 2008. Bacterial endophytes: recent developments and applications. FEMS Microbiol. Lett. 278, 1–9.

Shehzadi, M., Afzal, M., Khan, M.U., Islam, E., Mobin, A., Anwar, S., et al., 2014. Enhanced degradation of textile effluent in constructed wetland system using *Typha domingensis* and textile effluent-degrading endophytic bacteria. Water Res. 58, 152–159. Available from: https://doi.org/10.1016/j.watres.2014.03.064.

Song, X., Zhang, X., Han, Q., Li, X., 2013. Xanthone derivatives from *Aspergillus sydowii*, an endophytic fungus from the liverwort *Scapania ciliata* S. Lac and their immunosuppressive activities. Phytochem. Lett. 6, 318–321.

Srivastava, S., Singh, M., Paul, A.K., 2016. Arsenic bioremediation and bioactive potential of endophytic bacterium *Bacillus pumilus* isolated from *Pteris vittata L.* Inte. J. Adv. Biotechnol. Res. 7 (1), 77–92 (IJBR)ISSN 0976–2612, Online ISSN 2278-599X.

Stierle, A., Strobel, G., Stierle, D., 1993. Taxol and taxane production by *Taxomyces andreanae*, an endophytic fungus of Pacific yew. Science 260, 214–216.

Strobel, G., Daisy, B., 2003. Bioprospecting for microbial endophytes and their natural products. Microbiol. Mol. Biol. Rev. 67, 491–502.

Strobel, G.A., Miller, R.V., Condron, N.M., Teplow, R.V., Hess, W.M., 1999. Crytocandin, a potent antimycotic from the endophytic fungus Cryptosporipsis cf. quercina. Microbiology 145, 1919–1926.

Strobel, G.A., Hess, W.M., Li, J.Y., Ford, E., Sears, J., Sidhu, R.S., et al., 1997. *Pestaliopsis guepinii*, a taxol-producing endophyte of the Wollemi pine, *Wollemia nobilis*. Aust. J. Bot. 45, 1073–1082.

Sturz, A.V., Christie, B.R., Matheson, B.G., Nowak, J., 1997. Biodiversity of endophytic bacteria which colonize red clover nodules, roots, stems and foliage and their influence on host growth. Biol. Fert. Soils 25 (1), 13–19. Available from: https://doi.org/10.1007/s003740050273.

Sturz, A.V., Christie, B.R., Nowak, J., 2000. Bacterial endophytes: potential role in developing sustainable systems of crop production. Crit. Rev. Plant. Sci. 19 (1), 1–30.

Sulistiyani, Ardyati, T., Winarsih, S., 2016. Antimicrobial and antioxidant activity of endophytic bacteria associated with *Curcuma longa* rhizome. J. Exp. Life Sci. 6, 45–51.

Swarnalatha, Y., Saha, B., Choudary, L., 2015. Bioactive compound analysis and antioxidant activity of endophytic bacterial extract from *Adhatoda beddomei*. Asian J. Pharm. Clin. Res. 8, 70–72.

Tan, R.X., Zou, W.X., 2001. Endophytes: a rich source of functional metabolites. Nat. Prod. Rep. 18, 448–459.

Tang-um, J., Niamsup, H., 2012. Chitinase production and antifungal potential of enophytic *Streptomyces griseoflavus* P4. Maejo Int. J. Sci. Technol. 6, 95–104.

Taware, A.S., More, Y.W., Ghag, S.V., Rajukar, S.K., 2017. Screening of endophytic fungi isolated from *Azadiruchta indica* A. Juss for production of enzymes. Biosci. Discov. 8, 688–694.

Theantana, T., Hyde, K.D., Lumyong, S., 2009. Asparaginase production by endophytic fungi from Thai medicinal plants: cytotoxic properties. Int. J. Integr. Biol. 7, 1–8.

Uzma, R., Konappa, N.M., Chowdappa, S., 2016. Diversity and extracellular enzyme activities of fungal endophytes isolated from medicinal plants of Western Ghats, Karnataka. Egypt. J. Basic Appl. Sci. 3, 335–342.

Verlicchi, P., Zambello, E., 2014. How efficient are constructed wetlands in removing pharmaceuticals from untreated and treated urban wastewaters? A review. Sci. Total Environ. 470–471, 1281–1306.

Vetrivelkalai, P., Sivakumar, M., Jonathan, E.I., 2010. Biocontrol potential of endophytic bacteria on *Meloidogyne incognita* and its effect on plant growth in bhendi. J. Biopesticides 3, 452–457.

Visioli, G., Vamerali, T., Mattarozzi, M., Dramis, L., Sanangelantoni, A.M., 2015. Combined endophytic inoculants enhance nickel phytoextraction from serpentine soil in the hyperaccumulator *Noccaea caerulescens*. Front. Plant. Sci. 6, 638. Available from: https://doi.org/10.3389/fpls.2015.00638.

Vymazal, J., Březinová, T., 2015. The use of constructed wetlands for removal of pesticides from agricultural runoff and drainage: a review. Environ. Int. 75, 11–20.

Wani, M.C., Taylor, H.L., Wall, M.E., Coggon, P., McPhail, A.T., 1971. Plant antitumor agents. VI. The isolation and structure of taxol, a novel antileukemic and antitumor agent from *Taxus brevifolia*. J. Am. Chem. Soc. 93, 2325–2327.

Wellensiek, B.P., Ramakrishnan, R., Bashyal, B.P., Eason, Y., Gunatilaka, A.A.L., Ahmad, N., 2013. Inhibition of HIV-1 replication by secondary metabolites from endophytic fungi of desert plants. Open Virol. J. 7, 72–80.

Wisniewski-Dyé, F., Borziak, K., Khalsa-Moyers, G., Alexandre, G., Sukharnikov, L.O., Wuichet, K., et al., 2011. Azospirillum genomes reveal transition of bacteria from aquatic to terrestrial environments. PLoS Genet. 7, e1002430.

Zhang, B., Salituro, G., Szalkowski, D., Li, Z., Zhang, Y., 1999. Discovery of small molecule insulin mimetic with antibiabetic activity in mice. Science 284, 974–977.

Zhao, L.F., Xu, Y.J., Lai, X.H., 2018. Antagonistic endophytic bacteria associated with nodules of soybean (*Glycine max* L.) and plant growth-promoting properties. Braz. J. Microbiol. 49 (2), . Available from: https://doi.org/10.1016/j.bjm.2017.06.007São Paulo Apr.

Zheng, L., Li, G., Wang, X., Pan, W., Li, L., Lv, H., et al., 2008. Nematicidal endophytic bacteria obtained from plants. Ann. Microbiol. 58, 569–572.

Zin, N.M., Ng, K.T., Sarmin, N.M., Getha, K., Tan, G.Y., 2011. Anti-trypanosomal activity of endophytic Streptomycete. Curr. Res. Bacteriol. 4, 1–8.

Zou, W.X., Meng, J.C., Lu, H., Chen, G.X., Shi, G.X., Zhang, T.Y., et al., 2000. Metabolites of *Colletotrichum gloeosporioides*, an endophytic fungus in *Artemisia mongolica*. J. Nat. Prod. 63, 1529–1530.

Engineering bacterial aromatic dioxygenase genes to improve bioremediation

7

Vachaspati Mishra[1,2,*], *S. Veeranna*[1] *and Jitendra Kumar*[1]
[1]Bangalore Bio Innovation Centre, Helix Biotech Park, Electronics City Phase 1, Bengaluru, Karnataka, India, [2]Alberta Plant Health Lab, Crop Diversification Centre North, Alberta Agriculture and Forestry, Edmonton, AB, Canada
*Corresponding author

7.1 Introduction

There are numerous synthetic aromatic compounds released into the environment due to anthropogenic activities. A growing body of literature over the past two decades has helped to unravel that microorganisms are the primary agents responsible for degradation of xenobiotic compounds and hence their remediation from the contaminated sites can be possible by suitably applying microorganisms at the contaminated sites. However, the ability to naturally acquire degradative traits for aromatic compounds in such microbes takes time and consequently, the rate of degradation of these compounds in a natural environment is much slower. This emphasizes the significance of their genetic manipulation to generate in them the ability to degrade such compounds or to enhance their natural degradative abilities. Concerted research efforts in bacterial degradation processes and bioremediation trials have been made over the last few decades. This has led to significant advancement in our knowledge on the characterization and cloning of genes responsible for certain compounds that have been considered priority pollutants like 2,4-dichlorophenoxyacetic acid, 2,4,5-trichlorophenoxyacetic acid, polychlorinated biphenyls (PCBs), phenols, polycyclic aromatic hydrocarbons, etc. Recent molecular techniques have provided means to manipulate catabolic genes to induce and activate the degradation pathways, generate pathways with broadened efficiency, or create absolutely new hybrid pathways for degrading recalcitrant compounds. The systems biology approach, a recent trend in obtaining crucial preliminary information on metabolic pathways, can provide useful details on pathway engineering of microbes for their enhanced bioremediation capabilities (Dangi et al., 2018). The advent of gene editing tools like CRISPR (clustered regularly interspaced short palindromic repeats)–Cas, TALEN (transcription activator-like effector nucleases), and ZFNs (zinc finger nucleases) with a potential to aid in designing microbes having functional catabolic genes for biodegradation of a particular recalcitrant compound for improved

Bioremediation of Pollutants. DOI: https://doi.org/10.1016/B978-0-12-819025-8.00007-7

bioremediation processes, has given enormous potential to improve microbes for more efficient engineering for bioremediation applications (Jaiswal et al., 2019).

Metabolic diversity in bacterial degradation of aromatic compounds has been comprehensively reviewed, highlighting the importance of oxygenases and their role in the initiation of degradation (Phale et al., 2007). Oxygenases are considered as the principal agents responsible for the turnover of aromatic compounds in nature (Dagley, 1986). Thus, oxygenases appear to be the most logical target for genetic manipulation of the degradation pathway of aromatic compounds. A comprehensive information on various oxygenases linked with biodegradation processes can be had from databases, OxDBase: a database of oxygenases involved in biodegradation (Arora et al., 2009) and Rhobase from the Department of Microbiology and Center for Excellence in Bioinformatics, Bose Institute, Kolkata. Many aspects of genetic manipulations of catabolic genes apart from the manipulation of genes encoding oxygenases, have been reviewed previously (Lal et al, 1995; Johri et al., 1999; Mishra et al., 2001). The major emphasis of this review article is to discuss the importance of oxygenases in biodegradation and how intervening in their genes through molecular biology approaches could be useful in increasing degradation efficiencies of bacteria possessing such enzymes in order to perform better in bioremediation trials.

7.2 Oxygenases are the key enzymes in aromatic degradation

Oxygenases belong to the oxidoreductive group of enzymes (E.C. Class 1), which oxidize the substrates by incorporating oxygen from molecular oxygen (O_2) and utilize FAD/NADH/NADPH as the cosubstrate (Fig. 7.1). They are key players in the metabolism of aromatic compounds. They affect cleavage of the aromatic ring and eventually a complete degradation of the compound in question. Some important oxygenases with their kinetic properties and substrates worked out over the last two decades are listed in Table 7.1. Initial transformation of aromatic compounds normally proceeds through oxidative activation of the aromatic ring in a dioxygenation reaction. Depending on the substrate utilized, they form intermediates, such as catechol, protocatechuate, gentisate, or other 1,2- or 1,4-diols (Fig. 7.1). The reaction is called dioxygenation, because both O molecules of O_2 are incorporated into the aromatic intermediation, thereby creating a diol. Dioxygenation is an activation for two reasons. First, it inserts oxygen into an aromatic ring, which prepares it for later dehydrogenation and hydroxylation steps. Second, the reaction is an activation because it does not yield any net electron equivalents that can be used for energy generation by the microorganisms. Thus, the activation steps yield no energy benefit for the microorganisms, but yield such intermediates that eventually release electrons and energy.

Subsequent dioxygenation of the diols results in cleavage of the aromatic ring. The diols are transformed through either *ortho-* or *meta*-pathways (Fig. 7.1). The latter pathway appears to be more common and has relaxed substrate specificities. For example, catechols with alkyl substituents typically undergo *meta*-cleavage,

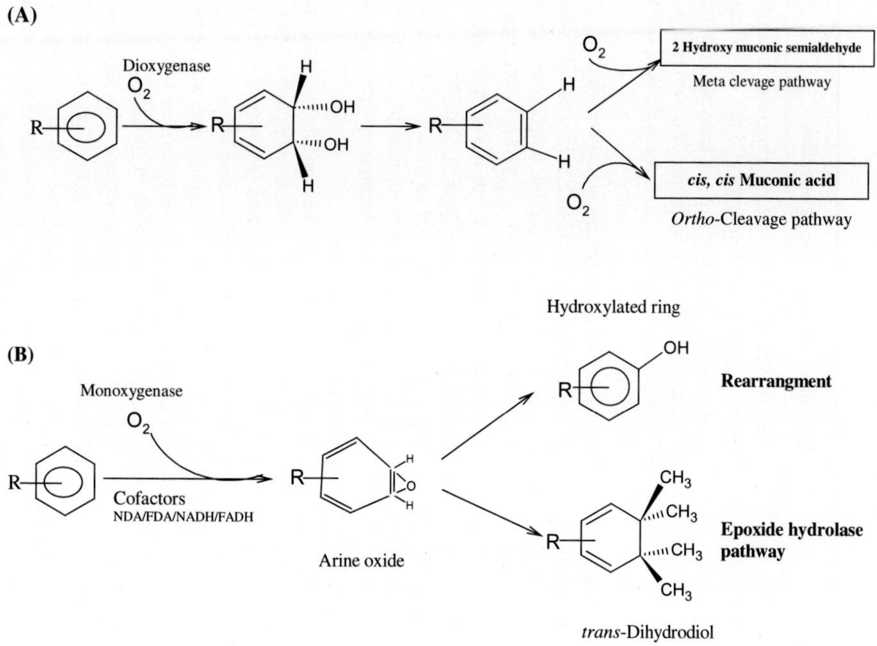

Figure 7.1 Aromatic structures showing action of mono- and dioxygenases leading to the formation of central metabolites for ring fission. (A) Aromatic ring cleavage takes place by the mediation of a dioxygenase leading into formation of either a semialdehyde or muconic acid; (B) Aromatic ring cleavage takes place by the mediation of a monooxygenase leading into dihydrodiol formation, which is cleaved eventually.

while catechols without substituents undergo either *ortho-* or *meta-*cleavage. For the same reasons, the subsequent dioxygenation reactions are oxidative activations. The ring-cleavage products are susceptible to a series of dehydrogenation and hydroxylation reactions that lead to and participate in the Krebs's cycle. Therefore, electrons and energy can be harvested by the reactions that follow ring cleavage.

The oxygenases have been well-characterized and are multicomponent proteins which form short electron transport chains with flavins and iron-sulfur clusters as redox mediators (Erickson and Mondello, 1993; Neidle et al., 1988). The oxygenase component of toluene dioxygenase in particular, from *Pseudomonas putida* F1 is an iron-sulfur protein (ISP) consisting of α and β subunits. The α subunit can accept electrons from reduced ferredoxin; however, both the substrates are required for the catalytic activity of the α subunit (Jiang et al., 1999). The resistance of some of the highly chlorinated or substituted derivatives of benzene or benzoic acids to oxygenation is primarily due to the inability of bacterial oxygenases to accept them as substrates (Reineke and Knackmuss, 1978). Such oxygenases cleave the aromatic ring between the dihydroxylated carbon atoms of the central metabolites, while the latter oxygenases, known as ring-cleavage dioxygenase, cleaves it at a point adjacent to one of the dihydroxylated carbon atoms. Ring-cleavage dioxygenases are myriad

Table 7.1 List of some aerobic mono- and dioxygenases utilizing various substrates through bacteria with their catalytic properties summarized, showing their usefulness for bioremediation applications.

Enzymes	Substrates	References	PDB ID	Remarks on catalytic properties
TOMO	*O*-Toluene	Cafaro et al. (2005)	DOI: 10.2210/pdb1T0Q/pdb	The TOMO is a very stable oxygenase as mutations in one of the enzyme subunit did not significantly change the efficiency of the enzyme and the k_{cat} values for benzene, toluene, and *o*-xylene were very similar to those measured for wild-type TOMO subunit
Monooxygenase	2C4NP	Min et al. (2018)	DOI: 10.2210/pdb6AIO/pdb	Enzyme kinetic study assay showed that biodegradation of 2C4NP followed Haldane substrate inhibition model, with the maximum specific growth rate (μ_{max}) of 0.148/h, half saturation constant (K_s) of 0.022 mM and substrate inhibition constant (K_i) of 0.72 mM
Monooxygenase phenol hydroxylase	Phenol	Sridevi et al. (2012) and Meyer et al. (2002)	DOI: 10.1016/S0969-2126(98)00062-8	Guaiacol as substrate led to increased V(max) from 0.22 to 0.43 U/mg protein and decreased the K_m from 588 to 143 μm, improving k(cat)/K(m) by a factor of 8.2
Naphthalene dioxygenase	Polycyclic aromatic hydrocarbons	Jin et al. (2012) and Jouanneau et al. (2006)	DOI: 10.1006/jmbi.1999.3462	Steady-state kinetic measurements revealed that the enzyme had a relatively low apparent Michaelis constant for naphthalene ($K_m = 0.92 \pm 0.15$ μM) and an apparent specificity constant of 2.0 ± 0.3 μM^{-1} s^{-1}. Naphthalene was converted to the corresponding 1,2-dihydrodiol with stoichiometric oxidation of NADH
Hydroquinone 1,2-dioxygenases	1,4-Dihydroxyphenols	Kolvenbach et al. (2011) and Enguita and Leitão (2013)	DOI: 10.1111/mmi.12204	The activity of 1 mg of enzyme with unsubstituted hydroquinone was 6.1 μmol/min, the apparent K_m was 2.2 μM. The enzyme lost activity upon exposure to oxygen, but could be reactivated by Fe (II) in the presence of ascorbate

Phenol monooxygenase	Naproxen	Domaradzka et al. (2015)	DOI: 10.1021/bi9619233	Phenol may possibly be responsible for enzyme induction for aromatic ring cleavage
Dioxygenase	Aromatic hydrocarbons	Ghosal et al. (2016) and Junca et al. (2004)	DOI: 10.2210/pdb6BDJ/pdb	Both enzymes exhibited $K_{app}^{mO_2}$ values of 8–10 mM in the presence of saturating concentrations of catechol
Catechol 2,3-dioxygenase	Toluene, xylene, naphthalene, and biphenyl derivatives	Cerdan et al. (1994) and Ajao et al. (2012)	DOI: 10.2210/pdb1MPY/pdb	Highly specific with a high specific activity of the wild-type enzyme exhibits only weak activity toward 3-methylcatechol but retained the ability to cleave catechol and 4-methylcatechol.
3NT dioxygenase	3NT	Mahan et al. (2015) and Kumari et al. (2017)	DOI: 10.2210/pdb5BOK/pdb	The enzyme had enhanced affinities for 3NT and were more catalytically efficient with 3NT than the wild-type enzyme. Amino acid substitutions in the closely related enzyme N12DO were detrimental to enzyme activity of N12DO
Benzoate dioxygenase	Benzene	Tavakoli et al. (2016)	DOI: 10.1007/s12223-017-0505-z	A marked concentration-dependence observed, the benzoate dioxygenase activity increased when the reaction mixture was incubated with successive portions of substrate. The highest benzoate dioxygenase activity was 6.54 U/mL, which closely matched to that of *Pseudomonas putida* benzoate dioxygenase activity (6.67 U/mL). The benzoate dioxygenase activity by *Pseudomonas arvilla* C-l was 4.2 U/mg. The reported benzoate dioxygenase activities of *Acinetobacter calcoaceticus* ADP (18.2 U/mg), *P. putida* ML2 (20 U/mg), and *P. putida* sp. (58 U/mg) were higher than that of *Rhodococcus ruber* UKMP-5M
Dioxygenase	Nitrophenols	Wojcieszynska et al. (2011) and Zhang et al. (2009)	DOI: 10.2210/pdb5M4O/pdb	Enzyme had a specific activity of 0.0004 U/mg in the presence of NADPH

(*Continued*)

Table 7.1 (Continued)

Enzymes	Substrates	References	PDB ID	Remarks on catalytic properties
Dioxygenase	Hydroxybenzenes and hydroxybenzoates	Tavakoli et al. (2016) and Nzila (2018)	DOI: 10.1021/bi00847a031	The highest benzoate dioxygenase activity was 6.54 U/mL, which was close to the benzoate dioxygenase activity by *P. putida* (6.67 U/mL). *P. arvilla* C-l was 4.2 U/mg. *A. calcoaceticus* ADP (18.2 U/mg), *P. putida* ML2 (20 U/mg), and *P. putida* sp. (58 U/mg)
Dioxygenase	Phenanthrene and pyrene (pyre)	Deveryshetty and Phale (2009)	DOI: 10.2210/pdb2HML/pdb	This enzyme from *Pseudomonas* sp. strain PPD had specific activity of 61.5 μmol/min/mg on phenanthrene and failed to show activity with gentisic, salicylic, and other hydroxynaphthoic acids
Dioxygenase	Naphthalene	Jouanneau et al. (2006) and Abo-State et al. (2018)	DOI: 10.2210/pdb1O7G/pdb	Measurements using steady-state kinetic parameters revealed that the enzyme had a relatively low apparent Michaelis constant for naphthalene ($K_m = 0.92 \pm 0.15$ μM) and an apparent specificity constant of 2.0 ± 0.3 μM^{-1} s^{-1}. Naphthalene got converted to the corresponding 1,2-dihydrodiol with stoichiometric oxidation of NADH
Dioxygenase	Nitrophenols	Zhang et al. (2009)	DOI: 10.2210/pdb5M4O/pdb	Enzyme assays indicated that cell extracts of strain WBC-3 oxidized PNP at a specific activity of 0.0004 U/mg with nitrite released in the presence of NADPH

Dioxygenase	Chloronitrophenols (2C4NP and PNP)	Min et al. (2016) and Min et al. (2019)	DOI: 10.2210/pdb2WL9/pdb	Substrate-induced dioxygenase were argued to be involved in 2C4NP and PNP catabolism. 2C4NP-induced cells of strain *Rhodococcus imtechensis* RKJ300 showed a lower rate of conversion of CHQ (4.3 mU/h OD600/cell) than the cells converting 2C4NP (100 mU/h OD600/cell), and PNP-induced cells also exhibited an evidently lower rate of conversion of HQ ($\sim$3 mU/h OD600/cell) than cells converting PNP (150 mU/h OD600/cell)
Monooxygenase	3M4NP	Leungsakul et al. (2006) and Min et al. (2019)	DOI: 10.2210/pdb4G5E/pdb	The purified wild-type DntB enzyme had apparent V_{max}, K_m, and k_{cat} values as 14 nmol/min/mg protein, 350 μM, and 0.014/S, respectively. For the purified variant M22L/L380I enzyme, the apparent V_{max}, K_m, and k_{cat} values were 44 nmol/min/mg protein, 100 μM, and 0.045/S, respectively. Therefore, the k_{cat}/K_m values of the purified wild-type DntB enzyme and purified variant M22L/L380I were 40 and 450 (per S per M), respectively, which shows that the M22L/L380I variant has 11-fold-higher efficiency than the wild-type enzyme for 4NP degradation
Monooxygenase	Regulation of *p*-nitrophenol degradation	Min et al. (2016) and Chen et al. (2016)	DOI: 10.2210/pdb4G5E/pdb	PNP 4-monooxygenase (PnpA) catalyzed rapid degradation of 3M4NP with a specific activity of 3.36 U/mg, together with consumption of NADPH. Kinetics assays revealed that the K_m value of H6-PnpA for 3M4NP was 20.3 ± 2.54 mM

(*Continued*)

Table 7.1 (Continued)

Enzymes	Substrates	References	PDB ID	Remarks on catalytic properties
Monooxygenase	Chlorophenols and their derivatives	Pimviriyakul et al. (2017) and Arora and Bae (2014)	DOI: 10.2210/pdb3RMK/pdb	The data indicated that this monooxygenase prefers to use 4-CP > 2,4-DCP > 2,4,5-TCP > 2,4,6-TCP
Monooxygenase	Aromatic compounds	Tao et al. (2004) and Arora et al. (2009)	DOI: 10.2210/pdb3RMK/pdb	Monooxygenase had no activity on 2-naphthol, and 1-naphthol. Engineered enzymes containing mutations in T4MO alpha hydroxylase TmoA acted on pure 1-naphthol or 2-naphthol. The wild monooxygenases were able to oxidize fluorine to different monohydroxylated products
Monooxygenase	Toluene, benzene	Cafaro et al. (2005) and Tao et al. (2004)	DOI: 10.2210/pdb1T0Q/pdb	The mutations in this enzyme; A subunit did not significantly change the efficiency of the enzyme in the first hydroxylation step; the k_{cat} values for benzene, toluene, and *o*-xylene were very similar to those measured for wild-type enzyme. The kinetic parameters of the enzyme on benzene as substrate with *Escherichia coli* cells showed K_{cat} of 0.36/S, K_m of 0.2 μM and K_{cat}/K_m of 1810 ($10^3\ S^{-1}\ \mu M^{-1}$)
Monooxygenase	Phenanthrene	Syed et al. (2010) and Ning et al. (2010)	DOI: 10.2210/pdb2HMK/pdb	A tighter substrate specificity was observed for phenanthrene, as only Pc-Pah4 (~65%) and Pc-Pah6 (~14%) could oxidize this compound
Monooxygenase	4-NT	Teramoto et al. (2004)	DOI: 10.2210/pdb3P3X/pdb	Various toluene derivatives were metabolized to their corresponding alcohol by *Phanerochaete chrysosporium*. 2-Nitrotoluene and 3NT were hydroxylated to 2-nitrobenzyl alcohol and 3-nitrobenzyl alcohol, respectively. Furthermore, 4-chlorotoluene was hydroxylated to 4-chlorobenzyl alcohol

Monooxygenase	Trichlorobiphenyl	Kamei et al. (2006)	DOI: 10.2210/pdb4G5E/pdb	Based on computational docking studies, Ala293 in TcpA (Ile292 in TftD) amino acid is possibly responsible for the differences in substrate specificity between the two monooxygenases
Dioxygenase	Phenanthrene	Deveryshetty and Phale (2009) and Chemerys et al. (2014)	DOI: 10.2210/pdb2HML/pdb	This enzyme (*Pseudomonas* sp. strain PPD) having a specific activity of 61.5 μmol/min/mg on phenanthrene failed to show activity with gentisic acid, salicylic acid, and other hydroxynaphthoic acids tested
Dioxygenase	Phenol, catechol	Mishra and Lal (2014) and Jones et al. (2004)	DOI: 10.2210/pdb3HJS/pdb	Broad substrate specificity; substrate's rate-limiting step of the reaction cycle is dependent on nucleophilic reactivity of the substrates concerned

The protein data bank (PDB) identification numbers of the listed enzymes are not necessarily their own, however, they are for the same enzyme studied for another organism. *2C4NP*, 2-Chloro-4-nitrophenol; *3M4NP*, 3-methyl-4-nitrophenol; *3NT*, 3-nitrotoluene; *4-NT*, 4-nitrotoluene; *N12DO*, nitrobenzene 1,2-dioxygenase; *TOMO*, toluene *o*-xylene monooxygenase.

depending on the substrates metabolized and the nature of the substrate which plays a very important role in induction of a particular enzyme for catabolism, as for instance, growth on toluene and phenol induced the *meta*-ring fission enzyme, catechol 2,3-dioxygenase, whereas growth on benzoate induced primarily *ortho*-ring-cleavage enzyme, catechol 1,2-dioxygenase (Heald and Jenkins, 1996). Without this contention, it is hard to understand the functioning of oxygenase, which paved the way for action of a particular ring-cleavage dioxygenase. Studies have revealed considerable homology among intradiol dioxygenases, the ring-cleavage dioxygenases (Hartnett et al., 1990; Ohlendorf et al., 1988) and their activity depends on ferric ion which is attached by two histidyl and two tyrosyl side chains within the catalytic subunit of the enzymes (Kohmueller and Howard, 1979; Ohlendorf et al., 1988). In a recent report, Carredano et al. (2000) have studied the substrate binding site of naphthalene 1,2-dioxygenase. They explained that there is electron density in the active site of the enzyme for the binding of a flat (planar) aromatic compound forming an indole adduct. This finding is further supported by crystallographic analysis, in which docking studies with indole, naphthalene, and biphenyl inside the substrate pocket of naphthalene dioxygenase (NDO) suggested the presence of sub-pockets where the one close to the active site iron is reserved for the binding of the aromatic ring which is hydroxylated upon catalysis (Carredano et al., 2000). The protocatechuate 3,4-dioxygenase has two nonidentical subunits, while the catechol 1,2-dioxygenase has identical subunits. However, Nakai et al. (1990) have reported an isozyme of catechol 1,2-dioxygenase which is a heterodimer. Oxygenases, such as catechol 1,2-dioxygenase from *Rhodococcus gordoniae* in crude cell extract with high specific activities (Mishra and Lal, 2014), could possibly be a reasonable candidate for in situ enzymatic studies that would provide a good insight into the behavior of similar bacteria in the environment. Furthermore, identification of microorganisms possessing such oxygenases is much necessitated even now when most of the microbial pathways of xenobiotic degradation has been worked out, because enzymes with a high specific activity in crude extract are very handy and cost effective as compared to purified enzyme.

Another versatile dioxygenase, biphenyl dioxygenase (BD) is a Resike-type, three-component enzyme, composed of a terminal dioxygenase and an electron transfer chain (ETC) (Furukawa and Miyazaki, 1986). The three-component enzyme of BD comprises a large subunit and a small subunit, associating as an $\alpha 3\beta 3$ heterohexamer (Broadus and Haddock, 1998; Maeda et al., 2001). The ETC consists of ferredoxin and reductase, which manifest electron transfer from NADH to the terminal dioxygenase for its reduction. The terminal dioxygenase triggers activated molecular oxygen to get incorporated into the biphenyl molecule at the 2,3 position to obtain a 2,3-dihydro-2,3-diol, which is then dehydrogenated to 2,3-dihydroxybiphenyl by dihydrodiol dehydrogenase (BphB). The other dioxygenase, 2,3-dihydroxybiphenyl dioxygenase (BphC), does not require any external reductant and cleaves the 2,3-dihydroxylated ring between carbon atoms 1 and 2 for yielding the product 2-hydroxy-6-oxo-6-phenylhexa-2,4-dienoic acid (the ring *meta*-cleavage product), which is subsequently hydrolyzed to benzoic acid and 2-hydroxypenta-2,4-dienoate by a hydrolase (BphD) enzyme. These are termed as upper pathway enzymes in biphenyl metabolism and are encoded by the bph gene clusters, in which bphA1 and bphA2 encode a large and a small subunit of the terminal dioxygenase, bphA3 encodes ferredoxin, and bphA4 encodes ferredoxin reductase (Erickson and Mondello, 1992; Hofer et al., 1994; Kikuchi et al., 1994; Masai et al., 1995; Mishra et al., 2001).

These enzymes are encoded either on chromosome or plasmid, which is subject to further discussions as follows:

7.2.1 Chromosome-mediated oxygenase genes

Oxygenases encoded by the Toluene (TOL)-degrading plasmid generally have broad substrate specificities and can even act on highly substituted compounds (Assinder and Williams, 1990). Some of the dioxygenases such as NDO, toluene dioxygenase, and BD, which exhibit relatively broader substrate specificities, are probably better suited for turnover (or oxygenation) of aromatic compounds. The structure and substrate specificity based on specific amino acid alteration at Phe-352 of NDO has been worked out in detail (Kauppi et al., 1998; Parales et al., 2000). The enzyme is a hexamer composed of three alpha subunits, and contains a Rieske (2Fe−2S) cluster and nonheme iron, which is bound to His 208, His 213, and Asp 362 by coordinate linkages. The alpha and beta subunits of dioxygenase have been implicated in determining substrate specificities in toluene and BDs (Kauppi et al., 1998; Parales et al., 2000). In spite of a relaxed substrate specificity, some of these enzymes maintain their enantiomeric preference. These properties have made such enzymes an attractive choice for catalysis in production of important industrial and pharmaceutical compounds (Gibson and Parales, 2000).

In the case of BD, the bphB, bphC, and bphD genes encode a dehydrogenase, a ring-cleavage dioxygenase, and a hydrolase, respectively. The large subunit of terminal dioxygenase is critical in being involved in the substrate specificity of BD (Kimura et al., 1997; Furukawa and Miyazaki, 1986). Molecular engineering has thus been suitably amenable to large-subunit genes of different origins, creating novel and evolved dioxygenases, which show enhanced and expanded degradation for not only PCBs, but also other related compounds and are highly effective for the synthesis of high-value organic molecules in the pharmaceutical industries (Furukawa and Miyazaki, 1986). Standfuss-Gabisch et al. (2012) have elucidated BD sequences and activities based on PCR amplification followed by cloning of DNA segment encoding the active site of the enzyme of extracted DNA from heavily contaminated soils with PCBs. The sequences of the amplicons thus obtained fell into three similarity clusters (I to III). Sequence identities were moderate or low between the clusters and amplicons from these three clusters were used by them to reconstitute and express complete BD operons.

7.2.2 Plasmid-mediated oxygenase genes

Catabolic enzymes of most aromatic hydrocarbons discussed currently, are encoded by the genes which are organized either on plasmid or on chromosomes (Burlage et al., 1990). TOL is an example of one such a plasmid, which when initially discovered was found to mediate degradation of toluene alone. Subsequently a number of catabolic plasmids, mostly derivatives of TOL, have been found to be responsible for the degradation of a wide range of xenobiotics and aromatic compounds such as phenol (Kasak et al., 1993), chlorobiphenyls (Lloyd-Jones et al., 1994), naphthalene (Dunn and Gunsalus, 1973), and salicylate (Chakrabarty, 1972). The TOL plasmid has been shown to confer the host bacterium with the ability to degrade not only toluene but also *m*- and *p*-xylene and other substituted benzenes. The genes encoding catabolic enzymes have been named xyl genes. These xyl genes of the TOL plasmid (pWWO) have been found to be organized into two operons referred to as upper and lower (*meta*) pathways. *Xyl*CAB, the upper pathway genes, encodes for enzymes involved in the degradation of toluene and xylene to benzoates and toluates, respectively and the lower pathway genes (*xyl*XYZLEGFJKIH) encode degradation of benzoate and toluate to acetaldehyde and pyruvate. The two operons (OP1 and OP2) have been found to be regulated by the elements known as regulatory elements XylR and XylS, respectively. Their regulatory genes are controlled by promoters Pr and Ps, respectively which are closely located but are functionally divergent.

The backbone genes of various catabolic plasmids have significant homologies existing amongst them, which is exemplified by the homologies in the genes of plasmids TOL, Naphthalene (NAH), Camphor (CAM), Octane (OCT), etc. (Burlage et al., 1990). It also appears that a horizontal flow of genes or plasmids might have occurred between different species or strains of *Pseudomonas* making them ubiquitous in distribution. A report on the presence of transposons on catabolic plasmids in *Pseudomonas* further explains the minor differences in gene organization and function of these plasmids (Assinder and Williams, 1990). One common feature of all the catabolic plasmids isolated from different strains of *Pseudomonas*, is that they are quite large in size varying from 117 kb to more than 500 kb (Burlage et al., 1990), and the catabolic genes are invariably clustered or organized to form distinct operons.

The genetic information within a single organism seems to be an outcome of rearrangement of plasmid-borne degradative genes, which might have resulted due to continuous cocultivation of different organisms possessing unique degradative genes. This has resulted in the development of new catabolic functions not shared by the parental strains (Smets et al., 1993). In the natural population, the plasmidless and plasmid bearing strains encounter at random with the frequency jointly proportional to the densities of both the population and only a fraction of these encounters results in the transmission of the plasmid (Smets et al., 1993).

7.3 Molecular interventions targeting dioxygenase genes to understand aromatic compound degradation

Majority of *Pseudomonas* catabolic genes occurring on chromosomes as individual gene(s) or gene clusters have shown characteristics similar to those found on TOL plasmids. Transposons seem to have played a major role in the transfer of these genes from chromosomes to the TOL plasmids and vice versa. Modifications in the catabolic operons or their regulatory genes through classical mutations or site-directed mutagenesis have been demonstrated to enhance the ability of these organisms or their dioxygenases to degrade the recalcitrant compounds (Erickson and Mondello, 1993). Furthermore, the studies on the regulatory mechanisms operative on the catabolic genes residing on chromosomes or plasmids, have been carried out mostly in *P. putida*, the organism in which the TOL plasmid was originally discovered. Some of these genes have also been cloned in *Escherichia coli* (Khan and Walia, 1989; Nakatsu and Wyndham, 1993). The organization of catabolic operons in biphenyl, toluene, and naphthalene degradation revealed significant homology (Zylstra and Gibson, 1991). Considerable homology in the sequences of aromatic ring-cleavage dioxygenase genes of phenol, chlorocatechol, catechol, and 2,4-D catabolism has been reported (Kivisaar et al., 1991; Ngai and Ornston, 1988).

Our current knowledge acquired over the last three decades on the mechanisms of aromatic degradation and their genetics, especially elucidation of the regulation of operons responsible for degradation of benzoate, has now made it possible to manipulate genes to increase the ability of dioxygenases to attack a wide array of xenobiotics including aromatic compounds. For instance, *xyl* operon of toluene and xylene degrading plasmid pWWO can be induced by the addition of benzoate or xylene or toluene to the medium, but a derivative of benzoate, 4-ethyl benzoate cannot induce or activate the *xyl* promoter. However, once activated, the same dioxygenase can convert 4-ethyl benzoate to 4-ethyl catechol (Cerdan et al., 1994). The latter being toxic, inhibits the production and action of a ring-cleavage enzyme, catechol 2,3-dioxygenase. This difficulty potentially can be overcome by mutating the target strain *Pseudomonas*, and selecting only those strains or mutants that could accept 4-ethyl catechol.

Another example where manipulation of catabolic genes have yielded fruitful results, is the identification and cloning of toluene dioxygenase genes from *Pseudomonas* and their transfer to a more suitable and convenient environment in *E. coli* (Furukawa et al., 1994; Zylstra et al., 1989). In *Pseudomonas* sp., toluene dioxygenase is represented by four genes *tod* C1, *tod* C2, *tod* B, and *tod* A. The *tod* operon dependent toluene dioxygenase can also degrade trichloroethylene (TCE) to vinyl chloride reductively. This degradation process is slow and also both these compounds are highly carcinogenic. Furthermore, the accumulation of this compound is highly toxic to the bacterium itself. This necessitated cloning of this operon in *E. coli* with its fusion with *tac* promoter, which resulted in the formation of recombinant *E. coli* strains that could degrade TCE and vinyl chloride more efficiently and without any side effects. This example illustrates that in some cases even after modification of dioxygenases, the metabolite produced could be toxic and block further degradation of the xenobiotics and therefore necessitates generating chimeric operons. Therefore, such problem once identified could also be effectively tackled by recombinant DNA technique available to us. A particular operon like *tod* can be expressed better in a different background of *E. coli*. The *tod* opeon has been further manipulated by Furukawa et al. (1994) by fusing the functional gene clusters of the *tod* with *bph* (biphenyl) operon. The new *tod–bph* operon constructed in this way was expressed in *E. coli*. The *E. coli* cells containing such twin operon degraded TCE much faster than original *tod* or *bph* operons (Furukawa et al., 1994). This is, in fact, one of the classical discoveries on application of genetic manipulation of catabolic genes in aromatic degradation.

Use of catechol 2,3-dioxygenase specific primers has recently been made to serve as a basis for identifying and qualifying bacteria in the petroleum contaminated soil (Mesarch et al., 2000). Direct field applications of molecular techniques for detecting and enumerating microbial genes involved in aromatic degradation are powerful tools for monitoring bioremediation and developing field evidence in support of natural attenuation (Mesarch et al., 2000). Parales et al (2000) have elaborately studied the three-dimensional structure of NDO enzyme system which carries out the first step in the aerobic degradation of naphthalene by *Pseudomonas* sp. Their study revealed that the hydrophobic amino acids at the active site of the oxygenase are necessary for the enzyme's preference for aromatic hydrocarbon substrates. Furthermore, the enzyme, NDO is highly region-specific even though it catalyzes *cis* dihydroxylation of a wide range of substrates. Their study reinforces that site-directed mutagenesis has led to the finding that several amino acid substitutions at the active site have little or no effect on product formation with naphthalene or biphenyl as substrates, but have significant effects on product formation from phenanthrene. A comparison of the two extensively studied PCB degraders, *Pseudomonas pseudoalcaligenes* KF 707 and *Burkholderia cepacia* LB 400 revealed that they differ with respect to their range of substrate specificity. KF 707 can catabolize seven *para* substituted PCBs while LB 400 can metabolize PCBs having a maximum of four chlorines. The two dioxygenases from the two bacterial systems with significant difference in their substrate preference, are otherwise nearly identical in their amino acid sequences. The site-directed mutagenesis has been useful in enhancing the structural range of dioxygenases (Erickson and Mondello, 1993). The studies revealed that about 20 residues of amino acids at the carboxyl terminal are involved in the recognition of chlorinated PCBs. Later, a hybrid strain KF 707-D34, which has the required capability to degrade a wide range of PCBs, was constructed. This strain retained its ability to oxidize 4—4′-chlorbenzene (CB) via 2,3-dioxygenation and acquired a novel activity to degrade 2,4,5-CB and 2,5,2′,5′-CB via 3,4 dioxygenation. This novel activity appears to be a consequence of a single amino acid change of threonine to aspartic acid at position 376 (Suenaga et al., 2002). Similarly, using DNA shuffling between the bphA1 gene of KF 707 and LB 400, Bruehlmann and Chen (1999) have created a novel oxygenase capable of recognizing both 2- and 4-substituted PCBs. In these cases, the common amino acid substitution was Ala in place of The 335 and Ile in place of Phe 336.

Mondello et al. (1997) based on an extensive comparison, identified four regions in which specific sequences were found to be associated with PCB substrate specificity. Other workers have also created hybrid dioxygenases by exchanging alpha- and beta-subunits between several PCB degrading strains (Hurtubise et al., 1998; Chebrou et al., 1999). Table 7.1 lists some of the important work done on oxygenases.

The bph gene cluster was very similar in *P. aeruginosa* KF 702 and *P. pseudoalcaligenes* KF 707 (Fujihara et al., 2015). However, Shumkova et al. (2015) identified a difference in the organization of the bph gene cluster in *Rhodococcus ruber*

P25 and other bph gene clusters known thus far. Though this article primarily highlights aerobic oxygenases, it is important to note here that Chakraborty and Das (2016a,b) demonstrated a marine dioxygenase possessor, *P. aeruginosa* JP-11 being able to utilize 98.86% and 2.29% of biphenyl within 72 hours and further an increased biphenyl stress led to 43.5-fold upregulation of bphA gene. A complete genome sequence of the PCB degrading *Rhodococcus* sp. WB1 was studied by Xu et al. (2016) and was found the gene sequence to occur as bphB, bphA3A2A1, bphC, bphD, and bphA4 along with other xenobiotic metabolic pathways and many related genes, such as PCBs degradation, metabolism of nicotinate and nicotinamide, degradation of polycyclic aromatic hydrocarbon (PAH), toluene and nitrotoluene, and atrazine. Bacterium degrading PCBs (*P. putida* KF715) was also found to contain salicylate catabolic *sal* genes in addition to *bph* and termed as bphsal element (Suenaga et al., 2017). This element consisted of the biphenyl metabolism cluster encoded on plasmid as bphR1A1A2A3A4BC, allowing bphsal element to spread by horizontal transfer between strains of KF in the natural environment (Suenaga et al., 2017). Another bacterium, *Pseudomonas alcaliphila* JAB1, was found to have bph genes similar to that of *P. pseudoalcaligenes* KF 707, however it lacks open reading frame possession in the bph operon for benzoate terminal dioxygenase, which is indispensable for the completion of biphenyl degradation (Ridl et al., 2018). Rather, the latter is found somewhere else in the JAB1 genome.

Homologies and attributes of catabolic genes of above mentioned enzymes and operons led biologists to design PCR-based primers to study their presence in the environment. Degenerate primers and PCR have been used to isolate a portion of the NDO ISP gene from strains of *Neptunomonas naphthovorans* and the ISP deduced amino acid sequences were found to be 97.6% homologous with that of *Pseudomonas* and *Burkholderia* strains (Hedlund et al., 1999).

Studies on the genetic aspects of some well-understood and important catabolic genes, such as BD, DDT 2,3-dioxygenase, and angular dioxygenase have been made (Chakraborty and Das, 2016a,b) which suggest degradation of biphenyl, organochlorine pesticides (OCPs), and dioxins/furans, respectively. Furthermore, some recent techniques like sequence and function-based screening into the metagenomic database are proficient in tracing out novel catabolic genes for utilization in enhanced biodegradation. Qu et al. (2015) reported a novel *Chryseobacterium* sp., which is able to degrade various OCP in different amounts, while enabling a complete degradation of DDT in highly contaminated soil. Qureshi et al. (2009) have studied on in situ bioremediation of OCP contaminated microcosm soil and evaluated the presence of microbial genes using gene probes. They found predominance of catabolic genes tfdC (catechol 1,2-dioxygenase) and cm genes (chlorophenol monoxygenase) in the samples.

7.4 Dioxygenase from organisms other than *Pseudomonas*

Multiple ring-hydroxylating oxygenase (RHO) genes have been found to occur in *Sphingobium* sp. PNB (Khara et al., 2014). *Arthrobacter* sp. known as a versatile aromatic degrader, was found to degrade seven aromatic compounds and the genetic basis encoded were further studied (Ren et al., 2018). A novel mechanism of benzene ring cleavage has been found to occur in the mediation of angular dioxygenase in addition to the ring-cleavage dioxygenase (Cai et al., 2017). A fresh water toluene degrading isolate, *Runella* sp. ABRDSP2, was found to possess many aromatic hydrocarbon degrading genes, such as monooxygenase, ring-cleaving dioxygenase, and catechol 1,2-dioxygenase (Kang et al., 2019). A rapid degradation of phenol is reported by possession of diverse dioxygenases (Tian et al., 2017), which might be worth trying for bioremediation purposes. *Paraburkholderia* strain BN5, capable of degrading naphthalene, BTEX, and short chain aliphatic hydrocarbons, possesses a genome encoding 29 monooxygenases and thus acquires diverse organic compounds degradation capabilities (Lee et al., 2019).

7.5 Ethical issues pertaining to bioremediation trials

A clear understanding of possible harm due to incorporation of genetically modified organisms (GMOs) at the bioremediation sites while evaluating their risks against the probable benefits, is a prerequisite for bioremediation trials. The GMOs are applied in bioremediation process for effective removal of contaminants that are not able to get degraded effectively by indigenous microbes. GMOs play a vital role in getting rid of industrial wastes, reduce the toxicity of some hazardous compounds, and have been immensely useful in removing pollution by hydrocarbons and petrol discharges. A variety of molecular tools, which include molecular cloning to gene transfer through conjugation and other methods, are available for the successful construction of GMOs. These GMOs can remove compounds such as xylene, toluene, octane, naphthalene, and salicylate when these traits are coded on their plasmids. A recent book by Kumar et al. (2018) extensively explains the pros and cons, ethical issues, and laws governing the application of GMOs. Essentially, proper laws need to be followed when GMOs are subjected to in situ applications at a site within our ecosystem. The importance of the risk assessment, management, and precautionary approaches in application of GMOs should be candidly analyzed prior to decision making regarding the use of GMOs.

7.6 Conclusion

Microbial aromatic oxygenases reveal significant homologies in structure and a small difference in the amino acid sequence can result in remarkable changes in the substrate specificities and catalytic activities. Their homologies have also been found at the genetic level. Since dioxygenases mediate the very first step in the initial conversion of many aromatic compounds, they are very crucial in being picked up for gene manipulations for performance enhancement in bioremediation trials. Novel dioxygenases are being created by employing recombinant DNA techniques like site-directed mutagenesis, DNA shuffling, domain exchange, etc., primarily targeting such enzymes. Bacterial strains carrying these novel dioxygenases together with complete catabolic pathways are subjected to field applications for bioremediation studies. Logically, it is assumed that bacteria possessing catabolic oxygenases with high specific activities will have a better chance to stay active in the environment and hence can effect an efficient bioremediation process. This promotes the idea of a continuous hunt for such bacteria from nature, conducting an adequate enzyme induction studies and their proper gene manipulations for further improvement. The recent advent of CRISPER–Cas9 technology combined with others may have better promise for paving the way for an efficient bioremediation by genome editing of the bacteria with the slightest ability to degrade aromatic compounds.

Acknowledgments

The authors are thankful to the technical staffs of Bangalore Bio Innovation Center for providing assistance in manuscript preparation.

Disclosure statement

Authors disclose no conflict of interest.

References

Abo-State, M.A.M., Riad, B.Y., Bakr, A.A., Abdel Aziz, M.F., 2018. Biodegradation of naphthalene by *Bordetella avium* isolated from petroleum refinery wastewater in Egypt and its pathway. J. Radiat. Res. Appl. Sci. 11 (1), 1–9. Available from: https://doi.org/10.1016/j.jrras.2017.10.001.

Ajao, A., Kannan, M., Yakubu, S., Vj, U., Jb, A., 2012. Homology modeling, simulation and molecular docking studies of catechol-2,3-dioxygenase from *Burkholderia cepacia*: involved in degradation of petroleum hydrocarbons. Bioinformation 8 (18), 848–854. Available from: https://doi.org/10.6026/97320630008848.

Arora, P.K., Bae, H., 2014. Bacterial degradation of chlorophenols and their derivatives. Microb. Cell Fact. 13 (1), 31. Available from: https://doi.org/10.1186/1475-2859-13-31.

Arora, P.K., Kumar, M., Chauhan, A., Raghava, G.P., Jain, R.K., 2009. OxDBase: a database of oxygenases involved in biodegradation. BMC Res. Notes 2, 67. Available from: https://doi.org/10.1186/1756-0500-2-67.

Assinder, S.J., Williams, P.A., 1990. The TOLplasmids: determinants of the catabolism oftoluene and the xylenes. Adv. Microbiol. Phys 31, 1.

Broadus, R.M., Haddock, J.D., 1998. Purification and characterization of the NADH: ferredoxinBPH oxidoreductase component of biphenyl 2,3-dioxygenase from *Pseudomonas* sp. strain LB400. Arch. Microbiol. 170 (2), 106–112.

Bruehlmann, F., Chen, W., 1999. Transformation of polychlorinated biphenyls by a novel BphA variant through the *meta*-cleavage pathway. FEMS Microbiol. Lett. 179 (2), 203–208. Available from: https://doi.org/10.1111/j.1574-6968.1999.tb08728.x.

Burlage, R.S., Bemis, L.A., Layton, A.C., Sayler, G.S., Larimer, F., 1990. Comparative genetic organization of incompatibility group P degradative plasmids. J. Bacteriol. 172 (12), 6818–6825. Available from: https://doi.org/10.1128/jb.172.12.6818-6825.1990.

Cafaro, V., Notomista, E., Capasso, P., Di Donato, A., 2005. Mutation of glutamic acid 103 of toluene *o*-xylene monooxygenase as a means to control the catabolic efficiency of a recombinant upper pathway for degradation of methylated aromatic compounds. Appl. Environ. Microbiol. 71 (8), 4744–4750. Available from: https://doi.org/10.1128/aem.71.8.4744-4750.2005.

Cai, S., Chen, L.-W., Ai, Y.-C., Qiu, J.-G., Wang, C.-H., Shi, C., et al., 2017. Degradation of diphenyl ether in *Sphingobium phenoxybenzoativorans* SC_3 is initiated by a novel ring cleavage dioxygenase. Appl. Environ. Microbiol. 83, e00104–e00117.

Carredano, E., Karlsson, A., Kauppi, B., Choudhury, D., Parales, R.E., Parales, J.V., et al., 2000. Substrate binding site of naphthalene 1,2-dioxygenase: functional implications of indole binding. J. Mol. Biol. 296 (2), 701–712. Available from: https://doi.org/10.1006/jmbi.1999.3462.

Cerdan, P., Wasserfallen, A., Rekik, M., Timmis, K.N., Harayama, S., 1994. Substrate specificity of catechol 2,3-dioxygenase encoded by TOL plasmid pWW0 of *Pseudomonas putida* and its relationship to cell growth. J. Bacteriol. 176 (19), 6074–6081. Available from: https://doi.org/10.1128/jb.176.19.6074-6081.1994.

Chakrabarty, A.M., 1972. Genetic basis of the biodegradation of salicylate in *Pseudomonas*. J. Bacteriol. 112 (2), 815–823. Retrieved from: <https://www.ncbi.nlm.nih.gov/pubmed/4628746>. <https://www.ncbi.nlm.nih.gov/pmc/articles/PMC251491/>.

Chakraborty, J., Das, S., 2016a. Characterization of the metabolic pathway and catabolic gene expression in biphenyl degrading marine bacterium *Pseudomonas aeruginosa* JP-11. Chemosphere 144, 1706–1714. Available from: https://doi.org/10.1016/j.chemosphere.2015.10.059.

Chakraborty, J., Das, S., 2016b. Molecular perspectives and recent advances in microbial remediation of persistent organic pollutants. Environ. Sci. Pollut. Res. Int. 23, 16883–16903.

Chebrou, H., Hurtubise, Y., Barriault, D., Sylvestre, M., 1999. Heterologous expression and characterization of the purified oxygenase component of *Rhodococcus globerulus* P6 biphenyl dioxygenase and of chimeras derived from it. J. Bacteriol. 181 (16), 4805–4811.

Chemerys, A., Pelletier, E., Cruaud, C., Martin, F., Violet, F., Jouanneau, Y., 2014. Characterization of novel polycyclic aromatic hydrocarbon dioxygenases from the bacterial metagenomic DNA of a contaminated soil. Appl. Environ. Microbiol. 80 (21), 6591–6600. Available from: https://doi.org/10.1128/aem.01883-14.

Chen, Q., Tu, H., Luo, X., Zhang, B., Huang, F., Li, Z., et al., 2016. The regulation of *para*-nitrophenol degradation in *Pseudomonas putida* DLL-E4. PLoS One 11 (5), e0155485. Available from: https://doi.org/10.1371/journal.pone.0155485.

Dagley, S., 1986. Biochemistry of aromatic hydrocarbon degradation in *Pseudomonas*. In: Sokatch, J.R. (Ed.), The Bacteria, vol X. The Biology of Pseudomonas. Academic Press, London, pp. 527–556.

Dangi, A.K., Sharma, B., Hill, R.T., Shukla, P., 2018. Bioremediation through microbes: systems biology and metabolic engineering approach. Crit. Rev. Biotechnol. 1–20. Available from: https://doi.org/10.1080/07388551.2018.1500997.

Deveryshetty, J., Phale, P.S., 2009. Biodegradation of phenanthrene by *Pseudomonas* sp. strain PPD: purification and characterization of 1-hydroxy-2-naphthoic acid dioxygenase. Microbiology 155 (Pt 9), 3083–3091. Available from: https://doi.org/10.1099/mic.0.030460-0.

Domaradzka, D., Guzik, U., Hupert-Kocurek, K., Wojcieszynska, D., 2015. Cometabolic degradation of naproxen by *Planococcus* sp. strain S5. Water Air Soil Pollut. 226 (9), 297. Available from: https://doi.org/10.1007/s11270-015-2564-6.

Dunn, N.W., Gunsalus, I.C., 1973. Transmissible plasmid coding early enzymes of naphthalene oxidation in *Pseudomonas putida*. J. Bacteriol. 114 (3), 974–979. Retrieved from: <https://www.ncbi.nlm.nih.gov/pubmed/4712575>. <https://www.ncbi.nlm.nih.gov/pmc/articles/PMC285353/>.

Enguita, F.J., Leitão, A.L., 2013. Hydroquinone: environmental pollution, toxicity, and microbial answers. Biomed. Res. Int. 2013, 14. Available from: https://doi.org/10.1155/2013/542168.

Erickson, B.D., Mondello, F.J., 1992. Nucleotide sequencing and transcriptional mapping of the genes encoding biphenyl dioxygenase, a multicomponent polychlorinated-biphenyl-degrading enzyme in *Pseudomonas* strain LB400. J. Bacteriol. 174 (9), 2903–2912. Available from: https://doi.org/10.1128/jb.174.9.2903-2912.1992.

Erickson, B.D., Mondello, F.J., 1993. Enhanced biodegradation of polychlorinated biphenyls after site-directed mutagenesis of a biphenyl dioxygenase gene. Appl. Environ. Microbiol. 59 (11), 3858–3862. Retrieved from: <https://www.ncbi.nlm.nih.gov/pubmed/8285689>. <https://www.ncbi.nlm.nih.gov/pmc/articles/PMC182541/>.

Fujihara, H., Yamazoe, A., Hosoyama, A., Suenaga, H., Kimura, N., Hirose, J., et al., 2015. Draft genome sequence of *Pseudomonas aeruginosa* KF702 (NBRC 110665), a polychlorinated biphenyl-degrading bacterium isolated from biphenyl-contaminated soil. Genome Announc. 3 (3). Available from: https://doi.org/10.1128/genomeA.00517-15.

Furukawa, K., Miyazaki, T., 1986. Cloning of a gene cluster encoding biphenyl and chlorobiphenyl degradation in *Pseudomonas pseudoalcaligenes*. J. Bacteriol. 166 (2), 392–398. Available from: https://doi.org/10.1128/jb.166.2.392-398.1986.

Furukawa, K., Hirose, J., Hayashida, S., Nakamura, K., 1994. Efficient degradation of trichloroethylene by a hybrid aromatic ring dioxygenase. J. Bacteriol. 176 (7), 2121–2123. Available from: https://doi.org/10.1128/jb.176.7.2121-2123.1994.

Ghosal, D., Ghosh, S., Dutta, T.K., Ahn, Y., 2016. Current state of knowledge in microbial degradation of polycyclic aromatic hydrocarbons (PAHs): a review. Front. Microbiol. 7, 1369. Available from: https://doi.org/10.3389/fmicb.2016.01369.

Gibson, D.T., Parales, R.E., 2000. Aromatic hydrocarbon dioxygenases in environmental biotechnology. Curr. Opin. Biotechnol. 11, 236–243.

Hartnett, C., Neidle, E.L., Ngai, K.L., Ornston, L.N., 1990. DNA sequences of genes encoding *Acinetobacter calcoaceticus* protocatechuate 3,4-dioxygenase: evidence indicating shuffling of genes and of DNA sequences within genes during their evolutionary divergence. J. Bacteriol. 172 (2), 956–966. Available from: https://doi.org/10.1128/jb.172.2.956-966.1990.

Heald, S.C., Jenkins, R.O., 1996. Expression and substrate specificity of the toluene dioxygenase of *Pseudomonas putida* NCIMB 11767. Appl. Microbiol. Biotechnol. 45 (1–2), 56–62.

Hedlund, B.P., Geiselbrecht, A.D., Bair, T.J., Staley, J.T., 1999. Polycyclic aromatic hydrocarbon degradation by a new marine bacterium, *Neptunomonas naphthovorans* gen. nov., sp. nov. Appl. Environ. Microbiol. 65 (1), 251. Retrieved from: <http://aem.asm.org/content/65/1/251.abstract>.

Hofer, B., Backhaus, S., Timmis, K.N., 1994. The biphenyl/polychlorinated biphenyl-degradation locus (bph) of *Pseudomonas* sp. LB400 encodes four additional metabolic enzymes. Gene 144 (1), 9–16. Available from: https://doi.org/10.1016/0378-1119(94)90196-1.

Hurtubise, Y., Barriault, D., Sylvestre, M., 1998. Involvement of the terminal oxygenase beta subunit in the biphenyl dioxygenase reactivity pattern toward chlorobiphenyls. J. Bacteriol. 180 (22), 5828–5835. Retrieved from: <https://www.ncbi.nlm.nih.gov/pubmed/9811638>. <https://www.ncbi.nlm.nih.gov/pmc/articles/PMC107654/>.

Jaiswal, S., Singh, D.K., Shukla, P., 2019. Gene editing and systems biology tools for pesticide bioremediation: a review. Front. Microbiol. 10 (87). Available from: https://doi.org/10.3389/fmicb.2019.00087.

Jiang, H., Parales, R.E., Gibson, D.T., 1999. The alpha subunit of toluene dioxygenase from *Pseudomonas putida* F1 can accept electrons from reduced FerredoxinTOL but is catalytically inactive in the absence of the beta subunit. Appl. Environ. Microbiol. 65 (1), 315–318. Retrieved from: <https://www.ncbi.nlm.nih.gov/pubmed/9872799>. <https://www.ncbi.nlm.nih.gov/pmc/articles/PMC91022/>.

Jin, H.M., Kim, J.M., Lee, H.J., Madsen, E.L., Jeon, C.O., 2012. Alteromonas as a key agent of polycyclic aromatic hydrocarbon biodegradation in crude oil-contaminated coastal sediment. Environ. Sci. Technol. 46 (14), 7731–7740. Available from: https://doi.org/10.1021/es3018545.

Johri, A.K., Dua, M., Singh, A., Sethunathan, N., Legge, R.L., 1999. Characterization and regulation of catabolic genes. Crit. Rev. Microbiol. 25 (4), 245–273. Available from: https://doi.org/10.1080/10408419991299248.

Jones, A.L., Brown, J.M., Mishra, V., Perry, J.D., Steigerwalt, A.G., Goodfellow, M., 2004. *Rhodococcus gordoniae* sp. nov., an actinomycete isolated from clinical material and phenol-contaminated soil. Int. J. Syst. Evol. Microbiol. 54 (Pt 2), 407–411. Available from: https://doi.org/10.1099/ijs.0.02756-0.

Jouanneau, Y., Meyer, C., Jakoncic, J., Stojanoff, V., Gaillard, J., 2006. Characterization of a naphthalene dioxygenase endowed with an exceptionally broad substrate specificity toward polycyclic aromatic hydrocarbons. Biochemistry 45 (40), 12380–12391. Available from: https://doi.org/10.1021/bi0611311.

Junca, H., Plumeier, I., Hecht, H.J., Pieper, D.H., 2004. Difference in kinetic behaviour of catechol 2,3-dioxygenase variants from a polluted environment. Microbiology 150 (Pt 12), 4181–4187. Available from: https://doi.org/10.1099/mic.0.27451-0.

Kamei, I., Kogura, R., Kondo, R., 2006. Metabolism of 4,4′-dichlorobiphenyl by white-rot fungi *Phanerochaete chrysosporium* and *Phanerochaete* sp. MZ142. Appl. Microbiol. Biotechnol. 72 (3), 566–575. Available from: https://doi.org/10.1007/s00253-005-0303-4.

Kang, H.K., Ryu, B.G., Choi, K.M., Jin, H.M., 2019. Complete genome sequence of *Runella* sp. ABRDSP2, a new mono-aromatic compounds degrading bacterium isolated from freshwater. Korean J. Microbiol. 55, 55–57.

Kasak, L., Horak, R., Nurk, A., Talvik, K., Kivisaar, M., 1993. Regulation of the catechol 1,2-dioxygenase- and phenol monooxygenase-encoding pheBA operon in *Pseudomonas putida* PaW85. J. Bacteriol. 175 (24), 8038–8042. Available from: https://doi.org/10.1128/jb.175.24.8038-8042.1993.

Kauppi, B., Lee, K., Carredano, E., Parales, R.E., Gibson, D.T., Eklund, H., et al., 1998. Structure of an aromatic-ring-hydroxylating dioxygenase-naphthalene 1,2-dioxygenase. Structure 6 (5), 571–586.

Khan, A., Walia, S., 1989. Cloning of bacterial genes specifying degradation of 4-chlorobiphenyl from *Pseudomonas putida* OU83. Appl. Environ. Microbiol. 55 (4), 798–805.

Khara, P., Roy, M., Chakraborty, J., Ghosal, D., Dutta, T.K., 2014. Functional characterization of diverse ring-hydroxylating oxygenases and induction of complex aromatic catabolic gene clusters in *Sphingobium* sp. PNB. FEBS Open Bio 4, 290–300.

Kikuchi, Y., Nagata, Y., Hinata, M., Kimbara, K., Fukuda, M., Yano, K., et al., 1994. Identification of the bphA4 gene encoding ferredoxin reductase involved in biphenyl and polychlorinated biphenyl degradation in *Pseudomonas* sp. strain KKS102. J. Bacteriol. 176 (6), 1689–1694. Available from: https://doi.org/10.1128/jb.176.6.1689-1694.1994.

Kimura, N., Nishi, A., Goto, M., Furukawa, K., 1997. Functional analyses of a variety of chimeric dioxygenases constructed from two biphenyl dioxygenases that are similar structurally but different functionally. J. Bacteriol. 179 (12), 3936–3943. Available from: https://doi.org/10.1128/jb.179.12.3936-3943.1997.

Kivisaar, M., Kasak, L., Nurk, A., 1991. Sequence of the plasmid-encoded catechol 1,2-dioxygenase-expressing gene, pheB, of phenol-degrading *Pseudomonas* sp. strain EST1001. Gene 98 (1), 15–20. Available from: https://doi.org/10.1016/0378-1119(91)90098-v.

Kohmueller, N.A., Howard, J.B., 1979. The primary structure of the alpha subunit of protocatechuate 3,4-dioxygenase. II. Isolation and sequence of overlap peptides and complete sequence. J. Biol. Chem. 254 (15), 7309–7315.

Kolvenbach, B.A., Lenz, M., Benndorf, D., Rapp, E., Fousek, J., Vlcek, C., et al., 2011. Purification and characterization of hydroquinone dioxygenase from *Sphingomonas* sp. strain TTNP3. AMB Express 1 (1), 8. Available from: https://doi.org/10.1186/2191-0855-1-8.

Kumar, N.M., Muthukumaran, C., Sharmila, G., Gurunathan, B., 2018. Genetically modified organisms and its impact on the enhancement of bioremediation. In: Varjani, S.J., Agarwal, A.K., Gnansounou, E., Gurunathan, B. (Eds.), Bioremediation: Applications for Environmental Protection and Management. Springer, Singapore.

Kumari, A., Singh, D., Ramaswamy, S., Ramanathan, G., 2017. Structural and functional studies of ferredoxin and oxygenase components of 3-nitrotoluene dioxygenase from *Diaphorobacter* sp. strain DS2. PLoS One 12 (4), e0176398. Available from: https://doi.org/10.1371/journal.pone.0176398.

Lal, R., Lal, S., Dhanaraj, P.S., Saxena, D.M., 1995. Manipulations of catabolic genes for the degradation and detoxification of xenobiotics. Adv. Appl. Microbiol. 41, 55–95.

Leungsakul, T., Johnson, G.R., Wood, T.K., 2006. Protein engineering of the 4-methyl-5-nitrocatechol monooxygenase from *Burkholderia* sp. strain DNT for enhanced degradation of nitroaromatics. Appl. Environ. Microbiol. 72 (6), 3933–3939. Available from: https://doi.org/10.1128/aem.02966-05.

Lee, Y., Lee, Y., Jeon, C.O., 2019. Biodegradation of naphthalene, BTEX, and aliphatic hydrocarbons by *Paraburkholderia aromaticivorans* BN5 isolated from petroleum-contaminated soil. Sci. Rep. 9 (1), 860.

Lloyd-Jones, G., de Jong, C., Ogden, R.C., Duetz, W.A., Williams, P.A., 1994. Recombination of the bph (biphenyl) catabolic genes from plasmid pWW100 and their deletion during growth on benzoate. Appl. Environ. Microbiol. 60 (2), 691–696.

Maeda, T., Takahashi, Y., Suenaga, H., Suyama, A., Goto, M., Furukawa, K., 2001. Functional analyses of Bph–Tod hybrid dioxygenase, which exhibits high degradation activity toward trichloroethylene. J. Biol. Chem. 276 (32), 29833–29838. Available from: https://doi.org/10.1074/jbc.M102025200.

Mahan, K.M., Penrod, J.T., Ju, K.S., Al Kass, N., Tan, W.A., Truong, R., et al., 2015. Selection for growth on 3-nitrotoluene by 2-nitrotoluene-utilizing *Acidovorax* sp. strain JS42 identifies nitroarene dioxygenases with altered specificities. Appl. Environ. Microbiol. 81 (1), 309–319. Available from: https://doi.org/10.1128/aem.02772-14.

Masai, E., Yamada, A., Healy, J.M., Hatta, T., Kimbara, K., Fukuda, M., et al., 1995. Characterization of biphenyl catabolic genes of gram-positive polychlorinated biphenyl degrader *Rhodococcus* sp. strain RHA1. Appl. Environ. Microbiol. 61 (6), 2079–2085.

Mesarch, M.B., Nakatsu, C.H., Nies, L., 2000. Development of catechol 2,3-dioxygenase-specific primers for monitoring bioremediation by competitive quantitative PCR. Appl. Environ. Microbiol. 66 (2), 678–683. Available from: https://doi.org/10.1128/aem.66.2.678-683.2000.

Meyer, A., Wursten, M., Schmid, A., Kohler, H.P., Witholt, B., 2002. Hydroxylation of indole by laboratory-evolved 2-hydroxybiphenyl 3-monooxygenase. J. Biol. Chem. 277 (37), 34161–34167. Available from: https://doi.org/10.1074/jbc.M205621200.

Min, J., Lu, Y., Hu, X., Zhou, N.Y., 2016. Biochemical characterization of 3-methyl-4-nitrophenol degradation in *Burkholderia* sp. strain SJ98. Front. Microbiol. 7, 791. Available from: https://doi.org/10.3389/fmicb.2016.00791.

Min, J., Wang, J., Chen, W., Hu, X., 2018. Biodegradation of 2-chloro-4-nitrophenol via a hydroxyquinol pathway by a Gram-negative bacterium, *Cupriavidus* sp. strain CNP-8. AMB Express 8 (1), 43. Available from: https://doi.org/10.1186/s13568-018-0574-7.

Min, J., Xu, L., Fang, S., Chen, W., Hu, X., 2019. Molecular and biochemical characterization of 2-chloro-4-nitrophenol degradation via the 1,2,4-benzenetriol pathway in a Gram-negative bacterium. Appl. Microbiol. Biotechnol. . Available from: https://doi.org/10.1007/s00253-019-09994-7.

Mishra, V.A., Lal, R., 2014. Enhanced degradation of phenol by a new species of *Rhodococcus*: *R. gordoniae* through *ortho*-pathway. Int. J. Basic. Appl. Biol. 1, 23–26.

Mishra, V., Lal, R., Srinivasan, 2001. Enzymes and operons mediating xenobiotic degradation in bacteria. Crit. Rev. Microbiol. 27 (2), 133–166. Available from: https://doi.org/10.1080/20014091096729.

Mondello, F.J., Turcich, M.P., Lobos, J.H., Erickson, B.D., 1997. Identification and modification of biphenyl dioxygenase sequences that determine the specificity of polychlorinated biphenyl degradation. Appl. Environ. Microbiol. 63 (8), 3096–3103.

Nakai, C., Horiike, K., Kuramitsu, S., Kagamiyama, H., Nozaki, M., 1990. Three isozymes of catechol 1,2-dioxygenase (pyrocatechase), alpha alpha, alpha beta, and beta beta, from *Pseudomonas arvilla* C-1. J. Biol. Chem. 265 (2), 660–665.

Nakatsu, C.H., Wyndham, R.C., 1993. Cloning and expression of the transposable chlorobenzoate-3,4-dioxygenase genes of *Alcaligenes* sp. strain BR60. Appl. Environ. Microbiol. 59 (11), 3625–3633.

Neidle, E.L., Hartnett, C., Bonitz, S., Ornston, L.N., 1988. DNA sequence of the Acinetobacter calcoaceticus catechol 1,2-dioxygenase I structural gene catA: evidence for evolutionary divergence of intradiol dioxygenases by acquisition of DNA sequence repetitions. J. Bacteriol. 170 (10), 4874–4880. Available from: https://doi.org/10.1128/jb.170.10.4874-4880.1988.

Ngai, K.L., Ornston, L.N., 1988. Abundant expression of *Pseudomonas* genes for chlorocatechol metabolism. J. Bacteriol. 170 (5), 2412–2413. Available from: https://doi.org/10.1128/jb.170.5.2412-2413.1988.

Ning, D., Wang, H., Ding, C., Lu, H., 2010. Novel evidence of cytochrome P450-catalyzed oxidation of phenanthrene in *Phanerochaete chrysosporium* under ligninolytic conditions. Biodegradation 21 (6), 889–901. Available from: https://doi.org/10.1007/s10532-010-9349-9.

Nzila, A., 2018. Current status of the degradation of aliphatic and aromatic petroleum hydrocarbons by thermophilic microbes and future perspectives. Int. J. Environ. Res. Public. Health 15 (12). Available from: https://doi.org/10.3390/ijerph15122782.

Ohlendorf, D.H., Lipscomb, J.D., Weber, P.C., 1988. Structure and assembly of protocatechuate 3,4-dioxygenase. Nature 336 (6197), 403–405. Available from: https://doi.org/10.1038/336403a0.

Parales, R.E., Lee, K., Resnick, S.M., Jiang, H., Lessner, D.J., Gibson, D.T., 2000. Substrate specificity of naphthalene dioxygenase: effect of specific amino acids at the active site of the enzyme. J. Bacteriol. 182 (6), 1641–1649. Available from: https://doi.org/10.1128/jb.182.6.1641-1649.2000.

Phale, P.S., Basu, A., Majhi, P.D., Deveryshetty, J., Vamsee-Krishna, C., Shrivastava, R., 2007. Metabolic diversity in bacterial degradation of aromatic compounds. Omics 11 (3), 252–279. Available from: https://doi.org/10.1089/omi.2007.0004.

Pimviriyakul, P., Thotsaporn, K., Sucharitakul, J., Chaiyen, P., 2017. Kinetic mechanism of the dechlorinating flavin-dependent monooxygenase HadA. J. Biol. Chem. 292 (12), 4818–4832. Available from: https://doi.org/10.1074/jbc.M116.774448.

Qu, J., Xu, Y., Ai, G.M., Liu, Y., Liu, Z.P., 2015. Novel *Chryseobacterium* sp. PYR2 degrades various organochlorine pesticides (OCPs) and achieves enhancing removal and complete degradation of DDT in highly contaminated soil. J. Environ. Manage. 161, 350–357.

Qureshi, A., Mohan, M., Kanade, G.S., Kapley, A., Purohit, H.J., 2009. In situ bioremediation of organochlorine-pesticide-contaminated microcosm soil and evaluation by gene probe. Pest Manage. Sci. 65 (7), 798–804.

Reineke, W., Knackmuss, H.J., 1978. Chemical structure and biodegradability of halogenated aromatic compounds. Substituent effects on dehydrogenation of 3,5-cyclohexadiene-1,2-diol-1-carboxylic acid. Biochim. Biophys. Acta 542 (3), 424–429. Available from: https://doi.org/10.1016/0304-4165(78)90373-2.

Ren, L., Jia, Y., Zhang, R., Lin, Z., Zhen, Z., Hu, H., et al., 2018. Insight into metabolic versatility of an aromatic compounds degrading *Arthrobacter* sp. YC-RL1. Front. Microbiol. 9, 2438.

Ridl, J., Suman, J., Fraraccio, S., Hradilova, M., Strejcek, M., Cajthaml, T., et al., 2018. Complete genome sequence of *Pseudomonas alcaliphila* JAB1 (=DSM 26533), a versatile degrader of organic pollutants. Stand. Genomic Sci. 13, 3. Available from: https://doi.org/10.1186/s40793-017-0306-7.

Shumkova, E.S., Voronina, A.O., Kuznetsova, N.V., Plotnikova, E.G., 2015. [Diversity of the key biphenyl destruction genes in the microbial community of the Anadyr Bay coastal sediments]. Genetika 51 (7), 841–846.

Smets, B.F., Rittmann, B.E., Stahl, D.A., 1993. The specific growth rate of *Pseudomonas putida* PAW1 influences the conjugal transfer rate of the TOL plasmid. Appl. Environ. Microbiol. 59 (10), 3430–3437.

Sridevi, V., Lakshmi, M., Manasa, M.R., Sravani, M.S., 2012. Metabolic pathways for the biodegradation of phenol. J. Eng. Sci. Adv. Technol. 2, 705.

Standfuss-Gabisch, C., Al-Halbouni, D., Hofer, B., 2012. Characterization of biphenyl dioxygenase sequences and activities encoded by the metagenomes of highly polychlorobiphenyl-contaminated soils. Appl. Environ. Microbiol. 78 (8), 2706–2715. Available from: https://doi.org/10.1128/aem.07381-11.

Suenaga, H., Watanabe, T., Sato, M., Ngadiman, Furukawa, K., 2002. Alteration of regiospecificity in biphenyl dioxygenase by active-site engineering. J. Bacteriol. 184 (13), 3682–3688. Available from: https://doi.org/10.1128/jb.184.13.3682-3688.2002.

Suenaga, H., Fujihara, H., Kimura, N., Hirose, J., Watanabe, T., Futagami, T., et al., 2017. Insights into the genomic plasticity of *Pseudomonas putida* KF715, a strain with unique biphenyl-utilizing activity and genome instability properties. Environ. Microbiol. Rep. 9 (5), 589–598. Available from: https://doi.org/10.1111/1758-2229.12561.

Syed, K., Doddapaneni, H., Subramanian, V., Lam, Y.W., Yadav, J.S., 2010. Genome-to-function characterization of novel fungal P450 monooxygenases oxidizing polycyclic aromatic hydrocarbons (PAHs). Biochem. Biophys. Res. Commun. 399 (4), 492–497. Available from: https://doi.org/10.1016/j.bbrc.2010.07.094.

Tao, Y., Fishman, A., Bentley, W.E., Wood, T.K., 2004. Altering toluene 4-monooxygenase by active-site engineering for the synthesis of 3-methoxycatechol, methoxyhydroquinone, and methylhydroquinone. J. Bacteriol. 186 (14), 4705–4713. Available from: https://doi.org/10.1128/jb.186.14.4705-4713.2004.

Tavakoli, A., Hamzah, A., Rabu, A., 2016. Expression, purification and kinetic characterization of recombinant benzoate dioxygenase from *Rhodococcus ruber* UKMP-5M. Mol. Biol. Res. Commun. 5 (3), 133–142.

Teramoto, H., Tanaka, H., Wariishi, H., 2004. Degradation of 4-nitrophenol by the lignin-degrading basidiomycete *Phanerochaete chrysosporium*. Appl. Microbiol. Biotechnol. 66 (3), 312–317. Available from: https://doi.org/10.1007/s00253-004-1637-z.

Tian, M., Du, D., Zhou, W., Zeng, X., Cheng, G., 2017. Phenol degradation and genotypic analysis of dioxygenase genes in bacteria isolated from sediments. Braz. J. Microbiol. 48 (2), 305–313.

Wojcieszynska, D., Guzik, U., Gren, I., Perkosz, M., Hupert-Kocurek, K., 2011. Induction of aromatic ring: cleavage dioxygenases in *Stenotrophomonas maltophilia* strain KB2 in cometabolic systems. World J. Microbiol. Biotechnol. 27 (4), 805–811. Available from: https://doi.org/10.1007/s11274-010-0520-6.

Xu, Y., Yu, M., Shen, A., 2016. Complete genome sequence of the polychlorinated biphenyl degrader *Rhodococcus* sp. WB1. Genome Announc. 4 (5). Available from: https://doi.org/10.1128/genomeA.00996-16.

Zhang, J.J., Liu, H., Xiao, Y., Zhang, X.E., Zhou, N.Y., 2009. Identification and characterization of catabolic *para*-nitrophenol 4-monooxygenase and para-benzoquinone reductase from *Pseudomonas* sp. strain WBC-3. J. Bacteriol. 191 (8), 2703–2710. Available from: https://doi.org/10.1128/jb.01566-08.

Zylstra, G.J., Gibson, D.T., 1991. Aromatic hydrocarbon degradation: a molecular approach. In: Setlow, J.K. (Ed.), Genetic Engineering, vol. 13. Plenum Press, New York, pp. 183–203.

Zylstra, G.J., Wackett, L.P., Gibson, D.T., 1989. Trichloroethylene degradation by *Escherichia coli* containing the cloned *Pseudomonas putida* F1 toluene dioxygenase genes. Appl. Environ. Microbiol. 55 (12), 3162–3166.

Further reading

Arora, P.K., Jain, R.K., 2012. Metabolism of 2-chloro-4-nitrophenol in a Gram negative bacterium, *Burkholderia* sp. RKJ 800. PLoS One 7 (6), e38676. Available from: https://doi.org/10.1371/journal.pone.0038676.

Dorn, E., Knackmuss, H.J., 1978. Chemical structure and biodegradability of halogenated aromatic compounds. Two catechol 1,2-dioxygenases from a 3-chlorobenzoate-grown pseudomonad. Biochem. J. 174 (1), 73–84. Available from: https://doi.org/10.1042/bj1740073.

Fielding, A.J., Lipscomb, J.D., Que Jr., L., 2014. A two-electron-shell game: intermediates of the extradiol-cleaving catechol dioxygenases. J. Biol. Inorg. Chem. 19 (4–5), 491–504. Available from: https://doi.org/10.1007/s00775-014-1122-9.

Hirose, J., Fujihara, H., Watanabe, T., Kimura, N., Suenaga, H., Futagami, T., et al., 2019. Biphenyl/PCB degrading bph genes of ten bacterial strains isolated from biphenyl-contaminated soil in Kitakyushu, Japan: comparative and dynamic features as integrative conjugative elements (ICEs). Genes (Basel) 10 (5). Available from: https://doi.org/10.3390/genes10050404.

Karishma, M., Trivedi, V.D., Choudhary, A., Mhatre, A., Kambli, P., Desai, J., et al., 2015. Analysis of preference for carbon source utilization among three strains of aromatic compounds degrading *Pseudomonas*. FEMS Microbiol. Lett. 362 (20). Available from: https://doi.org/10.1093/femsle/fnv139.

Lau, P.C., Garnon, J., Labbe, D., Wang, Y., 1996. Location and sequence analysis of a 2-hydroxy-6-oxo-6-phenylhexa-2,4-dienoate hydrolase-encoding gene (bpdF) of the biphenyl/polychlorinated biphenyl degradation pathway in *Rhodococcus* sp. M5. Gene 171 (1), 53–57. Available from: https://doi.org/10.1016/0378-1119(96)00025-x.

Resnick, S.M., Torok, D.S., Lee, K., Brand, J.M., Gibson, D.T., 1994. Regiospecific and stereoselective hydroxylation of 1-indanone and 2-indanone by naphthalene dioxygenase and toluene dioxygenase. Appl. Environ. Microbiol. 60 (9), 3323–3328.

Saby, S., Leroy, P., Block, J.C., 1999. *Escherichia coli* resistance to chlorine and glutathione synthesis in response to oxygenation and starvation. Appl. Environ. Microbiol. 65 (12), 5600–5603.

Suenaga, H., Sato, M., Goto, M., Takeshita, M., Furukawa, K., 2006. Steady-state kinetic characterization of evolved biphenyl dioxygenase, which acquired novel degradation ability for benzene and toluene. Biosci. Biotechnol. Biochem. 70 (4), 1021–1025. Available from: https://doi.org/10.1271/bbb.70.1021.

Microbial remediation progress and future prospects

8

Monu Jariyal[1], Manish Yadav[2,], Nitin Kumar Singh[3], Suman Yadav[4], Iti Sharma[5], Swati Dahiya[6] and Arti Thanki[7]*

[1]Regional Centre of Organic Farming, Ministry of Agriculture & Farmers Welfare, Government of India, Ghaziabad, India, [2]Central Mine Planning and design Institute Limited, Bhubaneswar, India, [3]Department of Environmental Science and Engineering, Marwadi University, Rajkot, India, [4]APS University, Rewa, India, [5]Advance Technology Development Centre, Indian Institute of Technology, Kharagpur, India, [6]Department of Civil Engineering, Indian Institute of Technology, Roorkee, India, [7]Department of Environmental Science and Engineering, Marwadi University, Rajkot, India

*Corresponding author

8.1 Introduction

Bioremediation is the process of reducing pollutant complexity through biodegradation in which complex compounds are degraded into simpler compounds by microorganisms (Marinescu et al., 2009). When the process is completed, it is called "biomineralization." However, generally the term biodegradation is used to describe almost any biologically mediated change in the substrate (Bennet et al., 2002). Hence, to get proper understanding of bioremediation, it is important to understand the biological processes. The microorganisms degrade organic pollutants by enzymatic or metabolic pathways, depending upon its growth and cometabolism. In this process, the chemical compound provides the source of carbon and energy to the tiny microorganism. This method results in a complete mineralization or degradation of organic pollutants. Cometabolism is the process of breakdown of pollutants in the presence of a growth substrate which is the primary source of carbon and energy (Fritsche and Hofrichter, 2008). Biodegradation is performed by microorganisms, such as bacteria, fungi, and yeasts whereas, there is less information available regarding the role of algae and protozoa involved in the process of biodegradation (Das and Chandran, 2011). Biodegradation methods vary with different microorganisms and diverse substrate, but often the final product of the process is carbon dioxide (Pramila et al., 2012). Organic pollutants can be transformed with oxygen or without oxygen (Fritsche and Hofrichter, 2008). Biodegradable material is usually organic matter such as plant and animal matter and other materials originating from living organisms, or synthetic materials that are alike to plant and animal matter. Microbes have the amazing natural catabolic diversity to decompose a large number of pollutants comprising of heavy metals, hydrocarbons (e.g., oil), polyaromatic hydrocarbons (PAHs), and polychlorinated biphenyls (PCBs) (Leitão, 2009).

Bioremediation of Pollutants. DOI: https://doi.org/10.1016/B978-0-12-819025-8.00008-9

The word biodegradation is frequently used in relation to ecology, waste management, and generally related with environmental remediation (bioremediation) (Marinescu et al., 2009). Biodegradation can be divided into three phases or levels. First, natural attenuation, pollutants are degraded by native microorganisms without human intervention. Second, utilization of oxygen and nutrients are used to increase their efficacy and to accelerate biodegradation (biostimulation). In the third, special microorganisms are added with the substrate to enhance the biodegradation process which is known as bioaugmentation. These add-on microorganisms must increase efficiency than native flora to transform the target pollutant (Diez, 2010). A realistic remedial strategy needs an adaptive microorganism that would use target pollutants within considerable time (Seo et al., 2009). Many factors such as genetic and certain environmental factors such as temperature, pH, and available nitrogen and phosphorus sources affect microbes to use contaminants as substrates or cometabolize to regulate the rate and the extent of the process (Fritsche and Hofrichter, 2008). Hence, applications of genetically engineered microorganisms (GEMs) in bioremediation have received a great deal of attention. These GEMs have higher degradative ability and have been used effectively for the degradation of various contaminants under defined conditions. However, there are certain concerns such as ecological, environmental, and regulatory rules for using GEMs in the biodegradation field (Menn et al., 2008).

In recent years, developments in next generation sequencing has allowed genomic, metagenomic, and bioinformatics studies of environmentally significant microbes, thereby giving a novel insight into degradative pathways. In the case of oil contaminated sites, microbes isolated from the different hydrocarbon-accumulated sites proposed that there are substantial taxonomic and functional dissimilarity in diverse geographically and spatially isolated sites in a study conducted by Mukherjee et al. (2017). This study provided an unprecedented insight into the ecological dynamics of contaminated sites.

Luo et al. (2014) reported, that microbial diversity varies in accordance to heavy metal contamination in soil. In a study of microbial metagenomes from ice sheets of Greenland, Hauptmann et al. (2017) isolated prospective microbial genes for the degradation and resistance to pollutants. A similar study was done by Joshi et al. (2014) from the metagenomes of petroleum muck. In recent times, there is still extensive ongoing research on individual genomic or transcriptomic methods for biodegradation of contaminated sites (George et al., 2011; Kumavath and Deverapalli, 2013; Bell et al., 2014; Ufarte et al., 2015; Czaplicki and Gunsch, 2016; Chistoserdova, 2017). Earlier remediation used to be carried out using single omics-approach; however, nowadays the integrative role of the multiomics methods is getting more attention in microbial mediated bioremediation.

In the present chapter, ecofriendly microbial remediation approach and various microorganisms that are involved in biodegradation of organic environmental pollutants are discussed. Other processes such as biostimulation and bioaugmentation along with factors governing biodegradation are discussed. The application of genetically modified microorganisms in the field of bioremediation for the removal of contaminants is the next level approach. However, due to several constraints for

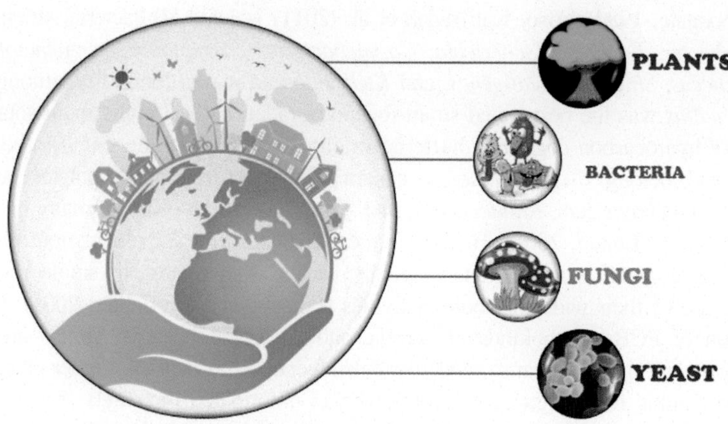

Figure 8.1 Factors responsible for the bioremediation.

using GEMs, this chapter also describes the application of multiomics which provide a better understanding of structural as well as functional characteristics of microbial consortia towards different environmental pollutants (Fig. 8.1).

8.2 Microorganisms associated with bioremediation

Biodegradation is the microorganisms-mediated biological process for the transformation of complex and harmful organic pollutants into nontoxic chemicals which can be used as a source of nutrients for metabolic pathways of the microbial community. A variety of bacterial population, yeast, and fungi can decompose or biodegrade the organic materials for their metabolic pathways. Several pollutants such as heavy metals, polyaromatic hydrocarbons (PAHs), hydrocarbons, and PCBs can be decomposed by catabolic activities of microbes by biotransformation and bioremediation methods.

8.2.1 Bacterial degradation

Different classes of bacteria are involved in the remediation of environmental contaminants using biological processes. There are number of bacteria that are specific to hydrocarbon degradation (Yakimov et al., 2007), they are generally known as hydrocarbon degrading bacteria and bacteria strains such as *Pseudomonas* sp. and *Brevibacillus* sp. which are isolated from petroleum contaminated soil and used for nitrate degradation. The process of bioremediation can occur aerobically or anaerobically that is with or without oxygen (Grishchenkov et al., 2000). Wiedemeier et al. (1995) reported that anaerobic biodegradation is also an effective method that enables us to harness resources and energy from the waste.

There are 25 genera of hydrocarbon degrading bacteria isolated from marine environment (Floodgate, 1984). Also, Kafilzadeh et al. (2011) isolated 80 bacterial strains such as *Enterobacter, Bacillus, Escherichia, Corynebacterium, Streptococcus, Acinetobacter, Staphylococcus, Shigella, Alcaligenes*, and *Klebsiella* and it was found that among these strains, *Bacillus* was the best suited strain for biodegradation. Soil is the main source for isolation of hydrocarbon degrading bacterial strains; most of them are found to be *Gram negative* and belong to *pseudomonas* genera. Generally microbial isolates such as *Bacillus, Corynebacterium, Rhodococcus*, and *Aermonas* are involved in many biodegradation processes (Leitão, 2009). However, microbial consortia are much more effective in degrading complex contaminants in polluted sites as compared to single microbial strain because of their wide metabolic activities (Fritsche and Hofrichter, 2005). For the degradation of PCBs, both kinds of aerobic and anaerobic bacterial strains are used. Anaerobic microorganisms cause reductive dehalogenation to degrade higher chlorinated PCBs and aerobic microorganisms oxidized lower chlorinated biphenyls (Seeger et al., 2001). Extensive research has been done on isolation of aerobic bacteria primarily on *Gram-negative* strains mainly belonging to genera such as *Ralstonia, Comamonas, Pseudomonas, Sphingomonas, Achromobacter*, and *Burkholderia.* Though, it has been found in several case studies that a large number of *Gram-positive* strains such *Bacillus, Microbacterium, Janibacter, Rhodococcus*, and *Paenibacillus* are also effective in PCB degradation on contaminated sites (Petric et al., 2007).

PCB degradation is usually a four step degradation process in the case of most aerobic bacteria which involves enzymes such as biphenyl dioxygenase (BphA), dihydrodiol dehydrogenase (BphB), 2, 3-dihydroxybihenyl dioxygenase (DHBD) (BphC), and hydrolase (BphD) (Taguchi et al., 2001).

Pesticides such as atrazine have also been successfully alleviated by bacteria mediated degradation (Struthers et al., 1998). Different bacterial strains such as *Stenotrophomonas, Bacillus, Providencia stuartii*, and *Staphylococcus* which have been isolated from agricultural and nonagricultural land, are able to degrade pesticides such as chlorpyrifos (Rani et al., 2008) and dichlorodiphenyltrichloroethane (Kanade et al., 2012).

There are numerous reports on removing azo dyes by aerobic and anaerobic bacterial strains (Dos Santos et al., 2007). And it has been found that aerobic bacteria are much more effective in transformation of azo dyes as compared to anaerobic bacteria which offend forming azo dyes into aromatic amine, which is toxic and harmful to humans and thus requires an aerobic degradation at later stages. Thus for an effective degradation of dye contaminated wastewater, a combination of aerobic and anaerobic strategy is required (Lodato et al., 2007; Chaube et al., 2010).

In many case studies, a pure culture of bacterium known as *Shewanella decolorations* has been used effectively for the removal of azo dyes (Hong et al., 2007). Pure culture has many advantages as compared to mixed consortia which can provide a better understanding of catabolic pathways with effective degradation and formation of nontoxic products in better controlled conditions. And by using molecular and microbial cultural techniques their activity can be observed which is helpful to identify bacterial density in a certain time period which would eventually enable to predict the biodegradation kinetic analysis (Khalid et al., 2010).

Heavy metals are difficult to degrade biologically, however, they can be converted into another chemical form by microorganism which can be easily processed further (Garbisu and Alkorta, 2001). Also, bacteria have been found to be most effective in the case of heavy metal degradation.

In the case of heavy metal toxicity, microbes have the capacity to protect them by adapting a number of strategies such as reduction, methylation, oxidation, uptake, and adsorption. Dissimilatory reactions can be used to reduce metals (Fernández et al., 2012). In the case of anaerobic respiration of bacteria, metals can be used as terminal electron acceptors. Besides these, bacteria are able to transform metals by reductive mechanisms which have been reported in several case studies. As methylated compounds are volatile in nature, it holds major significance in heavy metal degradation through methylation process. Such as, a number of bacterial strains have been reported to transform mercury, Hg(II) to gaseous methyl mercury (De et al., 2008). Environmental contaminated sites with high concentrations of Cd, Co, As, Zn, and Cu are easily recovered by bacterial strains like sulfur oxidizing bacteria (White et al., 1998) and *Acidithiobacillus ferrooxidans* (Takeuchi and Sugio, 2006). Sulfate reducing bacteria are also found to be effective in precipitation of heavy metals as insoluble sulfides (white et al., 1998). They are anaerobic heterotrophs using organic substrates with SO_4^{-2} as the terminal electron acceptor. Adsorption and entrapment are also the most common methods of heavy metal degradation by bacterial cells. This is also known as passive uptake and it is independent on microbial catabolic activities. Whereas, during active uptake of heavy metals, it is transported through cell membrane and metabolized inside the cell. *Pseudomonas* strains have been extensively studied for the removal of heavy metals such as uranium from the contaminated sites (Sar et al., 2004). Most of the research have been carried out by using single microbial strain, however, there are numerous benefits of using microbial consortia. Bioremediation is the best example of showing effective efficiency of mixed microbial strains. Adarsh et al. (2007) have used mixed microbial consortia for the effective recovery of heavy metal contaminated environmental sites. It has been observed that there are better chances of bacteria growth in microbial consortia as compared to pure culture for removing metals in heavily contaminated sites (Joutey et al., 2011). The bacteria, which have been isolated from the enrichment culture techniques, are able to degrade multiple metals in a contaminated site (Joutey et al., 2011). It is reported in a study that, *Pseudomonas* sp. consortia are able to degrade dye from contaminated wastewater as compared to pure culture (Jadhav et al., 2010).

8.2.2 *Plant growth promoting* Rhizobacteria *role in remediation*

For bioremediation of toxic contaminants from the environmental contaminated sites, plant endophytic and plant rhizospheric bacteria play an important role by contributing in the phytoremediation (Leigh et al., 2006). They are naturally found bacteria in the soil that colonize in plant roots aggressively and also supply plant growth nutrients and promote hormones to increase plant growth and yield (Saharan and Nehra, 2011). A few plants have the tendency to produce compounds that are structurally similar to phenols which will be eventually favorable for the growth and degradation of PAHs.

Among bacterial strains, *Pseudomonas* spp. was found to be best in plant-microbial interactions in the aspect of plant growth promoting Rhizobacteria (PGPR) and hydrocarbon degrading activity (Hontzeas et al., 2004). In a case study of exploration of vegetation in a contaminated area resulted in isolation of PAH degrading microbial strains, especially *Lysinibacillus* was found (Ma et al., 2010). Bacterial strains such as *Rhodococcus* spp. are capable of growing in the *rhizospheric* area of environmental contaminated sites and are potent to degrade PCB. This clearly indicates that *rhizobacterias* can be used as effective tools for onsite bioremediation (Leigh et al., 2006).

Among organophosphorus, marathon is one of the most commonly used pesticides, which is also degraded by *rhizobacteria, Azospirillum lipoferum*, a free living nitrogen fixing bacteria (Kanade et al., 2012). It has been observed in a several number of studies that PGPR are involved in the bioremediation of heavy metal contaminated soil. Even in the high concentration of metals, these *rhizobacteria* are able to enhance plants growth, and facilitate phytoremediation of metals (Glick, 2010). Thus rhizoremediation is an effective strategy of combining microbes and plants together to facilitate contaminant removal from environmental contaminated sites.

8.2.3 Degradation by micro fungi and mycorrhiza

Eukaryotic microorganisms from the single cell yeasts to multicellular molds are known as micro fungi (Rossman, 1995). Yeasts can be found from unicellular form to *pseudomycelia*, however, molds classically occur in mycelium form. Like bacteria, fungi are also known as one of the important decomposers of organic materials and they play a key role in the degradation of carbon present in the biosphere. However, unlike bacteria, fungi are able to grow in low pH and low moisture conditions which is helpful in the degradation of organic material (Spellman, 2008). Fungi have been found to be the most effective in decomposition of complex compounds due to its extracellular and multienzyme complexes. They are able to penetrate organic substrates easily by colonization and rapidly transport and distribute nutrients via the mycelium (Matavulj and Molitoris, 2009).

The association where a fungus and the roots of higher plants lived symbiotically is known as Mycorrhiza. The plant and roots association often occurs in two forms such as extracellular and intracellular. In intracellularly, fungus can colonize plant roots intracellularly, for example, arbuscular mycorrhizal fungi (AMF), whereas in extracellularly, colonization takes place extracellularly, for example, ectomycorrhizal fungi. Fungi play an important role in the soil dynamics. Mycorrhizo remediation is the process of removing contaminants by growth of mycorrhizobium (Khan, 2006). To remove harmful contaminants from the environment and to reuse xenobiotics complexes such as lignin, fungi play an important role in degradation because of catabolic enzymes carried by them (Fritsche and Hofrichter, 2005). In this chapter, a brief insight of contaminants degradation by different types of fungi has bccn elaborately described.

8.2.4 Degradation by yeasts

Yeasts have the ability to use complex substances as growth substrates as well as cometabolites. Some of the yeasts such as *Trichosporon cutaneum* found in the soil, have a special system for the uptake of phenol, an aromatic compound (Mörtberg and Neujahr, 1985).

Moreover, the role of yeast in biodegradation of hydrocarbon such as petroleum compounds has been studied thoroughly. Micro fungi are able to readily degrade simpler alkanes ranging from C10 and C20 as growth substrates (Bartha, 1986). Whereas, bioremediation of higher alkanes with a chain length of more than C24 has also been reported (Fritsche and Hofrichter, 2005). *Candida lipolytica*, and *Rhodoturularubra* are examples of yeasts that are able to degrade alkanes (Miranda et al., 2007). To degrade aniline, a product of azo dye degradation, yeast has been exploited (Mucha et al., 2010). There are numerous reports that observed bacteria as the most efficient in the degradation of hydrocarbon, but a few case studies showed that yeast is a better degrader of hydrocarbon as compared to bacterial degradation (Ijah, 1998).

Besides aromatic organopollutants, micro fungi are able to degrade other hydrocarbons also cometabolically such as pesticides, biphenyls, plasticizers, and nitro aromatics (Fritsche and Hofrichter, 2000). Strains of yeasts such as *C. lipolytica* have been reported to degrade PCB (Šašek et al., 1993). *Saccharomyces cerevisiae* have been also involved in the degradation of numerous pesticides and insecticides in aerobic fermentation (Cabras et al., 1988).

To degrade toxic heavy metal, yeasts play a key role. There are several reports; Yeasts are known to remove heavy metal by biosorption and accumulation. They are better metal accumulators unlike bacteria (Wang and Chen, 2006). There are many reports of yeasts involved in the removal of hexavalent chromium [Cr(VI)], such as *Candida tropicalis, Wickerhamomyces anomalus*, and *Cyberlindnera fabianii* which are reported to transform Cr(VI) by biosorption (Bahafid et al., 2013). Reduction of Cr(VI) to Cr(III) has been also reported by several others strains such as, *S. cerevisiae* and *Hansenula polymorpha* (Ksheminska et al., 2006). The degradation of Cr(VI) and Cr(III) by *Pichia guilliermondii* takes place through the chelation process (Ksheminska et al., 2008). Immobilized cells of *Schizosaccharomyces pombe* have been used effectively for removing copper (Subhashini et al., 2011).

8.2.5 Degradation by filamentous fungi

The fungi which are filamentous in nature are generally very good in the biodegradations process as compared to other microorganisms because of their filamentous nature of mycelium. Unlike yeasts and bacteria, these fungi have mycelial growth which has better penetration and colonization over the substrates. Fungi can easily spread over the substrates and have faster degradation by using numerous extracellular digestive enzymes. This will eventually result into better degradation as compared to other forms of microorganisms via deeper penetration. These fungi filaments have better interaction with the contaminants mainly due to high surface to area ratio. And the secretive enzymes which are extracellular in nature are better degraders of contaminants as compared to other intracellular enzymes.

The bioremediation processes generally take place through three ways. First the growth substrate is used as a carbon source. In another way, target contaminants are used as cometabolites and are attacked by enzymes. Alternatively, the target contaminant can be bioaccumulated within the cell, which is not being directly broken down but gets concentrated inside the cell. Though all these mechanisms have been found in the fungi biodegradation, the most common one is bioaccumulation and cometabolism as compared to, where contaminants are used as a carbon source (Abrusci et al., 2007). Fungi such as *Pseudeurotium zonatum* have been reported to use toluene as a source of carbon and whereas fungal strain such as *Leptodontium* are known as deuteromycetes (Prenafeta-Boldú et al., 2001).

For biodegradation of petroleum-related compounds, generally a mixed consortia of bacterial and fungal strains are preferred, as fungal strains individually are not able to completely degrade aromatic compounds but can reduce their toxicity and convert them into compounds which can be later easily decomposed by using bacterial strains (Atlas, 1981). To degrade aromatic hydrocarbon, fungal strains such as *Aspergillus* and *Penicillinum* have been used, whereas in case of aliphatic hydrocarbon, *Aspergillus* and *Cladosporium* have been used commonly (Steliga, 2012). In a large number of isolation studies, petroleum hydrocarbon sites are the most common source of fungal isolation and they have been found out to be potent hydrocarbon degraders (Chaillan et al., 2004).

To degrade crude hydrocarbon, *Pencillium* and *Aspergillus* were found to be the most commonly used fungal strains (Singh, 2006). It has been observed that there is less contribution of fungal strains to degrade PCBs, whereas in some cases, the role of filamentous fungi has been observed. For the aromatic contaminant's degradation, there has been a huge role for lignolytic filamentous fungi because of the secretion of the extracellular and specific oxidative and reductive enzymes. There have been reports of the role of ectomycorrhizal fungi and *Aspergillus niger* in the PCB degradation (Donnelly and Fletcher, 1995; Dmochewitz and Ballschmiter, 1988). For a very long time, fungi have been known to be a great cellolytic degrader and can degrade products such as fibers, woods materials, cellulose, paints, fuels, drugs, plastics, and other materials.

There are several processes by which fungi can transform metal from the contaminated sites such as active uptake, intra and extracellular precipitation, and valence transformation (Gadd, 1994). The mode of adsorption of metals such as zinc, cadmium, lead, copper, and mercury usually takes place through fungal spores and mycelium. At times, dead fungi can also bind effectively with the contaminants as compared to living fungi. There have been reports where fungi such as *Rhiloprzs arrhizus* have been used to treat thorium and uranium. In different case studies, it has been found that fungal strains are using their functional group present on the cell surfaces as a ligand to sequester metal in a liquid culture for example *A. niger*

AB10 for cadmium and *R. arrhizus* M1 for lead removal (Pal et al., 2010). Moreover, *Abruscular mycorhizal* fungi proteins present on the cell walls also have the ability to sequester toxic elements. It can be evidenced that these AMF can sequester these toxic elements by using glomalin present on the cell surfaces (Gonzalez-Chavez et al., 2004). These fungi play an important role in the ecology by removing these toxic elements from the environmental contaminated sites using sequestration.

Ligninolytic fungi have been researched extensively for the bioremediation of dye (Bumpus, 2004). Abrusci et al. (2007) has isolated nine filamentous fungi from the cinematographic film comprising of *Penicillium chrysogenum, Alternaria alternata, Trichoderma longibrachiatum, Phomaglomerata, Mucor racemosus, Aspergillus versicolor, Aspergillus ustus*, and *Aspergillus nidulans* which are also able to degrade gelatin emulsion by producing CO_2 production. Generally, there are two types of enzymes secreted by filamentous fungi, in case of extracellular it is laccases and peroxidases for the lignin degradation and intracellular degradation generally involved cytochromes P450. It is really important to control their metabolism, as these systems can be inhibited or activated using pesticides.

8.2.6 Degradation by algae and protozoa

Though algae and protozoa play an important role in the microbial communities of terrestrial and aquatic environment, there are scanty reports regarding their role in the degradation of hydrocarbon (Das and Chandran, 2011). *Prototheca zopfi*, an algae, can degrade or transform a number of hydrocarbon such as mixed hydrocarbon substrate, crude oil, aromatic hydrocarbons, and *n*-alkanes and iso-alkanes in a study by Walker et al. (1975); and naphthalene degradation is also reported by using various algae strains (two diatoms, one red alga, five green, one brown alga, and nine cyanobacteria) in a report by Cerniglia and Gibson (1977). Unlike bacteria and fungi, it has been found out that till date there is no report on degradation of hydrocarbon by Protozoa, instead they can negatively affect biodegradation process by reducing the population of bacterial strains which degrade hydrocarbons (Stapleton and Singh, 2002). Similar results were also found by Rogerson and Berger (1981) that protozoa which were cultured on hydrocarbon-utilizing bacteria and yeasts are not able to degrade crude oil. Generally, it has been found that algae and protozoa do not have an important role for the hydrocarbon degradation from the environmental contaminated sites (Bossert and Bartha, 1984). In several reports it has been observed that algae such as *Scenedesmus platydiscus* and *Chlorella vulgaris* are used to degrade and transform (Wang and Zhao, 2007). Algae are not only able to degrade pesticides but can also degrade other major environmental contaminants due to great interaction between algae and pesticides. Similarly, dyes can be used as a source of carbon and nitrogen, irrespective of their chemical structure, for example azo dye is degraded by *C vulgaris* (Jinqi and Houtian, 1992). The catabolic process involved in the biodegradation can be inducible. Different algae strains were found to decolorize and degrade a large number of dyes such as *Nostoc, Chlorella, Volvox*, and *Oscillatoria* which decolorize basic fuchsin, G-Red

(FN-3G), orange II, methyl red, and basic cationic (El-Sheekh et al., 2009). Several species of *Anabaena, Synechoccus*, and *Chlorella* have been reported to degrade heavy metals. However, there are certain limitations regarding commercial application of using these algae systems for the bioremediation of heavy metals because of the optimum working conditions (Dwivedi, 2012). Adsorption is one of the ways of metal transformation. In another way, algae also convert metals by chelation. Fora very long time, metal biosorption using brown algae is also known to be occurring; many cell wall constituents such as fucoidan and alginate are involved in the heavy metal sorption. This type of research is mainly found to be in soil and marine algae (Davis et al., 2003). In a continuous culture, removal of Cd, Cu, and Cr is reported to be performed by microalga Scenedesmus *incrassatulus* (Pena-Castro et al., 2004). For the heavy metal biodegradation, green algae such as *Chlorella* has been reported to transform Cr(III) (Akhtar et al., 2008). The interaction between degrading bacterial strains and protozoa would affect bioremediation process, due to the characteristics of protozoa engulfing degrading bacterial strains. In a case study of Mattison et al. (2005), they studied a food chain model to depict the overall effect of grazing bacteria of flagellated protozoa on the methylbenzene and benzene degradation. They concluded that active growing flagellated protozoa have a positive impact on the rate of degradation of benzene and methylbenzene. It has been found that protozoa infusoria is able to improve the rate of degradation by 8.5 as compared to the previous rate of degradation. Thus protozoa can speed up bioremediation of environmental contaminated sites.

It also has been noticed that similar results were found for the naphthalene degradation by increasing its degradation four times. There are many expected theories for the mechanism responsible for the contaminants biodegradation by protozoa which are as follows: nutrients breakdown which in turn increase nutrients availability; activation of bacteria which can control dead cells, secretion of active compound; selective grazing of unwanted bacteria which are the competition for the good degrading strains and provide enough food and space for them; increasing oxygen amount of degraded materials to accelerate degradation; secretion of selective enzymes responsible for the faster degradation; provide carbon and energy sources together for the growth of the bacteria and ultimately these processes collectively responsible to increase the rate of biodegradation (Chen et al., 2007a,b).

8.2.7 Major contaminants

In the last few years, extremely toxic chemical compounds have been synthesized and introduced into the environment rapidly such as dyes, hydrocarbon, polycyclic aromatic hydrocarbons (PAHs), pesticides, and PCBs (Diez, 2010). Whereas, it has been found that synthetic compounds such as heavy metals are difficult to degrade as compared to natural compounds using native microorganisms.

Dyes have a wide range of industrial applications. The aromatic dye having (−N=N−) group known as azo dyes, are most commonly used in industrial application for commercial purpose. Due to their structures, they are extremely resistant and thus treatment of contaminated wastewater having dyes involves many chemical and physical approaches. How effectively microorganisms remove dyes from contaminated water depends on its catabolic and degradative efficiency of microbes and factors regulating the process during the bioremediation of dyes (Fig. 8.2).

Hydrocarbons are the compounds comprising of hydrogen and carbon. Hydrocarbons are linear linked, branched, or cyclic molecules. They are usually aromatic, for example, benzene (C_6H_6) or aliphatic such as alkanes, alkenes, and alkynes (McMurry, 2000). Another important class of pollutant hydrocarbons is polycyclic aromatic hydrocarbons generally present in soil, sediments, and air. Industrial pollution is the main source of PAH pollution. They have been considered as an interesting topic of studies because of their toxicity, environmental resistance, and dominance (Okere and Semple, 2012). They can absorb to soil, water, and ultimately affecting food chain via aquatic organism. To restore PAH contaminated sites, microbial mediated remediation is one of the most feasible technology.

PCBs have a wide application in the industrial process such as in rubber, dye, and paint, etc. industries because of their physical and chemical properties. They are the most toxic synthetic organic chemical compounds that can cause disruption in the endocrine system and this may be one of the lead reasons for cancer. Hence, removal of PCB from the PCBs contaminated environment sites are a matter of major concern and need faster microbial mediated bioremediation (Seeger et al., 2010).

Pesticides are the mixture of compounds used for controlling any kind of pest. There are two types of persistent pesticides that are easily degraded and nonpersistent which are not easily degraded. In the soil contaminated with pesticides, the most used mode of microbial mediated remediation is generally carried out by bacteria and fungi (Vargas, 1975).

Heavy metals: For the bioremediation of heavy metals, it should be transformed into a stable form or any other form and cannot be destroyed. So the major mode of biodegradation of heavy metals is via transformation. The main mechanism behind the biodegradations of heavy metals involves bioleaching where mobilization of heavy metal occurs by the methylation and release of organic acids, secondly by biosorption where

Figure 8.2 List of some major contaminants.

by using physiochemical techniques heavy metals can be absorbed into microbial cell surfaces. In biomineralization, heavy metal is immobilized via the complexes formation, transformation using enzyme-based reactions (Lloyd and Lovley, 2001).

8.3 Biodegradation and bioremediation

In recent times, biodegradation is one of the emerging tools in the biotechnology field, as it has a huge potential to solve the dangers of rapidly spreading contaminants in the environment. Thus biodegradation holds a major significance in the microbiology field. Microbial mediated remediation has advantages over other bioremediation processes as they are much more effective for contaminants removal from the environmental surroundings (Demnerová et al., 2005). In addition, biodegradation is considered as the most effective method among bioremediation strategies to remove contaminants, though it is also a cost effective method which makes this process a more viable method for bioremediation. Biostimulation, bioaugmentation, and natural attenuation are the most commonly used methods for bioremediation (Olaniran et al., 2006).

Natural attenuation is one of the simplest ways of bioremediation, as in this method active conversion of contaminants takes place and pollutants concentration is observed in the soil (Kaplan and Kitts, 2004). Whereas, the area where native active microbial strains are scanty or absent in numbers, there pollutants degrading specialized microbial strains are added to transform the contaminants present in the environmental sites, also known as bioaugmentation (El Fantroussi and Agathos, 2005). In those cases, adaption of theses added microbial strains to the contaminants environmental sites is the major challenge, as they exhibit extra enzymatic activities for the degradation of contaminants (El Fantroussi et al., 1999). In addition to this, nutrients along with microbial strains are also added to speed up degradation of contaminants present in the environment, also called biostimulation (Olaniran et al., 2006).

Natural attenuation is the process of reducing the concentration of contaminants present in the environment through various processes such as physical (sorption/desorption, diffusion, dispersion, dilution and volatilization); chemical (abiotic transformation, ion exchange and complexation); and biological (through animal, plant, and anaerobic and aerobic process). However, natural bioattenuation's definition includes the terms biotransformation or intrinsic remediation. Though biodegradation is the most commonly used process in the natural attenuation, microbes are used to convert or degrade pollutants from one form to another form. Microbes can transform toxic pollutants into nontoxic forms or less toxic forms without any harmful effects under optimum conditions. Mainly, at the most contaminated sites, natural attenuation takes place. Though, to detoxify the contaminated sites, an optimum environmental condition is a major requirement. Otherwise lack of optimum conditions, makes the biodegradation process delayed. So these processes require scientific observation for better performance. This is also known as monitored natural attenuation (MNA). Therefore MNA is the process where one can observe the degradation process of contaminants in the water and soil. It can be used in combination with other bioremediation processes or as an independent bioremediation strategy for the faster degradation of the pollutants from contaminated sites. The remaining pollutants can be removed by the natural bioremediation; this reduction in contaminants can be monitored by the regular analysis of soil and groundwater (EPA, 2001). In nature, there are four strategies to detoxify environments from contaminants (Li et al., 2010). First, microorganisms use pollutants as a source of carbon and energy present in the environment, they completely transform these contaminants into CO_2 and H_2O. Second through adsorption, microbes can bind to the pollutants present in the environment. In this process, microbes do not destroy pollutants, but help to avoid further contamination. Third, If the pollutants spread into groundwater and soil, it would dilute the contamination by mixing the soil and clean water. Fourth, in some cases, such as volatile liquid, solvents and oil can easily evaporate from the ground surface to the air in the form of gases, and these contaminants would generally destroy in the presence of sunlight. Generally, bioaugmentation and biostimulation will enhance bioremediation process when the pollutants are not being degraded by the natural microflora. The process of addition of trace minerals, electron donors, electron acceptors, or soil nutrients enhance the conversion of a large number of harmful pollutants into the nontoxic form is known as biostimulation (Li et al., 2010). Indigenous microbes are much effective in the contaminant's removal by biostimulation. It has been found out that in the process of biostimulation, perchloroethene, and Trichloroethene are degraded into ethane within a short period of time by microbes with the addition of lactate (Shan et al., 2010). Humic substances as electron shuttles also play an important role in the bioconversion of contaminants by increasing the speed of electron transfer in the biodegradation process under anaerobic condition. Humic substances such as Anthraquinone-2,6-disulfonate (AQDS) act as electron shuttle for the conversion of chlorinated organic contaminants and iron oxides (Bond and Lovley, 2002). In a case study of Chen et al. (2007a,b) it has been reported that with the addition of AQDS and lactate, as biostimulator and dechlorinating and iron reducing bacteria will lead to the increased degradation of pentachlorophenol de-chlorination in the soils. Indigenous Rhizobium ralstonia taiwanensis, along with the glycerol, increase phenol degradation which clearly indicates that biostimulation will enhance biodegradation rates of pollutant removal (Chen et al., 2007a,b). It has been found that biostimulation is the effective technique in comparison to natural degradation of diesel contaminated sites (Meneghetti et al., 2012). Though Bento et al. (2005) reported in a comparative case study that among these bioremediation strategies, bioaugmentation is the most effective process, whereas the natural attenuation is more efficient as compared to biostimulation of pollutants from the environmental contaminated sites with diesel oil. Yu et al. (2005) reported that for recovery of phenanthrene and fluorine contaminated mangrove sediments, natural native microbial strains proved to be most effective; these indigenous microbes have better integral with other microbial consortia for effectively relieving contaminated sites from the pollutants as compare to other techniques where in the case of pyrene degradation, biostimulation appears to be the most effective bioremediation strategy.

Bioaugmentation is the process of addition of specific single or mixed microbial strains to reduce contamination. It is the process of enhancing bioremediation technique for better recovery of environmental contaminated sites (Lebeau et al., 2008).

The basic idea of bioaugmentation is to increase the enzymatic and catabolic activities of native microbial strains present in the environmental contaminated areas by adding other microbial strains and ultimately this will lead to the numerous metabolic reactions for faster bioremediation (El Fantroussi and Agathos, 2005). Although, to enhance the biodegradation process, genetically engineered microbes having higher degradation abilities also play a significant role which enable them to hold a good potential for the bioaugmentation process in the soil. When the bioattenuation and biostimulation are not able to degrade completely, then bioaugmentation is the most effective approach for clean-up (El Fantroussi and Agathos, 2005).

There are numerous important biotic and abiotic factors affecting the process of bioaugmentation such as pH, temperature, organic matter content, and moisture; the other abiotic factors such as nutrient content, soil types, and aeration also play an important role in determining the bioaugmentation efficiency. Whereas in the case of biotic factors, for limited growth substrates, there is a competition between native and added microbial strains. Bacteriophages and protozoa also show activities such as predation and antagonistic interactions which play a key role in the bioaugmentation process (Mrozik and Piotrowska-Seget, 2006). So, to enhance bioremediation, a combination of biostimulation and bioaugmentation is one of the effective strategies. Through the introduction of any nutrients and electron sources, both the native and added could gain benefits, making it a promising collaborative technique for bioremediation (El Fantroussi and Agathos, 2005). Phytoextraction along with bioaugmentation via AMF or PGPR is one of the effective methods to remove metal from environmental contaminated sites (Lebeau et al., 2008).

8.4 Factors involved in microbial remediation

Microbes can transform a large number of contaminants due to a wide range of metabolic activities and their ability to transform in the contaminated environment. Hence microbes play a major role within in situ remediation. But their degradation capacity depends on many things such as pollutants availability to microbes, their concentration, the chemical nature of the pollutant, and the physical and chemical environment (El Fantroussi and Agathos, 2005). Therefore main factors such as microorganisms, and its nutritional parameters and the environment associated with it are regulating factors of the biodegradation process.

8.4.1 Biotic factors

It involves the microbial metabolic process. The biological factors that influence rate of degradation by microbial strains generally involve the process of inhibition of microbial enzymatic activities and growth of microorganisms (Riser-Roberts, 1998). This is mainly due to the following reasons such as competition between two microbial strain for limited growth substrate; protozoa or bacteriophage predates microorganisms and antagonistic activities among microbes (ERD, 1998). Generally contaminant concentration and enzyme quantity determines the biodegradation rate. Thus the quantity of catalyst depicts the number of microbial strains present during the bioremediation process and also represents the enzyme concentration produced by each microbial strains. In addition, the rate of biodegradation of contaminants in the environmental sites is mainly due to interaction of specific enzymes with the contaminants and the availability of pollutants to the microbial strains present in the environment. Furthermore, a sufficient amount of nutrients and oxygen should be present in proper form for good microbial growth (ERD, 1998).

Factors such as pH, temperature, and moisture also play an important role in controlling the biodegradation rate by regulating the rate of enzyme catalytic reactions. The bio-enzymes have an optimum working temperature for better degradation, and the degradation rate will keep on changing with respect to changing temperature (Van der Heul, 2011). For every 10°C decrease in temperature, the biodegradation rate will be decreased by roughly one-half. Bioremediation can perform over wide pH, in terrestrial and aquatic environments, and the most preferred range of pH is from 6.5 to 8.5 for biodegradation. The biodegradation rate also depends on the moisture, as it affects the quantity of available soluble materials, pH, and osmotic pressure of the environmental sites (Cases and de Lorenzo, 2005).

8.4.2 Factors present in the environment

For better biodegradation, the type of soil and the amount of organic matter are essential parameters for good adsorption of contaminants on the soil surface. Whereas in the case of absorption, pollutants are used to penetrate into the matrix of soil. Both techniques are used to decrease the biodegradation rate as it reduces the availability of pollutants to most of the microbes (ERD, 1998). There is a variation in the pore sizes of the saturated and unsaturated aquifer matrix and that affects the pollutants and fluids movement in the groundwater. When the soil is saturated with the water and the tendency of matrix to release gases is also reduced, then it ultimately influences the rate of biodegradation (ERD, 1998). The electron density of the system can be measured by the oxidation–reduction potential of the soil. In the oxidation process, electron acceptors are used to accept electrons and release biological energy during the process. The aerobic conditions involve low electron density, whereas anaerobic conditions involve high electron density (ERD, 1998).

8.5 Genetically engineered microbes and their role in degradation

To enhance recovery of environmental contaminated polluted sites, methods such as biostimulation and bioaugmentation have been commonly employed. During

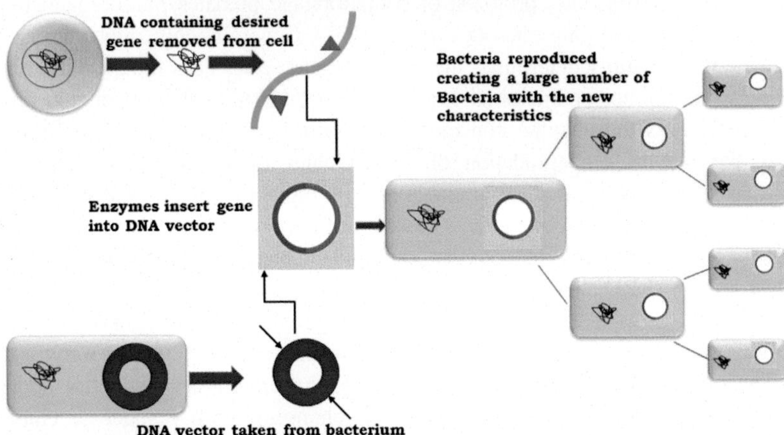
Genetically Modified Bacteria
DNA containing desired gene removed from cell
Bacteria reproduced creating a large number of Bacteria with the new characteristics
Enzymes insert gene into DNA vector
DNA vector taken from bacterium

Figure 8.3 Process of making genetically modified bacteria.

1980 microbial genes carrying metabolic enzymes have been characterized and cloned for faster degradation of xenobiotics compounds. Thereafter, extensive work has been done regarding exploration of the role of genetic engineered strains for biodegradation in the molecular biology and microbiology field (Cases and de Lorenzo, 2005). Naturally, there is an exchange of genetic material among microorganisms, which lay the basic foundation of genetic modification of microbes using genetic engineered tools. This process used to be called recombinant DNA technology. These GEMs have a great potential to clean-up environments polluted from various pollutants. They have a wide range of catabolic activities so can degrade numerous environmental contaminants (Sayler and Ripp, 2000). But many concerns such as risk analysis and biosafety have been raised regarding the use of these GEMs; extensive research is required to make this technique a real practical approach (Cases and de Lorenzo, 2005). For the application of bioremediation, there have been four major principals employed (Menn et al., 2008)——enzyme activity (specificity and affinity) which can be modified; regulation of the metabolic pathways; control, monitoring, and development of the bioprocesses; as a biosensor and bioreporter in a chemical reaction, reduction of toxicity and analysis of end point (Fig. 8.3).

8.5.1 Genetically modified microbes

To standardize the biodegradation activities of the microbes, genetic engineered tools have been used most commonly in the molecular biology field, which will lead to generation of new enzymatic activities and pathways by cloning genes of one microorganism into another one (Ramos et al., 1994). There are numerous biodegradative genes that have been identified for removing environmental contaminants such as halogenated pesticides, toluene, toxic wastes, and chlorobenzene acids. There is one specific plasmid for specific compound. It is not possible that one plasmid can detoxify all the environmental pollutants of the different groups. There are four major group of plasmids-- to destroy decane, octane, and hexane (OCT plasmid); to destroy toluene and xylene (XYL plasmid); to destroy camphor (CAM plasmid) and to transform naphthalene (NAH plasmid) (Ramos et al., 1994). Friello et al. reported that genetic engineering has a huge potential to degrade a large number of hydrocarbons by creating genetic modified microbes. They have successfully created a multiplasmid of Pseudomonas which is capable of degrading polyaromatic hydrocarbons, aromatic, terpenic, and aliphatic. To degrade multiple compounds such as naphthalene, octane, xylene, and camphor, a strain of *Pseudomonas putida* is genetic engineered in such a way that it would have XYL and NAH plasmid and a hybrid plasmid created by recombining CAM and OCT plasmid through conjugation (Sayler and Ripp, 2000) and also able to transform crude hydrocarbon in a more effective way as compared to other strains due to its capacity of degrading multiple hydrocarbons because of carrying multiplasmids (Markandey and Rajvaidya, 2004). Superbug was created to degrade oil using these genetic modified tools. It is one of the successful products of genetic engineering. There are number of plasmids found in *P. putida* for degradation of environmental pollutants such as for toluene (TOL), for 3-cne chlorobenzoate (pAC 25), for salicylate toluene (pKF439), and for 3,5-xylene (RA500). It was the first case of intellectual property rights involving a living being. Therefore it is concluded that molecular systems such as genetic engineering tools and plasmid conjugations are able to create microorganisms of higher degradation capacity, and ultimately efficiently detoxifying environmental contaminated sites (Sayler and Ripp, 2000). There are reports on environmental contaminants degradation by genetically engineered bacteria using genetic engineered tools such modification of pathway and change in substrate specificity (Hrywna et al., 1999).

Genetic engineering has also played a great role in the removal of heavy metals. The example of such processes is as to degrade chromium from industrial wastewater, a strain of *Alcaligenes eutrophus* AE104 (pEBZ141) has been used (Srivastava et al., 2010), whereas a genetically modified strain of *Rhodopseudomonas palustris* was used to detoxify heavy metal from wastewater (Deng and Jia, 2011). To degrade PCBs, catabolic genes of *Achromobacter* and *Ralstonia eutropha* have been used and in the case of heavy metal removal, *R. eutropha* strain is genetically modified using genes from *Achromobacter denitrificans* through conjugation method (Menn et al., 2008).

The bacteria present in the endophyte and rhizosphere are found to be most suitable for production of genetic engineered strains, which could be used for the degradation of toxic compounds present in the soil and considered to be an important strategy for the recovery of environmental contaminated sites (Divya and Kumar, 2011). There are three major principles that are used for the selection of a suitable microorganism for genetic engineering and introduction into the soil such as: (1) After cloning, microbial strains should have high stability and high expressions; (2) Microbial strains must be resistant to the environmental contaminants present in the soil; and (3) Survival of microbial strains to specific environmental conditions (Huang et al., 2004). Though some of the bacteria present in the soil have a lower degradation capacity, the rhizoremediation can be advanced by creating genetic modified bacteria with higher biodegradation capacity in the molecular biology field (Glick, 2010). A large number of studied have been carried out for the contaminants removal using molecular tools, for example, for degradation of PCBs and trichloroethylene (Sriprang et al., 2003). Genes such as phytochelatin synthase (PCS; PCSAt) have been introduced to the *Mesorhizobium huakuii* subsp. rengei strain B3 from the Arabidopsis thaliana strains for the detoxification of heavy metals by Sriprang et al. (2003) and later on create a symbiosis between *Astragalus sinicus* and *M. huakuii* subsp. rengei strain B3. The expressed gene used to produce phytochelatins and accumulate Cd^{2+}, under the regulation of bacteroid specific promoter, was the nifH gene (Sussman et al., 1988). Problems such as competition among microbial strains can be avoided by using genetic modified strains at the time of seeding during inoculation. Though, there are always many environmental concerns including the release of these genetic engineered strains into the environment which causes many biosafety and ethical issues. These GEMs cannot be released into the field before addressing the safety issues and assurance of no ecological damage to the environment (Wackett, 2004).

8.5.2 Problems involved in application of bioremediation using genetically engineered microorganisms

It has been found out that are numerous genetic engineered strains that only degrade pollutants compounds in a small amount and within in vitro conditions, whereas these strains are not been able to degrade pollutants in in situ conditions (Sayler and Ripp, 2000). Another major issue in the bioremediation process is the selection of the biodegrading strains; as the cultures expecting to be the most potent degrader in this process by enrichment techniques are not actual real microbial strains responsible for the bioremediation of contaminants in bulk condition in the natural environment. In microbial ecology, techniques such as stable isotope probing (SIP) have discovered that the strains can be grown aerobically for example *Rhodococcus* and *Pseudomonas* are the most efficient strains for the bioremediation, but these strains were found to be less effective in in situ conditions (Wackett, 2004).

In additions, microbes that are growing rapidly result in accumulation of a large amount of biomass. To overcome this problem, a microbial strain should be showing the highest degradation capacity with the production of minimum biomass, which is the most efficient one for optimum bioremediation. By using stationary phase promoters, the genes responsible for the biodegradation of pollutants can be expressed separately, by uncoupling them from the exponential phase (Matin, 1994). Furthermore, due to the recent advancement in the molecular biology field, "suicidal GEMs" (S-GEMs) have been created to reduce unwanted hazards and for efficient bioremediation of contaminated environmental sites (Pandey et al., 2005). Whereas in a few cases, the recombinant bacterium is not able to transform contaminants efficiently because of sensitivity to a foreign environment. If the bacterial strains are introduced into the existing environment, it would make protozoa spp. grow and prevent the bacterial spp. from growing further (Iwasaki et al., 1993). There are many indigenous approaches to avoid these problems, such as encapsulation of efficient microbial strains. The efficiency of the biodegradation activity generally depends on the presence of catabolic enzymes on the target site. The main functional enzyme should be present or in contact with the target pollutant. On the other hand, it is contributed by a small fraction of all the microbes cells present. So, their expression does not hold much significance in the bioremediation process (Sayler and Ripp, 2000). The *Pseudomonas fluorescens* strain has been successfully released into the field on a controlled and large-scale level, which makes bioremediation a more applicable approach (Ripp et al., 2000). But there are huge risks associated with the release of GEMs into the field conditions. So the future application of genetically engineered microbes is always under doubt. Hence, the future prospects of the use of the genetically engineered microbes into the in situ environmental condition will be always a topic of concern, and extensive research and safety assessment are required before releasing these genetically engineered microbes into the environment. The main concern associated with the commercialization of bioremediation technology belongs to their adaption in the natural environment. Moreover, the genetic engineering studies have been well characterized on only in a few bacterial spp. such as Pseudomonas, Bacillus, and *Escherichia coli*, while the rest of the bacterial strains need to be studied more for the application of genetic engineered strains in the bioremediation field. The challenges associated with the bioremediation process demand the introduction of more genetically engineered microbes in the concerned area. The major challenge is to make genetically engineered microbes with higher biodegradative capacity and environmental stability.

For the successful release of bioremediation technology, huge efforts are needed such as observation of survival and activity of genetically engineered microbes in field and transfer of any genetic material into the environment. However, other concerns such as public and environmental safety are also being raised by many scientific organizations. In a real scenario, the majority of the bacteria that are optimized for the bioremediation application work well only under laboratory conditions, but not in the environmental conditions. But as such, there are also not any significant reports on the adverse efforts of genetically engineered microbes into the environment. And hence, due to many safety and environmental concerns, extensive research is going on in the field of application of genetically engineered microbes in the bioremediation area. Though, the survival of genetically engineered microbes in in situ environment is still unclear, it needs more research to address this issue (Singh et al., 2011).

8.6 Optimization of bioremediation process

Overall, in recent times, extensive research is going on for in-situ and ex-situ bioremediation such as removal of dyes, chlorides, solvents, uranium, and explosives, etc. (Nizzetto et al., 2010; He and McMahon, 2011).

The bioremediation strategy is the most commonly employed method for removing pollutants from the environment; they also help us to understand the ecological system and microbe interaction under in in situ conditions. Furthermore, a lot of success has been gained in the use of biodegradative strains under lab conditions but in some cases, it has failed miserably under in situ conditions (De Lorenzo, 2009). Thus sometimes it is not possible to replicate lab-based experiments in field conditions and therefore novel strategies are necessary in implementing the bioremediation process under natural field conditions. The single use of microbial strains is not efficient for successfully removing pollutants, and integration of biology along with genetic engineering approaches will effectively make a difference. Extensive research is required to reduce the time required for the operationalization of the bioremediation process and for the requirement of rapid sensor to observe the process throughout (Van Der Meer and Belkin, 2010).

Recently, there has been great advancement in the field of microbial mediated remediation by using actively growing microbial cells with a higher degradation capacity. Though as only a few microbial spp. are culturable, and due to the lack of exploration of uncultured microbes and not being able to replicate the natural conditions, role of the biodegradation potential remains unexplored. Recently, "omics" approaches are being used to find out the unexplored biodegradation potential of microbial strain for the removal of pollutants in environmental contaminated sites. Thus there is an introduction of metagenomic approaches, which would pave a way for the exploration of new microbial strains in the field of microbial mediated remediation.

8.6.1 Future perspectives

Understanding the current importance of "omics" in the microbial remediation, the following areas need extensive research to better understand the process. Mining of the bioremediation data is required for understanding degradative pathways and with the help of novel algorithms, these data can be fit into simulation and numerical modeling. There is a need for standard protocols for assemblage, exploration, reposition, and transmission of data. Exploration of novel biomarkers may aid to define the usage of specific bioremediation processes. A complete detailed picture of microbial remediation process could be understood by merging all the "omics" data via genetic engineered tools.

8.7 Conclusion

Microbes play a key role in the restoration of our ecosystem and conservation of our ecological cycles. All these activities are comprised under microbial remediation. Most of the chemical pollutants are removed by microbial remediation, whereas some such as heavy metals and petroleum products have toxic effects on the ecosystem. Although the biodegradative efficiency of these microbial strains can be assessed by growing under ideal lab conditions, there are certain limitations in natural environments, which is why environmental genetic engineering has been introduced along with bioaugmentation or biostimulation strategies. However, there are many safety concerns associated using GEM. Recently, omics-approaches to bioremediation would help in tracing the potent microorganisms for effective removal of the pollutants from the sites. Currently, there is an urgent requirement for environmental cleanup and decontamination due to rapid and indiscriminate use of toxic chemicals and pollutants. Bioremediation is the best alternative to overcome these environmental contaminations. Since it is difficult to comprehend the complexity of these microbial and environmental interactions, for clear and precise understanding of the complexity of these microbial mediated bioremediation processes, novel techniques have been introduced such as integration of "omics" along with genetic engineering tools to explore new biodegradative pathways. The integrative methods of these practices in this time of "omics" have paved the way for the effective implementation of the bioremediation approaches. Since it is a beginning of the integration of novel approaches, the generated data need to be organized effectively for informative databases to depict microbial metabolic and catabolic activities within contaminated sites which would be helpful to speed up bioremediation processes.

References

Abrusci, C., Marquina, D., Del Amo, A., Catalina, F., 2007. Biodegradation of cinematographic gelatin emulsion by bacteria and filamentous fungi using indirect impedance technique. Int. Biodeterior. Biodegrad. 60 (3), 137–143.

Adarsh, V.K., Mishra, M., Chowdhury, S., Sudarshan, M., Thakur, A.R., Chaudhuri, S.R., 2007. Studies on metal microbe interaction of three bacterial isolates from East Calcutta Wetland, OnLine J. Biol. Sci., 7.

Akhtar, N., Iqbal, M., Zafar, S.I., Iqbal, J., 2008. Biosorption characteristics of unicellular green alga *Chlorella sorokiniana* immobilized in loofa sponge for removal of Cr(III). J. Environ. Sci. 20 (2), 231–239.

Atlas, R.M., 1981. Microbial degradation of petroleum hydrocarbons: an environmental perspective. Microbiol. Rev. 45 (1), 180.

Bahafid, W., Joutey, N.T., Sayel, H., Iraqui-Houssaini, M., El Ghachtouli, N., 2013. Chromium adsorption by three yeast strains isolated from sediments in Morocco. Geomicrobiol. J. 30 (5), 422–429.

Bartha, R., 1986. Biotechnology of petroleum pollutant biodegradation. Microb. Ecol. 12 (1), 155–172.

Bell, T.H., Joly, S., Pitre, F.E., Yergeau, E., 2014. Increasing phytoremediation efficiency and reliability using novel omics approaches. TRENDS Biotechnol. 32 (5), 271–280.

Bennet, J.W., Wunch, K.G., Faison, B.D., 2002. Use of fungi in biodegradation, Environmental Microbiology., *second ed. ASM Press, Washington, DC.*

Bento, F.M., Camargo, F.A., Okeke, B.C., Frankenberger, W.T., 2005. Comparative bioremediation of soils contaminated with diesel oil by natural attenuation, biostimulation and bioaugmentation. Bioresour. Technol. 96 (9), 1049–1055.

Bond, D.R., Lovley, D.R., 2002. Reduction of Fe(III) oxide by methanogens in the presence and absence of extracellular quinones. Environ. Microbiol. 4 (2), 115–124.

Bossert, I., Bartha, R., 1984. The fate of petroleum in soil ecosystems. In: Atlas, R.M. (Ed.), Petroleum Microbiology. Macmillan Publishing Co., New York, pp. 434–476.

Bumpus, J.A., 2004. Biodegradation of azo dyes by fungi. In: Arora, D.K. (Ed.), Fungal Biotechnology in Agricultural, Food and Environmental Applications. Marcel Dekker, New York, pp. 457–480.

Cabras, P., Meloni, M., Pirisi, F.M., Farris, G.A., Fatichenti, F., 1988. Yeast and pesticide interaction during aerobic fermentation. Appl. Microbiol. Biotechnol. 29 (2–3), 298–301.

Cases, I., de Lorenzo, V., 2005. Genetically modified organisms for the environment: stories of success and failure and what we have learned from them. Int. Microbiol. 8 (3), 213–222.

Cerniglia, C.E., Gibson, D.T., 1977. Metabolism of naphthalene by *Cunninghamella elegans*. Appl. Environ. Microbiol. 34 (4), 363–370.

Chaillan, F., Le Flèche, A., Bury, E., Phantavong, Y.H., Grimont, P., Saliot, A., et al., 2004. Identification and biodegradation potential of tropical aerobic hydrocarbon-degrading microorganisms. Res. Microbiol. 155 (7), 587–595.

Chaube, P., Indurkar, H., Moghe, S., 2010. Biodegradation and decolorisation of dye by mix consortia of bacteria and study of toxicity on *Phaseolus mungo* and *Triticum aestivum*. Asiat. J. Biotech. Res. 1, 45–56.

Chen, B.Y., Chen, W.M., Chang, J.S., 2007a. Optimal biostimulation strategy for phenol degradation with indigenous rhizobium *Ralstonia taiwanensis*. J. Hazard. Mater. 139 (2), 232–237.

Chen, X., Liu, M., Hu, F., Mao, X., Li, H., 2007b. Contributions of soil micro-fauna (protozoa and nematodes) to rhizosphere ecological functions. Acta Ecol. Sin. 27 (8), 3132–3143.

Chistoserdova, L., 2017. Application of omics approaches to studying methylotrophs and methylotroph communities. Curr. Issues Mol. Biol. 24, 119–142.

Czaplicki, L.M., Gunsch, C.K., 2016. Reflection on Molecular Approaches Influencing State-of-the-Art Bioremediation Design: Culturing to Microbial Community Fingerprinting to Omics (Doctoral dissertation). American Society of Civil Engineers.

Das, N., Chandran, P., 2011. Microbial degradation of petroleum hydrocarbon contaminants: an overview. Biotechnol. Res. Int. 2011.

Davis, T.A., Volesky, B., Mucci, A., 2003. A review of the biochemistry of heavy metal biosorption by brown algae. Water Res. 37 (18), 4311–4330.

De Lorenzo, V., 2009. Recombinant bacteria for environmental release: what went wrong and what we have learnt from it. Clin. Microbiol. Infect. 15, 63–65.

De, J., Ramaiah, N., Vardanyan, L., 2008. Detoxification of toxic heavy metals by marine bacteria highly resistant to mercury. Mar. Biotechnol. 10 (4), 471–477.

Demnerová, K., Mackova, M., Speváková, V., Beranova, K., Kochánková, L., Lovecká, P., et al., 2005. Two approaches to biological decontamination of groundwater and soil polluted by aromatics—characterization of microbial populations. Int. Microbiol. 8 (3), 205–211.

Deng, X., Jia, P., 2011. Construction and characterization of a photosynthetic bacterium genetically engineered for Hg^{2+} uptake. Bioresour. Technol. 102 (3), 3083–3088.

Diez, M.C., 2010. Biological aspects involved in the degradation of organic pollutants. J. Soil Sci. Plant Nutr. 10 (3), 244–267.

Divya, B., Kumar, M.D., 2011. Plant-microbe interaction with enhanced bioremediation. Res. J. Biotechnol. 6 (4), 72–79.

Dmochewitz, S., Ballschmiter, K., 1988. Microbial transformation of technical mixtures of polychlorinated biphenyls (PCB) by the fungus *Aspergillus niger*. Chemosphere 17 (1), 111–121.

Donnelly, P.K., Fletcher, J.S., 1995. PCB metabolism by ectomycorrhizal fungi. Bull. Environ. Contam. Toxicol. 54 (4), 507–513.

Dos Santos, A.B., Cervantes, F.J., van Lier, J.B., 2007. Review paper on current technologies for decolourisation of textile wastewaters: perspectives for anaerobic biotechnology. Bioresour. Technol. 98 (12), 2369–2385.

Dwivedi, S., 2012. Bioremediation of heavy metal by algae: current and future perspective. J. Adv. Lab. Res. Biol. 3 (3), 195–199.

El Fantroussi, S., Agathos, S.N., 2005. Is bioaugmentation a feasible strategy for pollutant removal and site remediation? Curr. Opin. Microbiol. 8 (3), 268–275.

El Fantroussi, S., Belkacemi, M., Top, E.M., Mahillon, J., Naveau, H., Agathos, S.N., 1999. Bioaugmentation of a soil bioreactor designed for pilot-scale anaerobic bioremediation studies. Environ. Sci. Technol. 33 (17), 2992–3001.

El-Sheekh, M.M., Gharieb, M.M., Abou-El-Souod, G.W., 2009. Biodegradation of dyes by some green algae and cyanobacteria. Int. Biodeterior. Biodegrad. 63 (6), 699–704.

EPA (U.S. Environmental Protection Agency), 2001. A Citizen's Guide to Monitored Natural Attenuation. EPA 542-F-01-004. U.S. Environmental Protection Agency, Washington, DC..

ERD: Environmental Response Division, 1998. Fundamental Principles of Bioremediation (An Aid to the Development of Bioremediation Proposals).

Fernández, P.M., Martorell, M.M., Fariña, J.I., Figueroa, L.I., 2012. Removal efficiency of Cr6 + by indigenous *Pichia* sp. isolated from textile factory effluent. Sci. World J. 2012.

Floodgate, G., 1984. The fate of petroleum in marine ecosystem. Petroleum Microbiology, Macmillan Publishing Co. 355–398.

Fritsche, W., Hofrichter, M., 2000. Aerobic degradation by microorganisms, Biotechnology, vol. 11. John Wiley & Sons, New York, pp. 146–164.

Fritsche, W., Hofrichter, M., 2005. Aerobic degradation of recalcitrant organic compounds by microorganisms. Environ. Biotechnol. Concepts Appl. .

Fritsche, W., Hofrichter, M., 2008. Aerobic degradation by microorganisms. In: Rehm, H.J., Reed, G. (Eds.), Biotechnology Set, *2nd Edn.*, pp. 145–155.

Gadd, G.M., 1994. Interactions of fungi with toxic metals. The Genus Aspergillus. Springer, Boston, MA, pp. 361–374.

Garbisu, C., Alkorta, I., 2001. Phytoextraction: a cost-effective plant-based technology for the removal of metals from the environment. Bioresour. Technol. 77 (3), 229–236.

George, I.F., Bouhajja, E., Agathos, S.N., 2011. 6.06 – Metagenomics for Bioremediation, second ed. Elsevier B.V.

Glick, B.R., 2010. Using soil bacteria to facilitate phytoremediation. Biotechnol. Adv. 28 (3), 367–374.

Gonzalez-Chavez, M.C., Carrillo-Gonzalez, R., Wright, S.F., Nichols, K.A., 2004. The role of glomalin, a protein produced by arbuscular mycorrhizal fungi, in sequestering potentially toxic elements. Environ. Pollut. 130 (3), 317–323.

Grishchenkov, V.G., Townsend, R.T., McDonald, T.J., Autenrieth, R.L., Bonner, J.S., Boronin, A.M., 2000. Degradation of petroleum hydrocarbons by facultative anaerobic bacteria under aerobic and anaerobic conditions. Process. Biochem. 35 (9), 889–896.

Hauptmann, A.L., Sicheritz-Pontén, T., Cameron, K.A., Bælum, J., Plichta, D.R., Dalgaard, M., et al., 2017. Contamination of the arctic reflected in microbial metagenomes from the Greenland ice sheet. Environ. Res. Lett. 12 (7), 074019.

He, S., McMahon, K.D., 2011. Microbiology of 'Candidatus Accumulibacter' in activated sludge. Microb. Biotechnol. 4 (5), 603–619.

Hong, Y., Xu, M., Guo, J., Xu, Z., Chen, X., Sun, G., 2007. Respiration and growth of Shewanella decolorationis S12 with an azo compound as the sole electron acceptor. Appl. Environ. Microbiol. 73 (1), 64–72.

Hontzeas, N., Zoidakis, J., Glick, B.R., Abu-Omar, M.M., 2004. Expression and characterization of 1-aminocyclopropane-1-carboxylate deaminase from the rhizobacterium *Pseudomonas putida* UW4: a key enzyme in bacterial plant growth promotion. Biochim Biophys Acta 1703 (1), 11–19.

Hrywna, Y., Tsoi, T.V., Maltseva, O.V., Quensen, J.F., Tiedje, J.M., 1999. Construction and characterization of two recombinant bacteria that grow on ortho-and para-substituted chlorobiphenyls. Appl. Environ. Microbiol. 65 (5), 2163–2169.

Huang, X.D., El-Alawi, Y., Penrose, D.M., Glick, B.R., Greenberg, B.M., 2004. Responses of three grass species to creosote during phytoremediation. Environ. Pollut. 130 (3), 453–463.

Ijah, U.J.J., 1998. Studies on relative capabilities of bacterial and yeast isolates from tropical soil in degrading crude oil. Waste Manage. 18 (5), 293–299.

Iwasaki, K., Uchiyama, H., Yagi, O., 1993. Survival and impact of genetically engineered *Pseudomonas putida* harboring mercury resistance gene in aquatic microcosms. Biosci. Biotechnol., Biochem. 57 (8), 1264–1269.

Jadhav, J.P., Kalyani, D.C., Telke, A.A., Phugare, S.S., Govindwar, S.P., 2010. Evaluation of the efficacy of a bacterial consortium for the removal of color, reduction of heavy metals, and toxicity from textile dye effluent. Bioresour. Technol. 101 (1), 165–173.

Jinqi, L., Houtian, L., 1992. Degradation of azo dyes by algae. Environ. Pollut. 75 (3), 273–278.

Joshi, M.N., Dhebar, S.V., Bhargava, P., Pandit, A.S., Patel, R.P., Saxena, A.K., et al., 2014. Metagenomic approach for understanding microbial population from petroleum muck. Genome Announce. 2 (3), e00533–14.

Joutey, N.T., Bahafid, W., Sayel, H., Abed, S.E., Ghachtouli, N.E., 2011. Remediation of hexavalent chromium by consortia of indigenous bacteria from tannery waste-contaminated biotopes in Fez, Morocco. Int. J. Environ. Stud. 68 (6), 901–912.

Kafilzadeh, F., Sahragard, P., Jamali, H., Tahery, Y., 2011. Isolation and identification of hydrocarbons degrading bacteria in soil around Shiraz Refinery. Afr. J. Microbiol. Res. 5 (19), 3084–3089.

Kanade, S.N., Ade, A.B., Khilare, V.C., 2012. Malathion degradation by *Azospirillum lipoferum* Beijerinck. Sci. Res. Rep. 2 (1), 94–103.

Kaplan, C.W., Kitts, C.L., 2004. Bacterial succession in a petroleum land treatment unit. Appl. Environ. Microbiol. 70 (3), 1777–1786.

Khalid, A., Arshad, M., Crowley, D., 2010. Bioaugmentation of azo dyes. Biodegradation of Azo Dyes. Springer, Berlin, Heidelberg, pp. 1–37.

Khan, A.G., 2006. Mycorrhizoremediation—an enhanced form of phytoremediation. J. Zhejiang Univ. Sci. B 7 (7), 503–514.

Ksheminska, H., Honchar, T., Gayda, G., Gonchar, M., 2006. Extra-cellular chromate-reducing activity of the yeast cultures. Open. Life Sci. 1 (1), 137–149.

Ksheminska, H., Fedorovych, D., Honchar, T., Ivash, M., Gonchar, M., 2008. Yeast tolerance to chromium depends on extracellular chromate reduction and Cr(III) chelation. Food Technol. Biotechnol. 46 (4), 419–426.

Kumavath, R.N., Deverapalli, P., 2013. Scientific swift in bioremediation: an overview. In: Patil, Y.B., Rao, P. (Eds.), Applied Bioremediation-Active and Passive Approaches. *Intech, Croatia*, pp. 375–388.

Lebeau, T., Braud, A., Jézéquel, K., 2008. Performance of bioaugmentation-assisted phytoextraction applied to metal contaminated soils: a review. Environ. Pollut. 153 (3), 497–522.

Leigh, M.B., Prouzová, P., Macková, M., Macek, T., Nagle, D.P., Fletcher, J.S., 2006. Polychlorinated biphenyl (PCB)-degrading bacteria associated with trees in a PCB-contaminated site. Appl. Environ. Microbiol. 72 (4), 2331–2342.

Leitão, A.L., 2009. Potential of Penicillium species in the bioremediation field. Int. J. Environ. Res. Public Health 6 (4), 1393–1417.

Li, C.H., Wong, Y.S., Tam, N.F.Y., 2010. Anaerobic biodegradation of polycyclic aromatic hydrocarbons with amendment of iron(III) in mangrove sediment slurry. Bioresour. Technol. 101 (21), 8083–8092.

Lloyd, J.R., Lovley, D.R., 2001. Microbial detoxification of metals and radionuclides. Curr. Opin. Biotechnol. 12 (3), 248–253.

Lodato, A., Alfieri, F., Olivieri, G., Di Donato, A., Marzocchella, A., Salatino, P., 2007. Azo-dye conversion by means of *Pseudomonas* sp. OX1. Enzyme Microb. Technol. 41 (5), 646–652.

Luo, J., Bai, Y., Liang, J., Qu, J., 2014. Metagenomic approach reveals variation of microbes with arsenic and antimony metabolism genes from highly contaminated soil. PLoS One 9 (10), e108185.

Ma, B., Chen, H.H., He, Y., Xu, J.M., 2010. Isolations and consortia of PAH-degrading bacteria from the rhizosphere of four crops in PAH-contaminated field. In: 19th World Congress of Soil Science.

Marinescu, M., Dumitru, M., Lăcătuşu, A.R., 2009. Biodegradation of petroleum hydrocarbons in an artificial polluted soil. Res. J. Agric. Sci. 41 (2), 157–162.

Markandey, D.K., Rajvaidya, N., 2004. Environmental Biotechnology, first ed. APH Publishing Corporation, p. 79.

Matavulj, M., Molitoris, H.P., 2009. Marine fungi: degraders of poly-3-hydroxyalkanoate based plastic materials. Zb. Mat. Srp. Prir. Nauk. 116, 253–265.

Matin, A., 1994. Starvation promoters of *Escherichia coli*: their function, regulation, and use in bioprocessing and bioremediation. Ann. N.Y. Acad. Sci. 721 (1), 277–291.

Mattison, R.G., Taki, H., Harayama, S., 2005. The soil flagellate *Heteromita globosa* accelerates bacterial degradation of alkylbenzenes through grazing and acetate excretion in batch culture. Microb. Ecol. 49 (1), 142–150.

McMurry, J., 2000. In aromatic hydrocarbons, Organic Chmistry, fifth ed. Brooks/Cole: Thomson Learning, New York, pp. 120–180.

Meneghetti, L.R., Thomé, A., Schnaid, F., Prietto, P.D., Cavelhão, G., 2012. Natural attenuation and biostimulation of biodiesel contaminated soils from southern Brazil with different particle sizes. J. Environ. Sci. Eng. B 1 (2), 155–162.

Menn, F.M., Easter, J.P., Sayler, G.S., 2008. Genetically engineered microorganisms and bioremediation. In: second ed. Rehm, H.-J., Reed, G. (Eds.), Biotechnology: Environmental Processes II, 11b. Wiley-VCH Verlag GmbH, Weinheim, Germany, <https://doi.org/10.1002/9783527620951.ch21>.

Miranda, R.D.C., Souza, C.S.D., Gomes, E.D.B., Lovaglio, R.B., Lopes, C.E., Sousa, M.D.F. V.D., 2007. Biodegradation of diesel oil by yeasts isolated from the vicinity of Suape Port in the state of Pernambuco-Brazil. Braz. Arch. Biol. Technol. 50 (1), 147–152.

Mörtberg, M., Neujahr, H.Y., 1985. Uptake of phenol by *Trichosporon cutaneum*. J. Bacteriol. 161 (2), 615–619.

Mrozik, A., Piotrowska-Seget, Z., 2006. Bioaugmentation as a strategy for cleaning up of soils contaminated with aromatic compounds. Microb. Res. 165 (2009). Available from: https://doi.org/10.1016/j.micres.2009.08.001.

Mucha, K., Kwapisz, E., Kucharska, U., Okruszek, A., 2010. Mechanism of aniline degradation by yeast strain *Candida methanosorbosa* BP-6. Pol. J. Microbiol. 59 (4), 311–315.

Mukherjee, A., Chettri, B., Langpoklakpam, J.S., Basak, P., Prasad, A., Mukherjee, A.K., et al., 2017. Bioinformatic approaches including predictive metagenomic profiling reveal characteristics of bacterial response to petroleum hydrocarbon contamination in diverse environments. Sci. Rep. 7 (1), 1108.

Nizzetto, L., Macleod, M., Borgå, K., Cabrerizo, A., Dachs, J., Guardo, A.D., et al., 2010. Past, present, and future controls on levels of persistent organic pollutants in the global environment. Environ. Sci. Technol. 2010, 44.

Okere, U.V., Semple, K.T., 2012. Biodegradation of PAHs in 'pristine' soils from different climatic regions. J. Bioremed. Biodegrad. 1 (2).

Olaniran, A.O., Pillay, D., Pillay, B., 2006. Biostimulation and bioaugmentation enhances aerobic biodegradation of dichloroethenes. Chemosphere 63, 600–608.

Pal, T.K., Sauryya, B., Arunabha, B., 2010. Cellular distribution of bioaccumulated toxic heavy metals in *Aspergillus niger* and *Rhizopus arrhizus*. Int. J. Pharm. Biol. Sci. 1 (2).

Pandey, G., Paul, D., Jain, R.K., 2005. Conceptualizing "suicidal genetically engineered microorganisms" for bioremediation applications. Biochem. Biophys. Res. Commun. 327 (3), 637–639.

Pena-Castro, J.M., Martınez-Jerónimo, F., Esparza-Garcıa, F., Canizares-Villanueva, R.O., 2004. Heavy metals removal by the microalga *Scenedesmus incrassatulus* in continuous cultures. Bioresour. Technol. 94 (2), 219–222.

Petric, I., Hrsak, D., Fingler, S., Voncina, E., Cetkovic, H., Kolar, A.B., et al., 2007. Enrichment and characterization of PCB-degrading bacteria as potential seed cultures for bioremediation of contaminated soil. Food Technol. Biotechnol. 45 (1), 11–20.

Pramila, R., Padmavathy, K., Ramesh, K.V., Mahalakshmi, K., 2012. Brevibacillus parabrevis, *Acinetobacter baumannii* and *Pseudomonas citronellolis*—potential candidates for biodegradation of low density polyethylene (LDPE). Afr. J. Bacteriol. Res. 4 (1), 9–14.

Prenafeta-Boldú, F.X., Andrea, K.U.H.N., Luykx, D.M., Heidrun, A.N.K.E., Van Groenestijn, J.W., 2001. Isolation and characterisation of fungi growing on volatile aromatic hydrocarbons as their sole carbon and energy source. Mycol. Res. 105 (4), 477–484.

Ramos, J.L., Díaz, E., Dowling, D., de Lorenzo, V., Molin, S., O'Gara, F., et al., 1994. The behavior of bacteria designed for biodegradation. Bio/Technology 12 (12), 1349.

Rani, M.S., Lakshmi, K.V., Devi, P.S., Madhuri, R.J., Aruna, S., Jyothi, K., et al., 2008. Isolation and characterization of a chlorpyrifos degrading bacterium from agricultural soil and its growth response. Afr. J. Microbiol. Res. 2 (2), 26–31.

Ripp, S., Nivens, D.E., Ahn, Y., Werner, C., Jarrell, J., Easter, J.P., et al., 2000. Controlled field release of a bioluminescent genetically engineered microorganism for bioremediation process monitoring and control. Environ. Sci. Technol. 34 (5), 846–853.

Riser-Roberts, E., 1998. Remediation of Petroleum Contaminated Soils: Biological, Physical, and Chemical Processes. CRC Press.

Rogerson, A., Berger, J., 1981. Effect of crude oil and petroleum-degrading micro-organisms on the growth of freshwater and soil protozoa. Microbiology 124 (1), 53–59.

Rossman, A.Y., 1995. Microfungi: molds, mildews, rusts, and smuts. In: Our Living Resources., US Department of Interior- National Biological Service, Washington, DC, pp. 190–194.

Saharan, B.S., Nehra, V., 2011. Plant growth promoting rhizobacteria: a critical review. Life Sci. Med. Res. 21 (1), 30.

Sar, P., Kazy, S.K., D'Souza, S.F., 2004. Radionuclide remediation using a bacterial biosorbent. Int. Biodeterior. Biodegrad. 54 (2–3), 193–202.

Šašek, V., Volfová, O., Erbanová, P., Vyas, B.R.M., Matucha, M., 1993. Degradation of PCBs by white rot fungi, methylotrophic and hydrocarbon utilizing yeasts and bacteria. Biotechnol. Lett. 15 (5), 521–526.

Sayler, G.S., Ripp, S., 2000. Field applications of genetically engineered microorganisms for bioremediation processes. Curr. Opin. Biotechnol. 11 (3), 286–289.

Seeger, M., Cámara, B., Hofer, B., 2001. Dehalogenation, denitration, dehydroxylation, and angular attack on substituted biphenyls and related compounds by a biphenyl dioxygenase. J. Bacteriol. 183 (12), 3548–3555.

Seeger, M., Hernández, M., Méndez, V., Ponce, B., Córdova, M., González, M., 2010. Bacterial degradation and bioremediation of chlorinated herbicides and biphenyls. J. Soil Sci. Plant Nutr. 10 (3), 320–332.

Seo, J.S., Keum, Y.S., Li, Q., 2009. Bacterial degradation of aromatic compounds. Int. J. Environ. Res. Public Health 6 (1), 278–309.

Shan, H., Kurtz Jr, H.D., Freedman, D.L., 2010. Evaluation of strategies for anaerobic bioremediation of high concentrations of halomethanes. Water Res. 44 (5), 1317–1328.

Singh, H., 2006. Mycoremediation: Fungal Bioremediation. John Wiley & Sons.

Singh, J.S., Abhilash, P.C., Singh, H.B., Singh, R.P., Singh, D.P., 2011. Genetically engineered bacteria: an emerging tool for environmental remediation and future research perspectives. Gene 480 (1–2), 1–9.

Spellman, F.R., 2008. Ecology for Non Ecologists. pp. 176.

Sriprang, R., Hayashi, M., Ono, H., Takagi, M., Hirata, K., Murooka, Y., 2003. Enhanced accumulation of Cd^{2+} by a *Mesorhizobium* sp. transformed with a gene from *Arabidopsis thaliana* coding for phytochelatin synthase. Appl. Environ. Microbiol. 69 (3), 1791–1796.

Srivastava, N.K., Jha, M.K., Mall, I.D., Singh, D., 2010. Application of genetic engineering for chromium removal from industrial wastewater. Int. J. Chem. Biol. Eng. 3 (3).

Stapleton Jr, R.D., Singh, V.P. (Eds.), 2002. Biotransformations: Bioremediation Technology for Health and Environmental Protection, vol. 36. Elsevier.

Steliga, T., 2012. Role of fungi in biodegradation of petroleum hydrocarbons in drill waste. Pol. J. Environ. Stud. 21 (2).

Struthers, J.K., Jayachandran, K., Moorman, T.B., 1998. Biodegradation of atrazine by *Agrobacterium radiobacter* J14a and use of this strain in bioremediation of contaminated soil. Appl. Environ. Microbiol. 64 (9), 3368–3375.

Subhashini, S.S., Kaliappan, S., Velan, M., 2011. Removal of heavy metal from aqueous solution using *Schizosaccharomyces pombe* in free and alginate immobilized cells. In: Second International Conference on Environmental Science and Technology, vol. 6, pp. 108–111.

Sussman, M., Collins, C.H., Skinner, F.A., Stewart-Tull, D.E., 1988. Release of Genetically-Engineered Micro-Organisms. Academic Press, London.

Taguchi, K., Motoyama, M., Kudo, T., 2001. PCB/biphenyl degradation gene cluster in *Rhodococcus rhodochrous* K37, is different from the well-known bph gene clusters in *Rhodococcus* sp. P6, RHA1, and TA421. RIKEN Rev. 23–26.

Takeuchi, F., Sugio, T., 2006. Volatilization and recovery of mercury from mercury-polluted soils and wastewaters using mercury-resistant *Acidithiobacillus ferrooxidans* strains SUG 2-2 and MON-1. Environ. Sci. 13 (6), 305–316.

Ufarte, L., Laville, E., Duquesne, S., Potocki-Veronese, G., 2015. Metagenomics for the discovery of pollutant degrading enzymes. Biotechnol. Adv. 33 (8), 1845–1854.

Van der Heul, R.M., 2011. Environmental Degradation of Petroleum Hydrocarbons (Master's Thesis).

Van Der Meer, J.R., Belkin, S., 2010. Where microbiology meets microengineering: design and applications of reporter bacteria. Nat. Rev. Microbiol. 8 (7), 511.

Vargas, J.M., 1975. Pesticide degradation. J. Arboric. 1 (12), 232–233.

Wackett, L.P., 2004. Stable isotope probing in biodegradation research. Trends Biotechnol. 22 (4), 153–154.

Walker, J.D., Colwell, R.R., Vaituzis, Z., Meyer, S.A., 1975. Petroleum-degrading achlorophyllous alga *Prototheca zopfii*. Nature 254 (5499), 423.

Wang, J., Chen, C., 2006. Biosorption of heavy metals by *Saccharomyces cerevisiae*: a review. Biotechnol. Adv. 24 (5), 427–451.

Wang, X.C., Zhao, H.M., 2007. Uptake and biodegradation of polycyclic aromatic hydrocarbons by marine seaweed. J. Coast. Res. 1056–1061.

White, C., Shaman, A.K., Gadd, G.M., 1998. An integrated microbial process for the bioremediation of soil contaminated with toxic metals. Nat. Biotechnol. 16 (6), 572.

Wiedemeier, T.H., Miller, R.N., Wilson, J.T., Significance of anaerobic processes for the Intrinsic bioremediation of fuel hydrocarbons. In: Proceedings of the Petroleum Hydrocarbons and Organic Chemicals in Groundwater – Prevention, Detection, and Remediation Conference, November 29–December 1, 1995, Houston TX.

Yakimov, M.M., Timmis, K.N., Golyshin, P.N., 2007. Obligate oil-degrading marine bacteria. Curr. Opin. Biotechnol. 18 (3), 257–266.

Yu, K.S.H., Wong, A.H.Y., Yau, K.W.Y., Wong, Y.S., Tam, N.F.Y., 2005. Natural attenuation, biostimulation and bioaugmentation on biodegradation of polycyclic aromatic hydrocarbons (PAHs) in mangrove sediments. Mar. Pollut. Bull. 51 (8–12), 1071–1077.

Development of biologically-based activated carbon for advanced water and wastewater treatment process

9

Ravi Kant Singh[1,*], *Santosh Kumar Mishra*[2],
Balasubramanian Velramar[1] and *Priya Ranjan Kumar*[2]
[1]Amity Institute of Biotechnology, Amity University Chhattisgarh, Raipur, India,
[2]Department of Biotechnology, IMS Engineering College, Ghaziabad, India
*Corresponding author

9.1 Introduction

Development of biological activated carbon (BAC) technology is based on activated carbon technology in association with its biological counterpart such as microorganisms. Activated carbon is used as an absorption medium and also plays a significant and important role in comparison to conventional treatment process for removing various contaminants present in water including removal of organic contaminants. BAC technology is a combination of degradation by efficient microbes and activated carbon-based adsorption process. In this process when the adsorptive efficiency of activated carbon gets exhausted then microbial action takes place in biological activated carbon. The BAC system protects the microorganisms due to immobilization on the surface of activated carbon which acts as a support material. In this system the adsorption process due to adsorbent and biodegradation process due to microorganisms takes place simultaneously. The development of biologically-based activated carbon (BAC) system is required for improving the adsorptive capacity of activated carbon and the degradation efficiency of microorganisms attached on the outer surface for the purpose of wastewater treatment. Nowadays, BAC technology is one of the most mature and effective processes for the complete removal of organic pollutants and its colors and odors. The first water treatment plant using granular activated carbon (GAC) was used for water treatment in 1930 and the large-scale process was established in Philadelphia, Pennsylvania, United States (Wenjie et al., 2006). Later, activated carbon technology was utilized at small scale for effective treatment of wastewater released from industries and other household activities. Since free chlorine inhibits the growth of microorganisms, therefore after prechlorination as the first step of activated carbon treatment process got removed, hence no obvious biological activity showed on the outer surface of activated carbon molecules. In 1961, the combination of ozonation and

Bioremediation of Pollutants. DOI: https://doi.org/10.1016/B978-0-12-819025-8.00009-0

adsorption was used for the first time to improve the removal in Germany (Weissenhorn, 1977). Later, the growth of microorganisms on the outer layer of activated carbon was first time encouraged by Parkhrust and his partners in 1967 (Parkhurst et al., 1976; Weiguang et al., 1999). During the early 1970s, the study and application of ozonation-(O_3-BAC) treatment process was developed for the large-scale wastewater treatment system. Presently the use of O_3-BAC technology in Germany is widely spread and is being used globally for wastewater treatment. Environmental Protection Agency of the United States legislated that activated carbons must be used in potable water treatment process as a small scale process in areas not having population over 150,000. The quality for potable water was improved by Japan in 1988 by using ozonation-(O_3-BAC) system (Naiyun et al., 2005). Recently, BAC process has been considered as an advanced water treatment process for several developing countries also, earlier this process was significant in developed countries only (Wenjie et al., 2006). Meanwhile, the BAC process is also commonly used in wastewater treatment and its reclamation process. Because of the serious concerns associated with the removal of pollutants present in potable water and the increasing importance of the requirements for potable water treatment unit, BAC process has emerged as a conventional process due to its ease in adsorption and biological-oxidation degradation properties (Yu et al., 1998). Recent studies reveal that BAC filter accomplished a maximum removal of 2-methylisoborneol and 2,4-dichlorophenoxyacetic acid (2,4-D) compared to nonacclimated GAC filter (Shimabuku Kyle et al., 2019). Complex organic pollutants rereleased from municipal reverse osmosis concentrate can be treated using UV/ H_2O_2 and BAC (Pradhan et al., 2020). Currently, the processing of pharmaceutical industrial waste is an important issue, especially with drug disposal. In this context, *Pseudoxanthomonas* sp. DIN-3 isolated from the BAC system is used for the efficient removal of nonsteroidal antiinflammatory drugs (NSAIDs) by a bench-scale BAC method. The biofilm of *Pseudoxanthomonas* sp. DIN-3 formed inside the BAC removed 23%–48% NSAIDs such as, diclofenac, ibuprofen, and naproxen (Zedong et al., 2019).

9.2 Biological activated carbon process

9.2.1 Basic principles of biological activated carbon technology

BAC process is developed to use the synergistic effect of adsorption on adsorbent and microorganisms for wastewater treatment effectively. Due to high surface area and pore size the activated carbon is characterized by its great effect on absorbing organic pollutants in wastewater. In this system, activated carbon is used as a support material: the microorganisms are reproduced on the surface of the adsorbents and finally form biologically activated carbon (BAC), which can perform the physiochemical and microbial roles simultaneously (Tian and Chen, 2006). The BAC system is based on the interaction among activated carbon particles, microorganisms, contaminants, dissolved oxygen, and other pollutants present in the soluble

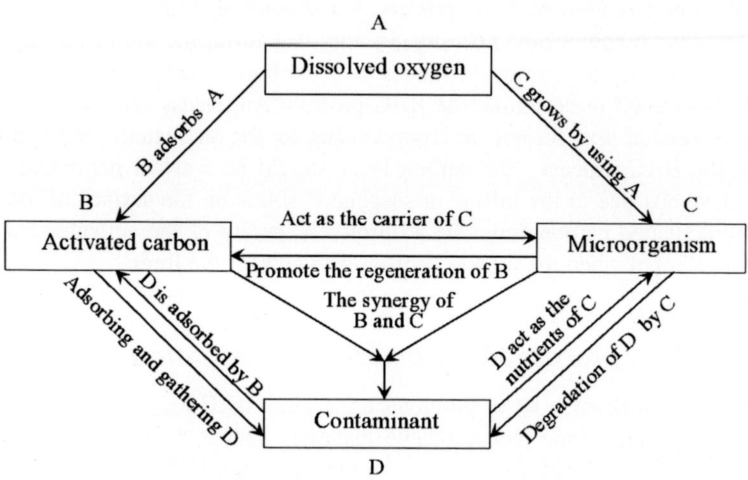

Figure 9.1 Schematic overview of interaction between activated carbon particles, microorganisms, contaminants, dissolved oxygen, and other pollutants present in the soluble state in the BAC process of wastewater treatment. *Source*: Adopted from Jin, P., Jin, X., Wang, X., Feng, Y., and Wang, X.C., 2013. Biological activated carbon treatment process for advanced water and wastewater treatment. Biomass Now-Cultivation and Utilization, 2013, IntechOpen London, 153-192.

state in the wastewater. The relationship between the activated carbon and pollutants can affect the adsorptive capacity of adsorbent, and the rate of reaction also depends on the properties of the adsorbents and level of pollutants. Activated carbon has the ability to absorb dissolve oxygen (DO) and microorganisms on the surface of activated carbon, and further feed of DO has the potential to biodegrade the pollutants. Briefly we can say that interaction of these agents is useful for removing pollutants from wastewater by adopting the BAC process (Lan, 2002). A simplified model of BAC process can be explain using Fig. 9.1.

The significant removal of disinfection by-products (DBPs) and their precursors is produced during the ozonation process by BAC (Chu et al., 2012, 2015; Liu et al., 2017; McKie et al., 2015; Wu et al., 2013; Yan et al., 2010). In the treatment of wastewater, after the BAC process, the O_3 treatment has been established in order to remove the DBPs (Chuang and Mitch, 2017; Reungoat et al., 2011).

9.2.2 Basic operational parameters of biological activated carbon process

Design and development of BAC system is necessary to comprehend the process of water quality and water treatment. Information about the adsorption performance and biodegradability of the wastewater needs an appropriate kind of activated carbon and can be selected on the basis of dynamic adsorption isotherms. The activated carbon is used in BAC process and should be uniform in size and processed to achieve appropriate porosity. Factors that influence the efficiency of the process such as filtering velocity, thickness of carbon layer, retention period, etc. must be considered in designing the BAC process. An excess amount of dissolved oxygen is needed for aerobic microorganisms in the activated carbon layer to improve the BAC process. The carbon layer should be washed periodically, as a biofilm develops due to the inflow of suspended solids on the surface of long-term operating activated carbon bed. The adsorptive capacity of the activated carbon is adversely affected when the thickness of biofilm crosses the limit.

9.2.3 O_3-BAC process and estimation of ozonation

During the experiments, a lot of problems arise when BAC technology is used separately; for example, some biodegradable materials cannot be removed completely and hence the life span of BAC system would be reduced. To ensure the safety of water distribution system. Disinfection is required after biological treatment process. If chlorinated water is used, an excess quantity of halogenated organic by-products is formed during the reaction between organic compounds and the chlorine, which are generally carcinogenic in nature. Therefore, preozonization for example, ozonization before BAC treatment is widely used. O_3-BAC process has three steps—ozonation, adsorptive efficiency of activated carbon, and biodegradation (Wang, 1999). In O_3-BAC process, the organic pollutants are first oxidized into small degradable molecules in the presence of strong oxidizing agents such as ozone, subsequently small degradable molecules gets adsorbed onto the surface of activated carbon and are degraded by microorganisms present on the outer surface of activated carbon. Simultaneously the oxygen decomposed from ozone enhances the level of DO, and leads to make DO in wastewater saturated, which further provides an optimized condition for biodegradation of waste materials (Wentworth et al., 2003; Kim and Nishijima, 1997; Nishijima and Speitel, 2004; Zhao et al., 1999; Anderson and Laurent, 2001).

9.3 Mechanisms of pollutants degradation in biological activated carbon filtration

9.3.1 Immobilization of microorganisms on the carriers

In ordinary BAC process, biofacies formed due to long-term operation causes several problems in the treatment process. Since biofacies are complex in nature and present on the surface, conventional BAC has some limitations. In the conventional BAC process, it is very hard for the dormant microflora to survive and grow. Due to technological advances in recent years in the area of immobilization techniques, nowadays immobilized biological activated carbon (IBAC) is being used. IBAC process majorly depends on the screening and acclimatization of dormant

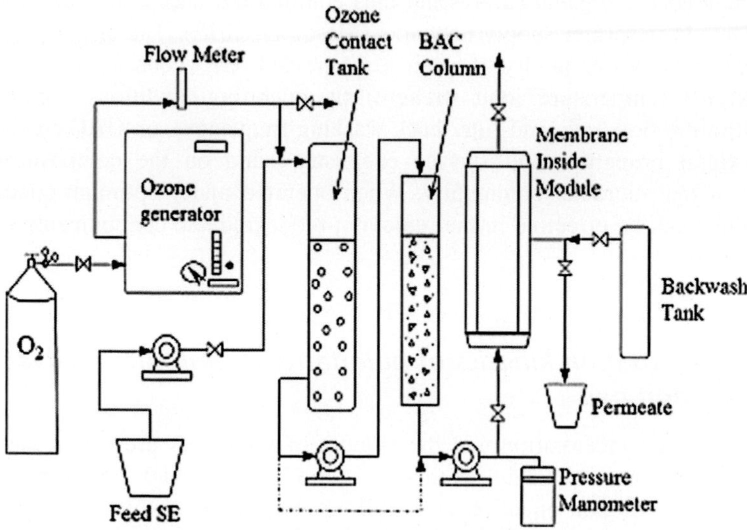

Figure 9.2 Diagrammatic illustration of biological activated carbon filtration unit with the ozonation for the treatment of wastewater. *Source*: Reproduced with permission from Hamid, K.I.A., Sanciolo, P., Gray, S., Duke, M., and Muthukumaran, S., 2017. Impact of ozonation and biological activated carbon filtration on ceramic membrane fouling. Water research, 126, 308-318. Elsevier Inc.

microflora from nature followed by immobilization on activated carbon for efficient degradation. Microflora used in IBAC should be nonpathogenic in nature. The biological community present in this system have strong antioxidant properties as well as enzyme activity which enables growth and reproduction of these microflora in poor nutrition conditions. Therefore, the efficiency of biodegradation is improved by using IBAC (Wang and Luo, 2003; Scholz and Martin, 1997). A simplified schematic diagram of BAC filtration unit is given in Fig. 9.2.

9.3.2 Synergy of activated carbon adsorption and biodegradation

GAC, besides being an excellent adsorbent material, is a good support media for the attachment and growth of microorganisms and biofilm development. Activated carbon biofilters take advantage of the accumulation of substrates in the GAC particle surface and of the roughness of the carbon granules which protect microbial cells from the fluid shear stress. Microorganisms established therein can biodegrade the adsorbed compounds. The synergy of the adsorption, desorption, biodegradation, and filtration processes determines the effectiveness of these systems. Biodegradation also promotes continuous bioregeneration of activated carbon and consequently increases filter lifetime. Furthermore, it allows removal of assimilable and biodegradable organic carbon and thus controls the undesirable development of biofilms in the water supply networks (Mesquita, 2012). The efficacy of BAC filters depends on the quality of water to be treated (e.g., concentration and type of NOM, pH, temperature, ionic strength), the operating conditions (e.g., contact time, filtration flow rate, and filter back washing frequency), on GAC characteristics (textural properties and surface chemistry), and on the composition and number of the microbial community. When operated under optimum conditions, BAC filters can be effective in the removing biodegradable contaminants such as cyanotoxins

9.3.3 Degradation kinetics of pollutants in biological activated carbon bed

The inoculums concentration and the amount of pollutants present in the water are key factors for establishing biodegradation kinetics model. For such biocatalytic reactions, Monod equation is a well-accepted model which is appreciable for nearly all such kinetics studies (Yanhui and Guowei, 1999). For wastewater treatment process the specific biodegradation rate of the substrate is considered as a key factor in comparison to specific growth rate of microbes. When pollutants present in the water are considered as substrate the specific degradation rate of the pollutants can be calculated by the equation below (Xiaochang and Chengzhong, 2010).

$$-dC/dt = V_{\max} * XC/(K_s + C) \tag{9.1}$$

wherein, dC/dt stands for the degradation rate of pollutant/substrate; C stand for residual pollutant/substrate conc. in mixed liquor after a reaction time t; K_s stand for saturated constant, X stands for biomass concentration; $V_{\max}$ stand for the max specific biodegradation rate of pollutant/substrate.

In the above equation, concentration of substrate, C is the key factor for developing biodegradation kinetic models.

When the saturation constant K_s of the microorganism is highly greater than the concentration of pollutant/substrate then the above equation will be found as below:

$$-dC/dt = k_1 XC \tag{9.2}$$

In the above equation, k_1 is the ratio of max specific biodegradation rate of pollutant/substrate and saturated constant ($V_{\max}/K_s$).

When the concentration of pollutant/substrate C is highly greater than the saturation constant K_s then the above Eq. (9.1) will be found as below:

$$-dC/dt = k_2 X \tag{9.3}$$

when the graph is plotted between biomass concentration X as horizontal coordinate and dC/dt as vertical coordinate, Eq. (9.3) shows linear behavior. In the above-mentioned conditions, the kinetic model is applicable for high matrix organic degradation, while if the graph is in exponential form then Eq. (9.2), is applicable for low matrix organic degradation.

9.3.4 Removal of micropollutants in biological activated carbon filters

Adsorption and biodegradation have predominant mechanisms to contribute to removing organic pollutants in biologically active carbon-based columns. The rate of bioremediation of pollutants can be attributed mostly to biodegradation (Yapsakli and Çeçen, 2010). BAC probably combines the adsorption with the biosorption on the biofilm and possible biodegradation for a longer extent with better efficacy. It potentially results in a more effective treatment system for aqueous solutions, with relevant concentrations of specific organic compounds. Unfortunately, it is difficult to identify their relative importance at different operational stages. The removal by adsorption alone can be determined under controlled conditions.

Development of biofilm in biologically active filters makes up a diverse microbial community (Zhang et al., 2011). The biofilm is capable of breaking down and removing adsorbed organic pollutants known as biodegradation which participates in the treatment of biodegradable organic waste, and may degrade in secondary wastewater treatment process. A similar effect has been observed for cyanotoxins in drinking water treated with biologically activated carbon (BAC) (Wang et al., 2007).

Table 9.1 Removal of organic micropollutants with biological activated carbon (BAC) filters (Snyder et al., 2006).

Compound	BAC removal (%)
Triclosan	99
Diclofenac	99
Fluoxetine	99
Atenolol	98
Naproxen	98
Trimethoprim	97
Carbamazepine	80
Diazepam	70
Norfluoxetine	66
Sulfamethoxazole	51
Dilantin	50
Meprobamate	29

BAC microbial biofilms are capable of bioregeneration, a mechanism known to regenerate activated carbon with the help of microbial activities (Aktas and Cecen, 2007). BAC biodegradation depends on a few factors besides reversibility of adsorption and biodegradability of the adsorbates, for example, the ability of the microbial population to biodegrade the substrates and mainly the micropollutants.

Bioregeneration is the hypothesis of desorption by enzymatic reactions, and indeed a biofilm containing bacteria in the macropores can be established on BAC system (Nath and Bhakhar, 2010; Scholz and Martin, 1997). These exo-enzymes are smaller than a bacterial cell, hence it can diffuse into the meso-and micropores of adsorbents. These enzymes are acted on the adsorbate which also induces its hydrolytic reactions. These products generated may have a lower affinity for the activated carbon and therefore desorbs, renewing adsorption capacity. Snyder et al. investigated the removal of individual organic micropollutants in a full-scale BAC filter (Table 9.1).

In BAC system, the microorganisms utilize electron acceptors which provide the maximum free energy during respiration. The organics that have a greater affinity for activated carbon have a low adsorption–desorption rate because the compounds are not released significantly in the aqueous phase (Saino et al., 2004). Therefore, the compounds most likely to be biodegraded are those that have the least affinity for GAC and therefore are the most likely to be found in the aqueous phase (Levine et al., 2000).

9.4 Conclusion and future remarks

In recent times regarding biodegradation of pollutants present in wastewater, BAC process is considered as the better alternative compared to processes where activated carbon were used for wastewater treatment. The clefts and macropores of activated carbon are serving as excellent support for the colonization of microorganisms by protection from the shear stress. Adsorption and biodegradation are simultaneously facilitated in BAC system and extend the life of activated carbon filter. O_3-BAC is developed to influence the oxidation of organic compounds present in effluent water. Recently, O_3-BAC is combined with microfilter and ultrafilter which enhance the total wastewater treatment process. Within the past few decades BAC system plays a significant role in wastewater treatment. Due to the heavy usage of groundwater there is scarcity throughout the world, for which wastewater treatment has been developed enormously to reuse wastewater after the treatment. The reuse of water is considered after the removal of organic matter in the treated effluent for different purposes. The quality of water has improved and reduced formaldehyde and NMDA-related health risk by BAC system. The recent research results support the BAC system development. The BAC process may be established in all type of chemical industries including pharmaceutical industries for wastewater treatment to remove pollutants and toxic chemicals.

References

Aktas, O., Cecen, F., 2007. Bioregeneration of activated carbon: a review. Int. Biodeterior. Biodegrad. 59 (4), 257–272.

Andersson, A., Laureent, P., 2001. Impact of temperature on nitrification in biological activated carbon (BAC) filters used for drinking water treatment. Water Res. 35 (12), 2923–2934.

Chu, W., Gao, N., Yin, D., Deng, Y., Templeton, M.R., 2012. Ozone-biological activated carbon integrated treatment for removal of precursors of halogenated nitrogenous disinfection by-products. Chemosphere 86, 1087–1091.

Chu, W., Yao, D., Gao, N., Bond, T., Templeton, M.R., 2015. The enhanced removal of carbonaceous and nitrogenous disinfection by-product precursors using integrated permanganate oxidation and powdered activated carbon adsorption pretreatment. Chemosphere 141, 1–6.

Chuang, Y., Mitch, W.A., 2017. Effect of ozonation and biological activated carbon treatment of wastewater effluents on formation of *N*-nitrosamines and halogenated disinfection byproducts. Environ. Sci. Technol. 51, 2329–2338.

Hamid, K.I.A., Sanciolo, P., Gray, S., Duke, M., and Muthukumaran, S., 2017. Impact of ozonation and biological activated carbon filtration on ceramic membrane fouling. Water research. 126, 308–318.

Jin, P., Jin, X., Wang, X., Feng, Y., and Wang, X.C., 2013. Biological activated carbon treatment process for advanced water and wastewater treatment. Miodrag Darko Matovic (Ed), *Biomass Now-Cultivation and Utilization*, IntechOpen, London, 153–192.

Kim, W.H., Nishijima, W., 1997. Micropollutant removal with saturated biological activated carbon (BAC) in ozonation-BAC process. Water Sci. Technol. 36 (12), 283–298.

Lan, S., 2002. Biological activated carbon technology and its application for wastewater treatment. Water Waste Water Eng. 28 (12), 125.

Levine, B.B., Madireddi, K., Lazarova, V., Stenstrom, M.K., Suffet, I.H., 2000. Treatment of trace organic compounds by ozone-biological activated carbon for wastewater reuse: the lake arrowhead pilot plant. Water Environ. Res. 72 (4), 388–396.

Liu, C., Olivares, C.I., Pinto, A.J., Lauderdale, C.V., Brown, J., Selbes, M., et al., 2017. The control of disinfection by-products and their precursors in biologically active filtration processes. Water Res. 124, 630–653.

McKie, M.J., Taylor-Edmonds, L., Andrews, S.A., Andrews, R.C., 2015. Engineered biofiltration for the removal of disinfection by-product precursors and genotoxicity. Water Res. 81, 196–207.

Mesquita, E., 2012. Remoção de cianotoxinas da água para consumo humano em filtros de carvão activado com actividade biológica. Algarve University, Faro, Portugal PhD thesis.

Naiyun G., Min Y., Yuesheng L. (2005) Strengthening Treatment Technology for Drinking Water.

Nath, K., Bhakhar, M., 2010. Microbial regeneration of spent activated carbon dispersed with organic contaminants: mechanism, efficiency, and kinetic models. Environ. Sci. Pollut. Res. 18 (4), 534–546.

Nishijima, W., Speitel, E.G., 2004. Fate of biodegradable dissolved organic carbon produced by ozonation on biological activated carbon. Chemosphere 56 (2), 113–119.

Parkhurst, J.P., et al., 1976. Pomona activated carbon pilot plant. J. WPCF 37 (1).

Pradhan, S., Fan, L., Roddick, F.A., Shahsavari, E., Ball, A.S., Zhang, X., 2020. A comparative study of biological activated carbon based treatments on two different types of municipal reverse osmosis concentrates. Chemosphere 240, 124925.

Reungoat, J., Escher, B.I., Macova, M., Argaud, F.X., Gernjak, W., Keller, J., 2011. Ozonation and biological activated carbon filtration of wastewater treatment plant effluents. Water Res. 46, 863–872.

Saino, H., Yamagata, H., Nakajima, H., Shigemura, H., Suzuki, Y., 2004. Removal of endocrine disrupters in wastewater by SRT control. J. Jpn. Soc. Water Environ. 27 (1), 61–67.

Scholz, M., Martin, R.J., 1997. Ecological equilibrium on biological activated carbon. Water Res. 31 (12), 2959–2968.

Shimabuku Kyle, K., Zearley, Thomas L., Dowdell, Katherine S., Scott Summers, R., 2019. Biodegradation and attenuation of MIB and 2,4-D in drinking water biologically active sand and activated carbon filters, Environ. Sci. Water Res. Technol., 5. pp. 849–860.

Snyder, S.A., Wert, E.C., Rexing, D.J., Zegers, R.E., Drury, D.D., 2006. Ozone oxidation of endocrine disruptors and pharmaceuticals in surface water and wastewater. Ozone: Sci. Eng. 28 (6), 445–460.

Tian, Q., Chen, J., 2006. Application of bioactivated carbon (BAC) process in water and wastewater treatment. Environ. Eng. 24 (1), 84–86.

Wang, B.Z., 1999. The efficacy and mechanism of removal of organic substances from water by ozone and activated carbon. Water Sci. Technol. 30 (1), 43–47.

Wang, L., Luo, Q., 2003. Study on degradation of immobilized microorganism on endocrine disruptor di-*n*-butyl phthalate. China J. Public Health 19 (11), 1302–1303.

Wang, H., Ho, L., Lewis, D.M., Brookes, J.D., Newcombe, G., 2007. Discriminating and assessing adsorption and biodegradation removal mechanisms during granular activated carbon filtration of microcystin toxins. Water Res. 41 (18), 4262–4270.

Weiguang, L., Fang, M., Xianji, Y., Qingliang, Z., Qingguo, W., 1999. A study on purification performance of biological activated carbon. J. Harbin Univ. Civ. Eng. Arch. 6 (32), 105–109.

Weissenhorn, F.J., 1977. The behavior of ozone in the system and its transformation. AMKBerlin 2, 51–57.

Wenjie, H., Weiguang, L., Xiaojian, Z., Tingling, H., Hongda, H., 2006. Novel Technology for Drinking Water Safety. China Architecture & Building Press.

Wentworth, P., Nieva, J., Akeuchi, T.C., et al., 2003. Evidence for ozone formation in human atherosclerotic arteries. Science 302 (7), 1053–1056.

Wu, Y., Zhu, G., Lu, X. (2013) Characteristics of DOM and removal of DBPs precursors across O3-BAC integrated treatment for the micro-polluted raw water of the huangpu river. Water (Switzerland) 5, 1472–1486.

Xiaochang, W., Chengzhong, Z., 2010. Environmental Engineering Science. Higher Education Press, Beijing.

Yan, M., Wang, D., Ma, X., Ni, J., Zhang, H., 2010. THMs precursor removal by an integrated process of ozonation and biological granular activated carbon for typical Northern China water. Separ. Purif. Technol. 72, 263–268.

Yanhui, G., Guowei, G., 1999. Water Pollution Control Engineering. Higher Education Press, Beijing.

Yapsakli, K., Çeçen, F., 2010. Effect of type of granular activated carbon on DOC biodegradation in biological activated carbon filters. Process. Biochem. 45 (3), 355–362.

Yu, T., Xiangrong, Z., Ding, Z., 1998. Development of coordinating technology for ozonation and biological activated carbon system. J. Harbin Inst. Technol. 30 (2), 21–25.

Zedong, Lu, Wenjun, Sun, Chen, Li, Xiuwei, Ao, Chao, Yang, Simiao, Li, 2019. Bioremoval of non-steroidal anti-inflammatory drugs by Pseudoxanthomonas sp. DIN-3 isolated from biological activated carbon process. Water Res. 161, 459–472.

Zhang, D., Li, W., Zhang, S., Liu, M., Zhao, X., Zhang, X., 2011. Bacterial community and function of biological activated carbon filter in drinking water treatment. Biomed. Environ. Sci. 24 (2), 122–131.

Zhao, X., Hickey, R.F., Voice, T.C., 1999. Long-term evaluation of adsorption capacity in biological activated carbon fluidized bed reactor system. Water Res. 33 (13), 2983–2991.

The role of microorganism in bioremediation for sustainable environment management

10

Arun Kumar Pal[1,†], *Jyotsna Singh*[1,†], *Ramendra Soni*[1,†], *Pooja Tripathi*[2], *Madhu Kamle*[3], *Vijay Tripathi*[1,*] *and Pradeep Kumar*[3,*]
[1]Department of Molecular and Cellular Engineering, JIBB, Sam Higginbottom University of Agriculture, Technology and Sciences, Prayagraj, India, [2]Department of Computational Biology and Bioinformatics, JIBB, Sam Higginbottom University of Agriculture, Technology and Sciences, Prayagraj, India, [3]Department of Forestry, North Eastern Regional Institute of Science and Technology, Nirjuli, India
*Corresponding authors

10.1 Introduction

Bioremediation is the process in which microorganisms are used to degrade and remove pollutants from the environment. In other words, bioremediation is a procedure for the eradication of environmental pollutants from the ecosystem through the application of microbes to restore the ecosystem to its original form (Ayangbenro and Babalola, 2017).

The native environment of microbial populations helped in the bioremediation property of the microorganism, but it cannot perform in a foreign environment (Ramírez et al., 2013). The native microbial communities which were exposed to hydrocarbons converted proverbial, and finally they were genetically modified. The microbial communities which are adapted in the hydrocarbon polluted area already take less time to degrade the pollutant in comparison to those microorganisms that are not exposed to the pollutant (Kannepalli and Farrish, 2016). Thus, home grown microorganisms have more pollutant degrading ability at the contaminated place, and serve as gifted for the pollutant degradation at the contaminated site. Finally it is proved that, this microorganism is the most suitable for pollutant degradation at the polluted area (Kannepalli and Farrish, 2016). For degradation of complex pollutants, they need to be more than a single species, because one species metabolizes only a limited number of pollutants. Therefore, mixed populations capable of broad enzymatic capacities are entailed to increase the rate and altitude of pollutant biodegradation further.

At this time it has been proven that the microbial consortium is the best technique for removing contaminants from the environment (Sharma, 2012). During

[†] Authors equally contributed.

Bioremediation of Pollutants. DOI: https://doi.org/10.1016/B978-0-12-819025-8.00010-7

bioremediation, microbial communities (consortia) on the polluted site where these microbial communities are use the contaminants as the nutrient, resulting in degradation of the pollutant (Singh et al., 2001). Nowadays various conventional physicochemical methods such as osmosis, evaporation, and sorption are used in removal of the contaminant, while bioremediation processes are working very efficiently even at very low levels of contaminants where physicochemical techniques cannot be performed efficiently (Coelho et al., 2015).

The marine environment is mostly polluted by heavy metals. Heavy metals interact with protein, and abort the metabolic process causing death of living beings. According to the World Health Organization heavy metals including cadmium, chromium, cobalt, copper, lead, nickel, mercury, and zinc contaminate the whole environment, that is, water, soil, sediments, which is extremely hazardous for the biological and ecological system (Priyalaxmi et al., 2014). Genetically engineered microorganisms (GEMs) are showing budding potential for biodegradation of pollutants in groundwater, activated sludge, and soil environments, exhibiting enhanced degradative capabilities surrounding a wide range of chemical contaminants. Recently, scientists focused on GEMs for bioremediation of pollutants through preparation of consortia of selected microorganisms. Microorganisms that are used in consortia preparation are ecofriendly; there is no side effect on human beings as well as the environment (Abatenh et al., 2017). Pollutants are introduced into the environment from industries, hospitals, municipal sewage treatment plants, agriculture lands, etc. which are dangerous for humans, animals, as well as plants and as the outcome, they imbalance the whole ecological system. These pollutants change the physical, chemical, and biological properties of water and soil. The requirement for alternative methods rather than traditional methods to clean polluted sites resulted in evolution of bioremediation techniques, that is, different microorganisms decontaminate different pollutants present in the soil and water (Das and Adholeya, 2012).

10.2 Types of bioremediation

Based on site of application, principles, advantages, limitations, and possible solutions, Bioremediation can be classified in following types:

- *Ex situ bioremediation techniques*: These techniques basically involved the excavation of the samples and the treatment is done away from the contaminated sites, resulting in the high cost and disruption of natural sites (Guang-Guo, 2018). The most common example for the ex situ bioremediation is composting which is the degradation of organic waste products in the presence of microorganisms typically at elevated temperatures ranging from 55°C to 65°C (Das and Dash, 2014).
- *In situ bioremediation techniques*: In situ bioremediation is the application of an onsite biological treatment to eliminate lethal compounds present in the environment. Some of the in situ bioremediation practices have been discussed below:

Biosparging: This strategy increases groundwater oxygen concentrations to enhance the rate of biological degradation and volatilization of contaminants with the help of naturally occurring bacteria by means of injecting air under pressure below the water table (Brusseau and Maier, 2004; Das and Dash, 2014).

Bioventing: Bioventing promotes the natural in situ biodegradation of any aerobically degradable compounds by supplying oxygen to the existing soil microorganisms with a low air flow rate to ensure enough oxygen to sustain microbial activity. Oxygen is directly injected into residual contamination in soil by means of wells. Henceforth, the adsorbed fuel residuals along with other contaminants are biodegraded and volatile compounds are also biodegraded as vapors move slowly through biologically active soil (Das and Dash, 2014).

Bioaugmentation: Bioaugmentation is a technique of bioremediation which involves a cluster of natural microbial strains or a genetically engineered strain for the treatment of contaminated soil or water. Generally, it is applied in municipal wastewater treatment plants to revive activated sludge bioreactors. The sites that are having soil and ground water contaminated with chlorinated ethanes, like tetrachloroethylene and trichloroethylene, are applied with bioaugmentation to make certain that the in situ microorganisms can absolutely break down these pollutants to ethylene and chloride, which are nontoxic in nature (Das and Dash, 2014).

Biopiling: Biopiling is a large scale technology in which excavated soils are heaped into piles, and further bioremediated under the robust aeration condition (Germaine et al., 2015; Li et al., 2004). The pollutants are further degraded to CO_2 and water. An elementary biopile system is consisted of a treatment bed, an aeration system, an irrigation system, and a leachate collection system (Das and Dash, 2014). Environmental factors like moisture, heat, nutrients, oxygen, and pH which can affect the bioremediation process are also monitored and controlled to augment biodegradation. The irrigation/nutrient system is fixed in fashion to be covered under the soil to pass the circulate air and nutrients either by vacuum or positive pressure.

10.3 Types of microorganisms used in bioremediation

Microorganisms play an important role as naturopaths in the bioremediation as biosorbents, biocatalysts, and bio-accumulators as shown in Fig. 10.1. Some of the categories are hereby discussed briefly.

1. *Bacteria*: Bacteria are very helpful for bioremediation as they are crucial biosorbents for the treatment of contaminated ecosystems since they carry the ability to propagate under controlled conditions and can also endure extreme environmental conditions. They exhibit good biosorbents activity against heavy metals in polluted environments (Srivastava and Dwivedi, 2015). Peptidoglycan layers containing alanine, glutamic acid, and teichoic acid, present in Gram-positive bacteria, while enzymes, glycoproteins, lipopolysaccharides, lipoproteins, and phospholipids present in Gram-negative bacteria cell walls (Ayangbenro and Babalola, 2017), serve as active sites for binding heavy metal ions which consequently results in the remediation of heavy metals from the polluted environment (Gupta et al., 2015). The various bacterial species such as *Bacillus cereus*, *Bacillus subtilis*, and *Pseudomonas putida* can be efficiently applied for the removal of Cr (VI) (Coelho et al., 2015; Balamurugan et al., 2014)

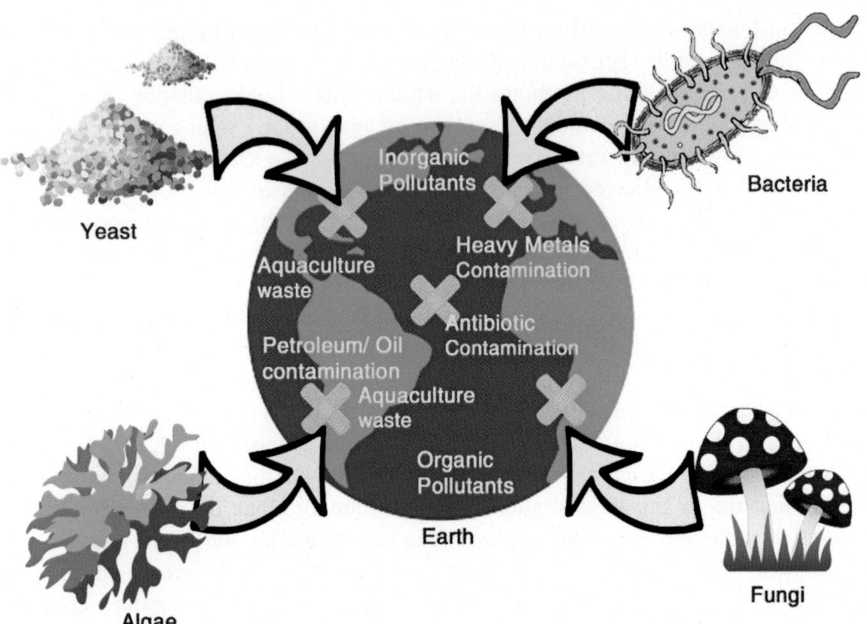

Figure 10.1 Bioremediation using microorganisms.

whereas *Sporosarcina ginsengisoli*, *Pseudomonas veronii*, and *Kocuria flava* have potential to eliminate As, Cu, Cd, and Zn from the polluted environment (Vullo et al., 2008; Achal et al., 2012a,b; Coelho et al., 2015). Consortia of bacterial strains is found to be more efficient than a single strain culture for bioremediation of heavy metals (Wang and Chen, 2009).

2. *Fungi (Mycoremediation)*: As fungi carry the capability to absorb heavy metals into their mycelium and spores, they act as robust biocatalysts for heavy metal bioremediation. Their modus operandi for bioremediation of metal ions is accumulation, intracellular, and extracellular precipitation, and valence transformation. Certainly, fungi are the only organisms on Earth that are capable of decaying wood. The mycelium of fungi emanates powerful extracellular enzymes and acids capable of decomposing lignin and cellulose (Gupta et al., 2017). Strain of *Rhizopus oryzae* CDBB-H-1877 has the biosorption ability of pentachlorophenol through the process of methylation and dechlorination (León-Santiesteban et al., 2016). Most of the fungi from class zygomycetes and *Aspergillus* sp. have been validated to transmit the property of decolorizing and detoxifying textile wastewaters. Several notable fungi like *Penicillium chrysogenum*, *Scedosporium apiospermum*, *Penicillium digitatum*, and *Fusarium solani* are also described for degradation capabilities of polychlorinated biphenyls (PCBs) (Tigini et al., 2009).
3. *Algae (cyanoremediation)*: Algae has been proven to be very efficient for the bioaccumulation of heavy metals and degradation of xenobiotics (Suresh and Ravishankar, 2004). Algae bioremediation process attracted the attention of researchers over the years as it plays a vital role in the carbon dioxide fixation and is also very advantageous biomass for biofuel production (Huang et al., 2010). Species like *Selenastrum capricornutum* and *Monoraphidium braunii* have the potential to remove bisphenol, benzene, toluene, and naphthalene from the polluted environment (Gavrilescu 2010; Gattullo et al., 2012).

The bioadsorption of organic and pharmaceutical contaminants (PC) by microalgae has been well reported (Xiong et al., 2018). Algae species like *Chlorella sorokiniana*, *Nannochloris* sp., and *Chlamydomonas mexicana*, have crucial application for the bioremediation of diclofenac (40–60), ibuprofen (100), paracetamol (100), metoprolol (100), carbamazepine (30), trimethoprim (40), Ibuprofen (40), trimethoprim (10), ciprofloxacin (100), carbamazepine (20), triclosan (100), and Carbamazepine (30–37) (de Wilt et al., 2016; Xiong et al., 2016; Bai and Acharya, 2017).

4. *Yeast*: Yeasts can be the most ubiquitous alternative for bioremediation of heavy metals from the polluted environment as they have the capability to withstand a wide range of environmental conditions (Abbas and Badr, 2015). Yeast cell walls carry a negative charge on it, and are capable to bind with heavy metal cations likely due to electrostatic interactions (Fukudome et al., 2002). *Saccharomyces cerevisiae* is the extensively applied yeast species for biosorption of heavy metals like Cr, Pb, Zn, Hg, Cd, and Ni (Ozer and Ozer, 2003; Mapolelo and Torto, 2004; Göksungur et al., 2005). Other than *S. cerevisiae, Candida tropicalis*, *Candida utilis*, and *Cyberlindnera fabianii* are also reported as bioremediation of Cr from polluted ecosystems (Bahafid et al., 2013).

10.4 Bioremediation of mined wasteland

Mining activities give rise to many negative environmental and socioeconomic impacts (Aznar-Sánchez et al., 2018). Blasting, drilling, construction of various auxiliary facilities, and storage of excess imposed dump materials, etc. drastically affect the land of mining areas (Sharma et al., 2013). Land deprivation due to mining has become a concern primarily due to disproportionate excavation and mismanagement of natural resources. Increased land degradation and decreased man to land ratio due to mining has utterly affected the environment in the forms of habitat destruction and biodiversity. It has also given rise to other environmental issues like groundwater alteration or contamination, affected workplace health and safety, the impact of metals toxicity, siltation, and changes in river regimes, effluent discharges, and acid drainage. Mining and related activities leave severe adverse effects on the soil environment as the green cover gets lost resulting in exposing more area to erosion, which consequently allows nutrients and contaminants to get leached into the soil and causes loss of productivity of the soil as well as the anchorage capacity of the soil for plants (Chaulya et al., 2000). Currently, there is a crucial demand for an effective solution to overcome the environmental issues caused by the mining activities. As physicochemical methods of remediation results in very high cost, some of the microorganisms can be an alternative tool and can play an important role in bioremediation of mined lands. Mwandira et al. (2017) first used *Pararhodobacter* sp. and achieved complete removal of 1036 mg/L of Pb^{2+} by using microbial-induced calcium carbonate precipitation (MICP) for bioremediation of Pb contaminated mining site. MICP is a very efficient and promising method for bioremediation of heavy metal (Cd, Ni, Pb) contaminated soils (Kang et al., 2014; Zhu et al., 2016; Achal et al., 2012b). *Pararhodobacter* sp. exhibits high urease activity which helps it to be buoyant when introduced in the ground and its enzyme activity retains for a stretched time duration to achieve biocementation and immobilization of the toxic ions (Danjo and Kawasaki, 2016; Li et al., 2016).

10.5 Bioremediations of landfill leachates

Landfilling is one of the most common and broadly accepted practices for waste dumping around the globe, consequently which gives rise to leachate. The term landfill leachate (LFL) can be simplified as the wastewater infiltrates and percolates through the degrading solid waste. It has become a significant threat to groundwater and surface water. Currently treatment of LFLs is a major challenge for the society, for which bioremediation can be a significant alternative approach as it comes up with a lower operation cost than the physicochemical methods. Bioremediation of LFLs may involve bacteria, fungi, algae, and other microorganisms. Morris et al. (2018) reported isolated bacterial strains belonging from phylum *Firmicutes*, *Actinobacteria*, and *Proteobacteria* from LFL and in their experiment, they achieved 90% removal rate of ammonia and 60% removal rate of nitrate and phosphate. Paskuliakova et al. (2018) obtained a significant decrease between 70% and 100% in ammonia nitrogen using the algal strain SW15aRL of *Chlamydomonas* sp., whereas Spina et al. (2018) assessed the bioremediation capacity of autochthonous fungi *Pseudallescheria boydii* MUT 721, *Phanerochaete sanguinea* MUT 1284 and achieved good and very fast decolorization percentage (60%), coupled by detoxification.

10.6 Bioremediation of aqua culture waste

After China and Indonesia, India has an output of about 5.7 million metric tons of aquaculture produced annually. Due to the lack of management, most of the aquaculture wastewater was discharged directly without being treated, and it may have caused significant damage to the receiving water. Ammonia is the main nitrogen-containing compound waste which was considered to be a primary contaminant of aquaculture. The heterotrophic bacteria degrade the wastes that are released in the aquatic environment through nitrification, denitrification, and hydrogen sulfide breakdown. However, in vigorous aquaculture, the naturally occurring bacteria are not sufficient to efficiently breakdown the waste. This is due to the growth of nitrifying bacteria which is relatively slow (Achuthan, 2000); and the bacteria that are more suitable for biodegrading the organic matter are generally found in the sediment, thus lacking in the water column (Antony and Philip, 2006). Therefore, for a proper aquaculture waste treatment, it should be necessary to accelerate the biodegradation process by either stimulating the microbe growth or by augmentation of microbes (Khan et al., 2004).

Among other chemicals, aquaculture wastes contain dissolved and suspended organic compounds which include mainly carbon chains and are abundantly available to microbes and algae. Microbial degraders require organic matter for their own growth (Wu et al., 2014). Microbes that bring metabolic capability for oxidizing organic matter, breakdown aquaculture wastes by their metabolic skills and transform them, either in a less reduced organic form or in the inorganic compounds CO_2 and H_2O, the end products of the complete oxidation process or mineralization, thus restoring the contaminated sites and controlling further pollution (Wu et al., 2012). Bacteria belonging to the Gram-positive genus *Bacillus* (e.g., *B. subtilis*, *Bacillus licheniformis*, *B. cereus*, and *Bacillus coagulans*) and of the genus *Paenibacillus* (e.g., *Paenibacillus polymyxa*) have been known to efficiently breakdown carbonaceous organic detritus to CO_2 by using enzymes that help to breakdown proteins and starch to small molecules (Antony and Philip, 2006). However, they are not usually present in the required quantities in the water column because their natural habitat is the sediment. The inorganic nutrients and CO_2 can be then used by the phytoplankton while the biomass and the cellular protein are eaten by zooplankton, prawns, shrimps, and fishes (Phillips et al., 2004). In the pond sediment, the microbial biomass bind with other dissolved matter to form detritus, which is a food source to the aquatic fish (Bowen, 1987). When certain *Bacillus* strains are added to the water in sufficient quantities, they can make an impact. They compete with the bacterial flora naturally present for the available organic matter, like leached or excess feed and shrimp feces (Sharma, 1999). Along with *Bacillus* to break down the organic detritus, *Lactobacillus* is also used by producing a variety of enzymes followed by a breakdown of proteins and starch to small molecules, which are then later consumed by other organisms as energy sources. The removal of abundant organic compounds reduces water turbidity (Haung, 2003).

Another type of bioremediation is through the breakdown of hydrogen sulfide (H_2S) under aerobic conditions. In aerobic conditions, the organic sulfur decomposes to sulfide later then be oxidized to sulfate which is highly soluble in water, and therefore it gradually disperses from sediments. This process is carried out by different microorganisms. Under anaerobic conditions, sulfate can be used as a terminal electron receptor of microbial metabolism, which leads to the production of H_2S gas. H_2S is soluble in water and has created an ill effect by causing damage of gills and other ailments in fish (Beveridge, 1987). which can later anaerobically metabolized by the photosynthetic purple and sulfur bacteria (Antony and Philip, 2006).

10.7 Microbial nitrification and denitrification in sediments

The nitrogen compounds (ammonia, nitrite, and nitrate) can lead to deterioration of water quality with toxicity to fish and shrimp. The ammonia nitrogen compounds are removed by autotrophic nitrification and sometimes by heterotrophic nitrification. The main autotrophic bacteria (*Nitrosomonas*, *Nitrosovibrio*, *Nitrosolobus*, *Nitrococcus*, *Nitrosococcus*, *Nitrospira*, and *Nitrobacter*) (Koops and Andreas, 2001), heterotrophic bacteria (*Nitrobacter*, *Nitrococcus*, and *Nitrospira*) (Nelson, 1929; Remacle, 1977), as well as some fungi such as *Aspergillus flavus* (Odu and Adeoye, 1970), are involved in the nitrification process. In the first step of the nitrification process ammonia is converted to nitrite through a complex process and is further oxidized to nitrate and the last phase is the denitrification where anaerobic bacteria reduce the accumulated nitrate to harmless nitrogen gas which bubbles out of the pond system. Sometimes the cycle ends before denitrification, as the nitrate produced by nitrification process is absorbed by the microbes and phytoplanktons thus limiting nitrate for the denitrification process (Chavez-Crooker and Obreque-Contreras, 2010). Apart from this there are some friendly bacteria which play the role as a bioremediator and act as biocontrol agents of aquatic diseases. These bacteria are applied as a probiotic to improve the health of the cultured animals to enhance and modify the intestinal microbiota balance when given as a food supplement (Parker, 1974). The most commonly used probiotic bacteria are *Lactobacillus*, *Carnobacterium*, *Vibrio alginolyticus*, *Bacillus*, and *Pseudomonas* (Singh et al., 2001).

10.8 Bioremediations of contaminated soil and water

Anthropogenic activities have led to environmental deterioration worldwide. These deteriorators can be either organic [e.g., PCBs, polycyclic aromatic hydrocarbons (PAHs), fertilizers, pesticides] or inorganic pollutants including various heavy metals (e.g., Cd, Cu, As, Zn, Hg, Pb). Augmentation of agriculture and industries render harm to humans, livestock, wildlife, flora, and fauna, causing ecological problems leading to disparity in nature by delivering a wide range of xenobiotics. Microorganisms play a necessary part in bioremediation of heavy metals. A few metals such as nickel, copper, iron, and arsenic are beneficial for the human body in minimal amounts but are toxic as well as mutagenic and carcinogenic in nature at a high concentration (Valko et al., 2016). Adverse health effects of some heavy metals are listed in Table 10.1. Prokaryotes and eukaryotes have a natural tendency to biosorb toxic heavy metal ions (Singh et al., 2014). Examples of various microorganisms including bacteria, algae, fungi, and yeast strategically used in bioremediation treatments for heavy metals are listed in Tables 10.2–10.5, respectively. The species *Enterococcus faecium*, *B. subtilis*, *Saccharomyces boulardii*, and *Staphylococcus aureus*, and *Escherichia coli* have also been used for removing heavy metals from water bodies (Bhakta et al., 2014). Some bacteria, such as *Geobacter metallireducens*, can remove uranium from drainage waters in mining operations and contaminated groundwater (Kumar et al., 2011). Numerous microorganisms can utilize oil as a source of food (Kumar et al., 2011). Bacteria that can degrade petroleum products include species of *Pseudomonas*, *Aeromonas*, *Moraxella*, *Beijerinckia*, *Flavobacteria*, *Chrobacteria*, *Nocardia*, *Corynebacteria*, *Modococci*, *Streptomyces*, *Bacilli*, *Arthrobacter*, *Aeromonas*, and cyanobacteria (Thapa et al., 2012) and some

Table 10.1 Adverse effects of heavy metals on human health.

Heavy metals	Adverse effects
Lead (Pb)	Acute or chronic damage to the nervous system/neurotoxic, affect children's brain development, anemia, hypertension, renal impairment, immunotoxicity, and toxicity to the reproductive organs
Cadmium (Cd)	Renal dysfunction, lung disease, lung cancer, and damage to respiratory systems
Copper (Cu)	Anemia at high dose, liver and kidney damage, stomach, and intestinal irritation
Mercury (Hg)	Blindness and deafness, brain damage, digestive problems, kidney damage, lack of coordination, mental retardation, nervous, immune systems, on lungs, kidneys, skin, and eyes
Arsenic (As)	Sensory changes, numbness, tingling and muscle tenderness, neuropathy, hyperpigmentation, hyperkeratosis, cancer, cardiovascular disease, and diabetes
Nickel (Ni)	Skin cancer, asthma, chronic bronchitis, reduced lung function, lung cancer, and nasal sinus
Chromium (Cr)	Irritation of skin, kidney, and liver damage and damage to circulatory and nerve tissues at long-term exposure
Zn	Abdominal pain, nausea, vomiting and diarrhea, gastric irritation, headache, irritability, lethargy, anemia

yeasts (Kumar et al., 2011). A recombinant strain of *Acinetobacter baumannii* was found to magnify degradation rates at crude oil contaminated sites (Paul et al., 2005). Hydrocarbon compounds such as petroleum are essential for life, but they can be hazardous. Fuel and lubricating oil spills have become a significant environmental hazard as of now for many countries. Petroleum hydrocarbon degraders include *Yokenella* sp., *Alcaligenes* sp., *Roseomonas* sp., *Stenotrophomonas* sp., *Acinetobacter* sp., *Flavobacter* sp., *Corynebacterium* sp., *Streptococcus* sp., *Providencia* sp., *Sphingobacterium* sp., *Capnocytophaga* sp., *Moraxella* sp., *Bacillus* sp., and *Enterobacter* sp. degrade a wide range of detoriators present in oily sludge (Eriksson et al., 1999; Barathi and Vasudevan, 2001; Mishra et al., 2001; Bhattacharya et al., 2003; Jain et al., 2010). Apart from bacteria, organisms such as fungi are also capable of degrading the engine oil hydrocarbons. However, they take more extended periods to grow as compared to their bacterial counterparts (Prenafeta-Boldu et al., 2001). Among fungi, genera of, *Neosartorya*, *Graphium*, *Talaromyces*, *Aspergillus*, and *Amorphoteca* potentially degrade hydrocarbon. Hydrocarbonoclastic bacteria (HCB) are of great importance since they belong to the key players in oil removal from contaminated marine sites. HCB such as *Alcanivorax*, *Marinobacter*, *Thallassolituus*, *Cycloclasticus*, and *Oleispira* play a vital role in the degradation of hydrocarbons as carbon and energy source for growth (Genovese et al., 2014). Among this, a group able to degrade PAHs, include *Pseudomonas*, *Marinomonas*, *Halomonas*, and *Micrococcus*. Moreover, *Acinetobacter* and *Sphingomonas* are well described in the degradation of aliphatic compounds

Table 10.2 Bacterial species employed in the bioremediation of heavy metals.

Heavy metals	Bioremediator	Reference
Cu	*Bacillus* sp.	Gunasekaran et al. (2003)
	Chlorella vulgaris	
	Pleurotus ostreatus	
	Pseudomonas aeruginosa	Philip et al. (2000)
	Phormidium valderianum	Chatterjee et al. (2008)
	Volvariella volvacea	
	Daedalea quercina	
	Micrococcus luteus	–
	Pseudomonas stutzeri	
	Pseudomonas cepacia	Savvaidis et al. (2003)
	Sphaerotilus natans	Beolchini et al. (2006)
	Kocuria flava	Achal et al. (2011)
	Stenotrophomonas sp.	Zaki and Farag (2010)
	Zooglea sp.	Sar and DSouza (2001)
	Citrobacter sp.	Sar and DSouza (2001)
	Enterobacter sp.	Bestawy et al. (2013)
Ni	*Zooglea* sp.	Sar and DSouza (2001)
	C. vulgaris	Gunasekaran et al. (2003)
	P. aeruginosa	Chatterjee et al. (2008)
	P. valderianum	
	Desulfovibrio desulfuricans	Ockjoo et al. (2015)
	Enterobacter cloacae	Banerjee et al. (2015)
Zn	*Bacillus* sp.	Chatterjee et al. (2008)
	C. vulgaris	
	D. quercina	
	P. ostreatus	Gunasekaran et al. (2003)
	Acinetobacter sp.	Tabaraki et al. (2013)
	Streptomyces rimosus	Mameri et al. (1999)
	Aphanothece halophytica	Incharoensakdi and Kitjaharn (2002)
	Thiobacillus ferrooxidans	Liu et al. (2004)
U	*P. aeruginosa*	Chatterjee et al. (2008)
	C. vulgaris	Gunasekaran et al. (2003)
	Zooglea sp.	Sar and DSouza (2001)
	Citrobacter sp.	
Co	*Zooglea* sp.	Sar and DSouza (2001)
	P. valderianum	Chatterjee et al. (2008)
Cd	*P. valderianum*	Chatterjee et al. (2008)
	Ganoderma applanatum	
	Zooglea sp.	Sar and DSouza (2001)
	Citrobacter sp.	Sar and DSouza (2001)
	P. ostreatus	Gunasekaran et al. (2003)
	Stereum hirsutum	Gunasekaran et al. (2003)
	D. desulfuricans	Ockjoo et al. (2015)
	E. cloacae	Banerjee et al. (2015)

(*Continued*)

Table 10.2 (Continued)

Heavy metals	Bioremediator	Reference
	Ochrobactrum anthropic	Ozdemir et al. (2003)
	Sphingomonas paucimobilis	Tangaromsuk et al. (2002)
	Stenotrophomonas sp.	Bestawy et al. (2013)
Pb	*Citrobacter* sp.	Chatterjee et al. (2008)
	G. applanatum	
	V. volvacea	
	D. quercina	
	C. vulgaris	Gunasekaran et al. (2003)
	Burkholderia sp.	Jiang et al. (2008)
	Corynebacterium glutamicum	Choi and Yun (2004)
	Bacillus firmus	Salehizadeh and Shojaosadati (2003)
	Bacillus sp.	Tunali et al. (2006)
	E. cloacae	Banerjee et al. (2015)
	P. aeruginosa	Ahmady-Asbchin et al. (2015)
	Pseudomonas putida	Uslu and Tanyol (2006)
Hg	*C. vulgaris*	Gunasekaran et al. (2003)
	Rhizopus arrhizus	
	V. volvacea	Chatterjee et al. (2008)
	Geobacter metallireducens	
Au	*C. vulgaris*	Gunasekaran et al. (2003)
	G. metallireducens	Chatterjee et al. (2008)
	Stenotrophomonas spp.	Song et al. (2008)
Ag	*R. arrhizus*	Chatterjee et al. (2008)
	G. metallireducens	
Cr	*Desulfovibrio vulgaris*	Chatterjee et al. (2008)
	Desulfuromonas acetoxidans	
	Desulfovibrio fructosovorans	
	Desulfovibrio norvegicus	
	D. desulfuricans	Ockjoo et al. (2015)
	Zoogloea ramigera	Nourbakhsh et al. (1994)
AS	*Sporosarcina ginsengisoli*	Achal et al. (2012)

(Harayama et al., 2004). *Bacillus thuringiensis* has been used for enhanced bacterial biodegradation of diesel oil (Mnif et al., 2017). Radioactive resistant bacteria (*Deinococcus radiodurans*) have updated to assimilate and utilize ionic mercury and toluene from overly radioactive waste (Manobala et al., 2018; Shukla and Rao, 2017).

10.9 Antibiotics bioremediation

From last few years, pharmaceutical companies have majorly contributed to pollution levels, therefore it became a serious concern for society and attracted the attention of whole research community for the dissemination of PC. But on the other

Table 10.3 Fungal spices employed in the bioremediation of heavy metals.

Heavy metals	Bioremediator	Reference
Cr	*Aspergillus* sp.	Zafar et al. (2007)
	Rhizopus sp.	
	Aspergillus lentulus	Mishra and Malik (2012)
	Aspergillus foetidus	Prasenjit and Sumathi (2005)
	Rhizopus oligosporus	Ariff et al. (1999)
	Termitomyces clypeatus	Fathima et al. (2015)
Zn	*Trametes versicolor*	Şahan et al. (2015)
	Rhizopus arrhizus	Fourest and Roux (1992)
	Penicillium spinulosum	Townsley and Ross (1985)
Cu	*R. arrhizus*	Fourest and Roux (1992)
	P. spinulosum	Townsley and Ross (1985)
	Penicillium chrysogenum	Niu et al. (1993)
	Ganoderma lucidum	Muraleedharan and Venkobachar (1990)
	Aureobasidium pullulans	Gadd and Mowll (1985)
	Aspergillus niger	Iskandar et al. (2011)
Ni	*R. arrhizus*	Fourest and Roux (1992)
	Rhizopus nigricans	
U	*Pleurotus mutilus*	Mezaguer et al. (2013)
	Aspergillus terreus	Tsezos and Volesky (1981)
Cd	*R. arrhizus*	Fourest and Roux (1992)
	R. nigricans	
	Phanerochaete chrysosporium	Gabriel et al. (1996)
	Pleurotus sapidus	Yalçinkaya et al. (2002)
	P. chrysogenum	Niu et al. (1993)
	Aspergillus sp.	Zafar et al. (2007)
Pb	*R. arrhizus*	Fourest and Roux (1992)
	R. nigricans	
	P. ostreatus	Zhang et al. (2016)
	P. chrysogenum	Niu et al. (1993)
	A. niger	Iskandar et al. (2011)
	A. lentulus	Mishra and Malik (2012)
Hg	*P. sapidus*	Yalçinkaya et al. (2002)
	Candida parapsilosis	Muneer et al. (2013)
Th	*A. terreus*	Tsezos and Volesky (1981)

side the use of low concentration antibiotics stimulate the antibiotic resistance in the microorganism. From the pharmaceutical industries and hospitals, pharmaceutical compounds are released into the environment. Consequently these compounds come in contact with surface water, groundwater, as well as wastewater treatment plant resulting in its widespread to the environment and finally stimulating the antibiotic resistance in the microorganism (Huerta et al., 2012). Excluding metabolites

Table 10.4 Algae spices employed in the bioremediation of heavy metals.

Heavy Metals	Bioremediator	Reference
Cr	*Spirogyra* sp.	Mane and Bhosle (2012)
	Spirulina sp.	
Cu	*Spirogyra* sp.	Mane and Bhosle (2012)
	Spirulina sp.	
	Chlorella vulgaris	Goher et al. (2016)
Pb	*C. vulgaris*	Kumaran et al. (2011)
	Nostoc sp.	Aung et al. (2013)
Cd	*C. vulgaris*	Kumaran et al. (2011)
	Nostoc sp.	Aung et al. (2013)
Zn	*Nostoc* sp.	Aung et al. (2013)
Ni	*C. vulgaris*	Wong et al. (2000)
	Nostoc sp.	Aung et al. (2013)
Fe	*Nostoc* sp.	Aung et al. (2013)

Table 10.5 Yeast as a microbial bioremediator.

Heavy metal	Bioremediator	Reference
Pb	*Saccharomyces cerevisiae*	Livia de and Benedito (2015)
	Streptomyces longwoodensis	Friss and Myers-Keith (1986)
Cd	*S. cerevisiae*	Livia de and Benedito (2015)
	Candida utilis	Kujan et al. (2006)
	Candida tropicalis	Mattuschka et al. (1993)
Cr	*S. cerevisiae*	Livia de and Benedito (2015)
	C. tropicalis	Mattuschka et al. (1993)
Ni	*S. cerevisiae*	Livia de and Benedito (2015)
	C. tropicalis	Mattuschka et al. (1993)
Cu	*Pichia guilliermondi*	Mattuschka et al. (1993)
	C. tropicalis	
Zn	*S. cerevisiae*	Livia de and Benedito (2015)
	C. tropicalis	Mattuschka et al. (1993)

and transformation products, there are more than 3500 pharmaceutical compounds that have been detected in different environmental compartments; they have been detected in aquatic environments, mainly in surface water and wastewater effluents. The nonsteroidal anti-inflammatory drugs represent a major class of detected pharmaceuticals (Marie-Pierre and Elena, 2016). At this time researchers are focusing on a biological tool for removing the pollutant. Toxic pharmaceutical compounds convert into a nontoxic compound with the help of biodegradation process in the presence and absence of oxygen (Mishra and Malik, 2014).

10.10 Biodegradability of antibiotics

Antibiotic degradation takes place in both biotic and abiotic conditions. Biotic degradation involves several microorganisms for the degradation of pollutants whereas abiotic degradation involves sorption, hydrolysis, photolysis, oxidation, and reduction reactions. The bioremediation process can be elucidated as a metabolic potential possessed by microorganisms. A web of interconnected metabolic pathways of microorganisms utilize various environmental substances, that is, petroleum, oil spill, pesticide, plastic, heavy metals as substrate, and release biodegradable byproducts (Braschi et al., 2013). In the case of antibiotics bioremediation, the microbial population attains antibiotic resistance genes (ARGs). Recent studies revealed that the comparison of ARG and antibiotic degradation is robable, but there is no similarity in ARGs and antibiotic degradation. Thus, these species would be the best ecofriendly option for engineering antibiotics degrading bacteria (Dangi et al., 2018).

10.11 Antibiotic degradation in soil

In the soil environment, antibiotics may be subjected to different abiotic and/or biotic processes, including transformation/degradation.

Beta-lactams antibiotic is more susceptible for hydrolysis process of degradation whereas macrolides and sulfonamides are less susceptible for hydrolysis process (Braschi et al., 2013). As many authors have suggested, environmental transformation or degradation depends on the molecular structure and physicochemical properties of antibiotics, which are the most important properties governing the fate of different antibiotics in soils (Chen et al., 2019). Since many abiotic and biotic factors affect the degradation of antibiotics, different groups of these pharmaceuticals differ in their rates of degradation in soils, as is evidenced by the large range of half-lives in soil, between <1 and 3466 days. Degradation of antibiotics is not totally dependent on catabolic activity of the soil, it also depends on physicochemical properties of the soil, for example, temperature, pH, moisture, soil texture, etc. (Cycoń et al., 2019).

10.12 Bioremediation of dye from textile industry

Synthetic dyes are widely used in many industries including paper printing, textile dyeing, pharmaceutical, color photography, cosmetics, food, and other industries. During textile dyeing process, the class of dye used is responsible for the toxicity to human and animals. Different industries discharge different dye effluent directly into the environment, which is a major source of pollution. Among these, azo dyes are the major pollutant because it has a more class of dye. This is the big challenge to reduce the toxicity of the effluent from textile industry, which is very dangerous for animals and humans. Because of its chemical nature (water soluble reactive, acidic) it can cross through conventional treatment systems unaffected.

10.13 Degradation of dye

The bacterial isolates, *Bacillus* sp. and *Pseudomonas* sp. were found to have significant potential to decolorize the dye effluent. Srinivasan et al. (2014) proved that, *Bacillus* sp., decolorization ability was 92%–97%, whereas, *Pseudomonas* sp., produced 87%–95%. *Bacillus* sp. has a better dye degradation ability in comparison to *Pseudomonas.* The reason behind this is that decolorization of the effluent by bacteria very effectively might be associated with the metabolic activities and interactions of these strains. For optimum decolorization of azo dyes the bacteria, *Lysinibacillus* sp. requires 96 hours and *Phanerochaete sordida* requires 48 hours (Srinivasan et al., 2014).

10.14 Conclusion

Natural and anthropogenic activities generate large quantities of aqueous effluents containing toxic metals. Several studies have been conducted in recent decades aimed at lowering metal concentrations derived from natural resources. Many microorganisms can break down metals naturally, but this is not a sufficient solution on a global scale. Therefore, as a means to resolve this problem, engineered microorganisms can be developed with the help of genetic engineering. This review provides an opportunity to reveal the role of microbial cell, biofilm, and their metabolites toward remediation of heavy metals and environmental research. Further research area needs to be extended on the focus of gene transfer within biofilms for heavy metal remediation.

Author contributions

Authors P.K., V.T. conceive and designed the paper, A.K.P. J.S., and R.S. wrote the manuscript. P.T. and M.K. helped in writing the manuscript. P.K. and V.T. critically reviewed the manuscript. All the authors read and approved the final manuscript.

Acknowledgment

All the authors would like to thank the higher authority of their respective department and university.

Conflicts of interest

The authors declare no conflict of interest.

References

Abatenh, E., Gizaw, B., Tsegaye, Z., Wassie, M., 2017. Application of microorganisms in bioremediation-review. J. Environ. Microbiol. 1 (1), 02–09.

Abbas, B.A., Badr, S.Q., 2015. Bioremediation of some types of heavy metals by *Candida spp*. IJETR 3, 2454–4698.

Achal, V., Pan, X., Zhang, D., 2011. Remediation of copper-contaminated soil by *Kocuria flava* CR1, based on microbially induced calcite precipitation. Ecol. Eng. 37 (10), 1601–1605.

Achal, V., Pan, X., Zhang, D., Fu, Q., 2012a. Bioremediation of Pb-contaminated soil based on microbially induced calcite precipitation. J. Microbiol. Biotechnol. 22, 244–247.

Achal, V., Pan, X., Fu, Q., Zhang, D., 2012b. Biomineralization based remediation of As (III) contaminated soil by *Sporosarcina ginsengisoli*. J. Hazard. Mater. 201–202, 178–184.

Achuthan, C., 2000. Development of Bioreactors for Nitrifying Water in Closed System Hatcheries of Penaeid and Non-Penaeid Prawns. PhD Thesis, Cochin University of Science and Technology, Kochi, India, 116 pp.

Ahmady-Asbchin, S., Safari, M., Tabaraki, R., 2015. Biosorption of Zn (II) by *Pseudomonas aeruginosa* isolated from a site contaminated with petroleum. Desalin. Water Treat. 54 (12), 3372–3379.

Antony, S.P., Philip, R., 2006. Bioremediation in shrimp culture systems. NAGA. World Fish. Cent. Q. 29, 62–66.

Ariff, A.B., Mel, M., Hasan, M.A., Karim, M.I.A., 1999. The kinetics and mechanism of lead (II) biosorption by powderized *Rhizopus oligosporus*. World J. Microbiol. Biotechnol. 15 (2), 291–298.

Aung, L.W., Hlaing, N.N., Aye, N.K., 2013. Biosorption of lead (Pb^{2+}) by using *Chlorella vulgaris*. Int. J. Chem. Environ. Biol. Sci. 1 (2), 2320–4087.

Ayangbenro, S.A., Babalola, O.O., 2017. A new strategy for heavy metal polluted environments: a review of microbial biosorbents. Int. J. Environ. Res. Public Health 14, 94.

Aznar-Sánchez, J.A., García-Gómez, J.J., Velasco-Muñoz, J.F., Carretero-Gómez, A., 2018. Mining waste and its sustainable management: advances in worldwide research. Minerals 8 (7), 284.

Bahafid, W., Tahri, N.J., Sayel, H., Iraqui-Houssaini, M., Ghachtouli, E.N., 2013. Chromium adsorption by three yeast strains isolated from sediments in Morocco. Geomicrobiol. J. 30, 422–429.

Bai, X., Acharya, K., 2017. Algae-mediated removal of selected pharmaceutical and personal care products (PPCPs) from Lake Mead water. Sci. Total Environ. 581–582, 734–740.

Balamurugan, D., Udayasooriyan, C., Kamaladevi, B., 2014. Chromium (VI) reduction by *Pseudomonas putida* and *Bacillus subtilis* isolated from contaminated soils. Int. J. Environ. Sci. 5, 522.

Banerjee, G., Pandey, S., Ray, A.K., Kumar, R., 2015. Bioremediation of heavy metals by a novel bacterial strain *Enterobacter cloacae* and its antioxidant enzyme activity, flocculant production, and protein expression in presence of lead, cadmium, and nickel. Water Air Soil Pollut. 226 (4), 1–9.

Barathi, S., Vasudevan, N., 2001. Utilization of petroleum hydrocarbons by *Pseudomonas fluorescence* isolated from a petroleum-contaminated soil. Environ. Int. 26, 413–416.

Beolchini, F., Pagnanelli, F., Toro, L., Veglio, F., 2006. Ionic strength effect on copper biosorption by *Sphaerotilus natans*: equilibrium study and dynamic modelling in membrane reactor. Water Res. 40 (1), 144–152.

Bestawy, E.E., Helmy, S., Hussien, H., Fahmy, M., Amer, R., 2013. Bioremediation of heavy metal-contaminated effluent using optimized activated sludge bacteria. Appl. Water Sci. 3 (1), 181–192.

Beveridge, M.C.M., 1987. Cage Aquaculture. Fishing News Books Ltd, Farnham, Surrey, UK.

Bhakta, J.N., Munekage, Y., Ohnishi, K., Jana, B.B., Balcazar, J.L., 2014. Isolation and characterization of cadmium and arsenic-absorbing bacteria for bioremediation. Water Air Soil. Pollut. 225, 2150–2159.

Bhattacharya, D., Sarma, P.M., Krishnan, S., Mishra, S., Lal, B., 2003. Evaluation of genetic diversity among *Pseudomonas citronellolis* strains isolated from oily sludge-contaminated sites. Appl. Environ. Microbiol 69, 1435–1441.

Bowen, S.H., 1987. Composition and Nutritional Value of Detritus. International Center for Living Aquatic Resources Management, Manila, Philippines, p. 385.

Braschi, I., Blasioli, S., Fellet, C., Lorenzini, R., Garelli, A., Pori, M., et al., 2013. Persistence and degradation of new b-lactam antibiotics in the soil and water environment. Chemosphere 93, 152–159.

Brusseau, M.L., Maier, R.M., 2004. Soil and groundwater remediation. In: Artiola, J.F., Pepper, I.A., Brusseau, M.L. (Eds.), Environmental Monitoring and Characterization. Elsevier. pp. 335–356.

Chatterjee, S., Chattopadhyay, P., Roy, S., Sen, S.K., 2008. Bioremediation: a tool for cleaning polluted environments. J. Appl. Biosci. 11, 594–601.

Chaulya, S.K., Singh, R.S., Chakraborty, M.K., Srivastava, B.K., 2000. Quantification of stability improvement of a dump through biological reclamation. Geotech. Geol. Eng. 18, 193–207.

Chavez-Crooker, P., Obreque-Contreras, J., 2010. Bioremediation of aquaculture wastes. Curr. Opin. Biotechnol. 21 (3), 313–317.

Chen, Q., Dong, J., Zhang, T., Yib, Q., Zhang, J., Chen, L.H.Q., et al., 2019. A method to study antibiotic emission and fate for data-scarce rural catchments. Environ. Int. 127, 514–521.

Choi, S.B., Yun, Y.S., 2004. Lead biosorption by waste biomass of *Corynebacterium glutamicum* generated from lysine fermentation process. Biotechnol. Lett. 26 (4), 331–336.

Coelho, L.M., Rezende, H.C., Coelho, L.M., deSousa, P.A., Melo, D.F., Coelho, N.M., 2015. Bioremediation of polluted waters using microorganisms. In: Shiomi, N. (Ed.), Advances in Bioremediation of Wastewater and Polluted Soil. InTech, Shanghai, China.

Cycoń, M., Mrozik, A., Piotrowska-Seget, Z., 2019. Antibiotics in the soil environment—degradation and their impact on microbial activity and diversity. Front. Microbiol. 10, 338.

Dangi, A., Sharma, B., Russell, T., Hill, Shukla, P., 2018. Bioremediation through microbes: systems biology and metabolic engineering approach. Crit. Rev. Biotechnol. 39, 79–98.

Danjo, T., Kawasaki, S., 2016. Microbially induced sand cementation method using *Pararhodobacter* sp. strain SO1, inspired by beachrock formation mechanism. Mater. Trans. 57, 428–437.

Das, M., Adholeya, A., 2012. Role of microorganisms in remediation of contaminated soil. In: Satyanarayana, T., et al. (Eds.), Microorganisms in Environmental Management: Microbes and Environment. Springer, Netherlands. pp. 81–111.

Das, S., Dash, H., 2014. Microbial bioremediation: a potential tool for restoration of contaminated areas. In: Das, S. (Ed.), Microbial Biodegradation and Bioremediation. Elsevier. pp. 1–21.

de Wilt, A., Butkovskyi, A., Tuantet, K., Leal, L.H., Fernandes, T.V., Langenhoff, A., et al., 2016. Micropollutant removal in an algal treatment system fed with source separated wastewater streams. J. Hazard. Mater 304, 84–92.

Díaz-Ramírez, I., Escalante-Espinosa, E., Adams Schroeder, R., Fócil-Monterrubio, R., Ramírez-Saad, H. 2013. Hydrocarbon biodegradation potential of native and exogenous microbial inocula in Mexican tropical soils. In: Chamy, R., Rosenkranz, F. (Eds.), Biodegradation of Hazardous and Special Products, Tech, Rijeka, Croatia, pp. 155–178.

Eriksson, M., Dalhammer, G., Borg-Karlson, A.K., 1999. Aerobic degradation of a hydrocarbon mixture in natural uncontaminated potting soil by indigenous microorganisms at 20°C and 6°C. Appl. Microbiol. Biotechnol 51, 532–535.

Fathima, A., Aravindhan, R., Rao, J.R., Nair, B.U., 2015. Biomass of *Termitomyces clypeatus* for chromium (III) removal from chrome tanning wastewater. Clean. Technol. Environ. Policy 17 (2), 541–547.

Fourest, E., Roux, J.C., 1992. Heavy metal biosorptionby fungal mycelial by-products: mechanisms and influence of pH. Appl. Microbiol. Biotechnol. 37 (3), 399–403.

Friss, N., Myers-Keith, P., 1986. Biosorption of uranium and lead by *Streptomyces longwoodensis*. Biotechnol. Bioeng. 28, 21–28.

Fukudome, K., Sato, M., Takata, Y., Kuroda, H., Watari, J., Takashio, M., 2002. Evaluation of yeast physiological state by Alcian blue retention. J. Am. Soc. Brew. Chem. 60, 149–152.

Gabriel, J., Vosahlo, J., Baldrian, P., 1996. Biosorption of cadmium to mycelial pellets of wood-rotting fungi. Biotechnol. Tech. 10 (5), 345–348.

Gadd, G.M., Mowll, J.L., 1985. Copper uptake by yeast-like cells, hyphae, and chlamydospores of *Aureobasidium pullulans*. Exp. Mycol. 9, 3.

Gattullo, C.E., Bährs, H., Steinberg, C.E.W., Loffredo, E., 2012. Removal of bisphenol A by the freshwater green alga *Monoraphidium braunii* and the role of natural organic matter. Sci. Total Environ. 416, 501–506.

Gavrilescu, M., 2010. Environmental biotechnology: achievements, opportunities and challenges. Dyn. Biochem. Process Biotech. Mol. Biol. 4, 1–36.

Genovese, M., Crisafi, F., Denaro, R., Cappello, S., Russo, D., Calogero, R., et al., 2014. Effective bioremediation strategy for rapid in situ cleanup of anoxic marine sediment in mesocosm oil spill simulation. Front. Microbiol. 5, 162.

Germaine, K.J., Byrne, J., Liu, X., Keohane, J., Culhane, J., Lally, R.D., et al., 2015. Ecopiling: a combined phytoremediation and passive biopiling system for remediating hydrocarbon impacted soils at field scale. Front. Plant Sci. 5, 1–6.

Goher, M.E., El-Monem, A.M.A., Abdel-Satar, A.M., Ali, M.H., Hussian, A.E.M., Napi orkowska-Krzebietke, A., 2016. Biosorption of some toxic metals from aqueous solution using nonliving algal cells of *Chlorella vulgaris*. J. Elementol. 21, 703–714.

Göksungur, Y., Uren, S., Güvenç, U., 2005. Biosorption of cadmium and lead ions by ethanol treated waste baker's yeast biomass. Bioresour. Technol. 96 (1), 103–109.

Guang-Guo, Y., 2018. Remediation and mitigation strategies. In: Integrated Analytical Approaches for Pesticide Management. Academic Press. pp. 207–217.

Gunasekaran, P., Muthukrishnan, J., Rajendran, P., 2003. Microbes in heavy metal remediation. Indian J. Exp. Biol. 41, 935–944.

Gupta, V.K., Nayak, A., Agarwal, S., 2015. Bioadsorbents for remediation of heavy metals: current status and their future prospects. Environ. Eng. Res. 20, 1–18.

Gupta, S., Wali, A., Gupta, M., Annepu, S.K., 2017. Fungi: an effective tool for bioremediation. In: Singh, D.P., et al. (Eds.), Plant-Microbe Interactions in Agro-Ecological Perspectives. Springer, Singapore. pp. 38–46.

Harayama, S., Kasai, Y., Hara, A., 2004. Microbial communities in oil-contaminated seawater. Curr. Opin. Biotechnol. 15 (3), 205–214.

Haung, H.J., 2003. Important tools to the success of shrimp aquaculture—aeration and the applications of tea seed cake and probiotics. Aqua Int. 13–16.

Huang, G., Chen, F., Wei, D., Zhang, X., Chen, G., 2010. Biodiesel production by microalgal biotechnology. Appl. Energy 87, 38–46.

Huerta, B., Rodríguez-Mozaz, S., Barceló, D., 2012. Pharmaceuticals in biota in the aquatic environment: analytical methods and environmental implications. Anal. Bioanal. Chem. 404, 2611–2624.

Incharoensakdi, A., Kitjaharn, P., 2002. Zinc biosorption from aqueous solution by a halotolerant cyanobacterium *Aphanothece halophytica*. Curr. Microbiol. 45 (4), 261–264.

Iskandar, N.L., Zainudin, N.A.I.M., Tan, S.G., 2011. Tolerance and biosorption of copper (Cu) and lead (Pb) by filamentous fungi isolated from a freshwater ecosystem. J. Environ. Sci. 23 (5), 824–830.

Jain, P.K., Gupta, V.K., Pathak, H., Lowry, M., Jaroli, D.P., 2010. Characterization of 2T engine oil degrading indigenous bacteria, isolated from high altitude (Mussoorie) India. World J. Microbiol. Biotechnol. 26, 1419–1426.

Jiang, C.Y., Sheng, X.F., Qian, M., Wang, Q.Y., 2008. Isolation and characterization of heavy metal resistant *Burkholderia* species from heavy metal contaminated paddy field soil and its potential in promoting plant growth and heavy metal accumulation in metal polluted soil. Chemosphere 72, 157–164.

Kang, C.H., Han, S.H., Shin, Y., Oh, S.J., So, J.S., 2014. Bioremediation of Cd by microbially induced calcite precipitation. Appl. Biochem. Biotechnol. 172, 2907–2915.

Kannepalli, S., Farrish, W.K., 2016. Effectiveness of various carbon amendments in the bioremediation of perchlorate contaminated soils. Int. J. Environ. Biorem. Biodegrad. 4 (3), 68–79.

Khan, F.I., Husain, T., Hejazi, R., 2004. An overview and analysis of site remediation technologies. Environ. Manage. 71, 95–122.

Koops, H.P., Pommerening-Röser, A., 2001. Distribution and ecophysiology of nitrifying bacteria emphasizing cultured species. FEMS Microb. Ecol. 1255, 1–9.

Kujan, P., Prell, A., Safar, H., Sobotka, M., Rezanka, T., Holler, P., 2006. Use of the industrial yeast *Candida utilis* for cadmium sorption. Folia Microbiol. 51 (4), 257–260.

Kumar, A., Bisht, B.S., Joshi, V.D., Dhewa, T., 2011. Review on bioremediation of polluted environment: a management tool. Int. J. Environ. Sci. 1 (6), 1079–1093.

Kumaran, N.S., Sundaramanicam, A., Bragadeeswaran, S., 2011. Adsorption studies on heavy metals by isolated cyanobacterialstrain (*Nostoc* sp.) from Uppanar estuarine water, southeast coast of India. J. Appl. Sci. Res. 7 (11), 1609–1615.

León-Santiesteban, H.H., Wrobel, K., Revah, S., Tomasini, A., 2016. Pentachlorophenol removal by *Rhizopus oryzae* CDBB-H-1877 using sorption and degradation mechanisms. J. Chem. Technol. Biotechnol. 91 (1), 65–71.

Li, L., Cunningham, C.J., Pas, V., Philp, J.C., Barry, D.A., Anderson, P., 2004. Field trial of a new aeration system for enhancing biodegradation in a biopile. Waste Manage. 24, 127–137.

Li, X., Peng, W., Jia, Y., Lu, L., Fan, W., 2016. Bioremediation of lead contaminated soil with *Rhodobacter sphaeroides*. Chemosphere 156, 228–235.

Liu, H.L., Chen, B.Y., Lan, Y.W., Cheng, Y.C., 2004. Biosorption of Zn (II) and Cu(II) by the indigenous *Thiobacillus thiooxidans*. Chem. Eng. J. 97, 195–201.

Livia de, C.F., Benedito, M.H., 2015. Potential application of modified *Saccharomyces cerevisiae* for removing lead and cadmium. J. Bioremed. Biodegrad. 6, 2.

Mameri, N., Boudries, N., Addour, L., Belhocine, D., Lounici, H., Grib, H., et al., 1999. Batch zinc biosorption by a bacterial nonliving *Streptomyces rimosus* biomass. Water Res. 33 (6), 1347–1354.

Mane, P.C., Bhosle, A.B., 2012. Bioremoval of some metals by living algae *Spirogyra* sp. and *Spirulina* sp. from aqueous solution. Int. J. Environ. Res. 6 (2), 571–576.

Manobala, T., Shukla, S., Rao, T.S., Kumar, M.D., 2018. A new uranium bioremediation approach using radio-tolerant *Deinocoocus radiodurans* biofilm. bioRxiv 503896.

Mapolelo, M., Torto, N., 2004. Trace enrichment of metal ions in aquatic environments by *Saccharomyces cerevisiae*. Talanta. 64, 39–47.

Marie-Pierre, H.-L., Elena, G., 2016. Pharmaceuticals in the environment. Environ. Sci. Pollut. Res. 23 (6), 4961–4963.

Mattuschka, B., Straube, G., 1993. Biosorption of metals by a waste biomass. J. Chem. Technol. Biotechnol. 58 (1), 57–63.

Mezaguer, M., El Hayet Kamel, N., Lounici, H., Kamel, Z., 2013. Characterization and properties of *Pleurotus mutilus* fungal biomass as adsorbent of the removal of uranium (VI) from uranium leachate. J. Radioanal. Nucl. Chem 295 (1), 393–403.

Mishra, A., Malik, A., 2012. Simultaneous bioaccumulation of multiple metals from electroplating effluent using *Aspergillus lentulus*. Water Res 46 (16), 4991–4998.

Mishra, A., Malik, A., 2014. Novel fungal consortium for bioremediation of metals and dyes from mixed waste stream. Bioresour. Technol 171, 217–226.

Mishra, S., Jyot, J., Kuhad, R.C., Lal, B., 2001. Evaluation of inoculum addition to stimulate *in situ* bioremediation of oily-sludge-contaminated soil. Appl. Environ. Microbiol. 67, 1675–1681.

Mnif, I., Sahnoun, R., Ellouz-Chaabouni, S., Ghribi, D., 2017. Application of bacterial biosurfactants for enhanced removal and biodegradation of diesel oil in soil using a newly isolated consortium. Process Saf. Environ. Prot. 109, 72–81.

Morris, S., Garcia-Cabellos, G., Enright, D., Ryan, D., Enright, A.M., 2018. Bioremediation of landfill leachate using isolated bacterial strains. IJEBB 6 (1), 26–35.

Muneer, B., Iqbal, M.J., Shakoori, F.R., Shakoori, A.R., 2013. Tolerance and biosorption of mercury by microbial consortia: potential use in bioremediation of waste water. Pak. J. Zool. 45 (1), 247–254.

Muraleedharan, T.R., Venkobachar, C., 1990. Mechanism of biosorption of Copper (Ii) by *Ganoderma lucidum*. Biotechnol. Bioeng. 35 (3), 320–325.

Mwandira, W., Nakashima, K., Kawasaki, S., 2017. Bioremediation of lead-contaminated mine waste by *Pararhodobacter* sp. based on the microbially induced calcium carbonate precipitation technique and its effects on strength of coarse and fine grained sand. Ecol. Eng. 109, 57–64.

Nelson, D.H., 1929. The isolation of some nitrifying organisms. Iowa State Collection 1. Science 3, 113–175.

Niu, H., Xu, X.S., Wang, J.H., Volesky, B., 1993. Removal of lead from aqueous solutions by *Penicillium* biomass. Biotechnol. Bioeng 42 (6), 785–787.

Nourbakhsh, M., Sag, Y., Ozer, D., Aksu, Z., Kutsal, T., Caglar, A., 1994. A comparative study of various biosorbents for removal of chromium (VI) ions from industrial waste waters. Process Biochem. 29 (1), 1–5.

Ockjoo, J., Choi, J.H., Kim, I.H., Kim, Y.K., Oh, B.K., 2015. Effective bioremediation of cadmium (II), nickel (II), and chromium (VI) in a marine environment by using Desulfovibrio desulfuricans. Biotechnol. Bioprocess Eng. 20 (5), 937–941.

Odu, C.T.I., Adeoye, K.B., 1970. Heterotrophic nitrification in soils – a preliminary investigation. Soil Biol. Biochem. 2, 41–45.

Ozdemir, G., Ozturk, T., Ceyhan, N., Isler, R., Cosar, T., 2003. Heavy metal biosorption by biomass of *Ochrobactrum anthropi* producing exopolysaccharide in activated sludge. Bioresour. Technol. 90 (1), 71–74.

Ozer, A., Ozer, D., 2003. Comparative study of the biosorption of Pb(II), Ni(II) and Cr(VI) ions onto *S. cerevisiae*: determination of biosorption heats. J. Hazard. Mater. 27 (1–3), 219–229.

Parker, R.B., 1974. Probiotics, the other half of the antibiotics story. Anim. Nutr. Health 29, 4–8.

Paskuliakova, A., McGowan, T., Tonry, S., Touzet, N., 2018. Microalgal bioremediation of nitrogenous compounds in landfill leachate – the importance of micronutrient balance in the treatment of leachates of variable composition. Algal Res. 32, 162–171.

Paul, D., Pandey, G., Jain, R.K., 2005. Suicidal genetically engineered microorganisms for bioremediation: need and perspectives. BioEssays 27 (5), 563–573.

Philip, L., Iyengar, L., Venkobacher, L., 2000. Site of interaction of copper on *Bacillus polymyxa*. Water Air Soil Pollut. 119, 11–21.

Phillips, P., Russell, A., Bender, J., Munoz, R., 2004. Management plan for utilization of a floating microbial mat with its associated detrital gelatinous layer as a complete tilapia *Oreochromis niloticus* feed system. Bioresour. Technol. 94, 229–238.

Prasenjit, B., Sumathi, S., 2005. Uptake of chromium by *Aspergillus foetidus*. J. Mater. Cycles Waste Manag. 7 (2), 88–92.

Prenafeta-Boldu, X.F., Kuhn, A., Mam, L.D., Anke, H., Van Groenestijin, J.W., De Bont, J. A.M., 2001. Isolation and characterization of fungi growing on volatile aromatic hydrocarbons as their sole carbon and energy source. Mycol. Res. 4, 477–484.

Priyalaxmi, R., Murugan, A., Raja, P., Raj, K.D., 2014. Bioremediation of cadmium by *Bacillus safensis* (JX126862), a marine bacterium isolated from mangrove sediments. Int. J. Curr. Microbiol. App. Sci 3 (12), 326–335.

Remacle, J., 1977. Microbial transformations of nitrogen in forests. Oecol. Plant. 7, 69–78.

Şahan, T., Ceylan, H., Aktaş, N., 2015. Optimization of biosorption of Zn (II) ions from aqueous solutions with low-cost biomass *Trametes versicolor* and the evaluation of kinetic and thermodynamic parameters. Desalin. Water Treat. 57, 1–12.

Salehizadeh, H., Shojaosadati, S.A., 2003. Removal of metal ions from aqueous solution by polysaccharide produced from *Bacillus firmus*. Water Res. 37 (17), 4231–4235.

Sar, P., DSouza, S.F., 2001. Biosorptive uranium uptake by *Pseudomonas* strain: characterization and equilibrium studies. J. Chem. Technol. Biotechnol. 76, 1286–1294.

Savvaidis, I., Hughes, M.N., Poole, R.K., 2003. Copper biosorption by *Pseudomonas cepacia* and other strains. World J. Microbiol. Biotechnol. 19 (2), 117–121.

Sharma, R., 1999. Probiotics: a new horizon in aquaculture. Fisheries World. 8-1.

Sharma, S., 2012. Bioremediation: features, strategies and applications. AJPLSC 2, 326–335.

Sharma, R., Rishi, M.S., Lata, R., 2013. Monitoring and assessment of soil quality near Kashlog limestone mine at Darlaghat district Solan, Himachal Pradesh, India. J. Environ. Earth Sci. 3, 1–40.

Shukla, S.K., Rao, T.S., 2017. The first recorded incidence of *Deinococcus radiodurans* R1biofilm formation and its implications in heavy metals bioremediation. bioRxiv 234781.

Singh, I.S., Jayaprakash, N.S., Somnath, P., 2001. Antagonistic bacteria as gut probiotics. Natl. Workshop Aquacul. Med. 55–59, 18–20.

Singh, R., Singh, P., Sharma, R., 2014. Microorganism as a tool of bioremediation technology for cleaning environment: a review. Proc. Int. Acad. Ecol. Environ. Sci. 4 (1), 1–6.

Song, H.P., Li, X.G., Sun, J.S., Xu, S.M., Han, X., 2008. Application of a magnetotactic bacterium, *Stenotrophomonas* sp. to the removal of Au (III) from contaminated wastewater with a magnetic separator. Chemosphere 72, 616–621.

Spina, F., Tigini, V., Romagnolo, A., Varese, G.C., 2018. Bioremediation of landfill leachate with fungi: autochthonous vs. allochthonous strains. Life (Basel) 8 (3), 27.

Srinivasan, V., Saravana, P., Krishnakumar, J., 2014. Bioremediation of textile dye effluent by bacillus and *Pseudomonas* species. Int. J. Sci. Environ. Technol. 3 (6), 2215–2224.

Srivastava, S., Dwivedi, A.K., 2015. Biological wastes the tool for biosorption of arsenic. J. Biorem. Biodegrad. 7, 2.

Suresh, B., Ravishankar, G.A., 2004. Phytoremediation - a novel and promising approach for environmental clean-up. Crit. Rev. Biotechnol. 24, 97–124.

Tabaraki, R., Ahmady-Asbchin, S., Abdi, O., 2013. Biosorption of Zn (II) from aqueous solutions by *Acinetobacter* sp. isolated from petroleum spilled soil. J. Environ. Chem. Eng. 1 (3), 604–608.

Tangaromsuk, J., Pokethitiyook, P., Kruatrachue, M., Upatham, E.S., 2002. Cadmium biosorption by *Sphingomonas paucimobilis* biomass. Bioresour. Technol. 85 (1), 103–105.

Thapa, B., Kumar, A., Ghimire, A., 2012. A Review on bioremediation of petroleum hydrocarbon contaminants in soil. Kathmandu Univ. J. Sci. Eng. Technol. 8 (1), 164–170.

Tigini, V., Prigione, V., Toro, S.D., Fava, F., Giovanna, C.V., 2009. Isolation and characterisation of polychlorinated biphenyl (PCB) degrading fungi from a historically contaminated soil. Microb. Cell Fact. 8, 1–14.

Townsley, C.C., Ross, I.S., 1985. Copper uptake by *Penicillium spinulosum*. Microbios 44, 125–134.

Tsezos, M., Volesky, B., 1981. Biosorption of uranium and thorium. Biotechnol. Bioeng. 23 (3), 583–604.

Tunali, S., Cabuk, A., Akar, T., 2006. Removal of lead and copper ions from aqueous solutions by bacterial strain isolated from soil. Chem. Eng. J. 115 (3), 203–211.

Uslu, G., Tanyol, M., 2006. Equilibrium and thermodynamic parameters of single and binary mixture biosorption of lead (II) and copper (II) ions onto *Pseudomonas putida*: effect of temperature. J. Hazard. Mater. 135 (1), 87–93.

Valko, M., Jomova, K., Rhodes, C.J., Kuca, K., Musilek, K., 2016. Redox-and non-redox-metal-induced formation of free radicals and their role in human disease. Arch. Toxicol. 90 (1), 1–37.

Vullo, D.L., Ceretti, H.M., Daniel, M.A., Ramírez, S.A., Zalts, A., 2008. Cadmium, Zinc and Copper biosorption mediated by *Pseudomonas veronii* 2e. Bioresour. Technol. 99, 5574–5581.

Wang, J., Chen, C., 2009. Biosorbents for heavy metals removal and their future. Biotechnol. Adv. 27, 195–226.

Wong, S.C., Li, X.D., Zhang, G., Qi, S.H., Min, Y.S., 2000. Heavy metals in agricultural soils of the Pearl River Delta, South China. Environ. Pollut. 119, 33–44.

Wu, Y., Li, T., Yang, L., 2012. Mechanisms of removing pollutants from aqueous solutions by microorganisms and their aggregates: a review. Bioresour. Technol. 107, 10–18.

Wu, Y., Xiaa, L., Yub, Z., Shabbira, S., Kerrd, P.G., 2014. In situ bioremediation of surface waters by periphytons. Bioresour. Technol. 151, 367–372.

Xiong, J.Q., Kurade, M.B., Abou-Shanab, R.A.I., Ji, M.K., Choi, J., Kim, J.O., et al., 2016. Biodegradation of carbamazepine using freshwater microalgae *Chlamydomonas mexicana* and *Scenedesmus obliquus* and the determination of its metabolic fate. Bioresour. Technol. 205, 183–190.

Xiong, J.Q., Kurade, M.B., Jeon, B.H., 2018. Can microalgae remove pharmaceutical contaminants from water? Trends Biotechnol. 36, 30–44.

Yalçinkaya, Y., Arica, M.Y., Soysal, L., Denizli, A., Genç, O., Bektaş, S., 2002. Cadmium and mercury uptake by immobilized *Pleurotus sapidus*. Turk. J. Chem. 26 (3), 441–452.

Zafar, S., Aqil, F., Ahmad, I., 2007. Metal tolerance and biosorption potential of filamentous fungi isolated from metal contaminated agricultural soil. Bioresour. Technol. 98 (13), 2557–2561.

Zaki, S., Farag, S., 2010. Isolation and molecular characterization of some copper biosorped strains. Int. J. Environ. Sci. Technol. 7 (3), 553–560.

Zhang, S., Zhang, X., Chang, C., Yuan, Z., Wang, T., Zhao, Y., et al., 2016. Improvement of tolerance to lead by filamentous fungus *Pleurotus ostreatus* HAU-2 and its oxidative responses. Chemosphere 150, 33–39.

Zhu, X., Li, W., Zhan, L., Huang, M., Zhang, Q., Achal, V., 2016. The large-scale process of microbial carbonate precipitation for nickel remediation from an industrial soil. Environ. Pollut. 219, 149–155.

Bioreactor and bioprocess technology for bioremediation of domestic and municipal wastewater

11

Nitin Kumar Singh[1,*], *Siddhartha Pandey*[2], *Rana Pratap Singh*[3], *Khalid Muzamil Gani*[4], *Manish Yadav*[5], *Arti Thanki*[1] and *Tarun Kumar*[6]
[1]Department of Environmental Science and Engineering, Marwadi University, Rajkot, India, [2]Chalapathi Institute of Technology, Guntur, India, [3]Department of Civil Engineering, Katihar Engineering College, Katihar, India, [4]Institute for Water and Wastewater Technology, Durban University of Technology, Durban, South Africa, [5]Central Mine Planning and Design Institute Limited, Bhubaneswar, India, [6]Department of Civil Engineering, Greater Noida Institute of Technology, Greater Noida, Uttar Pradesh, India
*Corresponding author

11.1 Background

The accelerated urbanization and unsustainable developmental activities have imposed huge pressures on environment, especially through the discharge of partially and untreated domestic and municipal wastewaters (Topare et al., 2011; Singh et al., 2015; Singh and Kazmi, 2017). These waste streams are typically laden with significant quantities of conventional and nonconventional pollutants such as organics, nutrients, and micropollutants such as metals and emerging contaminants, that is, pesticides, endocrine disrupting chemicals, etc. (Loupasaki and Diamadopoulos, 2013; Dixit et al., 2015; Grandclément et al., 2017). The presence of these pollutants in wastewater severely affects the water bodies, if discharged without appropriate treatment. In particular, the presence of organics reduces the dissolved oxygen in water bodies and it satisfies its biological oxygen demand (BOD), hence it disturbs the aquatic lives (Elsheikh and Al-Hemaidi, 2012; Singh and Kazmi, 2016; Singh et al., 2016a,b). It can also lead to the production of malodorous gases, produced from the anaerobic decomposition of organics (Ali et al., 2014). Along with this, micropollutants and metallic substances also accumulate in aquatic plants and animals, and ultimately affect the food chain of humans (Megharaj et al., 2011). Domestic and municipal wastewater streams usually contain a significant amount of pathogenic, or disease-causing microorganisms, thus threatening human health. Such wastewater may also have a certain amount of nutrients, which causes the eutrophication of lakes and other fresh water streams, and promotes the wild growth of algal blooms (Kesaano and Sims, 2014). Overall, the treatment of

Bioremediation of Pollutants. DOI: https://doi.org/10.1016/B978-0-12-819025-8.00011-9

domestic and municipal wastewater streams must be taken into account before leaving it in natural water bodies (Banyal et al., 2015; Dadrasnia et al., 2017).

11.2 Bioremediation: an overview

Bioremediation is an innovative and promising technology that utilizes the microorganism's consortia, microbial processes, and selective plants for the treatment of polluted waste streams and contaminated sites (Lynch and Moffat, 2005; Juwarkar et al., 2010). This biological approach of dealing with contaminants of soil, slurry, and liquid streams can be practiced in two ways, and are known as in situ and ex situ bioremediation, respectively (Head, 1998). Here, the term in situ refers to the treatment of pollutants/wastes streams at the place of contamination. In this approach, organic pollutants are degraded into less harmful or useful products such as carbon dioxide, water, and methane etc. under natural conditions. On the other hand, ex situ bioremediation involves the treatment of pollutants at a separate treatment facility—as in the case of decentralized and/or centralized wastewater treatment (Iwamoto and Nasu, 2001). Considering the practical uses of these approaches, low cost in situ bioremediation approaches such as soil remediation and natural attenuation, are significantly practiced by mainly developed countries. On the other hand, in spite of being a bit expensive developing countries focus more on ex situ technologies such as using wastewater treatment by bioreactors (Dixit et al., 2015). Further, these bioremediation technologies can also be divided into aerobic or anaerobic treatment units/approaches. Under an aerobic environment oxygen works as an electron acceptor and aerobic microbes such as *Pseudomonas*, *Alcaligenes*, *Sphingomonas*, *Rhodococcus*, and *Mycobacterium* perform biodegradation, whereas degradation of organics under anaerobic environment is achieved through reducing electron acceptors and specific heterotrophic microorganisms (Pieper and Reineke, 2000; Juwarkar et al., 2010; Pandey et al., 2016). To achieve these conditions, various bioremediation techniques such as bioaugmentation, biofilter, biostimulation, bioreactor, bioventing, composting, etc. are practiced in field (Perelo, 2010). Among these, bioreactors play an important role at large scale, especially for the treatment of domestic and municipal wastewaters (Singh et al., 2016a, b, 2017, 2018, 2019; Kumar et al., 2016; Bhatia et al., 2017). A brief summary of the advantages of bioremediation is given below:

- recycling of materials into the environment,
- low cost and environmentally viable, and
- less to negligible harmful products.

11.3 Need of the wastewater treatment

The wastewater generated from various activities can be categorized on the basis of sources of generation. These sources of municipal wastewater can be further classified effluent on the basis of their origins, that is, houses, apartments, residential buildings, institutions, commercial, and industrial hubs (Singh et al., 2015, 2017). Beside this, some small and medium scale industries also contribute to municipal wastewaters. Broadly, two types of approaches are used to deal with these wastewaters, known as decentralized and centralized treatment systems. The decentralized systems further can be categorized as onsite and cluster systems (Sharma et al., 2014, 2016; Singh et al., 2016a,b). The applicability of these two approaches depends on the technology type, up-scalability, budgetary requirements, treatment potential, and their robustness. Municipal wastewaters can also be treated at a remote facility through centralized wastewater treatment plants. The details of such systems is well documented in literature (Mace and Mata-Alvarez, 2002; Chan et al., 2009; Sun et al., 2010; Tran et al., 2018). The negative environmental impacts associated with untreated domestic and municipal wastewater, if discharged directly to fresh water sources or beyond the safe discharge limits, are as follows:

1. Depletion of dissolved oxygen of water bodies, hence it disturbs the aquatic lives and nearby biological communities.
2. Discharge of nutrients (N and P forms) into surface water bodies also promotes the eutrophication, while anaerobic decomposition of organic wastes produces huge quantities of malodorous gases such as methane and carbon dioxide.
3. Biomagnification of heavy metals and other micropollutants, thus it affects the food chain of humans and living organisms.
4. Increase in bacterial masses in water bodies, leads to the dispersion of pathogenic or disease-causing microorganisms; thus, affects the human health.

11.4 Wastewater treatment using biological processes

Biological processes have been used for over 100 years for the treatment of organics, nutrients, and micro-pollutant-bearing domestic and municipal wastewaters. To deal with such wastewaters, biological methods are one of the promising ways that deal with the applications of bacterial populations for the removal/conversion of pollutants into less harmful products such as carbon dioxide, water, and methane gas (Singh and Srivastava, 2011; Khan et al., 2011; Pandey and Sarkar, 2019a,b). These biological processes are typically performed in specially designed and operated bioreactors. During these processes, biodegradable matter (colloidal or dissolved) present in wastewater is converted in small environmental-friendly substances (Topare et al., 2011). Some of these substances may be in gaseous forms that can escape into the atmosphere, while biomass generated during the conversion/removal process can be removed effectively through the appropriate settling unit such as secondary sedimentation tank. Along this, to remove nitrogen and phosphorous from wastewater, aerobic and anaerobic combinations are used (Patel et al., 2006; Singh et al., 2019).

Conventionally, two approaches are used for municipal wastewater treatment, that is, physicochemical and biological/natural treatment. Between these, biological treatment systems have their own advantages such as low cost and environmental-friendly, when compared to expensive and unsustainable physicochemical processes (Eddy et al., 2013). Furthermore, biological wastewater treatment methods have been proven more cost-effective than engineered physicochemical systems. Broadly, biological wastewater treatment systems are categorized in the following ways:

1. aerobic and anaerobic systems and
2. natural and engineered systems.

Conventionally, aerobic and anaerobic (anoxic/anaerobic) biological processes are predominantly used for the treatment of domestic and/or municipal wastewater, and considered as cost-effective and environmental-friendly solutions for environmental protection. In particular, aerobic treatment plants have high energy requirements (50%–80% of total operation and maintenance cost), required mainly for pumping recycled and waste sludge streams, and aeration. On the other side, anaerobic systems are less expensive and capable to produce useable energy forms such as methane gas. In the last three decades, successful attempts have also been made to design a more efficient system in combination of anaerobic/anoxic processes and aerobic processes (Singh et al., 2018, 2019). Table 11.1 depicts the differences between these two types of systems. Besides, natural treatment systems are also used for municipal wastewater treatment, especially in rural areas and at those places where funds are very limited for their establishments (Crites et al., 2014; Almasi et al., 2019; Resende et al., 2019). The natural treatment systems allow degradation of pollutants under ambient conditions, however engineered systems (well controlled and maintained) can be used to enhance the removal rates of pollutants from the fed wastewaters.

11.4.1 Role of bioreactor and its classification

A bioreactor can be defined as an engineered system, deployed to facilitate the growth of biological mass through the transformation or degradation of material fed to the reactor. Such systems generally smoothen the way for biomechanical and/or

Table 11.1 Characteristics of aerobic and anaerobic systems.

Characteristic	Aerobic system	Anaerobic system
Treatment efficiency	High	Low to moderate
Quality of effluent	High	Low to moderate
Pollutant loading rate	Moderate	High
Biomass production	High	Low
Nutrient requirements	High	Low
Electricity requirement	Moderate to high	Low
Start-up time	3–4 Weeks	2–4 Months
Odor	Less odor	Potential odor

biochemical reactions, through which the conversion of feed compounds (present in wastewater) into desired biotransformation products or less harmful compounds takes place. A good bioreactor design is expected to address improved productivity, operational viability, capability to cope-up latest discharge standards, and appreciable removal efficiencies of pollutants in a cost-effective manner. Further, the design and operational strategies planned in bioreactors also depend on the growth of particular microbial communities, characteristics of feed wastewater, desired effluent quality, target pollutants, and its scalability. Along these, economic considerations are also critical factors in selecting technology as well as design of bioreactor.

With respect to reactor classifications on the basis of microbial growth mechanism, bioreactor systems are classified as suspended, attached, and hybrid bioreactors. Suspended growth systems are the ones that maintain the microbial biomass in the suspension in bioreactors. These systems can be further classified into aerobic (either aerated or nonaerated) and anaerobic units. Contrary, attached growth systems require some support media (either fixed or moving inside the reactor) for the growth of the microbial community. These attached growth systems can maintain high biomass concentration over packing material used within it, resulting in appreciable removal rates of contaminants. Some researchers used another classification which combined the features of both the process, and is known as hybrid growth systems (Singh et al., 2015; Singh and Kazmi, 2017). These reactors are practically employed along with appropriate primary and tertiary treatments units.

11.4.2 Algal biomass growth systems

One bioreactor of recent trend is algal biomass growth system, which gained interest especially through the development of microalgae-based bioreactors. In such reactors, algal biomass grows by using moisture and nutrients available in wastewater, either on the surfaces provided in the bioreactors or in suspension. At present, these systems are in infancy phase with limited applications especially at lab and pilot scale level. Besides, these systems are also facing several technical and operational challenges, probably due to limited information about the algal growth requirements, required operational inputs, performance potential for different pollutant types, and unstandardized operating procedures. However, these challenges can be overcome by conducting comprehensive research, especially through the detailed investigations on algal communities' reactors, at pilot or large-scale systems fed with actual municipal wastewaters (Kesaano and Sims, 2014; Delanka-Pedige et al., 2019).

11.4.3 Typical performance parameters of bioreactors

In order to analyze the performance potential of biological wastewater treatment plants, various physicochemical and biological parameters need to be monitored. A brief definition of these parameters is mentioned below:

pH: This parameter refers to the acidity or alkalinity of treated wastewater. The desirable limits for these parameters vary in the range of 6.5–8.5. If the pH value goes beyond this, an acidic or basic treatment system must be provided, before the safe discharge/reuse of effluent.

Alkalinity: It refers to levels of bicarbonates, carbonates, and hydroxides present in treated effluents and expressed as "mg/L of calcium carbonate ($CaCO_3$)." It also presents the scalability potential of treated effluents.

BOD: This is a well-known parameter used to quantify the residual/available organic fraction of treated and untreated wastewaters. It is estimated as dissolved oxygen requirement to degrade the organic matter in aerobic environment through the microbial populations.

Chemical oxygen demand (COD): It is a measure of the oxygen requirement of water sample needed to decompose the organic matter, recalcitrant compounds, and to oxidize the inorganic components such as ammonia, nitrate, and metals, etc.

Total suspended solids (TSS): This is the suspended solid concentration of water sample, and used to estimate the solid loading on water receiving bodies. The high values of suspended solids affect the aquatic life of water bodies. The volatile fraction of suspended solids, which generally comprises living or nonliving organic matter it helps in determining the microbial retaining capacity or clarification potential of bioreactors.

Total nitrogen (TN): This parameter clubs the concentrations of three commonly appearing nitrogen forms, known as ammonia nitrogen (NH_3–N), nitrate/nitrite nitrogen (NO_3/NO_2–N), and organic nitrogen. Another important parameter, which is used in performance assessment of bioreactor, is total Kjeldahl nitrogen. This parameter refers to the combined levels of ammonia and organic nitrogen. The increased levels of these forms are always expected to enhance the eutrophication activities in water receiving bodies.

Total phosphorous (TP): Like TN, this parameter quantifies the level of soluble as well as insoluble forms of phosphorous. The high levels of TP are reported to be responsible for eutrophication and algal blooms in water bodies, receiving treated wastewaters from bioreactors.

11.5 Common operational stresses of bioreactors

11.5.1 pH and temperature shocks

pH and temperature are common stresses faced by most bioreactors. A bioreactor must be sufficient enough to handle these fluctuations during wastewater treatment. A significant change in wastewater or reaction pH can alter the production rate of necessary enzymes, required for the effective conversion of degradation of selective pollutants present in wastewater. In particular, removing micropollutants such as metals may have a significant effect on its removal efficiencies by small changes in pH conditions. With respect to temperature, the growth rate of microorganisms is generally linked with temperature of the reactor. Higher temperature conditions generally favor thc growth, but these are limited to 35°C−40°C(Qian et al., 2019).

11.5.2 Hydraulic and organic shock loads

Hydraulic and organic shock loads are generally defined in terms of varying hydraulic retention times (HRT) and organic loadings, respectively. The HRT refers to the time spent by pollutants in the bioreactors and it is governed by the flow rate of the system. The HRT has a direct impact on the reaction volume and indirect impact on the substrate to biomass ratio. Although, the increasing HRT improves the treatability of bioreactors but simultaneously lowers down its treatment capacity too. On the other side, organic loading stress refers to load and characteristics of the pollutants entering in bioreactor. A significant change in organic loading and type of pollutants alters the degradation efficiencies of microbial communities in the reactor (Ruffino et al., 2019).

11.5.3 Biomass/sludge retention time

The retention of biomass/sludge is also one of the critical stresses that affects the operational and/or steady state performance of bioreactors. It is calculated as sludge retention time (SRT), also known as mean cell residence time or sludge age, which gives an estimation of the mean residence time of biomass in the reactor. This parameter is directly linked with growth of microbial communities as well as wastage of excess biomass from the reactor. This parameter is calculated as the ratio of the amount of biomass wasted from the reactor to the biomass present in the reactor. Bioreactors with high SRT values are typically desired for the degradation of conventional and emerging contaminants of the wastewater. Increased SRT values also provide an enriched population of slow growing bacteria such as nitrifiers. At high SRT values, microbes can also break down recalcitrant molecules such as aromatic rings (Solmaz and Işik, 2019).

11.5.4 Dissolved oxygen stress

Oxygen level plays an important role in biodegradation of pollutants. In particular, aerobic treatment systems need a specific control on dissolved conditions. A significant variation can affect the enzymatic activity as well as separability of biomass from treated wastewater. In general, the high dissolved oxygen (DO) rates enhance the pollutant removal rates in aerobic systems. However, a trade-off is also required between pollutant removal efficiency and required energy against it. Furthermore, higher DO rates may also affect the overall nitrogen removals from wastewaters. With respect to biofilm or attached growth reactors, various studies reported that high aeration rate intensity affects the sloughing of biomass from the biofilm, and thus promotes the formation of excess biomass against a reduced SRT. With respect to membrane applications in biofilm reactors, colloidal particles formed due to detachment of biofilm which created fouling problems in membranes (Singh et al., 2016a,b).

11.5.5 Feeding effects: batch, continuous, and intermittent

Under field conditions, a bioreactor may be operated in batch, continuous, or intermittent mode; depending on the wastewater loading, operational constraints, targeted pollutants, and legislative requirements. In batch feeding mode, pollutants are subjected to microbial degradation for a specific time interval. A simple example of such feeding strategy is operation of sequential batch reactor (SBR). Such feeding strategy is found to helpful in nutrient parameters. On the other hand, continuous feeding strategy is typically used for enhanced removal of organics/substrates pollutants such as in conventional activated sludge reactors. These strategies allow continuous treatment of wastewaters with appreciable removal efficiencies of pollutants. The intermittent feeding strategies are generally adopted intentionally, probably due to operational constraints or for removing targeted pollutants. Such feeding strategies are still under research trends (Torres et al., 2019).

11.6 Case studies on field-scale bioreactors

Wastewater is typically generated from water consuming sources/activities. These sources are broadly divided into municipal and industrial systems. Such wastewater is treated by adopting either physicochemical or biological strategies. Out of these two strategies, the biological methods have been proven effective for wastewater laden with conventional and nonconventional organic pollutants. As per the statistics given by Central Pollution Control Board (CPCB) (2016), only ~50% municipal wastewater is treated through centralized/decentralized treatment units, against a total amount of 61,000 millions litres per day (MLD) generated across India.

Till today various biological systems have been reported in literature along with their conventional and advanced operational strategies. However, their field experiences are always different than the lab scale models. Moreover, the field-scale systems also provide an actual scenario with a varying degree of pre- and posttreatment systems. A detailed analysis of various full-scale wastewater treatment systems is published by Singh and Kazmi (2017). Below are some case studies of full-scale treatment systems installed in different geographies of India, having conventional to advanced level of biological treatment strategies.

11.6.1 Waste stabilization ponds

Wastewater stabilization ponds (WSPs) are treatment systems with large surface area and shallow depth. It may include various ponds in series, having anaerobic to aerobic environment. In such a reactor, both physicochemical and biological processes work for the removal of organic matter, nutrients, pathogens, etc. In developing countries, wherever sufficient land is available, use of WSPs is the simplest and most sustainable way of wastewater treatment. These reactors treat the wastewater in a natural environment, with the help of natural growing microorganism, algae, and small aquatic lives. In addition to this, the plants grown on material fed to the reactor also play an important role in removing pollutants from the wastewater. Such systems require very less operation and maintenance needs; however, a timely sludge withdrawal is required to make the system operationally viable and environmentally sustainable. Furthermore, a good design and structure of ponds also helps in cultivating the necessary bacterial communities as well as for the growth of the required micro algae species. Such architecture control ensures the efficient and complete conversion of pollutants from the wastewater. The schematic of a 6 MLD waste stabilization pond based biological treatment plant, installed at Rishikesh, Uttarakhand, India, is shown in Fig. 11.1.

This wastewater treatment plant was established under Ganga Action Plan in 1990. This plant comprises of multiple ponds in series with anaerobic, anoxic, and facultative environment. The initial ponds of anaerobic nature are typically designed to remove suspended matter of wastewater with degradation of dissolved organic matter. In further stage, that is, facultative ponds, slowly and readily biodegradable organic compounds are broken down into smaller and stable compounds through the coordinated action of heterotrophic bacteria and algal biomass. This stage also helps

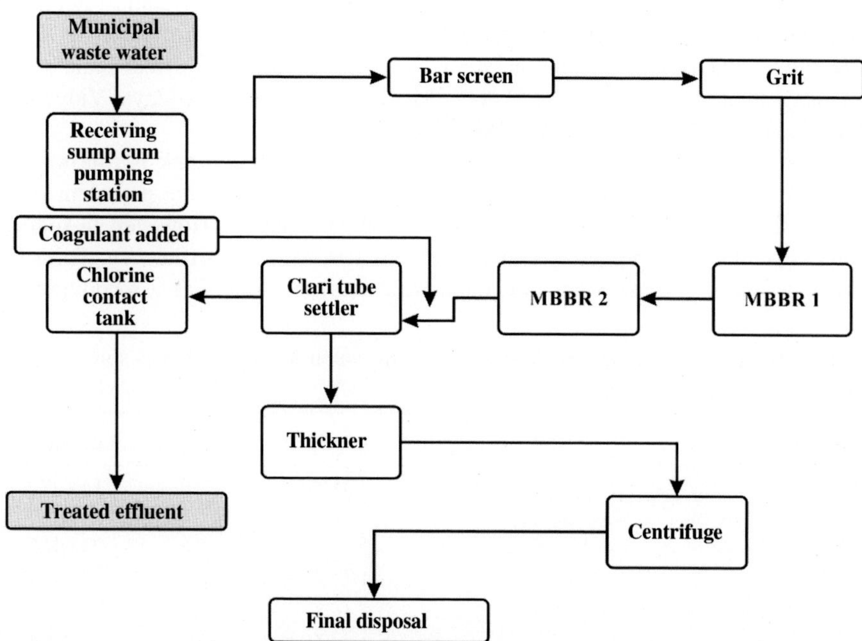

Figure 11.1 Schematic layout of 6 MLD waste stabilization pond, treating municipal wastewater in Rishikesh, Uttarakhand, India.

in attaining significant removal of nitrogen and phosphorous species through the growth of algae species. However, algal species grown through photosynthetic activity causes a diurnal variation in pH and dissolved oxygen levels of facultative ponds. Next to facultative pond(s), aerobic ponds are used as polishing units for the treated effluent received from facultative ponds. As compared to anaerobic ponds, aerobic ponds are generally of shallow depth (<0.6 m) and allow less stratification, as oxygen is expected to penetrate the depth of pond and the entire volume works under aerobic conditions. The typical performance of this waste stabilization system is shown in Table 11.2. The effluent results show a consistent quality that is quite difficult in such natural wastewater treatment systems.

11.6.2 Anaerobic package treatment plant

Anaerobic package treatment systems are typically known for their enhanced potentials, when compared to the conventional anaerobic systems such as septic tank. In general, the performance potential of these systems increases through the incorporation of some inert media in anaerobic tanks. Some of the examples of such media include small pieces of ceramic, raschig rings, bricks, glass, plastic, polypropylene, and other inert materials. The shape and size of these packing materials also play an important role in effective removal of pollutants present in wastewater. When wastewater laden with dissolved impurities is passed through the packed bed of reactor, biomass growth takes place through the consumption of organics, nutrients, and micropollutants. Further, the packed bed can be operated in up-flow or down flow mode, and with or without aeration. A typical example of a field-scale setup, having a treatment capacity of 0.0006 MLD, installed at Navodaya Vidyalaya, Roorkee, Uttarakhand, is shown in Fig. 11.2.

These systems present a two-stage reactor, in which septic tank is followed by anaerobic filter. It is well known that the septic tank is one of the prevalent treatment options in rural and urban areas of developing countries. However, in spite of simple operation, the biggest disadvantage is its low treatment efficiency. To tackle this issue, these systems were integrated with anaerobic filters as a recommended

Table 11.2 Typical treatment performance of waste stabilization pond bioreactor, analyzed through grab sampling.

Parameters	Unit	Influent	Effluent
BOD	mg/L	31	14–27
COD	mg/L	71	60–140
TSS	mg/L	56	23–116
Turbidity	NTU	27.8	22.8–32.1
NH_3–N	mg/L	22.4	9.8–17
NO_3–N	mg/L	3.9	4–4.1
Soluble P	mg/L	1.8	2.4–2.6

BOD, Biological oxygen demand; *COD*, chemical oxygen demand; *TSS*, total suspended solids.

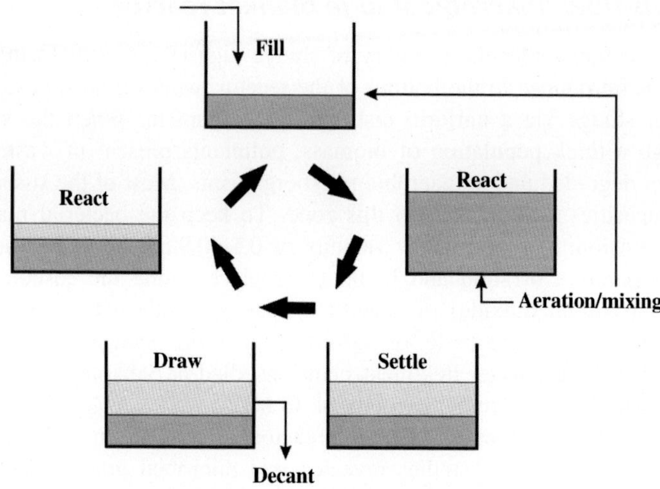

Figure 11.2 Schematic of anaerobic package treatment system, installed at Navodaya Vidyalaya, Roorkee, Uttarakhand, India.

Table 11.3 Treatment performance of anaerobic package plant.

Parameters	Unit	Influent	Effluent
pH	–	7.4	7–9
Alkalinity	mg/L as $CaCO_3$	380	486
BOD	mg/L	1178	581
COD	mg/L	1502	820
TSS	mg/L	1104	260
VSS	mg/L	950	100
NH_3–N	mg/L	33.2	19.4
NO_3–N	mg/L	90.2	35.1
TKN	mg/L	79	57
PO_4–P	mg/L	25.2	12.2

BOD, Biological oxygen demand; *COD*, chemical oxygen demand; *TKN*, total Kjeldahl nitrogen; *TSS*, total suspended solids; *VSS*, volatile suspended solids.

option. These package systems are typically made up of low linear density polyethylene plastic materials. In such a system, effluent of septic tank is passed through the anaerobic filter for the removal of residual pollutants. The first chamber, that is, settling tank allows the settling of settable impurities followed by anaerobic degradation. The second chamber, that is, anaerobic filter promotes the growth of biofilm over media (surface area, 100 m^2/m^3) through the degradation of residual concentrations of pollutants. The desired up-flow velocity and applied organic loading rates are typically less than 1 m/h and 15 kg COD/(m^3 day), respectively. The typical performance, analyzed through grab sampling, is shown in Table 11.3.

11.6.3 Up-flow anaerobic sludge blanket reactor

This process comes under the category of sludge digester process. In this process, wastewater is introduced to the bottom of the reactor, where from it passes through a blanket of sludge via a uniform distributing mechanism. When the wastewater rises through a thick population of biomass, pollutants present in wastewater are consumed or degraded under anaerobic microorganisms. Most of the suspended and colloidal impurities are removed in this zone. To keep the bacterial mass of this zone in suspension, and an up-flow velocity of 0.6–0.9 m/h is maintained. Above this zone, separation of solid and liquid takes place, while the gaseous products (methane and carbon dioxide) of anaerobic zone are collected in the gas dome. Fig. 11.3 shows the schematic of up-flow anaerobic sludge blanket (UASB) technology-based wastewater treatment plant, installed at Saharanpur, Uttarakhand, India. This plant has a treatment capacity of 38 MLD.

The typical performance of such bioreactors is shown in Table 11.4. The research results of previous studies revealed that microbial granulation also takes place in such reactor, which facilitates the higher removals of contaminants as compared to the conventional anaerobic systems such as septic tank. Considering their high potential, these bioreactors are used for the treatment of wastewaters generated from the small and medium scale industries.

11.6.4 Activated sludge process-based wastewater treatment plant

Activated sludge process is one of the oldest and well established technologies. This technology has been proven effective for municipal as well as industrial wastewater treatment. In this process, bacterial mass is grown through the degradation of pollutants under aerobic environment. This bioreactor allows simultaneous treatment of organics, nutrients, and other micropollutants. Various modifications of conventional activated sludge process are also being implemented over the field. In this reactor, municipal wastewater is mixed with microbial culture under aerobic conditions, and the growth of biomass takes place along with degradation of pollutants. After spending the required time in aerobic environment, the generated biomass along with treated wastewater is allowed to settle in clarifier. Some of the settled biomass is recycled back to aeration tank, to maintain the requisite microbial culture. Further, the overall efficiency of such a bioreactor is directly dependent on biomass settling properties. The typical schematic of such a bioreactor-based wastewater treatment plant is shown in Fig. 11.4, and installed in Srinagar, Jammu and Kashmir, India.

This wastewater treatment plant (17.08 MLD) is based on activated sludge process technology to treat the wastewater from adjoining areas of Srinagar. The type of wastewater received for treatment at the plant is purely domestic wastewater. This plant comprises of coarse and fine screens, grit chamber, aeration tank, clarifier, chlorination system, sludge concentrator, aerobic digester, sludge thickener, and sludge drying beds. Activated sludge processes essentially involve a phase in

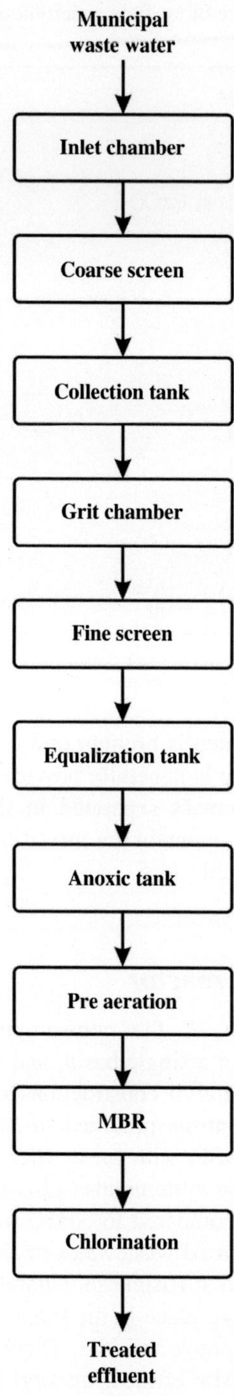

Figure 11.3 Schematic of the UASB-based municipal wastewater treatment plant. *UASB*, Up-flow anaerobic sludge blanket.

Table 11.4 Treatment performance of up-flow anaerobic sludge blanket reactor, analyzed through grab sampling.

Parameters	Units	Influent	Effluent
Temperature	°C	18.4	18.2
pH	–	7.65	7.85
Alkalinity	mg/L as $CaCO_3$	380	320
BOD	mg/L	200	27
COD	mg/L	405	65
TSS	mg/L	195	41
VSS	mg/L	104	25
Turbidity	NTU	210	22
NH_3-N	mg/L	41	12
NO_3-N	mg/L	3.1	5.1
TKN	mg/L	45	19
TN	mg/L	48.1	24.1
PO_4-P	mg/L	6.9	3.2
T–P	mg/L	7.3	5.5
SO_4	mg/L	203	140
Total coliforms	MPN/100 mL	22×10^8	21×10^5
Fecal coliforms	MPN/100 mL	41×10^7	13×10^4

BOD, Biological oxygen demand; *COD*, chemical oxygen demand; *TKN*, total Kjeldahl nitrogen; *TN*, total nitrogen; *TSS*, total suspended solids; *VSS*, volatile suspended solids.

which the wastewater to be treated is brought into contact with a bacterial biomass in the presence of oxygen which is generally provided by aeration. This is then followed by settling process. Biomass separated in the clarification tank is partly recycled to the aeration tank to maintain the mixed liquor. The performance data of such a bioreactor is shown in Table 11.5.

11.6.5 Sequential batch reactor

SBR is an advanced version of the conventional activated sludge process. Treatment in SBR takes place in a single basin, and it requires not only 35%–50% less land but also 40% reduced civil construction cost as compared to ASP plant. Furthermore, the automated controls provided in SBR also facilitates about 40% less power consumption than ASP, with lower chemical consumption and reduced manpower or labor costs. These systems also provide enhanced nutrient (nitrogen and phosphorous) removal, as compared to conventional ASP. Fig. 11.5 shows the operational layout of an SBR-based wastewater treatment plant, having a treatment capacity of 3 MLD and installed at Rishikesh, Uttarakhand, India.

The various stages, which take place in the same tank and in a sequence of time, are fill, react or aerate, settle, draw, and idle. These different stages can be modified, depending upon the requisite effluent quality. Whenever required, aerobic or anaerobic conditions can be endured in SBR to achieve both organic and nutrient

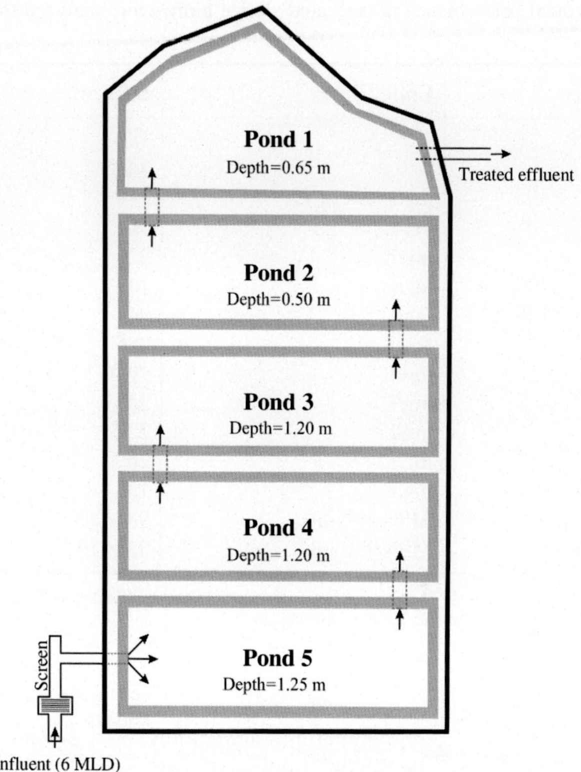

Figure 11.4 Schematic of activated sludge process-based bioreactor, treating municipal wastewater.

removal efficiently. The typical performance of this reactor is also shown in Table 11.6.

11.6.6 Membrane bioreactor

The membrane bioreactor (MBR) combines the features of membrane filtration (ultrafiltration or microfiltration) and conventional activated sludge process. Basically, this modification helps in retaining the biomass in aeration tank for longer times. Consequently, it promotes the growth of slow growing autotrophic bacteria leading to the enhanced removal of nutrients from the wastewater. Other added advantages of these bioreactors include less foot print and high-quality effluent. In this reactor, integration of membrane helps in retention of suspended impurities and maintain higher sludge/biomass retention times (>15 days). Several studies have also reported that along with conventional contaminants, MBR have significant potential (20%–50%)

Table 11.5 Typical performance of activated sludge bioreactor, analyzed through grab sampling.

Parameters	Units	Influent	Effluent
Temperature	°C	18.5	19.5
pH	–	7.95	7.64
Alkalinity	mg/L as $CaCO_3$	289	230
BOD	mg/L	104	91
COD	mg/L	222	176
TSS	mg/L	149	132
VSS	mg/L	94	72
Turbidity	NTU	286	119.6
NH_3–N	mg/L	19.9	8.8
NO_3–N	mg/L	3.8	3.7
TKN	mg/L	22.5	12
TN	mg/L	26.3	15.7
PO_4–P	mg/L	1.8	1.4
Total coliforms	MPN/100 mL	96×10^8	43×10^6
Fecal coliforms	MPN/100 mL	23×10^6	96×10^5

BOD, Biological oxygen demand; *COD*, chemical oxygen demand; *TKN*, total Kjeldahl nitrogen; *TN*, total nitrogen, *TSS*, total suspended solids; *VSS*, volatile suspended solids.

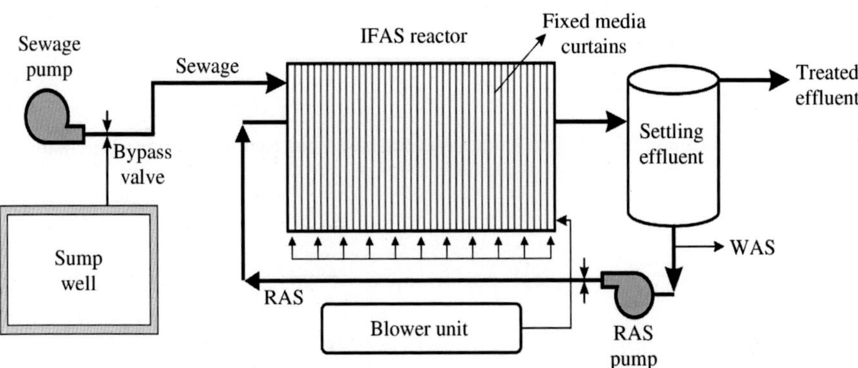

Figure 11.5 Schematic and operational layout of SBR plant treating municipal wastewater. *SBR*, Sequential batch reactor.

for removing the micropollutants and emerging contaminants. Fig. 11.6 shows the schematic of a full-scale MBR, installed in New Delhi, India.

The membrane used in these reactors can be defined as "shrill sheet of natural/synthetic compounds" that is porous/permeable to sludge. Such an arrangement of membrane helps to maintain the huge biomass concentration in reaction tank with a mixed liquor suspended solids (MLSS) concentration of 8000–10,000 mg/L. Such a high biomass concentration enables the easy separation of biosolids from treated effluent. Consequently, sludge processing steps are also minimized for such reactor.

Table 11.6 Typical performance of the sequential batch reactor plant, analyzed through grab sampling.

Parameters	Unit	Influent	Effluent
Temperature	°C	18.9	17.9
pH	–	8.92	7.8
COD	mg/L	331	10
COD	mg/L	63	6
TSS	mg/L	188	18
NH_3–N	mg/L	20.8	0
NO_3–N	mg/L	9.6	5.1
PO_4–P	mg/L	3.8	2.8
Total coliforms	MPN/100 mL	1.5×10^4	NIL
Fecal coliforms	MPN/100 mL	2.3×10^3	NIL

COD, Chemical oxygen demand; *TSS*, total suspended solids.

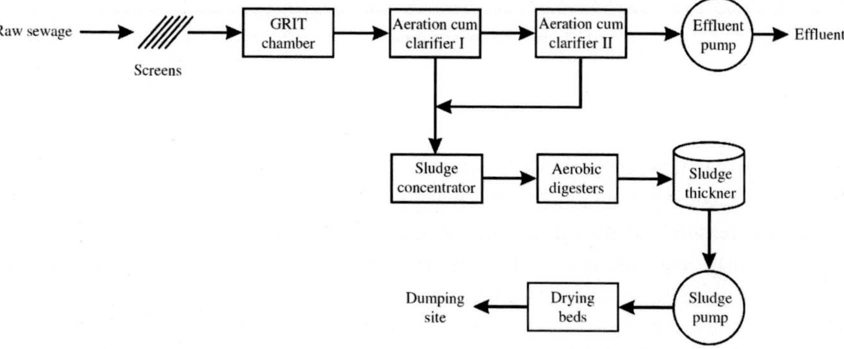

Figure 11.6 Schematic flow diagram of MBR technology-based wastewater treatment plant in New Delhi, India. *MBR*, Membrane bioreactor.

This progressive technology is far and widely used for municipal and industrial wastewater treatment. The actual field performance of such bioreactor is presented in Table 11.7.

11.6.7 Moving bed biofilm reactor

The term "moving bed" refers to a hybrid bioreactor that combines the features of suspended and attached growth processes in a single aerobic tank. The main idea behind these hybrid reactors is to have a continuous operating bioreactor with minimized head losses and high surface for biomass growth. Moving bed biofilm reactor (MBBR) is also another one of the advanced aerobic systems. In such reactor, along with suspended biomass, attached biomass is also grown on some specially

Table 11.7 Treatment performance of membrane bioreactor plant, analyzed through grab sampling.

Parameters	Unit	Influent	Effluent
Temperature	°C	31.4	30.4
pH	–	7.22	8.1
Alkalinity	mg/L as $CaCO_3$	450	143
DO	mg/L	0.20	5.7
BOD	mg/L	492	4.5
COD	mg/L	1050	22
TSS	mg/L	1572	03
VSS	mg/L	1035	02
Turbidity	NTU	689	0.6
NH_3–N	mg/L	77.0	0.3
NO_3–N	mg/L	8.4	47.8
TKN	mg/L	90.7	1.2
PO_4–P	mg/L	14.7	1.2
Total coliforms	MPN/100 mL	4300	11
Fecal coliforms	MPN/100 mL	230	2

BOD, Biological oxygen demand; *COD*, chemical oxygen demand; *TKN*, total Kjeldahl nitrogen; *TSS*, total suspended solids; *VSS*, volatile suspended solids.

designed biocarriers of high surface area. A known quantity of these carriers is placed into reactor for enhanced removal of pollutants. The media can be fluidized through the aeration or by mechanical mixing. In general, these media have specific gravity slightly less than one, but some recently invented media with enhanced features need to be kept in suspension through mechanical mixing/aeration. These systems have been reported to achieve organics removal efficiencies as 80%, at an applied loading of 4 kg COD/m^3 day on diluted wastewaters. The typical scheme of such a bioreactor is shown in Fig. 11.7. This plant is designed for approximately 2000 people and installed at Srinagar, Jammu and Kashmir, India.

This reactor incorporates some plastic or sponge media (usually polyethylene or polypropylene) with specific gravity ~1.0 and known packing ratio (30%–70%) which allows growth of biofilm over it, thus enhances the total biomass concentration in the reactor. The combination of various units used along with such reactor, like bar screen, grit chamber, MBBR, chlorination unit leads to significant removal of pollutants. The performance, analyzed through grab sampling, is presented in Table 11.8.

11.6.8 Integrated fixed film activated sludge system

This plant comprises the feature of suspended and attached growth processes. Along with suspended biomass growth, it promotes the growth of attached biomass on specially design fixed media, installed in the aeration tank of treatment plant.

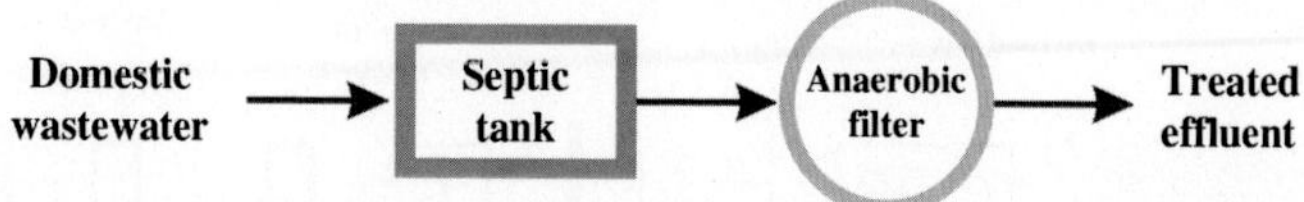

Figure 11.7 Actual schematic diagram for MBBR plant installed in Srinagar, Jammu and Kashmir, India. *MBBR*, Moving bed biofilm reactor.

Table 11.8 Analysis results of raw and treated wastewater collected from the moving bed biofilm reactor.

Parameters	Unit	Influent	Effluent
pH	–	7.23	7.38
Alkalinity	mg/L as $CaCO_3$	360	280
Total BOD	mg/L	173	19.4
Total COD	mg/L	398	96.3
TSS	mg/L	440	90
VSS	mg/L	203	26
Turbidity	NTU	243	17.4
NH_3–N	mg/L	40	38.6
NO_3–N	mg/L	5.5	3.6
PO_4–P	mg/L	4.6	1.2
Ortho-P	mg/L	2.7	0.3

BOD, Biological oxygen demand; *COD*, chemical oxygen demand; *TSS*, total suspended solids; *VSS*, volatile suspended solids.

As compared to MBBR, very less media filling fraction is applied. One example of such bioreactor, having a treatment capacity of 0.05 MLD and installed at Rishikesh, Uttarakhand, India, is shown in Fig. 11.8.

This media is basically, is a loop knitted polypropylene fabric in a rectangular geometry. The bioreactor was operated without any pre- and posttreatment units. The typical performance of this bioreactor is presented in Table 11.9. A detailed analysis of this bioreactors is presented in our recently published studies (Singh et al., 2016a,b).

11.7 Conclusions

The discharge of untreated or partially treated domestic and/or municipal wastewater causes significant harm to the water environment. This environment concern has led to the development of various bioremediation technologies. Among various bioremediation approaches, low cost bioreactor is experiencing rapid development and increasing deployment in practical applications, especially for the treatment of

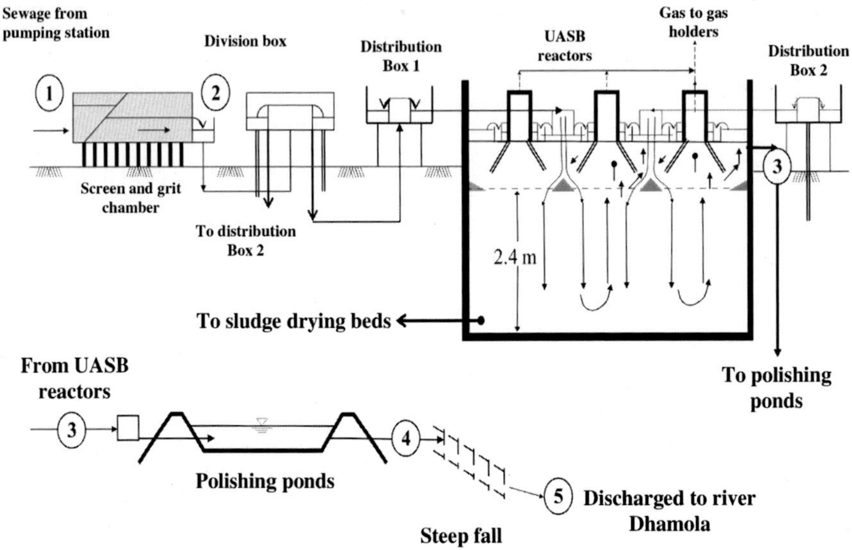

Figure 11.8 Schematic layout of IFAS technology-based bioreactor treating municipal sewage in Rishikesh, Uttarakhand, India. IFAS, Integrated fixed film activated sludge.

Table 11.9 Typical performance of integrated fixed film activated sludge reactor, achieved through yearly monitoring.

Parameters	Unit	Influent	Effluent
pH	–	7.3	7.26
BOD	mg/L	258	<10
COD	mg/L	432	<50
TSS	mg/L	275	<20
NH_3-N	mg/L	28	<5
TN	mg/L	38	<10
PO_4-P	mg/L	<2	<1

BOD, Biological oxygen demand; *COD*, chemical oxygen demand; *TN*, total nitrogen, *TSS*, total suspended solids.

municipal and domestic wastewater. Depending on the technology type, each bioreactor has its own features. In this chapter, along with a brief description of bioremediation approaches, various types of full-scale bioreactors are presented. Moreover, the typical performance is also mentioned in this chapter. Although, abundant literature is available on the applications of lab scale technologies, but very limited published studies are available for full-scale applications. Therefore further studies are required to develop field-scale biological treatment systems, particular in terms of technical and economic competitiveness.

References

Ali, M., Singh, N.K., Bhatia, A., Singh, S., Khursheed, A., Kazmi, A.A., 2014. Sulfide production control in UASB reactor by addition of iron salt. J. Environ. Eng. 141 (6), 06014008.

Almasi, A., Mahmoudi, M., Mohammadi, M., Dargahi, A., Biglari, H., 2019. Optimizing biological treatment of petroleum industry wastewater in a facultative stabilization pond for simultaneous removal of carbon and phenol. Toxin Rev. 1–9.

Banyal, P., Singh, N., Kazmi, A.A., 2015. Assessment of decentralized wastewater treatment systems for sanitation of small communities using a qualitative approach methodology: a case study from Northern India. Int. J. Eng. Adv. Technol. 4 (4), 32–39.

Bhatia, A., Singh, N.K., Bhando, T., Pathania, R., Kazmi, A.A., 2017. Effect of intermittent aeration on microbial diversity in an intermittently aerated IFAS reactor treating municipal wastewater: a field study. J. Environ. Sci. Health, A 52 (5), 440–448.

Central Pollution Control Board (CPCB), 2016. Status of Water Treatment Plants in India. Central Pollution Control Board, Ministry of Environment and Forests, India.

Chan, Y.J., Chong, M.F., Law, C.L., Hassell, D.G., 2009. A review on anaerobic–aerobic treatment of industrial and municipal wastewater. Chem. Eng. J. 155 (1–2), 1–18.

Crites, R.W., Middlebrooks, E.J., Bastian, R.K., 2014. Natural Wastewater Treatment Systems. CRC Press.

Dadrasnia, A., Usman, M.M., Lim, K.T., Velappan, R.D., Shahsavari, N., Vejan, P., et al., 2017. Microbial Aspects in Wastewater Treatment – A Technical Review, Vol. 2, Environmental Pollution and Protection.

Delanka-Pedige, H.M., Munasinghe-Arachchige, S.P., Cornelius, J., Henkanatte-Gedera, S. M., Tchinda, D., Zhang, Y., et al., 2019. Pathogen reduction in an algal-based wastewater treatment system employing Galdieria sulphuraria. Algal Res. 39, 101423.

Dixit, R., Malaviya, D., Pandiyan, K., Singh, U., Sahu, A., Shukla, R., et al., 2015. Bioremediation of heavy metals from soil and aquatic environment: an overview of principles and criteria of fundamental processes. Sustainability 7 (2), 2189–2212.

Eddy, M.A., Burton, F.L., Tchobanoglous, G., Tsuchihashi, R., 2013. Wastewater Engineering: Treatment and Resource Recovery. McGraw-Hill Education, New York, p. 2048.

Elsheikh, M.A., Al-Hemaidi, W.K., 2012. Approach in choosing suitable technology for industrial wastewater treatment. J. Civ. Environ. Eng. 2 (5), 2.

Grandclément, C., Seyssiecq, I., Piram, A., Wong-Wah-Chung, P., Vanot, G., Tiliacos, N., et al., 2017. From the conventional biological wastewater treatment to hybrid processes, the evaluation of organic micropollutant removal: a review. Water Res. 111, 297–317.

Head, I.M., 1998. Bioremediation: towards a credible technology. Microbiology 144 (3), 599–608.

Iwamoto, T., Nasu, M., 2001. Current bioremediation practice and perspective. J. Biosci. Bioeng. 92 (1), 1–8.

Juwarkar, A.A., Singh, S.K., Mudhoo, A., 2010. A comprehensive overview of elements in bioremediation. Rev. Environ. Sci. Biotechnol. 9 (3), 215–288.

Kesaano, M., Sims, R.C., 2014. Algal biofilm-based technology for wastewater treatment. Algal Res. 5, 231–240.

Khan, A.A., Gaur, R.Z., Tyagi, V.K., Khursheed, A., Lew, B., Mehrotra, I., et al., 2011. Sustainable options of post treatment of UASB effluent treating sewage: a review. Resour. Conserv. Recycl. 55 (12), 1232–1251.

Kumar, T., Hari Prasad, K.S., Singh, N.K., 2016. Substrate removal kinetics and performance assessment of a vermifilter bioreactor under organic shock load conditions. Water Sci. Technol. 74 (5), 1177–1184.

Loupasaki, E., Diamadopoulos, E., 2013. Attached growth systems for wastewater treatment in small and rural communities: a review. J. Chem. Technol. Biotechnol. 88 (2), 190–204.

Lynch, J.M., Moffat, A.J., 2005. Bioremediation—prospects for the future application of innovative applied biological research. Ann. Appl. Biol. 146 (2), 217–221.

Mace, S., Mata-Alvarez, J., 2002. Utilization of SBR technology for wastewater treatment: an overview. Ind. Eng. Chem. Res. 41 (23), 5539–5553.

Megharaj, M., Ramakrishnan, B., Venkateswarlu, K., Sethunathan, N., Naidu, R., 2011. Bioremediation approaches for organic pollutants: a critical perspective. Environ. Int. 37 (8), 1362–1375.

Pandey, S., Sarkar, S., 2019a. Performance evaluation and substrate removal kinetics of an anaerobic packed-bed biofilm reactor. Int. J. Environ. Res. 13, 1–11.

Pandey, S., Sarkar, S., 2019b. Spatial distribution of major bacterial species and different volatile fatty acids in a two-phase anaerobic biofilm reactor with PVA gel beads as biocarrier. Prep. Biochem. Biotechnol. 12, 1–14.

Pandey, S., Singh, N.K., Bansal, A.K., Arutchelvan, V., Sarkar, S., 2016. Alleviation of toxic hexavalent chromium using indigenous aerobic bacteria isolated from contaminated tannery industry sites. Prep. Biochem. Biotechnol. 46 (5), 517–523.

Patel, A., Zhu, J., Nakhla, G., 2006. Simultaneous carbon, nitrogen and phosphorous removal from municipal wastewater in a circulating fluidized bed bioreactor. Chemosphere 65 (7), 1103–1112.

Perelo, L.W., 2010. In situ and bioremediation of organic pollutants in aquatic sediments. J. Hazard. Mater. 177 (1–3), 81–89.

Pieper, D.H., Reineke, W., 2000. Engineering bacteria for bioremediation. Curr. Opin. Biotechnol. 11 (3), 262–270.

Qian, W., Ma, B., Li, X., Zhang, Q., Peng, Y., 2019. Long-term effect of pH on denitrification: high pH benefits achieving partial-denitrification. Bioresour. Technol. 278, 444–449.

Resende, J.D., Nolasco, M.A., Pacca, S.A., 2019. Life cycle assessment and costing of wastewater treatment systems coupled to constructed wetlands. Resour. Conserv. Recycl. 148.

Ruffino, B., Cerutti, A., Campo, G., Scibilia, G., Lorenzi, E., Zanetti, M., 2019. Improvement of energy recovery from the digestion of waste activated sludge (WAS) through intermediate treatments: the effect of the hydraulic retention time (HRT) of the first-stage digestion. Appl. Energy 240, 191–204.

Sharma, M.K., Khursheed, A., Kazmi, A.A., 2014. Modified septic tank-anaerobic filter unit as a two-stage onsite domestic wastewater treatment system. Environ. Technol. 35 (17), 2183–2193.

Sharma, M.K., Tyagi, V.K., Saini, G., Kazmi, A.A., 2016. On-site treatment of source separated domestic wastewater employing anaerobic package system. J. Environ. Chem. Eng. 4 (1), 1209–1216.

Singh, M., Srivastava, R.K., 2011. Sequencing batch reactor technology for biological wastewater treatment: a review. Asia-Pac. J. Chem. Eng. 6 (1), 3–13.

Singh, N.K., Kazmi, A.A., 2016. Environmental performance and microbial investigation of a single stage aerobic integrated fixed-film activated sludge (IFAS) reactor treating municipal wastewater. J. Environ. Chem. Eng. 4 (2), 2225–2237.

Singh, N.K., Kazmi, A.A., 2017. Performance and cost analysis of decentralized wastewater treatment plants in northern India: case study. J. Water Resour. Plann. Manage. 144 (3), 05017024.

Singh, N.K., Kazmi, A.A., Starkl, M., 2015. A review on full-scale decentralized wastewater treatment systems: techno-economical approach. Water Sci. Technol. 71 (4), 468–478.

Singh, N.K., Kazmi, A.A., Starkl, M., 2016a. Treatment performance and microbial diversity under dissolved oxygen stress conditions: Insights from a single stage IFAS reactor treating municipal wastewater. J. Taiwan Inst. Chem. Eng. 65, 197–203.

Singh, N.K., Banyal, P., Kazmi, A.A., 2016b. Techno-economic assessment of full scale MBBRs treating municipal wastewater followed by different tertiary treatment strategies: a case study from India. Nat. Environ. Pollut. Technol. 15 (4).

Singh, N.K., Bhatia, A., Kazmi, A.A., 2017. Effect of intermittent aeration strategies on treatment performance and microbial community of an IFAS reactor treating municipal waste water. Environ. Technol. 38 (22), 2866–2876.

Singh, N.K., Pandey, S., Singh, R.P., Dahiya, S., Gautam, S., Kazmi, A.A., 2018. Effect of intermittent aeration cycles on EPS production and sludge characteristics in a field scale IFAS reactor. J. Water Process Eng. 23, 230–238.

Singh, N.K., Yadav, M., Singh, R.P., Kazmi, A.A., 2019. Efficacy analysis of a field scale IFAS reactor under different aeration strategies applied at high aeration rates: a statistical comparative analysis for practical feasibility. J. Water Process Eng. 27, 185–192.

Solmaz, A., Işik, M., 2019. Effect of sludge retention time on biomass production and nutrient removal at an algal membrane photobioreactor. BioEnergy Res. 1–8.

Sun, S.P., Nàcher, C.P.I., Merkey, B., Zhou, Q., Xia, S.Q., Yang, D.H., et al., 2010. Effective biological nitrogen removal treatment processes for domestic wastewaters with low C/N ratios: a review. Environ. Eng. Sci. 27 (2), 111–126.

Topare, N.S., Attar, S.J., Manfe, M.M., 2011. Sewage/wastewater treatment technologies: a review. Sci. Rev. Chem. Commun. 1 (1), 18–24.

Torres, K., Álvarez-Hornos, F.J., Ferrero, P., Gabaldón, C., Marzal, P., 2019. Intermittent operation of UASB reactors treating wastewater polluted with organic solvents: process performance and microbial community evaluation. Environ. Sci.: Water Res. Technol. 5.

Tran, N.H., Reinhard, M., Gin, K.Y.H., 2018. Occurrence and fate of emerging contaminants in municipal wastewater treatment plants from different geographical regions-a review. Water Res. 133, 182–207.

Metagenomics approach for bioremediation: challenges and perspectives

12

*Indra Mani**
Department of Microbiology, Gargi College, University of Delhi, New Delhi, India
*Corresponding author

12.1 Introduction

An advancement in DNA sequencing technology like next generation sequencing (NGS) and availability of international nucleotide sequence database collaboration, and in silico tools (software) significantly help to rapidly generate the genome sequences and understand the functional genomics of any organism. Exploiting the NGS technology and databases in the microbiology, metagenomics molecular method is one of the powerful culture-independent approaches to understanding the microbial diversity and mining a specific gene from any sample such as environmental, human gut, and rumen. In addition, metagenomics also help to understand the bacterial species concept due to an acquisition of new genes through horizontal gene transfer (HGT) (McDaniel et al., 2010). Because of new genes acquisition, it may be responsible for the origin of new bacterial species (Gogarten et al., 2002).

Presently, microbial bioremediation is one of the effective methods to clean the environment with the help of microorganisms instead of conventional methods such as physical and chemical. Due to an emerging concept of metagenomics, it can be utilized to understand the active microbial species, beneficial genes, enzymes, and bioactive molecules from the particular environmental sample. Such microbial species or specific genes can be utilized for effective bioremediation of a particular biohazards compounds (Marco, 2008; Ju and Zhang, 2015). There are several metagenomics projects that have been completed in various environmental conditions like soil and marine environments (Tyson et al., 2004; Nealson and Venter, 2007; Vogel et al., 2009), and these studies have helped us better understand different fields like bioremediation, climate change, and agriculture (Delmont et al., 2011). In a typical hydrocarbon-polluted site, metagenomics molecular approaches have been utilized to characterize viruses and bacteria. Results have shown that dominance of bacteria and viruses in the polluted sites and various bacteriophages host populations including *Clostridia* and *Proteobacteria* have been identified (Costeira et al., 2019).

The leather industry is one of the main sources of aquatic environmental problems which disturbed the natural flora and fauna. A study reported from India, showed that various types of residual organic pollutants such as benzoic acid, resorcinol, dibutyl

Bioremediation of Pollutants. DOI: https://doi.org/10.1016/B978-0-12-819025-8.00012-0

phthalate benzeneacetamide, and benzene-1,2,4-triol were found in the industry wastewater samples. As these chemicals caused the genotoxicity, they are not safe to release into natural aquatic systems (Yadav et al., 2019). The cork boiling wastewater (CBW) is generated from the cork industry and is another challenge for the environment. Although, CBW has low toxicity and less biodegradability as compared to other types of wastewater. To analyze the toxicity effects of CBW on different microorganisms, various approaches, that is, classical to molecular methods have been utilized (Ponce-Robles et al., 2018). Such type of screening will help us to determine effective microorganisms for the treatment of wastewater containing a similar contaminant.

Metagenomics strategies have been utilized to predict and design the degradative mechanism for 2-chlorobenzoate, 4-chlorobenzoate, and 4-methylphenol molecules, which play major contributions in industrial wastewater. A group of different microbes were involved in the process which act as an active biomass (Jadeja et al., 2019). Moreover, metagenomic methods will help us to understand and isolate the genes that are involved in the treatment of industrial wastewater. Cabral et al. (2019) have utilized metagenomics molecular approach to study the genes responsible for degradation/biotransformation of metals and antibiotics resistance in the oil mangrove microbiome. The genes have been identified as lipopolysaccharide transport system, halo acid dehalogenase-like hydrolase, chromate transport protein ChrA, and 3-oxoacyl-(acyl-carrier-protein) reductase. Further, in the analysis, they have found more dominant genus of *Geobacter*, which has been shown tolerant against 24 heavy metals.

12.2 Metagenomics

General procedure of metagenomics methodology is shown in Fig. 12.1. The general steps of metagenomics involved are (1) isolation of metagenomic DNA from

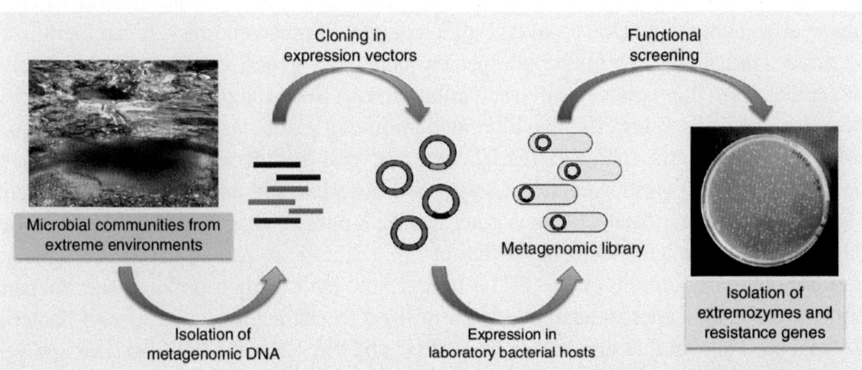

Figure 12.1 A general procedure of metagenomics to retrieve the particular gene from extreme environmental samples.
Source: Adapted from Mirete, S., Morgante, V., González-Pastor, J.E., 2016. Functional metagenomics of extreme environments. Curr. Opin. Biotechnol. 38, 143–149 with permission.

an extreme environmental sample, (2) cloning in expression vectors, (3) expression in laboratory bacterial hosts, (4) metagenomic library, (5) functional screening, and (6) isolation of extremozymes and resistance genes (Mirete et al., 2016). These steps are generally common for the mining of any genes through the metagenomic approach. A summary of microbial community analysis using 16S rRNA gene sequences is shown in Fig. 12.2. 16S rRNA gene sequences have been utilized for the census and characterization of bacteria (Schloss and Handelsman, 2004; Amaral-Zettler et al., 2010; The Human Microbiome Consortium, 2012; Hou et al., 2013; Gilbert et al., 2014; Schloss et al., 2016). In addition, the series of omics like metagenomics, metatranscriptomics, metaproteomics, metabolomics, and fluxomics are also utilized in the characterization, identification, and selection of particular strain of microbes (Schloss and Handelsman, 2004; Bharagava et al., 2019). Therefore multiomics approaches like metagenomics, metatranscriptomics, metaproteomics, and metabolomics provide an excellent way to understand the metabolic pathway and microbes, which are involved in bioremediation of particular contaminants in the environmental sites. Utilizing these approaches to establish microbial consortium may be useful to provide specific microbial strain for degradation of a particular contaminant from the environment.

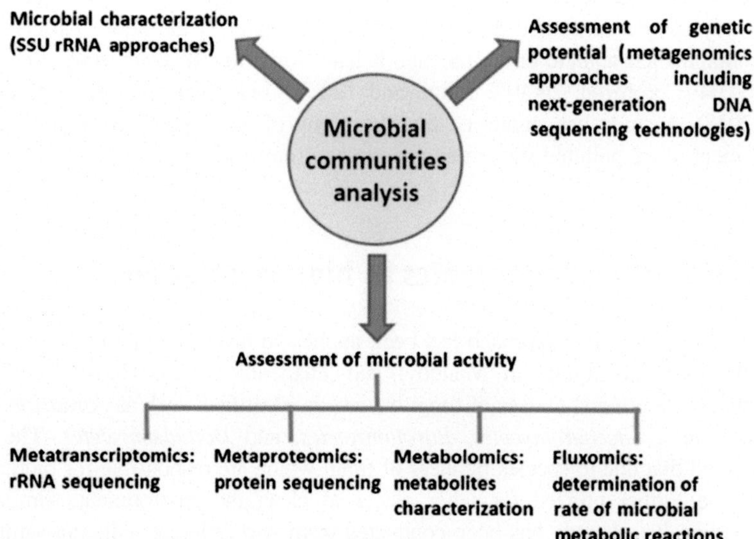

Figure 12.2 An outline of microbial community analysis using 16S rRNA gene sequences. In addition, the series of omics like metagenomics, metatranscriptomics, metaproteomics, metabolomics, and fluxomics are also utilized in the characterization, identification, and selection of particular strain of microbes.
Source: Adapted from Bharagava, R.N., Purchase, D., Saxena, G., Mulla, S.I., 2019. Applications of metagenomics in microbial bioremediation of pollutants: from genomics to environmental cleanup. Microb. Diversity Genomic Era. 459–477 with permission.

12.3 Use of shotgun metagenomics in bioremediation

A shotgun metagenomics method has been utilized to identify the microbes from uranium containing soil. A study has shown that the most abundant among bacteria are *Proteobacterial* phyla and *Burkholderia* (Jaswal et al., 2019). These microbes can be potent candidates for bioremediation of uranium in uraniferous soils. In another study, shotgun genome sequencing method has been applied to understand the role of microorganisms in plant biomass degradations. Different glycosyl hydrolases targeting lignocellulosic and starch-based polysaccharides bacteria have been characterized, namely, *Proteobacteria, Acidobacteria*, *Bacteroidetes*, and *Actinobacteria* (Kanokratana et al., 2011). These bacteria may play an important role in bioremediation of plant biomass.

A study has demonstrated that the identification and characterization of thioesterase coding gene construct from a metagenomic library. Thioesterase is involved in the degradation of phenylacetic acid. Through the phylogenetic analysis of thioesterase gene, it was shown that it belongs to *Actinobacterium* (Folch-Mallol et al., 2019). A metagenomics molecular approach has been utilized to identify the bacteria from bovine dung compost such as corn stover and rice straw. In this study, two new strains ET1 and GL4 have been isolated and were more closely related to *Tepidanerobacter* and *Tepidimicrobium*. These two strains are considered as noncellulolytic bacteria. Findings suggested that these two strains will help in lignocellulose degradation along with cellulolytic bacteria the *Herbivorax saccincola* (Shikata et al., 2018). A metagenomics-based study has demonstrated that the microbial diversity in the South African petroleum contaminated aquatic environments. It has shown that in their study, there were about 62.04% protozoa, 24.49% fungi, and 12.87% unknown organisms (Kachienga et al., 2018). A study has suggested that the group of microorganisms can explore in management of oil polluted sites in an aquatic environment.

12.4 Use of metagenomics in bioremediation

Metagenomics molecular approach has been applied to determine the analysis of fungi from the sediment sample in Mandovi and Zuari inlets, Goa, India (Haldar and Nazareth, 2019). Several classes of fungi have been identified, such as *Sordariomycetes*, *Agaricomycetes*, *Saccharomycetes*, *Eurotiomycetes*, and *Dothideomycetes*. The study hypothesized that due to species richness of fungi which are responsible for bioremediation can be further utilized for other places to clean the environment. Similarly, a metagenomics-based study has been conducted from two different sediment samples of mangrove forest, which is covered by the tropical and subtropical coasts. The study has shown that the metabolic ability of microorganisms in these areas were involved in the nitrogen fixation and methanogenesis (Li et al., 2019). Due to methanogenesis, increased emission of greenhouse gases has been observed from these areas, which indicated that induced bioremediation development occurred.

Ammonia-oxidizing archaea (AOA), is a potential candidate for nitrogen bioremediation from wastewater (Yin et al., 2018). A metagenomics technique has been utilized to compare the seven AOA genomes from the Pearl River estuary with *Nirosopumilus maritimus* SCM1 strain. After comparative genomic analysis of these organisms, four genomes have shown more similarity with *Nitrosomatrinus catalina* strain SPOT01. Interestingly, a comparative genomic analysis has indicated that such strain which has important genes for degradation of the contaminates can be utilized in eutrophic surroundings (Zou et al., 2019). Besides, a combined approach, that is, metagenomics, metranscriptomics, and metabolomics have been utilized to examine the microbial community involved in the degradation of bisphenol A (BPA) molecules. A coculture of *Sphingomonas* species and *Pseudomonas* species have shown increased degradation of BPA as compared to *Sphingomonas* species only (Yu et al., 2019). A study advocated that the microbial interaction plays a vital role in the biodegradation of environmental pollutants and understanding of such a type of interaction can be important to clean the environment. On the other hand, insertion sequence of bacteria may play an important role in the acquisition of xenobiotic degradation gene into a new species. A metagenomics study has demonstrated that an insertion sequence IS1071, which was flanked with xenobiotic degradation gene could help in the degradation of pesticides (Dunon et al., 2018). And after screening of such strain through metagenomics molecular approach, it can be utilized for the degradation of xenobiotic contaminants present in particular environments.

Naphthalene is ubiquitous in nature and a major source of the contaminant in anaerobic environments. Due to difficulty in the cultivation of naphthalene degrading bacteria, metagenomics and metabolomics play an important role in understanding the degradation mechanisms. A study has determined the role of *Desulfuromonadales* in the degradation of naphthalene (Toth et al., 2018). Due to a carcinogenic property of vinyl chloride (VC), it is a great challenge for its degradation. Metagenomics study has been utilized to identify the bacteria that involved in the degradation of VC. Several dominated microbial communities such as *Actinobacteria*, *Proteobacteria*, and *Bacteroide* have been determined in VC containing culture media (Liu et al., 2018b).

A comparative study has been performed through culture-dependent and culture-independent (metagenomics) way to analyze the bacterial diversity using 16S rDNA sequences. Two different samples have been utilized like oily seawater and soil samples. A study has examined the significant variation in bacterial diversity through metagenomics as compared with culture-dependent approach (Dashti et al., 2019). A bioaugmentation-based analysis has been established in total petroleum hydrocarbons (TPH) polluted groundwater. Due to the use of bioaugmentation, TPH was significantly reduced from 1564 to 89 mg/L after 32 days. Interestingly, metagenomics analysis has shown that there was a variation in the microbial community as compared to initial treatment periods (Poi et al., 2018). Metagenomics analysis has also been utilized to examine the microbes associated with oil sands process-affected water. A study has examined the more abundant of *Rhodococcus* bacteria from sand samples. In this study, 21.8% naphthenic acids reduction has been accomplished through biofilter systems (Zhang et al., 2018). As in significant reduction of naphthenic acids, *Rhodococcus* could play an important role.

To study the microorganisms associated with diesel bioremediation, metagenomics molecular approaches have been utilized to examine the diversity of microbes. Due to a mixture of hydrocabons (aliphatic and aromatic nature) present in diesel, there were several types of microorganisms that have been identified, namely, *Pseudomonas*, *Aquabacterium*, *Chryseobacterium*, and *Sphingomonadaceae* (Garrido-Sanz et al., 2019). These microbes further can be utilized to degrade the hydrocabons. Release of thiocyanate (SCN^-) from coal and gold mining, significantly polluted the environments globally. To analyze the degradation of SCN^-, microbiomes have been used in the bioreactor. Metagenomics investigation has shown that SCN^- degradation occurred through SCN^- hydrolase coding *gamma-proteobacteria* (Watts et al., 2019). Further, metagenomics molecular approach has been utilized to assess the *o*-xylene degrading genes from the methanogenic group of bacteria. A study has been shown that fumarate-adding enzyme, napthylmethylsuccinate synthase (nmsA), benzylsuccinate synthase, and alkylsuccinate synthase coding genes were involved in the biodegradation of *o*-xylene compound (Rossmassler et al., 2019). Such a construct can be utilized to degrade *o*-xylene in other contaminant sites into environments.

Metagenomics study has also revealed the microorganisms associated with tarballs (contain toxic hydrocarbons) degradation in Goa Beach, India. There were several microorganisms that have been identified, which were already known to the degradation of hydrocarbon (*Halomonas*, *Marinobacter*) and other human pathogeneic bacteria (*Acinatobacter*, *Klebsiella*, *Vibrio*, and *Staphylococcus*) (Fernandes et al., 2019). A dual approach of metagenomics and metatranscriptomics have been utilized to understand the effect of bioaugmentation in the removal of trichlorinated ethylene/*cis*-dichloroethylene (TCE/*c*DCE) from polluted sites. The study has shown that significantly increased of the *Rhodococcus* genus, including strain RHA1 from 0.1% to 76.6% of the total microbiomes. However, it has observed that other TCE/*c*DCE degrading bacteria decreased over time. Moreover, metatranscriptomics study has shown that high expression of aromatic compound degrading genes (bphA1-A4) occurred during the experimental periods (Watahiki et al., 2019).

HGT of antibiotic resistance gene (ARG) from one bacterium to another bacterium is a global problem for clinicians and society. Further, landfill plays an important role to enhance these problems, because it acts as a reservoir for these genes and remaining antibiotics. Metagenomics study has been utilized to understand the dynamic of ARGs in a bioreactor, and it has shown that elevated level of ARG (Liu et al., 2018a). Moreover, metagenome-based study has been performed for the bioreactor samples which demonstrate that the reduced microbial diversity occurred due to the presence of solids. It has also shown that thiobacilli also suppressed in the bioreactor, which is responsible for the degradation of thiocyanate (SCN^-) (Rahman et al., 2017).

A joint effort of metagenomics and metatranscriptomics has been made to understand the microbes and genes involved in the metabolic mechanisms of the nitrifying sludge. A number of bacterial species have been examined from the nitrifying sludge such as *Bacteroidetes*, *Alphaproteobacteria*, and *Betaproteobacteria*. In addition, metatranscriptomics have revealed significant upregulation of aromatic compound degrading genes (Sun et al., 2019). Metagenomics molecular approach has been utilized to estimate the polyaromatic hydrocarbon ring hydroxylating dioxygenases (PAH-RHDα) gene diversity in bacterial diversity in oilfield soils and mangrove sediments samples. Several bacteria have been identified, which has shown PAH-RHDα gene diversity such as *Pseudomonas, Rhodococcus*, *Polymorphum gilvum*, *Sciscionella marina*, *Mycobacterium*, and *Burkholderia* (Liang et al., 2019). A metagenomics sequence assembled based genome of *Methylocystis* species (HL 18 strain) of the bacteria has potent mercuric reductase genes which are involved in the bioremediation of Hg(II) and As(V) in contaminated aquatic environments. After sequence analysis, the mercuric reductase gene has shown more similarity with Hg(II) reducing bacteria such as *Bradyrhizobium* species strain CCH5-F6 and *Paracoccus halophilus* (Shi et al., 2019). For the bioremediation of heavy metals, such types of microorganisms can be used in the contaminated sites.

Genes responsible for biodegradation and pesticides degradation have been investigated in activated sludge from the wastewater treatments sites. The diversity of bacterial community was varied from 50.4% to 76.8% as per cyclical conditions and by environmental variables (Fang et al., 2018). On the other hand, metagenomics molecular approach has been utilized to analyze the degradation of terephthalate (TA) compound by microorganisms. The hypermesophilic methanogens have been characterized, which belongs to *Pelotomaculum* species and is involved in the degradation of TA (Lykidis et al., 2011). A gene encoding of 3,5,6-trichloro-2-pyridinol construct has screened through a metagenomic library. This enzyme can be utilized for degradation of chlorpyrofis insecticides, which has toxic effects against animals and humans (Math et al., 2010). Therefore this nondegradable insecticide is not safe to release into the natural environment.

Functional screening of extradiol dioxygenase diversity has been performed in the mixture of hydrocarbons with catechol, which acts as a substrate. In the study, there were three stains that have been characterized, which carry encoding genes such as TbuE of *Ralstonia pickettii* PK01, Ipbc of *Pseudomonas* species JR1or DbtC of *Burkholderia* species DBT1 respectively (Brennerova et al., 2009). Genes accountable for biodegradation and phenol degradation have been investigated in activated sludge from the wastewater treatments sites. The diversity of bacterial community could have demonstrated that there was 87 bacterial genus, which has the ability to degrade the contaminants. The more dominating microorganisms were *Proteobacteria* (59.8%), *Bacteroidetes* (17.2%), and *Actinobacteria* (9.2%) (Fang et al., 2013).

The combined efforts of metagenomics, metatranscriptomics, and metabolomics have been made to investigate the cholesterol degradation pathway. A study has proposed that C25 dehydrogenase could be the indicator for degradation of cholesterol in the environments (Wei et al., 2018). A metagenomic molecular approach has been utilized to understand the diversity of the microbial community, which are involved in the bioremediation of arsenic (As) and antimony (Sb) heavy metals. A study has demonstrated that there were five phyla of microorganisms, namely, Actinobacteria, Nitrspirae, Gemmatimonadetes, Fermicutes, and Tenericutes (Luo et al., 2014). These microbes were involved in the bioremediation of As and Sb contaminated soil which can further be utilized into another contaminated sites.

12.5 Conclusion, challenges, and future perspective

The industry is a major source of contaminants which are releasing waste materials either untreated or partial treated into the environment. To degrade pollutants from an environment is a great challenge. These pollutants come into the soil and aquatic systems, and consequently may reach into the human and animal body through food chains. The contaminants may be toxic, lethal, and cause genetic alterations. Natural groups of microorganisms are responsible to clean the environment from hydrocarbons, metals, and nonmetal pollutants. However, it is not well understood how microorganisms degrade most of the contaminants, but due to the availability of metagenomics and nucleotide database and in silico approach, they will help us to understand the metabolic pathways. Furthermore, through metagenomics, metatranscriptomics, metaproteomics, and metabolomics approach, we can screen the potential microbial strains, putative gens, enzymes, metabolites to degrade/biotransformation of these contaminants from the environment. Because of the development of metagenomics methodology, it is facilitating to study the culture-independent investigation of microorganisms either sequence-based or mining the biotechnological important genes. Construction of a metagenomic library is very important to understand the various genes, proteins, enzymes, and pathways involved in the metabolism, growth, and development of putative microbes during the bioremediation process.

Acknowledgments

The author appreciates anonymous reviewers of this book for their valuable comments and suggestions to improve the quality of the chapters.

References

Amaral-Zettler, L., Artigas, L.F., Baross, J., Loka Bharathi, P.A., Boetius, A., et al., 2010. A global census of marine microbes. In: McIntyre, A. (Ed.), Life in the World's Oceans: Diversity, Distribution, and Abundance. Wiley-Blackwell, Chichester, UK, pp. 221–245.

Bharagava, R.N., Purchase, D., Saxena, G., Mulla, S.I., 2019. Applications of metagenomics in microbial bioremediation of pollutants: from genomics to environmental cleanup. In: Das, S., Dash, H.R. (Eds.), Microbial Diversity in the Genomic Era. Elsevier, Academic Press, Oxford, UK, pp. 459–477.

Brennerova, M.V., Josefiova, J., Brenner, V., Pieper, D.H., Junca, H., 2009. Metagenomics reveals diversity and abundance of meta-cleavage pathways in microbial communities from soil highly contaminated with jet fuel under air-sparging bioremediation. Environ. Microbiol. 11 (9), 2216–2227.

Cabral, L., Noronha, M.F., de Sousa, S.T.P., Lacerda-Júnior, G.V., Richter, L., Fostier, A.H., et al., 2019. The metagenomic landscape of xenobiotics biodegradation in mangrove sediments. Ecotoxicol. Environ. Saf. 179, 232–240.

Costeira, R., Doherty, R., Allen, C.C.R., Larkin, M.J., Kulakov, L.A., 2019. Analysis of viral and bacterial communities in groundwater associated with contaminated land. Sci. Total. Environ. 656, 1413–1426.

Dashti, N., Ali, N., Salamah, S., Khanafer, M., Al-Shamy, G., Al-Awadhi, H., et al., 2019. Culture-independent analysis of hydrocarbonoclastic bacterial communities in environmental samples during oil-bioremediation. MicrobiologyOpen 8 (2), e00630.

Delmont, T.O., Malandain, C., Prestat, E., Larose, C., Monier, J.M., Simonet, P., et al., 2011. Metagenomic mining for microbiologists. ISME J. 5, 1837–1843.

Dunon, V., Bers, K., Lavigne, R., Top, E.M., Springael, D., 2018. Targeted metagenomics demonstrates the ecological role of IS1071 in bacterial community adaptation to pesticide degradation. Environ. Microbiol. 20 (11), 4091–4111.

Fang, H., Cai, L., Yu, Y., Zhang, T., 2013. Metagenomic analysis reveals the prevalence of biodegradation genes for organic pollutants in activated sludge. Bioresour. Technol. 129, 209–218.

Fang, H., Zhang, H., Han, L., Mei, J., Ge, Q., Long, Z., et al., 2018. Exploring bacterial communities and biodegradation genes in activated sludge from pesticide wastewater treatment plants via metagenomic analysis. Environ. Pollut. 243 (Pt B), 1206–1216.

Fernandes, C., Kankonkar, H., Meena, R.M., Menezes, G., Shenoy, B.D., Khandeparker, R., 2019. Metagenomic analysis of tarball-associated bacteria from Goa, India. Mar. Pollut. Bull. 141, 398–403.

Folch-Mallol, J.L., Zárate, A., Sánchez-Reyes, A., López-Lara, I.M., 2019. Expression, purification, and characterization of a metagenomic thioesterase from activated sludge involved in the degradation of acylCoA-derivatives. Protein Expression Purif. 159, 49–52.

Garrido-Sanz, D., Redondo-Nieto, M., Guirado, M., Pindado Jiménez, O., Millán, R., Martin, M., et al., 2019. Metagenomic insights into the bacterial functions of a diesel-degrading consortium for the rhizoremediation of diesel-polluted soil. Genes (Basel) 10 (6), E456.

Gilbert, J.A., Jansson, J.K., Knight, R., 2014. The Earth Microbiome Project: successes and aspirations. BMC Biol. 12, 69.

Gogarten, J.P., Doolittle, W.F., Lawrence, J.G., 2002. Prokaryotic evolution in light of gene transfer. Mol. Biol. Evol. 19, 2226–2238.

Haldar, S., Nazareth, S.W., 2019. Diversity of fungi from mangrove sediments of Goa, India, obtained by metagenomic analysis using Illumina sequencing. 3 Biotech 9 (5), 164.

Hou, W., Wang, S., Dong, H., Jiang, H., Briggs, B.R., Peacock, J.P., et al., 2013. A comprehensive census of microbial diversity in hot springs of tengchong, Yunnan Province China using 16S rRNA gene pyrosequencing. PLoS One 8 (1), e53350.

Jadeja, N.B., Purohit, H.J., Kapley, A., 2019. Decoding microbial community intelligence through metagenomics for efficient wastewater treatment. Funct. Integr. Genomics. 19.

Jaswal, R., Pathak, A., Edwards III., B., Lewis III., R., Seaman, J.C., Stothard, P., et al., 2019. Metagenomics-guided survey, isolation, and characterization of uranium resistant microbiota from the Savannah River site, USA. Genes (Basel) 10 (5), E325.

Ju, F., Zhang, T., 2015. Experimental design and bioinformatics analysis for the application of metagenomics in environmental sciences and biotechnology. Environ. Sci. Technol. 49 (21), 12628–12640.

Kachienga, L., Jitendra, K., Momba, M., 2018. Metagenomic profiling for assessing microbial diversity and microbial adaptation to degradation of hydrocarbons in two South African petroleum-contaminated water aquifers. Sci. Rep. 8 (1), 7564.

Kanokratana, P., Uengwetwanit, T., Rattanachomsri, U., Bunterngsook, B., Nimchua, T., et al., 2011. Insights into the phylogeny and metabolic potential of a primary tropical peat swamp forest microbial community by metagenomic analysis. Microb. Ecol. 61 (3), 518–528.

Li, Y., Zheng, L., Zhang, Y., Liu, H., Jing, H., 2019. Comparative metagenomics study reveals pollution induced changes of microbial genes in mangrove sediments. Sci. Rep. 9 (1), 5739.

Liang, C., Huang, Y., Wang, Y., Ye, Q., Zhang, Z., Wang, H., 2019. Distribution of bacterial polycyclic aromatic hydrocarbon (PAH) ring-hydroxylating dioxygenases genes in oilfield soils and mangrove sediments explored by gene-targeted metagenomics. Appl. Microbiol. Biotechnol. 103 (5), 2427–2440.

Liu, X., Wu, Y., Wilson, F.P., Yu, K., Lintner, C., Cupples, A.M., et al., 2018b. Integrated methodological approach reveals microbial diversity and functions in aerobic groundwater microcosms adapted to vinyl chloride. FEMS Microbiol. Ecol. 94 (9).

Liu, X., Yang, S., Wang, Y., Zhao, H.P., Song, L., 2018a. Metagenomic analysis of antibiotic resistance genes (ARGs) during refuse decomposition. Sci. Total Environ. 634, 1231–1237.

Luo, J., Bai, Y., Liang, J., Qu, J., 2014. Metagenomic approach reveals variation of microbes with arsenic and antimony metabolism genes from highly contaminated soil. PLoS One 9 (10), e108185.

Lykidis, A., Chen, C.L., Tringe, S.G., McHardy, A.C., Copeland, A., Kyrpides, N.C., et al., 2011. Multiple syntrophic interactions in a terephthalate-degrading methanogenic consortium. ISME J. 5 (1), 122–130.

Marco, D., 2008. Metagenomics and the niche concept. Theory Biosci. 127 (3), 241–247.

Math, R.K., Asraful Islam, S.M., Cho, K.M., Hong, S.J., Kim, J.M., Yun, M.G., et al., 2010. Isolation of a novel gene encoding a 3,5,6-trichloro-2-pyridinol degrading enzyme from a cow rumen metagenomic library. Biodegradation 21 (4), 565–573.

McDaniel, L.D., Young, E., Delaney, J., Ruhnau, F., Ritchie, K.B., Paul, J.H., 2010. High frequency of horizontal gene transfer in the oceans. Science 330, 50.

Mirete, S., Morgante, V., González-Pastor, J.E., 2016. Functional metagenomics of extreme environments. Curr. Opin. Biotechnol. 38, 143–149.

Nealson, K.H., Venter, J.C., 2007. Metagenomics and the global ocean survey: what's in it for us, and why should we care. ISME J. 1, 185–187.

Poi, G., Shahsavari, E., Aburto-Medina, A., Mok, P.C., Ball, A.S., 2018. Large scale treatment of total petroleum-hydrocarbon contaminated groundwater using bioaugmentation. J. Environ. Manage. 214, 157–163.

Ponce-Robles, L., Polo-López, M.I., Oller, I., Garrido-Cardenas, J.A., Malato, S., 2018. Practical approach to the evaluation of industrial wastewater treatment by the application of advanced microbiological techniques. Ecotoxicol. Environ. Saf. 166, 123–131.

Rahman, S.F., Kantor, R.S., Huddy, R., Thomas, B.C., van Zyl, A.W., Harrison, S.T.L., et al., 2017. Genome-resolved metagenomics of a bioremediation system for degradation of thiocyanate in mine water containing suspended solid tailings. MicrobiologyOpen 6 (3). Available from: https://doi.org/10.1002/mbo3.446.

Rossmassler, K., Snow, C.D., Taggart, D., Brown, C., De Long, S.K., 2019. Advancing biomarkers for anaerobic *o*-xylene biodegradation via metagenomic analysis of a methanogenic consortium. Appl. Microbiol. Biotechnol. 103 (10), 4177–4192.

Schloss, P.D., Girard, R.A., Martin, T., Edwards, J., Thrash, J.C., 2016. Status of the archaeal and bacterial census: an update. MBio 7 (3), e00201–e00216.

Schloss, P.D., Handelsman, J., 2004. Status of the microbial census. Microbiol. Mol. Biol. Rev. 68, 686–691.

Shi, L.D., Chen, Y.S., Du, J.J., Hu, Y.Q., Shapleigh, J.P., Zhao, H.P., 2019. Metagenomic evidence for a *Methylocystis* species capable of bioremediation of diverse heavy metals. Front. Microbiol. 9, 3297.

Shikata, A., Sermsathanaswadi, J., Thianheng, P., Baramee, S., Tachaapaikoon, C., Waeonukul, R., et al., 2018. Characterization of an anaerobic, thermophilic, alkaliphilic, high lignocellulosic biomass-degrading bacterial community, ISHI-3, isolated from biocompost. Enzyme Microb. Technol. 118, 66–75.

Sun, H., Narihiro, T., Ma, X., Zhang, X.X., Ren, H., Ye, L., 2019. Diverse aromatic-degrading bacteria present in a highly enriched autotrophic nitrifying sludge. Sci. Total Environ. 666, 245–251.

The Human Microbiome Consortium, 2012. Structure, function and diversity of the healthy human microbiome. Nature 486, 207–214.

Toth, C.R.A., Berdugo-Clavijo, C., O'Farrell, C.M., Jones, G.M., Sheremet, A., Dunfield, P.F., et al., 2018. Stable isotope and metagenomic profiling of a methanogenic naphthalene-degrading enrichment culture. Microorganisms 6 (3), pii: E65.

Tyson, G.W., Chapman, J., Hugenholtz, P., Allen, E.E., Ram, R.J., Richardson, P.M., et al., 2004. Community structure and metabolism through reconstruction of microbial genomes from the environment. Nature 428, 37–43.

Vogel, T.M., Simonet, P., Jansson, J.K., Hirsch, P.R., Tiedje, J.M., van Elsas, J.D., et al., 2009. TerraGenome: a consortium for the sequencing of a soil metagenome. Nat. Rev. Microbiol. 7, 2.

Watahiki, S., Kimura, N., Yamazoe, A., Miura, T., Sekiguchi, Y., Noda, N., et al., 2019. Ecological impact assessment of a bioaugmentation site on remediation of chlorinated ethylenes by multi-omics analysis. J. Gen. Appl. Microbiol. . Available from: https://doi.org/10.2323/jgam.2018.10.003.

Watts, M.P., Spurr, L.P., Lê Cao, K.A., Wick, R., Banfield, J.F., Moreau, J.W., 2019. Genome-resolved metagenomics of an autotrophic thiocyanate-remediating microbial bioreactor consortium. Water Res. 158, 106–117.

Wei, S.T., Wu, Y.W., Lee, T.H., Huang, Y.S., Yang, C.Y., Chen, Y.L., et al., 2018. Microbial functional responses to cholesterol catabolism in denitrifying sludge. mSystems 3 (5), e00113–e00118.

Yadav, A., Raj, A., Purchase, D., Ferreira, L.F.R., Saratale, G.D., Bharagava, R.N., 2019. Phytotoxicity, cytotoxicity and genotoxicity evaluation of organic and inorganic pollutants rich tannery wastewater from a common effluent treatment plant (CETP) in Unnao district, India using *Vigna radiata* and *Allium cepa*. Chemosphere 224, 324–332.

Yin, Z., Bi, X., Xu, C., 2018. Ammonia-oxidizing archaea (AOA) play with ammonia-oxidizing bacteria (AOB) in nitrogen removal from wastewater. Archaea 2018, 8429145.

Yu, K., Yi, S., Li, B., Guo, F., Peng, X., Wang, Z., et al., 2019. An integrated meta-omics approach reveals substrates involved in synergistic interactions in a bisphenol A (BPA)-degrading microbial community. Microbiome 7 (1), 16.

Zhang, L., Zhang, Y., Gamal El-Din, M., 2018. Degradation of recalcitrant naphthenic acids from raw and ozonated oil sands process-affected waters by a semi-passive biofiltration process. Water Res. 133, 310–318.

Zou, D., Li, Y., Kao, S.J., Liu, H., Li, M., 2019. Genomic adaptation to eutrophication of ammonia-oxidizing archaea in the Pearl River estuary. Environ. Microbiol. . Available from: https://doi.org/10.1111/1462-2920.14613.

Microbial bioremediation of industrial effluents and pesticides

13

Gargi Bhattacharjee[1], *Nisarg Gohil*[1], *Sachin Vaidh*[2], *Krunal Joshi*[1], *Gajendra Singh Vishwakarma*[2] *and Vijai Singh*[1,*]
[1]Department of Biosciences, School of Science, Indrashil University, Rajpur, Mehsana, Gujarat, India, [2]Department of Biological Sciences and Biotechnology, Institute of Advanced Research, Gandhinagar, India
*Corresponding author

13.1 Introduction

Rapid industrialization, urbanization, and various anthropogenic activities have led the world toward a nascent era of global warming, climate change, and high level of pollution. This reckless behavior of humans has not only driven different species to be endangered but also made their own existence troublesome. The ill-effects of pollutants have not only deracinated the biodiversity in its vicinity but have also impaired the proper nutrient cycling of the soil. Not just confined to that, the persistent harsh chemicals that penetrate deep into the soil often reach the water table and thereby enter into the aquatic environment (Fig. 13.1). It is also frightening to note that developing countries discharge 90% of their wastewater into the water bodies without any treatment (UNICEF and WHO, 2015). This wastewater may contain hazardous heavy metals, radionuclides, organic compounds, chemicals, and agricultural wastes. Reports show that approximately 730 megatons of sewage and other effluents are released into the water annually (WWDR, 2017). Industries alone discharge 300–400 million tons of waste into the streams every year (Boretti and Rosa, 2019).

Pesticides are immensely important in healthcare settings to kill the vectors of diseases, such as mosquitoes, while its rather important role is seen in the agricultural sector where it prevents the loss of crops and commodities due to pests. While pesticides are much needed to eradicate the growth of undesirable biological components, an undeniable fact is that these agents are potentially toxic to many nontarget organisms, including humans. Moreover, pesticides and fertilizers have left a serious impact on the soil quality, while the accessibility to potable water is plummeting with each passing day. It has been observed that about 98% of the pesticides used are toxic for crustaceans and fishes (Klemick and Lichtenberg, 2008; Lushchak et al., 2018), while the high amount of phosphorus in the fertilizers contribute greatly to eutrophication (Carpenter, 2008). Hence, it becomes obligatory to keep a check on the proper usage and disposal of these chemicals.

Bioremediation of Pollutants. DOI: https://doi.org/10.1016/B978-0-12-819025-8.00013-2

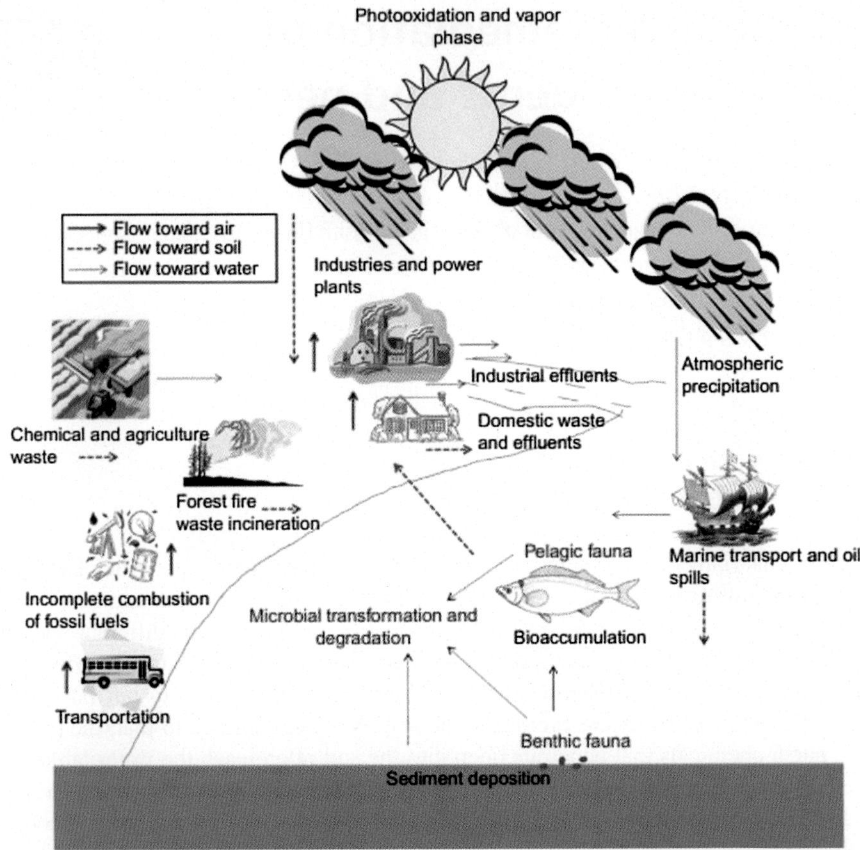

Figure 13.1 A generalized view of the discharge of industrial effluents and agricultural runoff and its repercussive effect on the environment.
Source: Reproduced with permission from Das, S., Raj, R., Mangwani, N., Dash, H.R., Chakraborty, J., 2014. Heavy metals and hydrocarbons: adverse effects and mechanism of toxicity. In: Das, S. (Ed.), Microbial Biodegradation and Bioremediation, Elsevier, London, pp. 23–54. Elsevier Inc.

The concept of using microorganisms to remediate polluted sites is nothing new and has been extensively studied to devise better ways to limit the spread of untreated waste matter. Pollutants-degrading microorganisms usually thrive at the sites where they are most exposed to the contaminants. With time, the constant encounter of the microorganisms with the pollutants endows them with some highly efficient characteristics, thereby allowing better restoration of contaminated sites (Ojuederie and Babalola, 2017). In this chapter, we highlight the hazards of industrial effluents and pesticides, and consortium of microorganisms that can be used for bioremediation for cleaning of polluted environments.

13.2 Microbial bioremediation of industrial effluents

The effluent discharged from various industries such as textile, pharmaceutical, chemical, food, petrochemical, distillery, and tannery contain different toxic organic and inorganic contaminants that can pose a serious impact on public health and the environment if discharged untreated. Therefore it is imperative to remediate and ensure that industrial wastewater meets water quality parameters before being released into nearby water resources (Ramírez-García et al., 2019). Ideally, industrial effluents and sewage should be shifted to wastewater treatment plants (WWTPs) before releasing into the environment. The waste in WWTPs usually goes through three stages, namely primary, secondary and tertiary treatment. Amongst them, the secondary treatment which uses activated sludge is crucial, wherein the biological processes for the decomposition of organic matters take place by employing various aerobic or anaerobic microorganisms (Ramírez-García et al., 2019).

The efficiency of degradation could be enhanced by altering the microorganisms according to the effluents. For example, the microorganisms isolated from brewery waste samples can efficiently degrade the brewery effluents as compared to the others (Oljira et al., 2018). In an attempt to provide information about the ecology and role of different microorganisms used in activated sludge, a database named MiDAS (Microbial Database for Activated Sludge, available at http://www.midasfieldguide.org/) has been generated (McIlroy et al., 2015, 2017). It provides researchers and WWTP practitioners an amalgamated data of microorganisms used in more than 667 WWTPs and 448 anaerobic digesters from more than 28 countries. The microorganisms are listed in detail with their phylum, morphology, functional importance, distribution, and abilities (McIlroy et al., 2015, 2017).

The effluent discharged from various industries has a distinct nature. As an example, the discharged water of textile industries contains different types of noxious synthetic dyes and phenols that make the soil or water bodies polluted by adding toxic synthetic dyes and metal ions which influence pH, biological oxygen demand (BOD), and chemical oxygen demand (COD) negatively. These types of pollutants can be degraded by incorporating microorganisms that have a rich source of degrading enzymes and the capability to grow in adverse conditions. *Aspergillus niger* has been used in numerous studies relating to the degradation of textile and pharmaceutical wastewater (Omar, 2016; Gulzar et al., 2017; Rana et al., 2017; Li et al., 2019), and removal of many different heavy metals contamination (Kapoor et al., 1999; Barros Júnior et al., 2003; Acosta-Rodríguez et al., 2018). It produces many significant ligninolytic enzymes such as laccase, manganese peroxidase, and lignin peroxidase which can break down the chromophoric groups of synthetic dye (Gulzar et al., 2017). Besides this, many bacterial strains such as *Pseudomonas*, *Nitrobacter*, *Nitrosomonas*, *Bacillus*, *Enterobacter*, *Streptomonas*, and *Rhodococcus* have also been widely used for the decolorization, reduction of BOD and COD, and degradation of organic and phenolic compounds

(Silveira et al., 2009; Singh et al., 2015; Tokiran et al., 2016; Guadie et al., 2017; Rana et al., 2017; Gupta et al., 2019).

Most pharmaceutical plants perform a wide range of treatments of wastewater, including pH neutralization, ozonation, sedimentation, Fenton's method, adsorption, precipitation, autoclaving, electrolysis, chlorination, ultrasonication, coagulation, distillation, membrane separation, irradiation, and various aerobic and anaerobic microbial treatments (Pal, 2018; Rana et al., 2017). Still, the pharmaceutical effluent majorly contains antibiotics, hormones, analgesics, surfactants, and other pharmaceutical ingredients. Amongst these, antibiotics and other drugs are of great concern as, if discharged untreated, the pace of multidrug resistance in bacteria would increase more rapidly (Gohil et al., 2018; Obayiuwana et al., 2018; Thai et al., 2018; Kumar et al., 2019). In this regard, numerous antibiotics degrading bacterial strains have been isolated and characterized that could be used in WWTPs, such as *Brevibacterium epidermidis* and *Castellaniella denitrificans* for degradation of sulfonamides (Levine, 2016); *Arthrobacter*, *Pseudomonas*, *Citrobacter youngae*, and *Enterobacter hormaechei* for a mixture of diclofenac, ibuprofen, and sulfamethoxazole (Aissaoui et al., 2017); *Bacteroides fragilis* for tetracycline (Park and Levy, 1988); *Microbacterium* for sulfamethazine (Hirth et al., 2016); *Nocardia* for erythromycin, midecamycin, and rokitamycin (Yazawa et al., 1994) and *Streptomyces lividans* for erythromycin and other macrolides (Wright, 2005) (Table 13.1). Similarly, various hormones, such as estradiol, diclofenac, progesterone, norgestrel, and testosterone are also engaged with many environmental problems, majorly, endocrine disruption in aquatic animals (Suresh and Abraham, 2018). Many species including *Rhodobacter*, *Rhodococcus*, *Nocardia*, *Pseudomonas*, *Comamonas*, and *Nitrosomonas* were found to degrade steroid hormones (Suresh and Abraham, 2018). As an alternative method, ultrafiltration, activated carbon adsorption, photolysis, radiation, ultrasonic treatment, ozonation (Koh et al., 2008; Eckert et al., 2012; Chi et al., 2013.13; Schröder et al., 2016;

Table 13.1 Different drug degrading bacteria.

Degrading bacteria	Drug	References
Brevibacterium epidermidis Castellaniella denitrificans	Sulfonamide	Levine (2016)
Arthrobacter, *Pseudomonas*, *Citrobacter youngae*, and *Enterobacter hormaechei*	Mixture of diclofenac, ibuprofen, and sulfamethoxazole	Aissaoui et al. (2017)
Bacteroides fragilis	Tetracycline	Park and Levy (1988)
Microbacterium	Sulfamethazine	Hirth et al. (2016)
Nocardia	Erythromycin, midecamycin, and rokitamycin	Yazawa et al. (1994)
Streptomyces lividans	Erythromycin and other macrolides	Wright (2005)

Roudbari and Rezakazemi, 2018), and mineralization (Layton et al., 2000) can also be applied. Nevertheless, it is important to analyze the final degraded or converted product, as sometimes the final product may even become more toxic or difficult to treat. For COD reduction in industrial effluent, the aforementioned bacterial strains and/or fungal strains such as *Pseudallescheria boydii* (over 95% reduction), *Fusarium solani* (89.4%), *Fusarium udum* (89.4%), *Candida inconspicua* (76.6%), *Galactomyces pseudocandidum* (76.6%), *Rhodotorula mucilaginosa* (76.6%), *Trichosporon asahii* (76.6%), and *Trypodendron domesticum* (76.6%) have been well explored (Rozitis and Strade, 2015; Aissaoui et al., 2017).

Oil and petroleum industrial effluents have different types of bioremediation challenges as they contain numerous saturates, aromatics, phenols, fatty acids, ketones, esters, porphyrins, and resins (Das and Chandran, 2011; Yuniati, 2018). Oil degrading microbial consortium having capabilities of bioaugmentation and biostimulation are proven to be the best approach for such bioremediation (Das and Chandran, 2011). Different bacterial (*Acinetobacter*, *Arthrobacter*, *Pseudomonas*, *Bacillus*, *Burkholderia*, *Flavobacterium*, *Mycobacterium*, *Rhodococcus*, *Sphingomonas*, *Alcaligenes*, *Micrococcus*) and fungal (*Candida*, *Pichia*, *Yarrowia*, *Talaromyces*, *Amorphoteca*, *Aspergillus*, *Geomyces*, *Graphium*, and *Neosartorya*) species isolated from contaminated sites have been successfully employed to degrade complex pollutants of oil and petroleum industries (Das and Chandran, 2011; Raju and Scalvenzi, 2017; Mahjoubi et al., 2018). Similar approaches can also be used for tannery, chemical, dairy, mining industrial effluents.

13.3 Microbial bioremediation of pesticides

Pests can find a niche in every nook and corner of the surrounding and may cause a whooping economic loss owing to their festering and blighting nature. Any substance or biological agent that limits the spread of the nocuous intruders such as weeds, fungi, insects, rodents, or any other harmful creatures is called a pesticide. Commonly used pesticides comprise of herbicides, fungicides, insecticides, and rodenticides. Besides these, there are several other lesser-known agents which include plant defoliants, growth regulators, surface disinfectants, certain chemicals to disinfect swimming pools and other water bodies, and even the floor cleaner solutions used in daily life are a form of pesticides.

The use of pesticides rocketed shortly after World War II with the advent of DDT (dichlorodiphenyltrichloroethane), benzene hexachloride, endrin, dieldrin, and many others. Though being effective against a large array of pests, the reckless use of these chemicals gave rise to some highly resistant organisms and the traces of pesticides began to appear at unexpected sites (Damalas, 2009). How much a certain pesticide can affect an organism depends on the relative toxicity of the chemical as well as the sensitivity of organisms toward it. Prolong usage and bulk handling of pesticides leads to its accumulation in soil and water bodies through leaching, runoff, and seepage. Additionally, these also tend to enter into the food chain resulting in biomagnification of the same in higher organisms (Bernardes et al., 2015). Despite imposing a ban on the use of pesticides which are hard to degrade, some countries still continue to use them unhesitantly.

Pesticides are predominantly classified under three broad categories based on composition (organochlorine, organophosphates, carbamates, nitrogen-based pesticides, etc.), applicative sector (domestic, agriculture, public health), and target-organism (herbicide, bactericide, fungicide, insecticide, rodenticide, etc.). Other than that, pesticides can also be classified according to their toxicity and mode of entry (Jayaraj et al., 2016). According to the World Health Organization, a pesticide can be classified into four categories, that is, extremely dangerous, highly dangerous, moderately dangerous, and slightly dangerous (WHO, 2009).

Of the numerous organisms residing in the soil, a small section of the soil microbiota has the ability to decompose and biotransform certain pesticides. The chemicals in pesticides supply the required amount of carbon and other growth factors to promote the growth of these microbes. This, in turn, establishes a means to treat the pesticide-contaminated sites (Qiu et al., 2007; Aktar et al., 2009). Microorganisms are known to possess degrading enzymes that can act on pesticides. However, merely the presence of enzymes does not ensure that the degradation is bound to happen. Other than the enzymes, several other ecological, biochemical, physiological, and molecular parameters play a key role in biodegradation and biotransformation of the noxious chemicals (Iranzo et al., 2001; Karigar and Rao, 2011; Ortiz-Hernández et al., 2013). Predominantly, pesticides degrading microbes usually thrive at sites where they are the most exposed to the contaminants. In 1973 the first reports of an organism with the ability to degrade organophosphorus came into light and the organism was later reported to be a bacterium belonging to *Flavobacterium* species (Singh and Walker, 2006). Since then, a number of microorganisms including bacteria, algae, fungi, actinomycetes, and several other species have been reported to degrade pesticides (Parte et al., 2017; Kumar et al., 2018). *Pseudomonas* sp., *Klebsiella* sp., *Bacillus subtilis*, etc. are common examples of bacteria that possess the ability to degrade pesticides. Similarly, marine *Chlorella* belonging to the *Chlorellaceae* algal family and *Trichoderma* spp., *Aspergillus* spp., and white-rot fungi are some of the widely known pesticide degraders (Huang et al., 2018).

Organophosphates are broad-spectrum pesticides used for controlling the spread of nematodes and insects. These pesticides act as potent nerve poisons against the target organisms. In higher organisms, these pesticides delineate their toxicity by acting as a cholinesterase inhibitor, thus hampering the conduction of nerve impulse across the synapse. This leads to rapid convulsions or spasms in the voluntary muscles, ultimately leading to paralysis and even death in some cases (Yadav and Devi, 2017). Cadusafos is one such organothiophosphate nematicide used extensively to protect crops. Persistent use of cadusafos is not just lethal to mammals but also presents a great deal of groundwater and surface contamination. *Pseudomonas* is the most efficient bacterial strain involved in cadusafos degradation (Abo-Amer, 2012). Besides *Pseudomonas*, other examples of organophosphate degrading bacteria include *Brevibacterium* sp. (Jariyal et al., 2018; Sidhu et al., 2019), *Microbacterium esteraromaticum* (Ou et al., 1994; Megharaj et al., 2003; Cáceres et al., 2009; Cabrera et al., 2010), *Rhodococcus* sp. (Singh and Walker, 2006), *Bacillus* sp. (Madhuri and Rangaswamy, 2009; Tang and You, 2012; Aziz et al., 2014), *Serratia marcescens* (Abo-Amer, 2011; Cycoń et al., 2013); *Sphingomonas paucimobilis* (Karpouzas et al., 2005), etc. Other than the bacterial genera, fungal species such as *Phanerochaete chrysosporium* (Eissa et al., 2006), *Cladosporium cladosporioides* (Gao et al., 2012), *Aspergillus fumigatus* (Pandey et al., 2014), *A. niger* (Liu et al., 2001a; Pandey et al., 2014), *Penicillium raistrickii*, and *Aspergillus sydowii* (Alvarenga et al., 2014) have the inherent ability to readily degrade organophosphates from the contaminated environment samples. Active organophosphate degraders belonging to the algae genera include *Chlorella*, *Scenedesmus*, and *Stichococcus* (Megharaj et al., 1987; Cáceres et al., 2009); and those belonging to cyanobacteria include *Anabaena* (Cáceres et al., 2008; Ibrahim and Essa, 2010), *Oscillatoria* (Salman and Abdul-Adel, 2015) and *Nostoc* (Megharaj et al., 1987; Cáceres et al., 2008; Ibrahim et al., 2014).

Organochlorides or chlorinated hydrocarbon-pesticides were amongst the very first pesticides to be synthesized and used in both healthcare settings and farmlands. Similar in action to the organophosphates, these pesticides have long been banned in most developed countries because of its highly toxic nature and long-lasting residual effect in the environment, however, several developing countries still make use of it to limit the spread of vectors (e.g., vectors of malarial parasites). Common examples of organochlorines are DDT, lindane, endosulfan, aldrin, dieldrin, and chlordane. Although few of the aforementioned bacterial strains can break down the pesticide to some degree, the sequential metabolites produced following the degradation does not fail to leave its baneful footprints on the environment. Devising solutions to curb the spread of noxious pollutants in the form of pesticides is on priority for researchers and environmentalists working in this field.

13.4 Microbial bioremediation of hydrocarbons

Contamination of hydrocarbons in soil, water, and marine ecosystem is one of the serious problems amongst other environmental threats. It comes from petroleum, pesticides, or other toxic effluents generated from different industries. Pollution associated with petroleum hydrocarbon is of great concern due to its high abundance and widespread toxicity in the world. The major sources of hydrocarbons are gasoline, diesel, lubricants, surfactants, etc. They are released into the environment via accidental spillage and leaks during fuel exploration, production, transport, and storage. As per the data provided by the US Energy Information Administration, the annual consumption of petroleum products in 2017 was 19.95 million barrels per day. Approximately 200,000 metric tons of crude oil gets spilled every year polluting aquatic and terrestrial ecosystems around the world (Kvenvolden and Cooper, 2003). Apart from that, it also seeps into soil columns and ultimately reaches groundwater aquifers, contaminating them. In addition, the residues of polyaromatic or petroleum hydrocarbons also persist in soil for a long time (Pawlak et al., 2008).

Hydrocarbons are classified into either open chain (aliphatic) or closed chain (cyclic). Most of the pollutants belong to the cyclic hydrocarbons, that is, aromatic and alicyclic hydrocarbon. Compounds belonging to the polycyclic aromatic hydrocarbons (PAHs) cover a major part of toxic hydrocarbons. PAHs generally spread into the environment either through incomplete combustion of wood, coal, oil, and gasoline or accidental spills of crude oils. US Environmental Protection Agency have identified 16 PAHs carcinogens and/or mutagens that are hazardous to human health and passed legislative restrictions over the release of these PAHs into the environment (Liu et al., 2001b).

Numerous physical, chemical, and biological methods can be applied for the remediation of hydrocarbon contamination, of which, biological approaches mediated by microbes are considered as economical, sustainable, ecofriendly, and inclined toward the total degradation of hydrocarbons (Logeshwaran et al., 2018). For oil degradation, bacterial genera *Gammaproteobacteria*, *Pseudoalteromonas*, *Marinobacter*, *Oleispira*, and *Alcanivorax*, have been used, and for other hydrocarbons such as naphthalene and phenanthrene (PHE), *Phingopyxis*, *Rhodobacter*, and *Hyphomonas* species have been used (Mahjoubi et al., 2018). A mangrove sediment-derived bacterial consortium collected from Thailand has shown enormous potential to degrade pyrene, in which, two *Mycobacterium* strains were found to contribute to the initiation process of pyrene degradation, while *Novosphingobium* and *Ochrobactrum* were capable of degrading the intermediates of pyrene, and *Bacillus* sp. enhanced the bioavailability of the pyrene via producing biosurfactant (Wanapaisan et al., 2018). Likewise, 150 bacterial strains were isolated from the contaminated soil of a refinery in Arzew, Algeria. The 16SrRNA study revealed that most the bacterial strains were affiliated with *Gammaproteobacteria* class. Among them, *Pseudomonas* species had the ability to metabolize high molecular weight hydrocarbon compounds such as pristane (C_{19}) and benzo[*a*]pyrene (C_{20}) (GuermoucheM'rassi et al., 2015). Other microbial consortia of *B. pumilus* KS2 and *B. cereus* R2 isolated from crude oil fields (Assam, India) represented a higher advantage when compared to single strains (Patowary et al., 2016). In addition to that, the strain LGM2 of *Pseudomonas* sp. could efficiently degrade the linear and branched alkanes as well as low and high PAHs (GuermoucheM'rassi et al., 2015). In the case of endophytic bacterial isolates, two bacterial strains *Stenotrophomonas* sp. and *Pseudomonas* sp. were found to degrade more than 90% of PHE within 7 days (Zhu et al., 2016). A wide range of hydrocarbon-degrading isolates *Rhodococcus*, *Achromobacter*, *Oerskovia paurometabola*, *Pantoea*, *Sejongia*, *Microbacterium*, and *Arthrobacter* sp. collected from the aged, weathered PAH-contaminated soil have been reported to grow on at least one of the three model PAHs (naphthalene, PHE, and pyrene) (Haleyur et al., 2018).

Bioremediation of hazardous PAHs using different phyla of fungi has also been extensively studied. Fungi are well adapted for terrestrial habitats. The fungal mycelium can easily penetrate in the soil and reach quickly to xenobiotic compounds in soil. Fungal strains also have different enzymes that have the potential to breakdown the PAHs. In this regard, the cytochrome p450 monooxygenase and ligninolytic enzymes present in fungus may work efficiently in the PAH degradation (Črešnar and Petrič, 2011). A group of prominent fungal enzymes such as catalases, laccases, and peroxidases have been found to be involved in the bioremediation of PAHs. White-rot fungi are considered as chief fungi due to its strong ligninolytic enzymes (Deshmukh et al., 2016). The majority of the studies have concluded that white-rot fungi *P. chrysosporium*, *Trametes versicolor*, *Bjerkandera adusta*, and *Pleurotus* sp., may efficiently clean-up the hydrocarbon material due to the presence of different ligninolytic enzymes such as laccases and peroxidases (dos Santos Bazanella et al., 2013). The group of marine fungi have been reported as an apt candidate for the hydrocarbon bioremediation due to production of secondary metabolites, biosurfactants, novel enzymes, polysaccharides, and polyunsaturated fatty acids (Damare et al., 2012). An extreme acting enzyme isolated from *Pestalotiopsis palmarum* was reported for the utilization of carbon from the crude oil (Naranjo-Briceño et al., 2019). The selective preference for *n*-alkanes low molecular weight or high molecular weight also varies among the microbes (Paniagua-Michel and Fathepure, 2018). Due to the complex nature of some environmental pollutants, individual populations and monocultures cannot perform the biodegradation. Additionally, factors associated with the local environmental conditions and site-specific conditions like pollutant flow, migration, and seepage also determine the rate of degradation.

13.5 Conclusion and future remarks

The world is experiencing a serious climatic change which is a matter of grave concern. With the insurmountable amount of waste that is being generated on a daily basis and the repercussions of it being clearly evident on the environment, it goes without saying that a mass extinction of biodiversity follows if immediate and strong concerted actions are not taken forthwith. A handful of microbes in nature are endowed with the ability to degrade pollutants coming from domestic sewage, industries, and pesticide runoffs from agricultural lands, at affordable rates and without compromising environmental aspects. However, the proficiency with which they do so is relatively slow and the efficiency with which these microbes delineate their degradation activity is subjected to alter the change in prevalent conditions. Resultantly, it is important to conduct fine studies and develop a clear understanding of ways to eliminate toxic pollutants by bringing about a few but necessary improvements in the existing pollutants-degrading microorganisms. Advanced genome editing techniques (Singh et al., 2017, 2018; Bhattacharjee et al., 2020; Khan, 2019), directed evolution (Copley, 2009; Cobb et al., 2013) and metabolic pathway engineering (Haro and de Lorenzo, 2001; Singh et al., 2008; Gohil et al., 2017; Dvořák et al., 2017) approaches could be applied for engineering microorganisms for enhancing the degradation capabilities (Stein et al., 2018; Jaiswal et al., 2019). There are some crucial factors that need to be looked after, which include temperature, pH, and availability of oxygen and nutrients, as they tend to negatively affect the operation of microbial degradation.

Another plausible solution is to inoculate mixed consortia of highly potent microbes that can significantly relieve the soil by acting on a diverse range of chemical compounds present at the polluted sites. It is the basic understanding and advancements of the underlying sciences of analytical geochemistry and environmental microbiology that drives the progress of biodegradation on the principles of biotechnology. Proper management of effluents could lead toward zero discharge of waste.

Nature has been offering several alarming evidences indicating that we may have already crossed or are on the verge of crossing several tipping points of the ecosystem and the planetary climatic system, reversal of which is nearly impossible. Loss of ice sheets, melting of glaciers, destruction of biodiversity in ecosystems as diverse as the Arctic tundra and Amazon rainforest all have been a result of the rapid warming of the planet. It is time that we shift toward a more sustainable approach and limit the use of hydrocarbon-based resources and ensure proper mitigation of the generated wastes. Tools in the form of pollutant-degrading microbes are already available, just an immediate and large-scale application awaits. Although it seems tough to achieve, every prudent steps toward mitigation of toxic wastes and the encouragement of waste valorization can reduce the possible environmental and health risks. The way in which we respond to this ongoing crisis will decide the fate of the current as well as coming generations of not just humans but all other species.

Acknowledgements

GB, NG, KJ, and VS thank Indrashil University for providing infrastructure and support for preparation of this chapter. SV and GSV gratefully acknowledge the Gujarat State Biotechnology Mission, Gandhinagar (GSBTM Project ID: L1Y5SU) for financial support during the preparation of this chapter.

References

Abo-Amer, A., 2011. Biodegradation of diazinon by *Serratia marcescens* DI101 and its use in bioremediation of contaminated environment. J. Microbiol. Biotechnol. 21, 71–80.

Abo-Amer, A.E., 2012. Characterization of a strain of *Pseudomonas putida* isolated from agricultural soil that degrades cadusafos (an organophosphorus pesticide). World J. Microbiol. Biotechnol. 28, 805–814.

Acosta-Rodríguez, I., Cárdenas-González, J.F., Rodríguez Pérez, A.S., Oviedo, J.T., Martínez-Juárez, V.M., 2018. Bioremoval of different heavy metals by the resistant fungal strain *Aspergillus niger*. Bioinorg. Chem. Appl. 2018, 3457196.

Aissaoui, S., Ouled-Haddar, H., Sifour, M., Beggah, C., Benhamada, F., 2017. Biological removal of the mixed pharmaceuticals: diclofenac, ibuprofen, and sulfamethoxazole using a bacterial consortium. Iran. J. Biotechnol. 15, 135–142.

Aktar, W., Sengupta, D., Chowdhury, A., 2009. Impact of pesticides use in agriculture: their benefits and hazards. Interdiscip. Toxicol 2, 1–12.

Alvarenga, N., Birolli, W.G., Seleghim, M.H., Porto, A.L., 2014. Biodegradation of methyl parathion by whole cells of marine-derived fungi *Aspergillus sydowii* and *Penicillium decaturense*. Chemosphere 117, 47–52.

Aziz, M.W., Sabit, H., Tawakkol, W., 2014. Biodegradation of malathion by *Pseudomonas* spp. and *Bacillus* spp. isolated from polluted sites in Egypt. Am.-Eurasian J. Agric. Environ. Sci. 14, 855–862.

Barros Júnior, L.M., Macedo, G.R., Duarte, M.M.L., Silva, E.P., Lobato, A.K.C.L., 2003. Biosorption of cadmium using the fungus *Aspergillus niger*. Braz. J. Chem. Eng. 20, 229–239.

Bernardes, M.F.F., Pazin, M., Pereira, L.C., Dorta, D.J., 2015. Impact of pesticides on environmental and human health. In: Andreazza, A.C. (Ed.), Toxicology Studies-Cells, Drugs and Environment. InTech, Croatia, pp. 195–233.

Bhattacharjee, G., Mani, I., Gohil, N., Khambhati, K., Braddick, D., Panchasara, H., et al., 2020. CRISPR technology for genome editing. In: Faintuch, J., Faintuch, S. (Eds.), Precision Medicine for Investigators, Practitioners and Providers. Academic Press, London, pp. 59–69.

Boretti, A., Rosa, L., 2019. Reassessing the projections of the World Water Development Report. npj Clean Water 2, 15.

Cabrera, J.A., Kurtz, A., Sikora, R.A., Schouten, A., 2010. Isolation and characterization of fenamiphos degrading bacteria. Biodegradation 21, 1017–1027.

Cáceres, T.P., Megharaj, M., Naidu, R., 2008. Biodegradation of the pesticide fenamiphos by ten different species of green algae and cyanobacteria. Curr. Microbiol. 57, 643–646.

Cáceres, T.P., Megharaj, M., Malik, S., Beer, M., Naidu, R., 2009. Hydrolysis of fenamiphos and its toxic oxidation products by *Microbacterium* sp. in pure culture and groundwater. Bioresour. Technol. 100, 2732–2736.

Carpenter, S.R., 2008. Phosphorus control is critical to mitigating eutrophication. Proc. Natl. Acad. Sci. U.S.A. 105, 11039–11040.

Chi, G.T., Churchley, J., Huddersman, K.D., 2013. Pilot-scale removal of trace steroid hormones and pharmaceuticals and personal care products from municipal wastewater using a heterogeneous Fenton's catalytic process. Int. J. Chem. Eng. 2013, 760915.

Cobb, R.E., Chao, R., Zhao, H., 2013. Directed evolution: past, present, and future. AIChE J. 59, 1432–1440.

Copley, S.D., 2009. Evolution of efficient pathways for degradation of anthropogenic chemicals. Nat. Chem. Biol. 5, 559–566.

Črešnar, B., Petrič, S., 2011. Cytochrome P450 enzymes in the fungal kingdom. Biochim. Biophys. Acta 1814, 29–35.

Cycoń, M., Żmijowska, A., Wójcik, M., Piotrowska-Seget, Z., 2013. Biodegradation and bioremediation potential of diazinon-degrading *Serratia marcescens* to remove other organophosphorus pesticides from soils. J. Environ. Manage. 117, 7–16.

Damalas, C.A., 2009. Understanding benefits and risks of pesticide use. Sci. Res. Essay 4, 945–949.

Damare, S., Singh, P., Raghukumar, S., 2012. Biotechnology of marine fungi. Prog. Mol. Subcell. Biol. 53, 277–297.

Das, N., Chandran, P., 2011. Microbial degradation of petroleum hydrocarbon contaminants: an overview. Biotechnol. Res. Int. 2011, 941810.

Das, S., Raj, R., Mangwani, N., Dash, H.R., Chakraborty, J., 2014. Heavy metals and hydrocarbons: adverse effects and mechanism of toxicity. In: Das, S. (Ed.), Microbial Biodegradation and Bioremediation. Elsevier, London, pp. 23–54.

Deshmukh, R., Khardenavis, A.A., Purohit, H.J., 2016. Diverse metabolic capacities of fungi for bioremediation. Indian J. Microbiol. 56, 247–264.

dos Santos Bazanella, G.C., de Souza, D.F., Castoldi, R., Oliveira, R.F., Bracht, A., Peralta, R.M., 2013. Production of laccase and manganese peroxidase by *Pleurotus pulmonarius* in solid-state cultures and application in dye decolorization. Folia Microbiol. 58, 641–647.

Dvořák, P., Nikel, P.I., Damborský, J., de Lorenzo, V., 2017. Bioremediation 3.0: engineering pollutant-removing bacteria in the times of systemic biology. Biotechnol. Adv. 35, 845–866.

Eckert, V., Bensmann, H., Zegenhagen, F., Weckenmann, J., Sörensen, M., 2012. Elimination of hormones in pharmaceutical waste water. Pharm. Ind. Pharmind. 74, 487–492.

Eissa, F.I., Mahmoud, H.A., Zidan, N.A., Belal, E.B.A., 2006. Microbial, thermal and photodegradation of cadusafos and carbofuran pesticides. J. Pestic. Contam. Environ. Sci. 14, 107–130.

Gao, Y., Chen, S., Hu, M., Hu, Q., Luo, J., Li, Y., 2012. Purification and characterization of a novel chlorpyrifos hydrolase from *Cladosporium cladosporioides* Hu-01. PLoS One 7, e38137.

Gohil, N., Panchasara, H., Patel, S., Ramírez-García, R., Singh, V., 2017. Book review: recent advances in yeast metabolic engineering. Front. Bioeng. Biotechnol. 5, 71.

Gohil, N., Ramírez-García, R., Panchasara, H., Patel, S., Bhattacharjee, G., Singh, V., 2018. Book review: quorum sensing vs. quorum quenching: a battle with no end in sight. Front. Cell. Infect. Microbiol. 8, 106.

Guadie, A., Tizazu, S., Melese, M., Guo, W., Ngo, H.H., Xia, S., 2017. Biodecolorization of textile azo dye using *Bacillus* sp. strain CH12 isolated from alkaline lake. Biotechnol. Rep. 15, 92–100.

GuermoucheM'rassi, A., Bensalah, F., Gury, J., Duran, R., 2015. Isolation and characterization of different bacterial strains for bioremediation of n-alkanes and polycyclic aromatic hydrocarbons. Environ. Sci. Pollut. Res. 22, 15332–15346.

Gulzar, T., Huma, T., Jalal, F., Iqbal, S., Abrar, S., Kiran, S., et al., 2017. Bioremediation of synthetic and industrial effluents by *Aspergillus niger* isolated from contaminated soil following a sequential strategy. Molecules 22, E2244.

Gupta, R., Sati, B., Gupta, A., 2019. Treatment and recycling of wastewater from pharmaceutical industry. In: Singh, R.L., Singh, R.P. (Eds.), Advances in Biological Treatment of Industrial Waste Water and their Recycling for a Sustainable Future. Springer, Singapore, pp. 267–302.

Haleyur, N., Shahsavari, E., Taha, M., Khudur, L.S., Koshlaf, E., Osborn, A.M., et al., 2018. Assessing the degradation efficacy of native Pah-degrading bacteria from aged, weathered soils in an Australian former gasworks site. Geoderma 321, 110–117.

Haro, M.A., de Lorenzo, V., 2001. Metabolic engineering of bacteria for environmental applications: construction of *Pseudomonas* strains for biodegradation of 2-chlorotoluene. J. Biotechnol. 85, 103–113.

Hirth, N., Topp, E., Dörfler, U., Stupperich, E., Munch, J.C., Schroll, R., 2016. An effective bioremediation approach for enhanced microbial degradation of the veterinary antibiotic sulfamethazine in an agricultural soil. Chem. Biol. Technol. Agric. 29, 2–11.

Huang, Y., Xiao, L., Li, F., Xiao, M., Lin, D., Long, X., et al., 2018. Microbial degradation of pesticide residues and an emphasis on the degradation of cypermethrin and 3-phenoxy benzoic acid: a review. Molecules 23, 2313.

Ibrahim, M.W., Essa, M.A., 2010. Tolerance and utilization of organophosphorus insecticide by nitrogen fixing Cyanobacteria. Egypt. J. Bot. 27, 225–240.

Ibrahim, W.M., Karam, M.A., El-Shahat, R.M., Adway, A.A., 2014. Biodegradation and utilization of organophosphorus pesticide malathion by cyanobacteria. BioMed Res. Int. 2014, 392682.

Iranzo, M., Sainz-Pardo, I., Boluda, R., Sanchez, J., Mormeneo, S., Hemery, G.E., et al., 2001. The use of microorganisms in environmental remediation. Ann. Microbiol. 51, 135–143.

Jaiswal, S., Singh, D.K., Shukla, P., 2019. Gene editing and systems biology tools for pesticide bioremediation: a review. Front. Microbiol. 10, 87.

Jariyal, M., Jindal, V., Mandal, K., Gupta, V.K., Singh, B., 2018. Bioremediation of organophosphorus pesticide phorate in soil by microbial consortia. Ecotoxicol. Environ. Saf. 159, 310–316.

Jayaraj, R., Megha, P., Sreedev, P., 2016. Organochlorine pesticides, their toxic effects on living organisms and their fate in the environment. Interdiscip. Toxicol 9, 90–100.

Kapoor, A., Viraraghavan, T., Cullimore, D.R., 1999. Removal of heavy metals using the fungus *Aspergillus niger*. Bioresour. Technol. 70, 95–104.

Karigar, C.S., Rao, S.S., 2011. Role of microbial enzymes in the bioremediation of pollutants: a review. Enzyme Res. 2011, 805187.

Karpouzas, D.G., Fotopoulou, A., Menkissoglu-Spiroudi, U., Singh, B.K., 2005. Non-specific biodegradation of the organophosphorus pesticides, cadusafos and ethoprophos, by two bacterial isolates. FEMS Microbiol. Ecol 53, 369–378.

Khan, S.H., 2019. Genome-editing technologies: concept, pros, and cons of various genome-editing techniques and bioethical concerns for clinical application. molecular therapy. Mol. Ther. Nucleic Acids 16, 326–334.

Klemick, H., Lichtenberg, E., 2008. Pesticide use and fish harvests in Vietnamese rice agroecosystems. Am. J. Agric. Econ. 90, 1–14.

Koh, Y.K.K., Chiu, T.Y., Boobis, A., Cartmell, E., Scrimshaw, M.D., Lester, J.N., 2008. Treatment and removal strategies for estrogens from wastewater. Environ. Technol. 29, 245–267.

Kumar, M., Jaiswal, S., Sodhi, K.K., Shree, P., Singh, D.K., Agrawal, P.K., et al., 2019. Antibiotics bioremediation: perspectives on its ecotoxicity and resistance. Environ. Int. 124, 448–461.

Kumar, S., Kaushik, G., Dar, M.A., Nimesh, S., Lopez-Chuken, U.J., Villarreal-Chiu, J.F., 2018. Microbial degradation of organophosphate pesticides: a review. Pedosphere 28, 190–208.

Kvenvolden, K.A., Cooper, C.K., 2003. Natural seepage of crude oil into the marine environment. Geo-Marine Lett. 23, 140–146.

Layton, A.C., Gregory, B.W., Seward, J.R., Schultz, T.W., Sayler, G.S., 2000. Mineralization of steroidal hormones by biosolids in wastewater treatment systems in Tennessee USA. Environ. Sci. Technol. 34, 3925–3931.

Levine, R., 2016. Microbial degradation of sulfonamide antibiotics: a thesis, 89. Available from: <https://digitalcommons.unl.edu/cgi/viewcontent.cgi?article = 1093&context = civilengdiss> (accessed 25.10.19.).

Li, S., Huang, J., Mao, J., Zhang, L., He, C., Chen, G., et al., 2019. *In vivo* and *in vitro* efficient textile wastewater remediation by *Aspergillus niger* biosorbent. Nanoscale Adv. 1, 168–176.

Liu, K., Han, W., Pan, W.P., Riley, J.T., 2001b. Polycyclic aromatic hydrocarbon (PAH) emissions from a coal-fired pilot FBC system. J. Hazard. Mater. 84, 175–188.

Liu, Y.H., Chung, Y.C., Xiong, Y., 2001a. Purification and characterization of a dimethoate-degrading enzyme of *Aspergillus niger* ZHY256, isolated from sewage. Appl. Environ. Microbiol. 67, 3746–3749.

Logeshwaran, P., Megharaj, M., Chadalavada, S., Bowman, M., Naidu, R., 2018. Petroleum hydrocarbons (PH) in groundwater aquifers: an overview of environmental fate, toxicity, microbial degradation and risk-based remediation approaches. Environ. Technol. Innov. 10, 175–193.

Lushchak, V.I., Matviishyn, T.M., Husak, V.V., Storey, J.M., Storey, K.B., 2018. Pesticide toxicity: a mechanistic approach. EXCLI J. 17, 1101–1136.

Madhuri, R., Rangaswamy, V., 2009. Biodegradation of selected insecticides by *Bacillus* and *Pseudomonas* sps in groundnut fields. Toxicol. Int. 16, 127.

Mahjoubi, M., Cappello, S., Souissi, Y., Jaouani, A., Cherif, A., 2018. Microbial bioremediation of petroleum hydrocarbon-contaminated marine environments. In: Zoveidavianpoor, M. (Ed.), Recent Insights in Petroleum Science and Engineering. IntechOpen, London, <https://doi.org/10.5772/intechopen.72207>.

McIlroy, S.J., Kirkegaard, R.H., McIlroy, B., Nierychlo, M., Kristensen, J.M., Karst, S.M., et al., 2017. MiDAS 2.0: an ecosystem-specific taxonomy and online database for the organisms of wastewater treatment systems expanded for anaerobic digester groups. Database 2017.

McIlroy, S.J., Saunders, A.M., Albertsen, M., Nierychlo, M., McIlroy, B., Hansen, A.A., et al., 2015. MiDAS: the field guide to the microbes of activated sludge. Database 2015.

Megharaj, M., Venkateswarlu, K., Rao, A.S., 1987. Metabolism of monocrotophos and quinalphos by algae isolated from soil. Bull. Environ. Contam. Toxicol. 39, 251–256.

Megharaj, M., Singh, N., Kookana, R.S., Naidu, R., Sethunathan, N., 2003. Hydrolysis of fenamiphos and its oxidation products by a soil bacterium in pure culture, soil and water. Appl. Microbiol. Biotechnol. 61, 252–256.

Naranjo-Briceño, L., Pernía, B., Perdomo, T., González, M., Inojosa, Y., De Sisto, Á., et al., 2019. Potential role of extremophilic hydrocarbonoclastic fungi for extra-heavy crude oil bioconversion and the sustainable development of the petroleum industry. In: Tiquia-Arashiro, S.M., Grube, M. (Eds.), Fungi in Extreme Environments: Ecological Role and Biotechnological Significance. Springer, Cham, pp. 559–586.

Obayiuwana, A., Ogunjobi, A., Yang, M., Ibekwe, M., 2018. Characterization of bacterial communities and their antibiotic resistance profiles in wastewaters obtained from pharmaceutical facilities in Lagos and Ogun States, Nigeria. Int. J. Environ. Res. Public Health 15, 1365.

Ojuederie, O.B., Babalola, O.O., 2017. Microbial and plant-assisted bioremediation of heavy metal polluted environments: a review. Int. J. Environ. Res. Public Health 14, E1504.

Oljira, T., Muleta, D., Jida, M., 2018. Potential applications of some indigenous bacteria isolated from polluted areas in the treatment of brewery effluents. Biotechnol. Res. Int. 2018, 9745198.

Omar, S.A., 2016. Decolorization of different textile dyes by isolated *Aspergillus niger*. J. Environ. Sci. Technol. 9, 149–156.

Ortiz-Hernández, M.L., Sánchez-Salinas, E., Dantán-González, E., Castrejón-Godínez, M.L., 2013. Pesticide biodegradation: mechanisms, genetics and strategies to enhance the process. In: Chamy, R. (Ed.), Biodegradation—Life of Science. IntechOpen, London, pp. 251–287.

Ou, L.T., Thomas, J.E., Dickson, D.W., 1994. Degradation of fenamiphos in soil with a history of continuous fenamiphos applications. Soil. Sci. Soc. Am. J. 58, 1139–1147.

Pal, P., 2018. Treatment and disposal of pharmaceutical wastewater: toward the sustainable strategy. Sep. Purif. Rev. 47, 179–198.

Pandey, B., Baghel, P.S., Shrivastava, S., 2014. To study the bioremediation of monocrotophos and to analyze the kinetics effect of Tween 80 on fungal growth. Indo Am. J. Pharm. Res. 4, 925–930.

Paniagua-Michel, J., Fathepure, B.Z., 2018. Microbial consortia and biodegradation of petroleum hydrocarbons in marine environments. In: Kumar, V., Kumar, M., Prasad, R. (Eds.), Microbial Action on Hydrocarbons. Springer, Singapore, pp. 1–20.

Park, B.H., Levy, S.B., 1988. The cryptic tetacycline resistance determinant on Tn4400 mediates tetracycline degradation as well as tetracycline efflux. Antimicrob. Agents Chemother. 32, 1797–1800.

Parte, S.G., Mohekar, A.D., Kharat, A.S., 2017. Microbial degradation of pesticide: a review. Afr. J. Microbiol. Res. 11, 992–1012.

Patowary, K., Patowary, R., Kalita, M.C., Deka, S., 2016. Development of an efficient bacterial consortium for the potential remediation of hydrocarbons from contaminated sites. Front. Microbiol. 7, 1092.

Pawlak, Z., Rauckyte, T., Oloyede, A., 2008. Oil, grease and used petroleum oil management and environmental economic issues. J. Achieve. Mater. Manuf. Eng. 26, 11–17.

Qiu, X., Zhong, Q., Li, M., Bai, W., Li, B., 2007. Biodegradation of *p*-nitrophenol by methyl parathion-degrading *Ochrobactrum* sp. B2. Int. Biodeterior. Biodegrad. 59, 297–301.

Raju, M.N., Scalvenzi, L., 2017. Petroleum degradation: promising biotechnological tools for bioremediation. In: Zoveidavianpoor, M. (Ed.), Recent Insights in Petroleum Science and Engineering. IntechOpen, London.

Ramírez-García, R., Gohil, N., Singh, V., 2019. Recent advances, challenges, and opportunities in bioremediation of hazardous materials. In: Pandey, V.C., Bauddh, K. (Eds.), Phytomanagement of Polluted Sites. Elsevier, Amsterdam, pp. 517–568.

Rana, R.S., Singh, P., Kandari, V., Singh, R., Dobhal, R., Gupta, S., 2017. A review on characterization and bioremediation of pharmaceutical industries' wastewater: an Indian perspective. Appl. Water Sci. 7, 1–12.

Roudbari, A., Rezakazemi, M., 2018. Hormones removal from municipal wastewater using ultrasound. AMB Express 8, 91.

Rozitis, D.Z., Strade, E., 2015. COD reduction ability of microorganisms isolated from highly loaded pharmaceutical wastewater pre-treatment process. J. Mater. Environ. Sci. 6, 507–512.

Salman, J.M., Abdul-Adel, E., 2015. Potential use of cyanophyta species *Oscillatoria limnetica* in bioremediation of organophosphorus herbicide glyphosate. Mesop. Environ. J. 1, 15–26.

Schröder, P., Helmreich, B., Škrbić, B., Carballa, M., Papa, M., Pastore, C., et al., 2016. Status of hormones and painkillers in wastewater effluents across several European states—considerations for the EU watch list concerning estradiols and diclofenac. Environ. Sci. Pollut. Res. 23, 12835–12866.

Sidhu, G.K., Singh, S., Kumar, V., Dhanjal, D.S., Datta, S., Singh, J., 2019. Toxicity, monitoring and biodegradation of organophosphate pesticides: a review. Crit. Rev. Environ. Sci. Technol. 49, 1135–1187.

Silveira, E., Marques, P.P., Silva, S.S., Lima-Filho, J.L., Porto, A.L.F., Tambourgi, E.B., 2009. Selection of *Pseudomonas* for industrial textile dyes decolourization. Int. Biodeterior. Biodegrad. 63, 230–235.

Singh, A.L., Chaudhary, S., Kayastha, A.M., Yadav, A., 2015. Decolorization and degradation of textile effluent with the help of *Enterobacter asburiae*. Ind. J. Biotechnol. 14, 101–106.

Singh, B.K., Walker, A., 2006. Microbial degradation of organophosphorus compounds. FEMS Microbiol. Rev. 30, 428–471.

Singh, S., Kang, S.H., Mulchandani, A., Chen, W., 2008. Bioremediation: environmental clean-up through pathway engineering. Curr. Opin. Biotechnol. 19, 437–444.

Singh, V., Braddick, D., Dhar, P.K., 2017. Exploring the potential of genome editing CRISPR-Cas9 technology. Gene 599, 1–18.

Singh, V., Gohil, N., Ramírez García, R., Braddick, D., Fofié, C.K., 2018. Recent advances in CRISPR-Cas9 genome editing technology for biological and biomedical investigations. J. Cell. Biochem. 119, 81–94.

Stein, H.P., Navajas-Pérez, R., Aranda, E., 2018. Potential for CRISPR genetic engineering to increase xenobiotic degradation capacities in model fungi. In: Prasad, R., Aranda, E. (Eds.), Approaches in Bioremediation. Springer, Cham, pp. 61–78.

Suresh, A., Abraham, J., 2018. Bioremediation of hormones from waste water. In: Hussain, C.M. (Ed.), Handbook of Environmental Materials Management. Springer, Cham, pp. 1–31.

Tang, M., You, M., 2012. Isolation, identification and characterization of a novel triazophos-degrading *Bacillus* sp. (TAP-1). Microbiol. Res. 167, 299–305.

Thai, P.K., Binh, V.N., Nhung, P.H., Nhan, P.T., Hieu, N.Q., Dang, N.T., et al., 2018. Occurrence of antibiotic residues and antibiotic-resistant bacteria in effluents of pharmaceutical manufacturers and other sources around Hanoi, Vietnam. Sci. Total Environ. 645, 393–400.

Tokiran, S., Maniyam, M.N., Yaacob, N.S., Ibrahim, A.L., 2016. Decolourization of textile dyes by Malaysian *Rhodococcus* strains. Indian J. Fundam. Appl. Life Sci. 6, 14–20.

UNICEF and WHO, 2015. Progress on sanitation and drinking water: 2015 update and MDG assessment, The United Nations Children's Fund (UNICEF) and World Health Organization (WHO), New York. <https://data.unicef.org/wp-content/uploads/2015/12/Progress-on-Sanitation-and-Drinking-Water_234.pdf> (accessed 25.09.19.).

Wanapaisan, P., Laothamteep, N., Vejarano, F., Chakraborty, J., Shintani, M., Muangchinda, C., et al., 2018. Synergistic degradation of pyrene by five culturable bacteria in a mangrove sediment-derived bacterial consortium. J. Hazard. Mater. 342, 561–570.

WHO, 2009. The WHO recommended classification of pesticides by hazard and guidelines to classification, World Health Organization (WHO), Geneva. <https://www.who.int/ipcs/publications/pesticides_hazard_2009.pdf> (accessed 25.10.19.).

Wright, G.D., 2005. Bacterial resistance to antibiotics: enzymatic degradation and modification. Adv. Drug Deliv. Rev. 57, 1451–1470.

WWDR, 2017. The United Nations world water development report, 2017: wastewater: the untapped resource, The United Nations Educational, Scientific and Cultural Organization (UNESCO), Paris. <https://www.unesdoc.unesco.org/ark:/48223/pf0000247153> (accessed 28.09.19.).

Yadav, I.S., Devi, N.L., 2017. Pesticides classification and its impact on human and environment. Environ. Sci. Eng. 6, 140–158.

Yazawa, K., Mikami, Y., Sakamoto, T., Ueno, Y., Morisaki, N., Iwasaki, S., et al., 1994. Inactivation of the macrolide antibiotics erythromycin, midecamycin, and rokitamycin by pathogenic *Nocardia* species. Antimicrob. Agents Chemother. 38, 2197–2199.

Yuniati, M.D., 2018. Bioremediation of petroleum-contaminated soil: a review. IOP Conf. Ser.: Earth Environ. Sci. 118, 012063.

Zhu, X., Ni, X., Waigi, M.G., Liu, J., Sun, K., Gao, Y., 2016. Biodegradation of mixed PAHs by PAH-degrading endophytic bacteria. Int. J. Environ. Res. Public. Health 13, E805.

Synthetic biology approaches for bioremediation

14

*Gargi Bhattacharjee, Nisarg Gohil and Vijai Singh**
Department of Biosciences, School of Science, Indrashil University, Rajpur, Mehsana, Gujarat, India
*Corresponding author

14.1 Introduction

The world population is surging day-by-day and with modernization, the demand for industrial goods is rising. The rate at which the industries have been exploiting natural resources and discharging the harmful effluents into the environment has left a huge impact on the earth's natural balance. The extent of damage has been such that the footprints can be seen not just on terra firma but also in the deep ocean waters or seafloor sediments. The world is experiencing an acute shortage of clean water to drink, fertile lands to cultivate, and a trembling level of fresh air to breathe which is further expected to culminate into a more worrisome state. Therefore it is time to conceive effective eco-friendly strategies to rehabilitate the human-driven fouled state of the environment.

Bioremediation is an approach to manage and treat waste and pollutants in a sustainable manner. Bioremediation can be simply described as the use of microbes to detoxify and degrade environmental contaminants in order to maintain the natural stability and improve the condition of the surrounding (Atlas and Hazen, 2011; Ramírez-García et al., 2019). The metabolic versatility of the genetically engineered microbes and plants can be employed for this purpose or even the wild-type microorganisms can be used with little modifications to adapt to the environmentally unfriendly conditions, thereby restoring the natural balance. These tailor-made microbes are competent to either degrade the toxic compounds (Karigar and Rao, 2011; Ojuederie and Babalola, 2017) or assimilate a large amount of it within the cell (Ayangbenro and Babalola, 2017; Igiri et al., 2018). Bioremediation is not just a strategy or an option to curb the damage that humans have been relentlessly causing on the environment; it has become obligatory if we want our natural resources to sustain for a few more generations. Undoubtedly, the effects of bioremediation are worthwhile in that it comes with several benefits such as the water quality can be improved or say the barren land can be turned into a productive field suitable for cultivation.

Since bioremediation stands on the pillars of biodegradation, the intention is to catabolize the xenobiotics and eliminate them completely. The microorganism chosen to serve this purpose can adapt and thrive on the waste matter or contaminants by using it as a substrate. Autochthonous microbes residing at polluted sites have been instrumental in solving some very critical challenges linked with the

Bioremediation of Pollutants. DOI: https://doi.org/10.1016/B978-0-12-819025-8.00014-4

hazardous pollutants (Verma and Jaiswal, 2016). Moreover, when proceeding with bioremediation, designing budget-friendly in situ markers to monitor the detection and degradation of environmental contaminants is advisable or rather requisite when the concentration of the contaminant is too low to be detected. That is when biosensors come into the picture which can fulfill both laboratory and on-site analytical applications. Using living organisms to sense pollution is not new, the instance of which can be drawn from the days when birds were used to detect the emission of toxic gases over coal mines.

Synthetic biology has emerged as a speedily evolving interdisciplinary area linking principles of biology and engineering, aimed at restructuring the existing lifeforms primarily by reprogramming their genetic framework. In simpler terms, synthetic biology is the design and construction of novel genetic parts, devices, and systems, or to redesign the present biological systems that lead towards betterment. Redesigning or de novo synthesis of the living system by employing principles of mathematical computations, computer science, chemistry, and other associated fields of engineering, may generate potent organisms carrying high utilitarian biological parts aimed to function in a preprogrammed manner as compared to their natural counterparts. For instance, synthetic biology may serve as a platform to develop high throughput sensors in microbes which endows them with the ability to respond and detect very low concentrations and to differentiate between closely related or highly similar compounds (Park et al., 2013; Carpenter et al., 2018).

With expeditious progression in molecular technologies, it is anticipated that synthetic biology would overcome the hurdles faced due to the natural evolution and alongside generate organisms endowed with desired functions. In this chapter, we highlight how synthetic biology can be amalgamated with the bioremediation techniques to move toward a greener and cleaner future.

14.2 Overview of synthetic biology

Synthetic biology provides an armamentarium of genetic tools to meticulously transform the molecular settings inside a living cell. Over the past decade, the leaping strides toward designing newer biological parts and devices have accelerated the journey toward building a cell with better physiological and biochemical properties. A pragmatic aspect of such developments is that technologies such as genome sequencing have become accessible and affordable for a larger section of the scientific community. Synthetic biology includes genetic parts (synthetic promoter, ribosome binding site (RBS), transcription factors, transcription terminator), synthetic devices (small noncoding RNAs such as riboregulators, riboswitches, synthetic oscillator, toggle switch, biologic gates), and many such cloning techniques such as clustered regularly interspaced short palindromic repeats with associated proteins (CRISPR–Cas), genome assembly and synthesis, cell-free protein synthesis, and microfluidics platform (Singh, 2014; Patel et al., 2018; Khambhati et al., 2019).

The very first genetic circuits, a genetic toggle switch (Gardner et al., 2000) and synthetic oscillator (Elowitz and Leibler, 2000), were based upon the knowledge of basic mathematical computations and involved remodeling of some usual DNA regulatory elements. Basically, a genetic toggle switch is an artificially built bistable genetic network that can switch between two steady states when induced by input signals such as a thermal or chemical inducer. Synthetic oscillators, on the other hand, are programmed genetic circuits designed to mediate functions in a time-dependent manner. Ever since their discovery, these developments have been the stepping stone for further advancements in the synthetic biology field (McAdams and Arkin, 2000).

The International Genetically Engineered Machine (iGEM) Foundation held its first competition in 2004. iGEM had earlier started as an independent, nonprofit organization at the Massachusetts Institute of Technology in 2003 to encourage students to create engineered biological systems. In the following year, this competition was opened up to all across the globe for whoever was interested in raising ideas to solve global challenges using synthetic biology techniques. Subsequently, databases such as the Registry of Standard Biological Parts (Endy, 2005) and OpenWetWare (Shee et al., 2010) came into existence. Based on these, a number of complex and effective genetic circuits began to be developed. Soon after, scientists began to work on engineering genetic logic gates to incorporate the mathematical Boolean functions that were expected to play some very important role in synthetic biology applications (Moon et al., 2012). Just as in mathematics, biological gates include AND, OR, NOR, NOT, NAND, XOR, and other such gates. The synthetic biology logic gate database (SynBioLGBD) is a user-friendly database that has over 189 registered logic circuits with functional evidence in three species including humans, *Escherichia coli* and *Bacillus clausii* (Wang et al., 2015).

CRISPR–Cas is another revolutionizing genome-editing technique widely used in biotechnology, synthetic biology and metabolic engineering (Jinek et al., 2012; Cong et al., 2013; Mali et al., 2013; Bikard et al., 2014; Singh et al., 2017, 2018; Gohil et al., 2017; Bhattacharjee et al., 2020). This technique is originally based on the idea of inherent RNA-mediated adaptive immune response in bacteria and archaea (Barrangou et al., 2007; Horvath and Barrangou, 2010; Barrangou and Marraffini, 2014). The most widely used Cas protein is Cas9 which delineates its endonuclease activity in a gRNA dependent manner. A mutated version of Cas9 called the dCas9 is widely used in a variety of synthetic circuits, which in itself does not show cleavage property but has the ability to specifically bind to the DNA and perform the desired functions (Bikard et al., 2013). It is used for gene repression and activation.

Besides these, other synthetic parts such as the transcriptional promoters or RNA-based transcription regulators are used to manage the rate of gene expression. Reports suggest that multiplying the copy number of the promoter or by expanding the upstream and downstream regions flanking the promoter sequence can achieve a tight control over the gene transcription (Ajikumar et al., 2010; Lubliner et al., 2015). The 5′ untranslated region (5′-UTR) and RBS play determinant roles in translation of an mRNA. Synthetic biology has helped to design tools such as RBS and UTR calculators which help determine the translation efficacy for a number of genes. Scaffolds can be designed such that the intermediates generated during metabolic conversions can be redirected toward the metabolic flux (Seo et al., 2013). The possibilities for generating novel parts using synthetic biology are immense and designing them with an insight of redirecting them toward bioremediation is something that the world needs in recent times.

14.3 Prospects of synthetic biology in bioremediation

Wastewater treatment plants (WWTP) exist in almost every part of the world and have been really effective in lessening the amount of pollutant released into the streams. Upon its initial collection at the WWTP, the contaminated water is first screened for the presence of bulky floating debris like chunks of paper, plastics, leaves and branches, rags, grits, and many such things that could clog and damage the pumps. Following their removal, the wastewater is directed through the subsequent treatment steps. In the primary treatment, the wastewater is passed through the quiescent settling tanks in which metal salts are added that tend to bind with waste materials to form complex and encourage their sedimentation or flotation. This process is called flocculation. The solid matter (sludge) settles at the bottom while the lipidic matter such as oil and grease float to the surface. The sludge settled at the bottom of the tank is scraped and removed using mechanical scrapers and the floating grease and dirt are removed using mechanical surface-skimming devices. Following this, the predissolved organic effluents that escape the preliminary treatment are treated in the next phase. The wastewater from the primary stage is made to pass through the aeration tank and microbes are added into it. The microbes utilize the organic impurities by breaking them into a simpler form and using them as a source of energy for their own growth and survival. The continuous pumping of air provides a favorable niche for the microbes to thrive upon. Elimination of dissolved organic waste is particularly important to maintain the dissolved oxygen level of the wastewater recipient stream as it readily reduces the biochemical oxygen demand of the sewage.

Quite a few water recovery systems are used in WWTPs to minimize the biological and chemical oxygen demands (BOD and COD), pathogen clearance, and removal of foul odour. These include biotrickling filters (McLamore et al., 2008), tubular reactors (Körbahti and Tanyolac, 2009), membrane-aerated bioreactors (MBR) (Pankhania et al., 1999; Wei et al., 2012), and a few other bioengineered immobilizing surfaces. Of the above-listed bioreactors, MBR comes with several merits over the conventional reactors such as its compact design, low biomass output, better quality of released effluent, transient hydraulic retention times, and extended solid retention times (Iorhemen et al., 2016). The conventional digesters require relatively large active sludge tanks with large sedimentation surface to allow longer retention time for the wastewater to sediment. For these reasons, MBR is well suited to be employed in urban areas where the discharged effluent have to be stringently regulated and that too in a very limited space and cost. However, a major drawback of using MBR is attributed to membrane fouling which is described as the excessive deposition of sludge cake onto the membrane filters, causing the membraneous pores to clog and reduce membrane permeability. Microorganisms play a very crucial role in biosorption and bioaccumulation of xenobiotics. Immobilization of certain biofilm-producing bacteria onto the hollow fiber membrane reactor helps to trap and degrade organic and inorganic pollutants from the wastewater (Iorhemen et al., 2016).

The repeated splashing of wastewater in the reactor tanks leads to unrestricted detaching of sessile microbes clinging to the membrane which is yet another limiting factor that impairs the bioreactor design. A plausible solution to restrict the detachment of the cells is by encapsulating the layered cells with a thin film of porous silica (biosilicification). The Purdue iGEM Team (2012) expressed the adhesion proteins in *E. coli* that in turn assists biofilm formation. The bacteria were engineered to produce a silica binding protein called silicatein alpha to encapsulate the biofilm-producing *E. coli* (Purdue iGEM Team, 2012). The silica entrapped cells were not just aimed to prevent the cells from unloosing but they also acted as a sturdy membrane capable of withstanding mechanical aberrations.

The bioreactors in WWTP function on the principles of biology and so the amount of pressurized air needs to meet the high oxygen demand for aerobic treatment which is energy-intensive and highly-priced. Another irrefutable point about water treatment in WWTP is that the sewage often begins to ferment even before the treatment proceeds. This is where synthetic biology may come into the picture and drive the organic detritus into a renewable energy source in the form of hydrogen or methane. This may readily bring down the BOD and COD of the wastewater.

Contrary to the popular belief, water pollution does not always arise from human-centric activities (domestic or industrial wastewater release, anthropogenic activities, groundwater contamination) but may also result due to natural phenomenon such as leaching of metals from their ores due to weathering and erosion. Groundwater contamination is actually a preordained action of soil pollution that stems due to continuous seeping of filthy water from polluted sites. Undeniably, domestic, sewage, and industrial wastewater are ineluctable consequences of the human existence. Water treatment and its management is a worrisome global concern (Shannon et al., 2008; Cosgrove and Loucks, 2015). The inaccessibility to clean water is already becoming a nightmare in some parts of the world. With nearly no clean water to function, people unintentionally turn to use contaminated water such as the arsenic-rich water in India and Bangladesh (Ahmad et al., 2018) and the poor quality of water in Africa (Yongsi, 2010; Pande et al., 2018), which brings along a myriad of diseases.

Desalinating the seawater is one probable solution for overcoming this shortage as it is an abundant, available resource. Conventionally, desalinating employs methods like distillation or reverse osmosis which is both costly as well as energy-driven with regards to the large-scale desalination required on a regular basis. One feasible approach is to use synthetic biology for improvising the wild-type *Halobacteria* to use their ion channels effectively to channelize the movement of different ions across membranes. This may even help to cut down on energy inputs as the functioning of these ion channels of these bacteria is fueled by adenosine triphosphate (ATP). Efforts are being taken to develop well-structured halorhodopsin pumps, which are light-driven ion pumps found in *Halobacteria* meant for chloride ions (Amezaga et al., 2014; Paul and Mormile, 2017). In the future, synthetic biology may help to devise novel treatment strategies to steer the water pollution issues which with the amalgamation of present knowledge may surpass the existing technologies.

World War I, II, and the Industrial Revolution which lasted for about two centuries not just played havoc with the air quality but also left quite a portion of the fecund land, derelict and barren which is regarded as the brownfields. The implications of the past are being felt after about 250 years of vandalism and the current rate of human-driven activities is just adding to it. Therefore a pressing need has arisen to regenerate the existing brownfields while slowing down the extensive use of the greenfields. For this to happen, newer ways of remediation have to be fostered, of which synthetic biology is one such field that can readily contribute to it. The soil pollutants are largely heavy metals or organic compounds such as polyaromatic hydrocarbons, mineral oil hydrocarbons, chlorinated hydrocarbons, and benzene derivatives (Bisht et al., 2015; Lasota and Błońska, 2018). These soil contaminants mix with the constituents of the soil and deracinate the natural balance. The most convenient way to dispose of the solid municipal waste is to shove it into landfills that subsequently hamper the quality of life around it (Stojić et al., 2019). In fact, the landfill leachate is amongst the predominant reasons behind groundwater and surface water pollution if the waste in the landfills goes untreated or unwatched (El-Salam and Abu-Zuid, 2015). Although bioremediation turns out to be a feasible approach for waste management, the organic compounds often tend to escape the normal treatment procedure either due to their low bioavailability or their toxicity which the wild-type microorganisms fail to resist. In such cases, little manipulations in the genome of these microbes through synthetic biology may help futuristic organisms to sustain toxic compounds and release enzymes to solubilize and degrade the same (Thornton et al., 2008).

Though this field is in its early days of conception, measures involving synthetic biology may accelerate any form of bioremediation processes leaving a positive impact on the environment. The effects would not just be limited to abandoned sites but may also help to revive the industrially polluted lands making them suitable for lucrative in situ applications.

14.4 Conclusion and future remarks

The advances in synthetic biology have diminished the demarcation between what is natural and what is artificial, which in turn have provided hope to unravel certain enigmas of life which were difficult to be studied in its natural niche. The ease with which synthetic biology enables to design constructive tools can ameliorate the current waste disposal techniques which in the future may be diverted to yield valuable resources and energy from waste at an economic cost. Another cogent application of synthetic biology in bioremediation is the formulation of biosensors to monitor the presence of contaminants even in trace amounts or to detect the extent of the hazards. Synthetic biology is expected to find maximum acceptance in the sector dealing with wastewater management and water desalination considering the fact that many countries would soon be running out of freshwater fit for drinking or even for agricultural purpose. The amount of wastewater and solid waste that is generated each day is mammoth and it is illogical to consider it being treated in closed chambers, simply because it would be difficult to manage such a large amount of waste in small sealed spaces. A serious impediment of this is that the engineered organisms may unceasingly spread in the environment. This limitation can be overcome by utilizing synthetic biology for tailoring a sustainable yet nonproliferative biological system which may include enzymes, genetic parts, or protocells.

However, what comes with so many benefits does have certain loopholes too. Firstly, the advancements in molecular biology have extensively aided the identification and characterization of pathogenic microorganisms (Gohil et al., 2019), the genetic sequences of which are now available in a number of databases such GenBank and EMBL and can be downloaded for free. Though it may be seen as an advantage that the genomes can now be commercially synthesized at a very low price than ever before, this can be a source to promote bioterrorism given the technical impediment for synthesizing artificial parts have already been eased with the number of scientific papers available online unfolding the technicalities. Moreover, conventional regulations laid for the use of pathogens under laboratory conditions does not suffice when it comes to the present challenges that synthetic biology brings along. It is very important to keep a check on the release of the organisms carrying synthetic parts in the environment. In a very short span of development, the field has been able to arouse some serious global debate amongst the governing bodies regarding its ethical concerns and the general population about its reach. Nevertheless, it is important to ensure that the newer strategies devised to detect and remediate pollutants are democratized and not just remain restricted to a specific geographical area so as to curb pollution at the global level.

Acknowledgment

The authors gratefully acknowledge the Gujarat State Biotechnology Mission, Gandhinagar (GSBTM Project ID: 5LY45F) for providing financial support and Indrashil University for providing infrastructure and facility.

References

Ahmad, S.A., Khan, M.H., Haque, M., 2018. Arsenic contamination in groundwater in Bangladesh: implications and challenges for healthcare policy. Risk Manag. Healthc. Policy 11, 251–261.

Ajikumar, P.K., Xiao, W.H., Tyo, K.E., Wang, Y., Simeon, F., Leonard, E., et al., 2010. Isoprenoid pathway optimization for Taxol precursor overproduction in *Escherichia coli*. Science 330 (6000), 70–74.

Amezaga, J.M., Amtmann, A., Biggs, C.A., Bond, T., Gandy, C.J., Honsbein, A., et al., 2014. Biodesalination: a case study for applications of photosynthetic bacteria in water treatment. Plant Physiol. 164 (4), 1661–1676.

Atlas, R.M., Hazen, T.C., 2011. Oil biodegradation and bioremediation: a tale of the two worst spills in US history. Environ. Sci. Technol. 45 (16), 6709–6715.

Ayangbenro, A., Babalola, O., 2017. A new strategy for heavy metal polluted environments: a review of microbial biosorbents. Int. J. Environ. Res. Public Health 14 (1), 94.

Barrangou, R., Marraffini, L.A., 2014. CRISPR-Cas systems: prokaryotes upgrade to adaptive immunity. Mol. Cell 54 (2), 234–244.

Barrangou, R., Fremaux, C., Deveau, H., Richards, M., Boyaval, P., Moineau, S., et al., 2007. CRISPR provides acquired resistance against viruses in prokaryotes. Science 315 (5819), 1709–1712.

Bhattacharjee, G., Mani, I., Gohil, N., Khambhati, K., Braddick, D., Panchasara, H., et al., 2020. CRISPR technology for genome editing. In: Faintuch, J., Faintuch, S. (Eds.), Precision Medicine for Investigators, Practitioners and Providers. Academic Press, London, pp. 59–69.

Bikard, D., Jiang, W., Samai, P., Hochschild, A., Zhang, F., Marraffini, L.A., 2013. Programmable repression and activation of bacterial gene expression using an engineered CRISPR-Cas system. Nucleic Acids Res. 41 (15), 7429–7437.

Bikard, D., Euler, C.W., Jiang, W., Nussenzweig, P.M., Goldberg, G.W., Duportet, X., et al., 2014. Exploiting CRISPR-Cas nucleases to produce sequence-specific antimicrobials. Nat. Biotechnol. 32 (11), 1146–1150.

Bisht, S., Pandey, P., Bhargava, B., Sharma, S., Kumar, V., Sharma, K.D., 2015. Bioremediation of polyaromatic hydrocarbons (PAHs) using rhizosphere technology. Braz. J. Microbiol. 46 (1), 7–21.

Carpenter, A., Paulsen, I., Williams, T., 2018. Blueprints for biosensors: design, limitations, and applications. Genes 9 (8), E375.

Cong, L., Ran, F.A., Cox, D., Lin, S., Barretto, R., Habib, N., et al., 2013. Multiplex genome engineering using CRISPR/Cas systems. Science 339 (6121), 819–823.

Cosgrove, W.J., Loucks, D.P., 2015. Water management: current and future challenges and research directions. Water Resour. Res. 51 (6), 4823–4839.

Elowitz, M.B., Leibler, S., 2000. A synthetic oscillatory network of transcriptional regulators. Nature 403 (6767), 335–338.

El-Salam, M.M.A., Abu-Zuid, G.I., 2015. Impact of landfill leachate on the groundwater quality: a case study in Egypt. J. Adv. Res. 6 (4), 579–586.

Endy, D., 2005. Foundations for engineering biology. Nature 438 (7067), 449–453.

Gardner, T.S., Cantor, C.R., Collins, J.J., 2000. Construction of a genetic toggle switch in *Escherichia coli*. Nature 403 (6767), 339–342.

Gohil, N., Panchasara, H., Patel, S., Ramírez-García, R., Singh, V., 2017. Book review: Recent advances in yeast metabolic engineering. Front. Bioeng. Biotechnol. 5, 71.

Gohil, N., Panchasara, H., Patel, S., Singh, V., 2019. Molecular biology techniques for the identification and genotyping of microorganisms. In: Tripathi, V., Kumar, P., Tripathi, P., Kishore, A. (Eds.), Microbial Genomics in Sustainable Agroecosystems. Springer, Singapore, pp. 203–226.

Horvath, P., Barrangou, R., 2010. CRISPR/Cas, the immune system of bacteria and archaea. Science 327 (5962), 167–170.

Igiri, B.E., Okoduwa, S.I., Idoko, G.O., Akabuogu, E.P., Adeyi, A.O., Ejiogu, I.K., 2018. Toxicity and bioremediation of heavy metals contaminated ecosystem from tannery wastewater: a review. J. Toxicol. 2018, 2568038.

Iorhemen, O., Hamza, R., Tay, J., 2016. Membrane bioreactor (MBR) technology for wastewater treatment and reclamation: membrane fouling. Membranes 6 (2), E33.

Jinek, M., Chylinski, K., Fonfara, I., Hauer, M., Doudna, J.A., Charpentier, E., 2012. A programmable dual-RNA–guided DNA endonuclease in adaptive bacterial immunity. Science 337 (6096), 816–821.

Karigar, C.S., Rao, S.S., 2011. Role of microbial enzymes in the bioremediation of pollutants: a review. Enzyme Res. 2011, 805187.

Khambhati, K., Bhattacharjee, G., Gohil, N., Braddick, D., Kulkarni, V., Singh, V., 2019. Exploring the potential of cell-free protein synthesis for extending the abilities of biological systems. Front. Bioeng. Biotechnol. 7, 248.

Körbahti, B.K., Tanyolac, A., 2009. Continuous electrochemical treatment of simulated industrial textile wastewater from industrial components in a tubular reactor. J. Hazard. Mater. 170 (2–3), 771–778.

Lasota, J., Błońska, E., 2018. Polycyclic aromatic hydrocarbons content in contaminated forest soils with different humus types. Water Air Soil Pollut. 229 (6), 204.

Lubliner, S., Regev, I., Lotan-Pompan, M., Edelheit, S., Weinberger, A., Segal, E., 2015. Core promoter sequence in yeast is a major determinant of expression level. Genome Res. 25 (7), 1008–1017.

Mali, P., Yang, L., Esvelt, K.M., Aach, J., Guell, M., DiCarlo, J.E., et al., 2013. RNA-guided human genome engineering via Cas9. Science 339 (6121), 823–826.

McAdams, H.H., Arkin, A., 2000. Gene regulation: towards a circuit engineering discipline. Curr. Biol. 10 (8), R318–R320.

McLamore, E., Sharvelle, S., Huang, Z., Banks, K., 2008. Simultaneous treatment of graywater and waste gas in a biological trickling filter. Water Environ. Res. 80 (11), 2096–2103.

Moon, T.S., Lou, C., Tamsir, A., Stanton, B.C., Voigt, C.A., 2012. Genetic programs constructed from layered logic gates in single cells. Nature 491 (7423), 249–253.

Ojuederie, O., Babalola, O., 2017. Microbial and plant-assisted bioremediation of heavy metal polluted environments: a review. Int. J. Environ. Res. Public Health 14 (12), E1504.

Pande, G., Kwesiga, B., Bwire, G., Kalyebi, P., Riolexus, A., Matovu, J.K., et al., 2018. Cholera outbreak caused by drinking contaminated water from a lakeshore water-collection site, Kasese District, south-western Uganda, June-July 2015. PLoS One 13 (6), e0198431.

Pankhania, M., Brindle, K., Stephenson, T., 1999. Membrane aeration bioreactors for wastewater treatment: completely mixed and plug-flow operation. Chem. Eng. J. 73 (2), 131–136.

Park, M., Tsai, S.L., Chen, W., 2013. Microbial biosensors: engineered microorganisms as the sensing machinery. Sensors 13 (5), 5777–5795.

Patel, S., Panchasara, H., Braddick, D., Gohil, N., Singh, V., 2018. Synthetic small RNAs: current status, challenges, and opportunities. J. Cell. Biochem. 119 (12), 9619–9639.

Paul, V.G., Mormile, M.R., 2017. A case for the protection of saline and hypersaline environments: a microbiological perspective. FEMS Microbiol. Ecol. 93 (8).

Purdue iGEM Team, 2012. Purdue biomakers. <http://2012.igem.org/Team:Purdue> (accessed 31.07.18).

Ramírez-García, R., Gohil, N., Singh, V., 2019. Recent advances, challenges, and opportunities in bioremediation of hazardous materials. In: Pandey, V.C., Bauddh, K. (Eds.), Phytomanagement of Polluted Sites. Elsevier, Amsterdam, pp. 517–568.

Seo, S.W., Yang, J., Min, B.E., Jang, S., Lim, J.H., Lim, H.G., et al., 2013. Synthetic biology: tools to design microbes for the production of chemicals and fuels. Biotechnol. Adv. 31 (6), 811–817.

Shannon, M.A., Bohn, P.W., Elimelech, M., Georgiadis, J.G., Marinas, B.J., Mayes, A.M., 2008. Science and technology for water purification in the coming decades. Nature 452 (7185), 301–310.

Shee, K., Strong, M., Guido, N.J., Lue, R.A., Church, G.M., Viel, A., 2010. Research, collaboration, and open science using web 2.0. J. Microbiol. Biol. Educ. 11 (2), 130–134.

Singh, V., 2014. Recent advancements in synthetic biology: current status and challenges. Gene 535 (1), 1–11.

Singh, V., Braddick, D., Dhar, P.K., 2017. Exploring the potential of genome editing CRISPR-Cas9 technology. Gene 599, 1–18.

Singh, V., Gohil, N., Ramírez García, R., Braddick, D., Fofié, C.K., 2018. Recent advances in CRISPR-Cas9 genome editing technology for biological and biomedical investigations. J. Cell. Biochem. 119 (1), 81–94.

Stojić, N., Štrbac, S., Prokić, D., 2019. Soil pollution and remediation. In: Hussain, C.M. (Ed.), Handbook of Environmental Materials Management. Springer, Cham, pp. 583–616.

Thornton, I., Farago, M.E., Thums, C.R., Parrish, R.R., McGill, R.A., Breward, N., et al., 2008. Urban geochemistry: research strategies to assist risk assessment and remediation of brownfield sites in urban areas. Environ. Geochem. Health 30 (6), 565–576.

Verma, J.P., Jaiswal, D.K., 2016. Book review: advances in biodegradation and bioremediation of industrial waste. Front. Microbiol. 6, 1555.

Wang, L., Qian, K., Huang, Y., Jin, N., Lai, H., Zhang, T., et al., 2015. SynBioLGDB: a resource for experimentally validated logic gates in synthetic biology. Sci. Rep. 5, 8090.

Wei, X., Li, B., Zhao, S., Wang, L., Zhang, H., Li, C., et al., 2012. Mixed pharmaceutical wastewater treatment by integrated membrane-aerated biofilm reactor (MABR) system—a pilot-scale study. Bioresour. Technol. 122, 189–195.

Yongsi, H.B.N., 2010. Suffering for water, suffering from water: access to drinking-water and associated health risks in Cameroon. J. Health Popul. Nutr. 28 (5), 424–435.

Microbial indicators and biosensors for bioremediation

*Ankita Chaurasia[1], Nihal Mohammed[2], Molka Feki Tounsi[3] and Heykel Trabelsi[4],**

[1]Centre for Bioinformatics, Institute of Interdisciplinary Sciences (IIDS), University of Allahabad, Allahabad, India, [2]Department of Biology, College of Science, Mosul University, Mosul, Iraq, [3]Centre of Biotechnology of Sfax, University of Sfax, Sfax, Tunisia, [4]Micalis Institute, INRA, AgroParisTech, Paris-Saclay University, Jouy-en-Josas, France

*Corresponding author.

15.1 Introduction

Pollution is a serious drawback of the industrial and technological progress of our civilization. For many long decades, the race for more wealth has neglected the environment. Despite the regulatory directives and the remediation policies to control the number and yields of pollutants released in the ecosystem, the results are still alarming and the adopted strategies remain inefficient. The growing awareness of the sensitivity of the current pollution situation has led decision-makers and scientists to prospect and search for more sustainable solutions. Conventional remediation methods for environmental contaminants are generally based on physical, chemical, and biological approaches, which may be used in combination to reach an acceptable and safe level (Khalid et al., 2017). Specialists strongly believe that a viable remediation method should detect the pollutant and promote adequate machinery for cleaning up. Through the course of evolution, microorganisms have faced the emergence of a huge number of chemical and biological compounds and have thus developed processes to degrade most of them for their survival. Inspired by nature, humans have designed and developed the biological remediation of pollutants or more precisely bioremediation which is microorganism mediated transformation or degradation of contaminants into nonhazardous or less hazardous substances. Bioremediation is being used to solve pollution problems in different domains including agriculture, textile, medicine, and all types of oil or chemical industries (Salleh et al., 2003; Mehta, 2012; Kulkarni et al., 2013; Baez-Rogelio et al., 2017; Kumar et al., 2019). Different strategies were adopted to respond to different pollution cases by using plants, algae, fungi, and bacteria (Kaur and Bhatnagar, 2002; Rosser et al., 2007; Saier and Trevors, 2016; Pal et al., 2016; de Lima et al., 2018; Quintella et al., 2019; Yin et al., 2019). Each strategy is specific to one compound and its success depends on the intrinsic conditions, the local environment, and the

Bioremediation of Pollutants. DOI: https://doi.org/10.1016/B978-0-12-819025-8.00015-6

regulations applied (Sivasubramanian, 2016). Microorganism based remediation is now well established and has shown a strong ability to treat complex pollution cases, but the increase of pollutants requests more sophisticated tools.

Microbial indicators have been traditionally used to suggest the presence of pathogens but nowadays three groups are recognized (Ashbolt et al., 2001). These indicator groups are highlighted in Table 15.1.

Microbial indicators, as their name suggests, could only indicate the presence of pollutants but are not able to produce a quantitative measurement or eliminate the contaminants. The recent advances in systems and synthetic biology have opened doors for new innovative approaches to solve this urgent need by introducing more sensitive, specific, and ecofriendly solutions (de Lorenzo, 2011; Nunes-Alves, 2015). These microbial tools are divided into two modules (1) detection and (2) remediation. The first module is based on the efficient and specific detection of the contaminants. An increasing number of scientific papers have shown successful synthetic biology-based methods able to detect molecules at very low concentrations (Fernandez-López et al., 2015; Libis et al., 2016a,b; Mannan et al., 2017; Zhang et al., 2017; Lin et al., 2018; Trabelsi et al., 2018). The generic name of the first module is a microbial indicator or more commonly known as biosensor. The second module contains all the elements responsible for the removal of the pollutant or involved in reducing its concentration below the authorized limit. These elements are mainly bio-bricks encoding complicated biodegradation pathways. Usually, a cascade of biochemical reactions is needed to reach the target. In this chapter, the development of such biosensors and their applications in detecting, monitoring, and remediation will be discussed and some cases will serve as examples for a deeper understanding of these valuable tools.

15.2 Biosensors development

The significant progress in synthetic biology, mainly in genetic network circuit design, through iterative cycles of "design-build-test," has provided new tools and has boosted the emergence of a large number of biosensors with diverse applications. Biosensors are becoming the workhorse of analytics thanks to their ability to rapidly and specifically detect an increasing number of molecules (Carpenter et al., 2018). Synthetic biologists have followed different approaches to develop different types of biosensors mainly including designs based on nucleic acids (Palchetti and

Table 15.1 Microbial indicator groups.

Indicators group	Target	Example of indicators
Process indicator	Process efficiency	Heterotrophic bacteria
Fecal indicator	Fecal contamination	Thermotolerant coliforms
Index and model	Pathogen presence and behavior	Coliphages

Mascini, 2008; Sassolas et al., 2008; Song et al., 2008), transcription factors (Zhang et al., 2012; Eggeling et al., 2015; Trabelsi et al., 2018), and transcription independent proteins (Nagai et al., 2001).

15.2.1 Nucleic acid-based biosensors

Nucleic acid-based biosensors are built using DNA or RNA sequences with structural conformations allowing binding affinity to a target ligand. The binding of the ligand to the nucleic acid sequence alters its structure. This structural change could be used to generate an output signal. The biosensors of this family could be aptamers or riboswitches. Aptamers are single-stranded DNA or RNA showing affinity to a target ligand (Tuerk and Gold, 1990; Ellington and Szostak, 1992). The binding of the aptamer to the target ligand provokes a conformational change in the aptamer (linear to stem-looped or hairpin to ligand-coordinated structure). This change in the secondary or tertiary structures could be used to generate a signal (Song et al., 2008). Riboswitches are single-stranded RNA comprised of two joined domains. The first domain binds to the target ligand and the second generates a signal after ligand binding (Mehdizadeh et al., 2016; Findeiß et al., 2017).

15.2.2 Transcription factor-based biosensors

The transcription factor-based biosensors are the fruit of labor from several years of combined research studies on transcription and regulation of protein expression. The association of target ligand and a transcription factor generate a specific gene transcriptional response. It is relatively easy to engineer and any natural transcription factor could be transformed into a biosensor (Fernandez-López et al., 2015; Mahr and Frunzke, 2016). The biosensor is a genetic circuit relying on the combination of a transcriptional regulator and promoter/operator pair with a signal output for the desired assay. The system is modular and heterologous components could be assembled together. Optimizing the dose–response, predicting the behavior of the biosensor, and controlling the specificity of the binding ligand are the main steps in the validation process. To extend the finite available number of transcription factor promoter pairs, scientists have used protein engineering tools to create de novo transcription factor which is able to bind new ligands (Machado et al., 2019; Snoek et al., 2019). Another computational approach named sensipath was also used to extend the scope of detection of molecules. A nondetectable compound could be transformed into a detectable one by adding one or two steps of catalytic reactions (Delpine et al., 2016; Libis et al., 2016a). The use of transcription factors from different species could be challenging. In fact, a given response in one species could be completely different when the same system is transplanted into another organism.

15.2.3 Transcription-independent protein-based biosensors

This family of biosensors is very diverse and offers plenty of mechanisms of detection and output responses. Generally, a receptor domain for a target ligand is expressed as a fusion protein with a response domain. When the ligand bind to the receptor domain, a conformational change is transmitted to the response domain inducing a signal output (Nagai et al., 2001; Guntas et al., 2004). A deep knowledge of the structural information for both the ligand-binding domain and the response domain is needed for successful design. However, when the information is missing, the design could be very challenging.

15.3 Pollution monitoring

Pollution monitoring refers to the quantitative or qualitative measure of the presence, effect, or level of any polluting substance in a defined environment (air, water, or soil). The accuracy of the measurements is mandatory in order to generate reliable data allowing pollution risk prediction and management. The current monitoring applied methods range from benthic algal communities (Whitton, 2013) to satellite missions equipped with high-performance single and multipolarization synthetic aperture radars (Migliaccio et al., 2015). The biological tools available to monitor environmental pollution are based on biomarkers that are generally native of the site of investigation and exposed to local environmental conditions for long periods of time. The biomarker is a biological response measured in an organism naturally exposed to the site under study that serves as an indicator of the presence and/or the effect of environmental pollutants (Beiras, 2018). Theoretically, the biological response measured should be quantitative, sensitive, and specific. However, technically the three requirements are rarely met at once. Despite the recorded success of the previous methods, and under the pressure of increasing needs for less costly, efficient, and accurate method, researchers were oriented to biosensors as a realistic and reliable alternative. Thus, their involvement is constantly increasing due to their high specificity and sensitivity. Table 15.2 is adopted from the review of Justino et al. (2017) and presents a brief summary of some successful implementations of biosensors in pollution monitoring.

To take advantage of the large possibilities brought by biosensors, synthetic biology approaches to engineer metabolite-actuated transcription factors that respond to environmental pollutants were also successful in designing new modular protein-based biosensors able to bind specific chemicals and regulate expression of a genetic program integrated into the microbial host.

15.4 Case studies

15.4.1 Context

Together with the massive industrial advancement and boom in urbanization, especially over the past half-century, it has not only proven detrimental to the global environment, as a whole but exclusively has left the world's water resources under crisis. Polluted water has now become the second biggest existential challenge of

Table 15.2 Examples of biosensors for environmental monitoring.

Pollutants	Example	Recognition element	Biosensor type	References
Pesticides	Paraoxon	Enzyme	Electrochemical	Justino et al. (2017)
	Atrazine	Antibodies (monoclonal)	Electrochemical	Liu et al. (2014) and Belkhamssa et al. (2016b)
	Acetamiprid	Aptamers	Electrochemical	Fei et al. (2015) and Jiang et al. (2015)
Pathogens	*Legionella pneumophila*	Nucleic acids	Optical	Foudeh et al. (2015)
	Bacillus subtilis	Antibodies	Electrochemical	Yoo et al. (2017)
	Escherichia coli	Antibodies (polyclonal)	Optical	Chen et al. (2017)
Toxins	Microcystin	Enzyme	Electrochemical	Catanante et al. (2015)
	Saxitoxin	Aptamers	Optical	Gao et al. (2017)
	Brevetoxin-2	Cardiomyocyte cells	Electrochemical	Wang et al. (2015)
Endocrine disrupting chemicals	Bisphenol A	Aptamers	Optical	Ragavan et al. (2013)
	Nonylphenol	Antibodies	Electrochemical	Belkhamssa et al. (2016a)
	17β-Estradiol	Antibodies	Electrochemical	Dai and Liu (2017)

the present era (Landrigan et al., 2018). The byproducts from the anthropogenic activities such as mining, manufacturing, and agricultural industries have contributed towards extensive release of several organic and inorganic pollutants into water bodies, which has not only severely impacted the aquatic ecosystem alone but also has placed considerable stress on the entire environment (Fu and Wang, 2011). Among a wide range of water pollutants, industrial wastewater particularly contains toxic levels of heavy metals such as Pb, Hg, Cu, Cd, etc. Most heavy metals are naturally found in the earth's crust. Though some of these under certain lower concentration levels are considered essential to human health, for example, Fe, Mg, Zn, etc. (Chitturi et al., 2015), t most of them are toxic and have adverse effects even at lower concentrations (Tchounwou et al., 2012). Additionally, their nondegradable nature leading to their accumulation in the food chain, has imposed a serious risk to all ecological niches and has now become a prominent concern to human health. Recently researches have shown that heavy metals particularly like Pb and Hg, widely used in power plants, automobile, construction, printing, and cosmetic industries, cause significant damage to human vital organs like the kidney, lungs, and brain and are reported to be carcinogenic (Cocârţă et al., 2016). Consumption of contaminated seafood has instigated fetal developmental abnormalities (Iwai-Shimada et al., 2019). These alarming findings have directed extensive research focusing on innovative, accurate metal detection, and elimination strategies, in order to minimize the impact of heavy metals on the ecosystem and to limit human exposure. To date several techniques have been implemented to tackle the issue namely: (1) analytical techniques like, atomic florescence spectroscopy, atomic absorption spectroscopy, and inductively coupled plasma mass spectrometry which are still being in use to detect Pb ion content (Iyengar and Woittiez, 1988; Townsend et al., 1998; Li and Lu, 2000; Verbych et al., 2004; Pyrzyńska, 2005). But despite their accuracy and sensitivity, these traditional procedures are lengthy, complex, and cost-inefficient; (2) relatively recent chemical-based procedures, for example, chemical filtration methods and nano-particles sensors have their own restraints in terms of their restrictive implementation (Li and Lu, 2000; Chen et al., 2005; He et al., 2006; Li et al., 2010; Lin et al., 2011; Kim et al., 2012); (3) adsorption techniques; and (4) ion exchange methods. The adsorption and ion exchange methods are not scalable with reference to their global implementation and are considerably expensive (Verbych et al., 2004; Namasivayam and Sangeetha, 2006).

Keeping in mind the pros and cons of the above techniques, in qualitative and qualitative measurements of the heavy metal ions and other critical factors such as toxicity and bioavailability of heavy metal pollutants require biosensors for their detection. A biosensor by definition is a sensor that can detect a specific chemical component inside the living cell/tissue. It comprises of two units: a genetically engineered microorganism acting as a recognition site and a physical or chemical transducer/detector. A bacterium-based whole-cell biosensor (WCB) not only exploits the power of the living cell by expressing a specific set of genes under particular environmental conditions but also these cells can be genetically modified to incorporate detector capabilities. Hence they, are able to respond to a wide array of biological and chemical conditions. Another study has reported bioluminescent bacteria for the detection of heavy metal ions namely: Hg, As, Cd and Pb (Jouanneau et al., 2011). Mainly focusing on modular systems involving bacteria-based WCB, the following section presents case studies that are potentially novel approaches and are efficient alternatives over the conventional methods of bioremediation.

15.4.2 Unifying gold-specific whole- (Escherichia coli) *cell biosensor system and adsorption*

Ongoing research and advancement in the field of biosensor techniques have additionally directed their implementation towards the detection of noble and rare heavy metals like gold (Au). Recent studies have shown the importance of gold nanoparticles in the field of biomedical applications, detection, and potential treatment of cancer (Messori et al., 2000; Casini et al., 2008) and nano-electronics (Homberger and Simon, 2010). Hence, there has been a growing scientific interest towards the detection and retrieval of natural gold in a much more efficient and environmentally friendly manner.

Principally, the assay involves the fusion of a reporter to a specific gene promoter, which gets activated in the presence of a target heavy metal (Belkin, 2003). Wei et al. have first reported a whole-cell bacterial system by assembling the gold-sensory regulator protein GolS from *Salmonella*, a gold-specific binding-protein GolB and its promoter with a red fluorescent protein (Wei et al., 2012; Yan et al., 2018). The reported fluorescence activity was highly sensitive as well as specific for the gold ions and also visibly detected by the naked eye. Based on these results, Yan et al. (2018), have presented a novel integrated system. Using a synthetic-biology technique BioBrick (Knight, 2003), they successfully unified a gold-ion specific detection technique using the GolS regulon and adsorption technique based on surface-display of GolB protein. The gold ions present in the cell bind with protein GoIS, whose expression is controlled by a promoter pgolTS, leading to the formation of a metalloprotein complex that on binding with promoter pgol1B further modifies the expression of GolB. Integration of newly synthesized regulon into *Escherichia coli*, the fluorescence results showed a much higher sensitivity towards gold ions, hence provided a much more precise and structured way of detecting heavy metal ions.

15.4.3 Lead specific whole-cell based biosensors and bioremediation system

Lead (Pb) is one of the most toxic heavy metals. Human-induced dissemination of lead into the environment is due to its use in several large-scale industries like batteries, petroleum, water pipes, paint, and cosmetics. In recent years, several measures have been taken to restrict the use of toxic lead but due to its nondegradable nature and subsequent accumulation in the food chain, there is still a need for an efficient system for its detection and remediation. Humans are exposed to the Pb toxicity mainly via the ingestion of contaminated food and water. Infants and young children are more susceptible to lead poisoning (Schell et al., 2004; Warniment et al., 2010). As a byproduct from the industrial wastes, lead pollution often accompanies other heavy metals like Cd, Zn, and Hg, which therefore restricts the use of simple detection and adsorption techniques. He et al. developed a fluorescent lead-probe, using exceptionally specific Pb^{2+} regulatory protein from *Cupriavidus metallidurans* CH34 (Chen et al., 2005). This fluorescent biosensor showed that the protein had high binding selectivity and sensitivity towards Pb^{2+}. Taking advantage from the above biosensor, Wei et al. have developed a lead-specific WCB and a remediation system, by engineering pbr promoter from the pbr operon and lead-specific binding protein (PbrR), into *E. coli*. This engineered *E. coli* along with a red fluorescent protein led to highly selective and sensitive recognition of Pb^{2+} (Wei et al., 2014). Their strategy was based on MerR family of proteins (Julian et al., 2009), which can efficiently distinguish Pb^{2+} from other heavy metals like Hg, Cd, Ni, Co, and Zn ions.

Plants being at the base of the food chain are exposed to heavy metal toxicity through polluted soil and water. For the bioremediation of heavy metals, several tools have been developed recently using transgene and bioengineering technologies. Zhao et al. reported the effect of Pb and Cd ions on the seed germination rates of *Arabidopsis thaliana* (Wei et al., 2014). The study has also reported the extent of effectiveness of the technique mentioned above based on engineered *E. coli* system and provided a comparison with the other conventional techniques. During the germination experiment under high-concentration Pb solution, a significant amount of the Pb ions were adsorbed on the PbrR-displayed *E. coli* cells, consequently decreasing lead toxicity in the *A. thaliana* seeds.

15.4.4 Transcription factor-based biosensors for remediation

Similarly, like other heavy metals, the unnatural rise in the level of Hg in the environment is widely due to its release as a byproduct from the industrial processes like power generation by coal, mining, and its extensive use in dental amalgam, production of batteries, fluorescent light bulbs, and in cosmetic industries. Consequently, mercury gets accumulation in the food chains, through biomagnification process. Due to its high toxicity and potential human health risk, there is an immediate need for an accurate and effective bioremediation system. Biological approaches like biosensors based on Hg specific transcription regulators detect and sequester Hg-ion, and hence provide cost-effective and efficient solutions (O'Halloran and Walsh, 1987; Wang et al., 2019). Hg-specific biosensors utilize transcription regulator MerR. The mercury resistance mer-operon of bacteria consists of the gene coding for the regulatory protein, merR, and enzymes that detoxify mercury. This method has gained special attention because it transforms toxic mercury (either ionic or organic) form into a less toxic elemental form (Hg(0)) (Summers, 1992). These engineered bacterial circuits for detection and absorption of mercury (Virta et al., 1995; Chen et al., 1998; Hakkila et al., 2002; Bae et al., 2003; Lin et al., 2010) have promising potential for remediation of mercury polluted ecosystems. However, the above sequestration techniques have their own limitations. Many studies have shown an alternative approach by using the extracellular material of biofilms which absorbs the heavy metal ions (François et al., 2012; Meliani and Bensoltane, 2016). Currently, synthetic materials containing self-assembling amyloid fibers, which are known to interact with specific heavy metals, are in great demand over biomaterials like keratin and cellulose. For the improved sequestration of Hg, Joshi et al. investigated the programmed synthetic biofilms and reported an engineered gene circuit for Hg bioremediation (Tay et al., 2017). Curli are the self-assembling extracellular amyloids present in *E. coli* and *Salmonella* biofilms (Chapman et al., 2002) and can be reengineered to provide a programmable template for detection of heavy metals contaminants via integration to specific biosensors.

15.4.5 Enzyme based remediation

Enzymes are proteic catalysts that facilitate the conversion of substrates into products by providing favorable conditions that lower the activation energy of the chemical reaction. Through evolution and the multiple evolutive bottlenecks, microorganisms have shaped plenty of enzymatic functions to survive harsh conditions. This rich toolbox of enzymatic activities continues to be valuable for customized bioremediation. Oxidoreductases (oxygenases, laccases, peroxidases) and hydrolases (lipases, cellulases, carboxylesterases, and haloalkane dehalogenases) have shown bioremediation capabilities and were successfully used. Table 15.3, adopted from Sharma et al. (2018), highlights some recorded bioremediation examples of enzymes.

In this context, Poirier et al. (2017) have shown the potential of the phosphotriesterase-like lactonase SsoPox from the archaea *Sulfolobus solfataricus* in degrading organophosphorus insecticides and have given an assessment of the degradation products shown to be less toxic. However, degradation of contaminants or detoxification using microorganisms is a slow process that requires optimized and controlled parameters to keep bioremediation feasible and efficient (Ghosh et al., 2017). In the majority of cases, the cell growth within the contaminated areas or sites is compromised due to the lack of oxygen and nutrients. Despite the relative success, the yields of produced enzymes are still very low and cannot fulfill all the needs. Screening for microbial enzymes in contaminated sites has offered new opportunities but further improvements are required (Baweja et al., 2016). The trade-off balance involving the yield, the stability, and the efficiency of an enzyme, in this kind of nonconventional environment, is very fragile, and any physical or chemical change could alter the whole process. To overcome all these limitations, enzymes were separated from their cells and were used purified or partially purified (Ruggaber and Talley, 2006; Thatoi et al., 2014). Moreover, the progress in genetic and protein engineering has boosted the design of customized enzymes with improved characteristics (Rayu et al., 2012; Dhanya, 2014). This large number of enzymes with high remediation capabilities could be combined in artificial pathways under the control of a detection module (biosensor) to generate "à la carte" responses. Synthetic biology could provide a plug and play strategy where different degrading or detoxifying pathways (remediation modules) could be merged with different microbial indicator bio-bricks (biosensor modules) in the required bacterial chassis to specifically respond to pollutants.

15.4.6 Plug and play strategy

One of the promising strategies inspired by synthetic genetic networks is merging different pathways from different microbial backgrounds. Fig. 15.1 schematically showcases a biosensor circuit that could be used for detection or/and remediation.

Table 15.3 Enzymes used in bioremediation of some chemicals.

Enzymes	Producing microorganisms	Pollutants	References
Chromium reductase	*Pseudomonas, Bacillus, Enterobacter, Deinococcus, Shewanella, Agrobacterium, Esherichia, Thermus*	Chromium	Thatoi et al. (2014)
Atrazine dechlorinase, triazine hydrolase	*Nocardioides* sp. C190, *Pseudomonas, Rhodococcus erythropolis*	Triazine herbicides	Scott et al. (2010)
Alkane hydroxylases	*Arthrobacter, Burkholderia, Mycobacterium, Pseudomonas, Sphingomonas, Rhodococcus*	Hydrocarbon	Das and Chandran (2011)
Laccase	*Trametes versicolor, Pleurotus ostreatus, Pycnoporus sanguineus*	Polychlorinated biphenyls, dyes	Mayer and Staples (2002) and Dodor et al. (2004)
Carboxylesterases	*Pseudomonas aeruginosa* PA1	Malathion and parathion	Qiao et al. (2003) and Singh et al. (2012)
Peroxidases	*Phanerochaete chrysosporium, Horseradish (Aromatica Rusticana)*	Nitroaromatic compounds, chlorophenol, phenol	Flock et al. (1999), Cameron et al. (2000), and Bilal et al. (2017)
Phytase	*Aspergillus niger* NCIM 563	Organophosphate	Shah et al. (2017)

In this context, Tay et al. (2017) were able to set up a synthetic circuit for mercury bioremediation. The authors have used two distinct pathways. The *mer* operon encodes enzymes for mercury detoxification under the control of the regulator MerR and the amyloid based biofilm pathway responsible for the synthesis of Curli nanofibers known to sequester mercury ions in an extracellular matrix. The circuit was designed in a manner in which the MerR regulator is constitutively expressed and repress the *Curli* operon cloned to replace the *mer* operon enzymes. When present, mercury ions bind to MerR to trigger an allosteric change and promote transcription and expression of *Curli* operon. The curliated cultures were able to retain mercury for over 10 days even after several washes. To extrapolate this encouraging

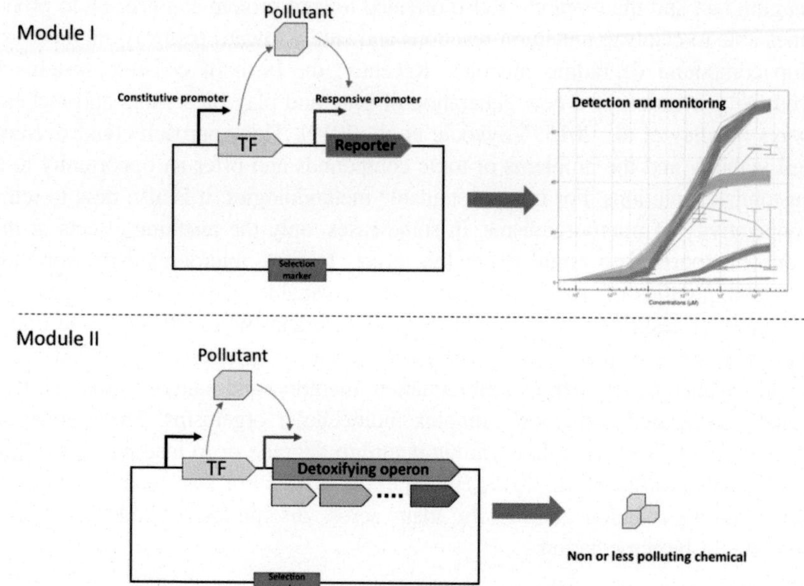

Figure 15.1 A schematic view of transcription factor-based biosensor. Module I: detection and monitoring, Module II: remediation. *TF*, Transcription factor.

success toward other metals, the authors have evaluated the cross selectivity of MerR. It was then concluded that this synthetic circuit could serve as a template for further fine-tuning to sequester other divalent metals. This work illustrates the potential of using engineered circuits and chassis in developing customized biosensor-based solutions for the increasing number of pollutants.

15.5 Perspectives

The increasing number of papers highlighting the successful impact of synthetic biology tools in bioremediation is a key indicator of the evolution of this field toward smart and surgical solutions. The synthetic biology community is growing and the recorded advances are promising. This interdisciplinary field has the potential to put on the table more innovative and out of the box methods. In the near future, the success or failure of any tool will depend on numerous factors including the choice of methodology, the used microorganism, the architecture of the remediation process and the regulations applied for each case. The development of new biosensors and enlarging their scoop of detection is mandatory to solve the problem of some new emerging molecules. The new generation of biosensors should have a large dynamic range and high specificity and affinity toward pollutants to detect them even at very low concentrations. On the other hand, technological and industrial activities will remain to produce compounds that are resistant to biological degradation. The community is aware of this challenging fact and the research is also oriented toward protein engineering to produce enzymes able to catalyze unnatural reactions and surely toward pathway engineering to develop compound degrading modules. Recently, the birth of cell-free systems has allowed the generation of a new generation of plug and play biosensors and metabolic pathways (Dudley et al., 2015; Voyvodic et al., 2019). This approach could overcome the cell viability and the problems of toxic compounds and offer an opportunity to tinker customized solutions. For more sustainable methodologies, it is also time to rethink the cooperativity of microorganisms. In some cases, only the resulting effects of more than one microorganism could reach the target. In fact, microorganisms commonly exist as communities of multiple species that are capable of performing more varied and complicated tasks than clonal populations (Brune and Bayer, 2012). These findings have inspired some studies focusing on microbial consortia to engineer clonal populations with characteristics such as differentiation, memory, and pattern formation, which are usually associated with more complex multicellular organisms. These genetically engineered populations are able to mimic nature to degrade or to inactivate or to transform pollutants (Dashti et al., 2019; Huang et al., 2019). For sustainable and efficient solutions, the assimilation of scientific ideas across disciplines is needed to optimize the potential of bioremediation.

References

Ashbolt, N., Grabow, W., Snozzi, M., 2001. Indicators of microbial water quality. Water Quality: Guidelines, Standard Health 289–316. Available from: https://doi.org/10.4324/9781315693606.

Bae, W., et al., 2003. Enhanced mercury biosorption by bacterial cells with surface-displayed MerR. Appl. Environ. Microbiol. 69 (6), 3176–3180. Available from: https://doi.org/10.1128/AEM.69.6.3176-3180.2003.

Baez-Rogelio, A., et al., 2017. Next generation of microbial inoculants for agriculture and bioremediation. Microb. Biotechnol. 10 (1), 19–21. Available from: https://doi.org/10.1111/1751-7915.12448.

Baweja, M., et al., 2016. Current technological improvements in enzymes toward their biotechnological applications. Front. Microbiol. 7. Available from: https://doi.org/10.3389/fmicb.2016.00965.

Beiras, R., 2018. Biological tools for monitoring: biomarkers and bioassays. In: Beiras, R.B. T.-M.P. (Ed.), Marine Pollution. Elsevier, pp. 265–291. Available from: http://doi.org/10.1016/B978-0-12-813736-9.00016-7.

Belkhamssa, N., da Costa, J.P., et al., 2016a. Development of an electrochemical biosensor for alkylphenol detection. Talanta 158, 30–34. Available from: https://doi.org/10.1016/j.talanta.2016.05.044.

Belkhamssa, N., Justino, C.I.L., et al., 2016b. Label-free disposable immunosensor for detection of atrazine. Talanta 146, 430–434. Available from: https://doi.org/10.1016/j.talanta.2015.09.015.

Belkin, S., 2003. Microbial whole-cell sensing systems of environmental pollutants. Curr. Opin. Microbiol. 206–212. Available from: https://doi.org/10.1016/S1369-5274(03)00059-6.

Bilal, M., Iqbal, H., Hu, H., et al., 2017. Development of horseradish peroxidase-based cross-linked enzyme aggregates and their environmental exploitation for bioremediation purposes. J. Environ. Manage. 188, 137–143. Available from: https://doi.org/10.1016/j.jenvman.2016.12.015.

Brune, K.D., Bayer, T.S., 2012. Engineering microbial consortia to enhance biomining and bioremediation. Front. Microbiol. 3 (JUN), 203. Available from: https://doi.org/10.3389/fmicb.2012.00203.

Cameron, M.D., Timofeevski, S., Aust, S.D., 2000. Enzymology of *Phanerochaete chrysosporium* with respect to the degradation of recalcitrant compounds and xenobiotics. Appl. Microbiol. Biotechnol. 54 (6), 751–758. Available from: https://doi.org/10.1007/s002530000459.

Carpenter, A.C., Paulsen, I.T., Williams, T.C., 2018. Blueprints for biosensors: design, limitations, and applications. Genes 9 (8), 375. Available from: https://doi.org/10.3390/genes9080375.

Casini, A., Hartlinger, C., Gabbiani, C., et al., 2008. Gold(III) compounds as anticancer agents: relevance of gold-protein interactions for their mechanism of action. J. Inorg. Biochem. 102 (3), 564–575. Available from: https://doi.org/10.1016/j.jinorgbio.2007.11.003.

Catanante, G., Espin, L., Marty, J.L., 2015. Sensitive biosensor based on recombinant PP1α for microcystin detection. Biosens. Bioelectron. 67, 700–707. Available from: https://doi.org/10.1016/j.bios.2014.10.030.

Chapman, M.R., Robinson, L.S., Pinkner, J.S., et al., 2002. Role of *Escherichia coli* curli operons in directing amyloid fiber formation. Science 295 (5556), 851–855. Available from: https://doi.org/10.1126/science.1067484.

Chen, P., et al., 2005. An exceptionally selective lead(II)-regulatory protein from *Ralstonia metallidurans*: development of a fluorescent lead(II) probe. Angew. Chem. Int. Ed. 44 (18), 2715–2719. Available from: https://doi.org/10.1002/anie.200462443.

Chen, S., Kim, E., Shuler, M.L., et al., 1998. Hg^{2+} removal by genetically engineered *Escherichia coli* in a hollow fiber bioreactor. Biotechnol. Prog. 14 (5), 667–671. Available from: https://doi.org/10.1021/bp980072i.

Chen, S., Chen, X., Zhang, L., et al., 2017. Electrochemiluminescence detection of *Escherichia coli* O157:H7 based on a novel polydopamine surface imprinted polymer biosensor. ACS Appl. Mater. Interfaces 9 (6), 5430–5436. Available from: https://doi.org/10.1021/acsami.6b12455.

Chitturi, R., Baddam, V.R., Prasad, L.K., et al., 2015. A review on role of essential trace elements in health and disease. J. Dr. NTR. Univ. Health Sci. 4 (2), 75. Available from: https://doi.org/10.4103/2277-8632.158577.

Cocârţă, D.M., Neamţu, S., Reşetar Deac, A.M., 2016. Carcinogenic risk evaluation for human health risk assessment from soils contaminated with heavy metals. Int. J. Environ. Sci. Technol. 13 (8), 2025–2036. Available from: https://doi.org/10.1007/s13762-016-1031-2.

Dai, Y., Liu, C.C., 2017. Detection of 17 β-estradiol in environmental samples and for health care using a single-use, cost effective biosensor based on differential pulse voltammetry (DPV). Biosensors 7 (2), 15. Available from: https://doi.org/10.3390/bios7020015.

Das, N., Chandran, P., 2011. Microbial degradation of petroleum hydrocarbon contaminants: an overview. Biotechnol. Res. Int. 13. Available from: https://doi.org/10.4061/2011/941810.

Dashti, N., Ali, N., Khanafer, M., et al., 2019. Plant-based oil-sorbents harbor native microbial communities effective in spilled oil-bioremediation under nitrogen starvation and heavy metal-stresses. Ecotoxicol. Environ. Saf. 181, 78–88. Available from: https://doi.org/10.1016/j.ecoenv.2019.05.072.

Delpine, B., Libis, V., Carbonell, P., et al., 2016. SensiPath: computer-aided design of sensing-enabling metabolic pathways. Nucleic Acids Res. 44 (W1), W226–W231. Available from: https://doi.org/10.1093/nar/gkw305.

Dhanya, M.S., 2014. Advances in microbial biodegradation of chlorpirifos. J. Environ. Res. Dev. 9 (1), 232–240.

Dodor, D.E., Hwang, H.M., Ekunwe, S.I.N., 2004. Oxidation of anthracene and benzo[*a*]pyrene by immobilized laccase from *Trametes versicolor*. Enzyme Microb. Technol. 35 (2–3), 210–217. Available from: https://doi.org/10.1016/j.enzmictec.2004.04.007.

Dudley, Q.M., Karim, A.S., Jewett, M.C., 2015. Cell-free metabolic engineering: biomanufacturing beyond the cell. Biotechnol. J. 69–82. Available from: https://doi.org/10.1002/biot.201400330.

Eggeling, L., Bott, M., Marienhagen, J., 2015. Novel screening methods-biosensors. Curr. Opin. Biotechnol. 30–36. Available from: https://doi.org/10.1016/j.copbio.2014.12.021.

Ellington, A.D., Szostak, J.W., 1992. Selection in vitro of single-stranded DNA molecules that fold into specific ligand-binding structures. Nature 355, 850–852. Available from: https://doi.org/10.1038/355850a0.

Fei, A., Liu, Q., Huan, J., et al., 2015. Label-free impedimetric aptasensor for detection of femtomole level acetamiprid using gold nanoparticles decorated multiwalled carbon nanotube-reduced graphene oxide nanoribbon composites. Biosens. Bioelectron. 15 (70), 122–129. Available from: https://doi.org/10.1016/j.bios.2015.03.028.

Fernandez-López, R., Ruiz, R., De la cruz, F., et al., 2015. Transcription factor-based biosensors enlightened by the analyte. Front. Microbiol. . Available from: https://doi.org/10.3389/fmicb.2015.00648.

Findeiß, S., Etzel, M., Will, S., et al., 2017. Design of artificial riboswitches as biosensors. Sens. (Switz.) 17 (9). Available from: https://doi.org/10.3390/s17091990.

Flock, C., Bassi, A., Gijzen, M., 1999. Removal of aqueous phenol and 2-chlorophenol with purified soybean peroxidase and raw soybean hulls. J. Chem. Technol. Biotechnol. 74 (4), 303–309.

Foudeh, A.M., Trigui, H., Mendis, N., et al., 2015. Rapid and specific SPRi detection of *L. pneumophila* in complex environmental water samples. Anal. Bioanal. Chem. 407 (18), 5541–5545. Available from: https://doi.org/10.1007/s00216-015-8726-y.

François, F., Lambard, C., et al., 2012. Isolation and characterization of environmental bacteria capable of extracellular biosorption of mercury. Appl. Environ. Microbiol. 78 (4), 1097–1106. Available from: https://doi.org/10.1128/aem.06522-11.

Fu, F., Wang, Q., 2011. Removal of heavy metal ions from wastewaters: a review. J. Environ. Manage. 92 (3), 407–418. Available from: https://doi.org/10.1016/j.jenvman.2010.11.011.

Gao, S., Zheng, X., Wu, J., 2017. A biolayer interferometry-based competitive biosensor for rapid and sensitive detection of saxitoxin. Sens. Actuators, B: Chem. 246, 169–174. Available from: https://doi.org/10.1016/j.snb.2017.02.078.

Ghosh, A., Dastidar, M.G., Sreekrishnan, T.R., 2017. Bioremediation of chromium complex dyes and treatment of sludge generated during the process. Int. Biodeterior. Biodegrad. 119, 448–460. Available from: https://doi.org/10.1016/j.ibiod.2016.08.013.

Guntas, G., Mitchell, S.F., Ostermeier, M., 2004. A molecular switch created by in vitro recombination of nonhomologous genes. Chem. Biol. 11 (11), 1483–1487. Available from: https://doi.org/10.1016/j.chembiol.2004.08.020.

Hakkila, K., Maximow, M., et al., 2002. Reporter genes lucFF, luxCDABE, gfp, and dsred have different characteristics in whole-cell bacterial sensors. Anal. Biochem. 301 (2), 235–242. Available from: https://doi.org/10.1006/abio.2001.5517.

He, Q., Miller, E.W., Wong, A.P., et al., 2006. A selective fluorescent sensor for detecting lead in living cells. J. Am. Chem. Soc. 128 (29), 9316–9317. Available from: https://doi.org/10.1021/ja063029x.

Homberger, M., Simon, U., 2010. On the application potential of gold nanoparticles in nanoelectronics and biomedicine. Philos. Trans. R. Soc. A: Math. Phys. Eng. Sci. 1405–1453. Available from: https://doi.org/10.1098/rsta.2009.0275.

Huang, J., Yang, X., Wu, Q., et al., 2019. Application of independent immobilization in benzo [*a*]pyrene biodegradation by synthetic microbial consortium. Environ. Sci. Pollut. Res. 26 (20), 21052–21058. Available from: https://doi.org/10.1007/s11356-019-05477-4.

Iwai-Shimada, M., Kameo, S., Nakai, K., et al., 2019. Exposure profile of mercury, lead, cadmium, arsenic, antimony, copper, selenium and zinc in maternal blood, cord blood and placenta: the Tohoku Study of Child Development in Japan. Environ. Health Prev. Med. 24 (1). Available from: https://doi.org/10.1186/s12199-019-0783-y.

Iyengar, V., Woittiez, J., 1988. Trace elements in human clinical specimens: evaluation of literature data to identify reference values. Clin. Chem. 34, 474–481.

Jiang, D., Du, X., Liu, Q., et al., 2015. Silver nanoparticles anchored on nitrogen-doped graphene as a novel electrochemical biosensing platform with enhanced sensitivity for aptamer-based pesticide assay. Analyst 140 (18), 6404–6411. Available from: https://doi.org/10.1039/c5an01084e.

Jouanneau, S., Durand, M.J., Courcoux, P., et al., 2011. Improvement of the identification of four heavy metals in environmental samples by using predictive decision tree models coupled with a set of five bioluminescent bacteria. Environ. Sci. Technol. 45 (7), 2925–2931. Available from: https://doi.org/10.1021/es1031757.

Julian, D.J., Kershaw, C.J., Brown, N.L., et al., 2009. Transcriptional activation of MerR family promoters in *Cupriavidus metallidurans* CH34. Antonie Van Leeuwenhoek 96 (2 SPEC. ISS.), 149–159. Available from: https://doi.org/10.1007/s10482-008-9293-4.

Justino, C.I.L., Duarte, A.C., Rocha-Santos, T.A.P., 2017. Recent progress in biosensors for environmental monitoring: a review. Sensors (Switzerland) . Available from: https://doi.org/10.3390/s17122918.

Kaur, I., Bhatnagar, A.K., 2002. Algae-dependent bioremediation of hazardous wastes. Prog. Ind. Microbiol. 36 (C), 457–516. Available from: https://doi.org/10.1016/S0079-6352(02)80024-8.

Khalid, S., Shahid, M., Niazi, N.K., et al., 2017. A comparison of technologies for remediation of heavy metal contaminated soils. J. Geochem. Explor. 182, 247–268. Available from: https://doi.org/10.1016/J.GEXPLO.2016.11.021.

Kim, H.N., Ren, W.X., Kim, J.S., et al., 2012. Fluorescent and colorimetric sensors for detection of lead, cadmium, and mercury ions. Chem. Soc. Rev. 3210–3244. Available from: https://doi.org/10.1039/c1cs15245a.

Knight, T., 2003. Idempotent Vector Design for Standard Assembly of Biobricks. MIT Synthetic Biology Working Group, pp. 1–11. Available from: http://hdl.handle.net/1721.1/21168.

Kulkarni, S.V., Palande, A.S., Deshpande, M.V., 2013. Bioremediation of petroleum hydrocarbons in soils. Microorganisms in Environmental: Microbes and Environment 589–606. Available from: https://doi.org/10.1007/978-94-007-2229-3_26.

Landrigan, P.J., Fuller, R., Acosta Jr, N., et al., 2018. The Lancet Commission on pollution and health. Lancet (London, Engl.) 391 (10119), 462–512. Available from: https://doi.org/10.1016/S0140-6736(17)32345-0.

Li, J., Lu, Y., 2000. A highly sensitive and selective catalytic DNA biosensor for lead ions [9]. J. Am. Chem. Soc. 10466–10467. Available from: https://doi.org/10.1021/ja0021316.

Li, T., Dong, S., Wang, E., 2010. A lead(II)-driven DNA molecular device for turn-on fluorescence detection of lead(II) ion with high selectivity and sensitivity. J. Am. Chem. Soc. 132 (38), 13156–13157. Available from: https://doi.org/10.1021/ja105849m.

Libis, V., Delépine, B., Faulon, J.-L.L., 2016a. Expanding biosensing abilities through computer-aided design of metabolic pathways. ACS Synth. Biol. 5 (10), . Available from: https://doi.org/10.1021/acssynbio.5b00225, p. acssynbio.5b00225.

Libis, V., Delépine, B., Faulon, J.L., 2016b. Sensing new chemicals with bacterial transcription factors. Curr. Opin. Microbiology 105–112. Available from: https://doi.org/10.1016/j.mib.2016.07.006.

de Lima, D.P., Dos Santos, E., Marques, M.R., et al., 2018. Fungal bioremediation of pollutant aromatic amines. Curr. Opin. Green. Sustain. Chem. 34–44. Available from: https://doi.org/10.1016/j.cogsc.2018.03.012.

Lin, C., Jair, Y.C., Chou, Y.C., et al., 2018. Transcription factor-based biosensor for detection of phenylalanine and tyrosine in urine for diagnosis of phenylketonuria. Anal. Chim. Acta 1041, 108–113. Available from: https://doi.org/10.1016/j.aca.2018.08.053.

Lin, K.H., Chien, M.F., Hsieh, J.L., et al., 2010. Mercury resistance and accumulation in *Escherichia coli* with cell surface expression of fish metallothionein. Appl. Microbiol. Biotechnol. 87 (2), 561–569. Available from: https://doi.org/10.1007/s00253-010-2466-x.

Lin, Y.W., Huang, C.C., Chang, H.T., 2011. Gold nanoparticle probes for the detection of mercury, lead and copper ions. Analyst 863–871. Available from: https://doi.org/10.1039/c0an00652a.

Liu, X., Li, W.J., Li, L., et al., 2014. A label-free electrochemical immunosensor based on gold nanoparticles for direct detection of atrazine. Sens. Actuators, B: Chem. 191, 408–414. Available from: https://doi.org/10.1016/j.snb.2013.10.033.

de Lorenzo, V., 2011. Systems biology approaches to bioremediation. Comprehensive Biotechnology 15–24. Available from: https://doi.org/10.1016/B978-0-08-088504-9.00460-8. second ed..

Saier Jr., M.H., Trevors, J.T., 2016. Phytoremediation. Encycl. Appl. Plant. Sci. 3 (May 2008), 327–331. Available from: https://doi.org/10.1016/B978-0-12-394807-6.00016-2.

Kumar, M., Jaiswal, S., et al., 2019. Antibiotics bioremediation: perspectives on its ecotoxicity and resistance. Environ. Int. 448–461. Available from: https://doi.org/10.1016/j.envint.2018.12.065.

Machado, L.F.M., Currin, A., Dixon, N., 2019. Directed evolution of the PcaV allosteric transcription factor to generate a biosensor for aromatic aldehydes. bioRxiv 689232. Available from: https://doi.org/10.1101/689232.

Mahr, R., Frunzke, J., 2016. Transcription factor-based biosensors in biotechnology: current state and future prospects. Appl. Microbiol. Biotechnol. 100 (1), 79–90. Available from: https://doi.org/10.1007/s00253-015-7090-3.

Mannan, A.A., Liu, D., Zhang, F., et al., 2017. Fundamental design principles for transcription-factor-based metabolite biosensors. ACS Synth. Biol. 6 (10), 1851–1859. Available from: https://doi.org/10.1021/acssynbio.7b00172.

Mayer, A.M., Staples, R.C., 2002. Laccase: new functions for an old enzyme. Phytochemistry 60 (6), 551–565. Available from: https://doi.org/10.1016/S0031-9422(02)00171-1.

Mehdizadeh Aghdam, E., Hejazi, M.S., Barzegar, A., 2016. Riboswitches: from living biosensors to novel targets of antibiotics. Gene 592 (2), 244–259. Available from: https://doi.org/10.1016/j.gene.2016.07.035.

Mehta, R., 2012. Bioremediation of textile waste water. Colourage 46–48.

Meliani, A., Bensoltane, A., 2016. Biofilm-mediated heavy metals bioremediation in PGPR *Pseudomonas*. J. Bioremed. Biodegrad. 7 (5). Available from: https://doi.org/10.4172/2155-6199.1000370.

Messori, L., Abbate, F., Marcon, G., et al., 2000. Gold(III) complexes as potential antitumor agents: solution chemistry and cytotoxic properties of some selected gold(III) compounds. J. Med. Chem. 43 (19), 3541–3548. Available from: https://doi.org/10.1021/jm990492u.

Migliaccio, M., Nunziata, F., Buono, A., 2015. SAR polarimetry for sea oil slick observation. Int. J. Remote. Sens. 3243–3273. Available from: https://doi.org/10.1080/01431161.2015.1057301.

Nagai, T., Sawano, A., Park, E.S., et al., 2001. Circularly permuted green fluorescent proteins engineered to sense Ca^{2+}. Proc. Natl. Acad. Sci. U.S.A. 98 (6), 3197–3202. Available from: https://doi.org/10.1073/pnas.051636098.

Namasivayam, C., Sangeetha, D., 2006. Recycling of agricultural solid waste, coir pith: removal of anions, heavy metals, organics and dyes from water by adsorption onto $ZnCl_2$ activated coir pith carbon. J. Hazard. Mater. 135 (1–3), 449–452. Available from: https://doi.org/10.1016/j.jhazmat.2005.11.066.

Nunes-Alves, C., 2015. Synthetic biology: GMOs in lockdown. Nat. Rev. Microbiol. 125. Available from: https://doi.org/10.1038/nrmicro3443.

O'Halloran, T., Walsh, C., 1987. Metalloregulatory DNA-binding protein encoded by the merR gene: Isolation and characterization. Science 235 (4785), 211–214. Available from: https://doi.org/10.1126/science.3798107.

Pal, R., Bhattacharya, P., Chakraborty, N., 2016. Bioremediation of toxic metals using algae. Algal Biorefinery: An Integrated Approach 439–462. Available from: https://doi.org/10.1007/978-3-319-22813-6_19.

Palchetti, I., Mascini, M., 2008. Nucleic acid biosensors for environmental pollution monitoring. Analyst . Available from: https://doi.org/10.1039/b802920m.

Poirier, L., Brun, L., Jacquet, P., et al., 2017. Enzymatic degradation of organophosphorus insecticides decreases toxicity in planarians and enhances survival. Sci. Rep. 7 (1), 19–21. Available from: https://doi.org/10.1038/s41598-017-15209-8.

Pyrzyńska, K., 2005. Recent developments in the determination of gold by atomic spectrometry techniques. Spectrochim. Acta, B: At. Spectrosc. 60 (9–10), 1316–1322. Available from: https://doi.org/10.1016/j.sab.2005.06.010.

Qiao, C.L., Huang, J., Li, X., et al., 2003. Bioremediation of organophosphate pollutants by a genetically-engineered enzyme. Bull. Environ. Contam. Toxicol. 70 (3), 455–461. Available from: https://doi.org/10.1007/s00128-003-0008-2.

Quintella, C.M., Mata, A.M.T., Lima, L.C.P., 2019. Overview of bioremediation with technology assessment and emphasis on fungal bioremediation of oil contaminated soils. J. Environ. Manage. 156–166. Available from: https://doi.org/10.1016/j.jenvman.2019.04.019.

Ragavan, K.V., Selvakumar, L.S., Thakur, M.S., 2013. Functionalized aptamers as nano-bioprobes for ultrasensitive detection of bisphenol-A. Chem. Commun. 49 (53), 5960–5962. Available from: https://doi.org/10.1039/c3cc42002g.

Rayu, S., Karpouzas, D.G., Singh, B.K., 2012. Emerging technologies in bioremediation: constraints and opportunities. Biodegradation 23 (6), 917–926. Available from: https://doi.org/10.1007/s10532-012-9576-3.

Rosser, S.J., French, C.E., Bruce, N.C., 2007. Engineering plants for the phytodetoxification of explosives. Vitro Cell. Dev. Biol. Plant. 37 (3), 330–333. Available from: https://doi.org/10.1007/s11627-001-0059-1.

Ruggaber, T.P., Talley, J.W., 2006. Enhancing bioremediation with enzymatic processes: a review. Pract. Period. Hazard., Toxic, Radioact. Waste Manage. 10 (2), 73–85. Available from: https://doi.org/10.1061/(asce)1090-025x(2006)10:2(73).

Salleh, A.B., Ghazali, F.M., Abd Rahman, R., et al., 2003. Bioremediation of petroleum hydrocarbon pollution. Indian J. Biotechnol. 411−425.

Sassolas, A., Leca-Bouvier, B.D., Blum, L.J., 2008. DNA biosensors and microarrays. Chem. Rev. 108 (1), 109−139. Available from: https://doi.org/10.1021/cr0684467.

Schell, L.M., Denham, M., Stark, A.D., et al., 2004. Relationship between blood lead concentration and dietary intakes of infants from 3 to 12 months of age. Environ. Res. 96 (3), 264−273. Available from: https://doi.org/10.1016/j.envres.2004.02.008.

Scott, C., Lewis, S.E., Milla, R., et al., 2010. A free-enzyme catalyst for the bioremediation of environmental atrazine contamination. J. Environ. Manage. 91 (10), 2075−2078. Available from: https://doi.org/10.1016/j.jenvman.2010.05.007.

Shah, P.C., Kumar, V.R., Dastager, S., et al., 2017. Phytase production by *Aspergillus niger* NCIM 563 for a novel application to degrade organophosphorus pesticides. AMB Express . Available from: https://doi.org/10.1186/s13568-017-0370-9.

Sharma, B., Dangi, A.K., Shukla, P., 2018. Contemporary enzyme based technologies for bioremediation: a review. J. Environ. Manage. . Available from: https://doi.org/10.1016/j.jenvman.2017.12.075.

Singh, B., Kaur, J., Singh, K., 2012. Biodegradation of malathion by *Brevibacillus* sp. strain KB2 and *Bacillus cereus* strain PU. World J. Microbiol. Biotechnol. 28 (3), 1133−1141. Available from: https://doi.org/10.1007/s11274-011-0916-y.

Sivasubramanian, V., 2016. Chapter 17—Phycoremediation and business prospects. In: Prasad, M.N.V. (Ed.), Bioremediation and bioeconomy. Elsevier, pp. 421−456. , <https://doi.org/10.1016/B978-0-12-802830-8.00017-4>.

Snoek, T., Chaberski, E.K., Ambri, F., et al., 2019. Evolution-guided engineering of small-molecule biosensors. bioRxiv 601823. Available from: https://doi.org/10.1101/601823.

Song, S., Wang, L., Li, J., et al., 2008. Aptamer-based biosensors. TrAC Trends Anal. Chem. 27 (2), 108−117. Available from: https://doi.org/10.1016/j.trac.2007.12.004.

Summers, A.O., 1992. Untwist and shout: a heavy metal-responsive transcriptional regulator. J. Bacteriol. 3097−3101. Available from: https://doi.org/10.1128/jb.174.10.3097-3101.1992.

Tay, P.K.R., Nguyen, P.Q., Joshi, N.S., 2017. A synthetic circuit for mercury bioremediation using self-assembling functional amyloids. ACS Synth. Biol. 6 (10), 1841−1850. Available from: https://doi.org/10.1021/acssynbio.7b00137.

Tchounwou, P.B., Yedjou, C.G., Patlolla, A.K., et al., 2012. Heavy metal toxicity and the environment. EXS 133−164. Available from: https://doi.org/10.1007/978-3-7643-8340-4_6.

Thatoi, H., Das, S., Mishra, J., et al., 2014. Bacterial chromate reductase, a potential enzyme for bioremediation of hexavalent chromium: a review. J. Environ. Manage. 146, 383−399. Available from: https://doi.org/10.1016/j.jenvman.2014.07.014.

Townsend, A.T., Miller, K.A., McLean, S., et al., 1998. The determination of copper, zinc, cadmium and lead in urine by high resolution ICP-MS. J. Anal. At. Spectrom. 13 (11), 1213−1219. Available from: https://doi.org/10.1039/a805021j.

Trabelsi, H., Koch, M., Faulon, J.-L., 2018. Building a minimal and generalizable model of transcription-factor based biosensors: showcasing flavonoids. Biotechnol. Bioeng. 115 (9), 2292−2304. Available from: https://doi.org/10.1002/bit.26726.

Tuerk, C., Gold, L., 1990. Systematic evolution of ligands by exponential enrichment: RNA ligands to bacteriophage T4 DNA polymerase. Science 249 (4968), 505−510. Available from: https://doi.org/10.1126/science.2200121.

Verbych, S., Hilal, N., Sorokin, G., et al., 2004. Ion exchange extraction of heavy metal ions from wastewater. Sep. Sci. Technol. 39 (9), 2031−2040. Available from: https://doi.org/10.1081/SS-120039317.

Virta, M., Lampinen, J., Karp, M., 1995. A luminescence-based mercury biosensor. Anal. Chem. 67 (3), 667–669. Available from: https://doi.org/10.1021/ac00099a027.

Voyvodic, P.L., Pandi, A., Koch, M., et al., 2019. Plug-and-play metabolic transducers expand the chemical detection space of cell-free biosensors. Nat. Commun. 10 (1). Available from: https://doi.org/10.1038/s41467-019-09722-9.

Wang, D., Zheng, Y., Fan, X., et al., 2019. Visual detection of Hg^{2+} by manipulation of pyocyanin biosynthesis through the Hg^{2+}-dependent transcriptional activator MerR in microbial cells. J. Biosci. Bioeng. 129 (2), 223–228. Available from: https://doi.org/10.1016/j.jbiosc.2019.08.005.

Wang, Q., Fang, J., Cao, D., et al., 2015. An improved functional assay for rapid detection of marine toxins, saxitoxin and brevetoxin using a portable cardiomyocyte-based potential biosensor. Biosens. Bioelectron. 72, 10–17. Available from: https://doi.org/10.1016/j.bios.2015.04.028.

Warniment, C., Tsang, K., Galazka, S.S., 2010. Lead poisoning in children. Am. Fam. Physician 6 (81), 751–757. Available from: http://www.cdc.gov/.

Wei, W., Zhu, T., Wang, Y., et al., 2012. Engineering a gold-specific regulon for cell-based visual detection and recovery of gold. Chem. Sci. 3 (6), 1780–1784. Available from: https://doi.org/10.1039/c2sc01119k.

Wei, W., Liu, X., Sun, P., et al., 2014. Simple whole-cell biodetection and bioremediation of heavy metals based on an engineered lead-specific operon. Environ. Sci. Technol. 48 (6), 3363–3371. Available from: https://doi.org/10.1021/es4046567.

Whitton, B.A., 2013. Use of benthic algae and bryophytes for monitoring rivers. J. Ecol. Environ. 95–100. Available from: https://doi.org/10.5141/ecoenv.2013.012.

Yan, L., Sun, P., Xu, Y., et al., 2018. Integration of a gold-specific whole *E. coli* cell sensing and adsorption based on BioBrick. Int. J. Mol. Sci. 19 (12), 3741. Available from: https://doi.org/10.3390/ijms19123741.

Yin, K., Wang, Q., Lv, M., et al., 2019. Microorganism remediation strategies towards heavy metals. Chem. Eng. J. 1553–1563. Available from: https://doi.org/10.1016/j.cej.2018.10.226.

Yoo, M.S., Shin, M., Kim, Y., et al., 2017. Development of electrochemical biosensor for detection of pathogenic microorganism in Asian dust events. Chemosphere 175, 269–274. Available from: https://doi.org/10.1016/j.chemosphere.2017.02.060.

Zhang, F., Carothers, J.M., Keasling, J.D., 2012. Design of a dynamic sensor-regulator system for production of chemicals and fuels derived from fatty acids. Nat. Biotechnol. 30 (4), 354–359. Available from: https://doi.org/10.1038/nbt.2149.

Zhang, J., Barajas, J.F., Burdu, M., et al., 2017. Development of a transcription factor-based lactam biosensor. ACS Synth. Biol. 6 (3), 439–445. Available from: https://doi.org/10.1021/acssynbio.6b00136.

Biosurfactant-based bioremediation

16

Simranjeet Singh[1,2,3,†], Vijay Kumar[4,†], Satyender Singh[3], Daljeet Singh Dhanjal[1], Shivika Datta[5], Deepansh Sharma[6], Nitin Kumar Singh[7] and Joginder Singh[1,]*
[1]Department of Biotechnology, Lovely Professional University, Phagwara, India, [2]Punjab Biotechnology Incubators, Mohali, India, [3]Regional Advanced Water Testing Laboratory, Mohali, India, [4]Regional Ayurveda Research Institute for Drug Development, Gwalior, India, [5]Department of Zoology, Doaba College, Jalandhar, India, [6]Amity Institute of Microbial Technology, Amity University, Jaipur, India, [7]Department of Environmental Science and Engineering, Marwadi University, Rajkot, India
*Corresponding author

16.1 Introduction

Over the years, modern lifestyle and industrialization have led to the contamination of soil and water. The most common environmental pollutants are heavy metals and hydrocarbon (Makombe and Gwisai, 2018). The primary concern of these molecules is that they readily bind with the soil particles, due to their high interfacial tension (IFT) and low water solubility, which makes their remediation from the contaminated location difficult (Saichek and Reddy, 2005; Ayangbenro and Babalola, 2017; Bhati et al., 2019; Kapoor et al., 2019). Various remediation approaches involving biological, chemical, and physical activities have been developed. Because of the advancement in sustainable technologies, the exploration of natural methods has increased for the remediation of hydrocarbon contamination from water and soil (Pichtel, 2016; Singh et al., 2019a,b; Kumar et al., 2019a). Hence, bioremediation emerged as the targeted approach for remediating the contaminated site. It is a process that involves the action of microbes and their enzymes to degrade the contaminants from the contaminated site (Karigar and Rao, 2011; Kumar et al., 2019b; Sidhu et al., 2019). In this process, different microbes like bacteria, fungi, and yeasts are isolated from the polluted site, and they are used to clean up contaminated soil and water (Das and Chandran, 2011; Kumar et al., 2019c; Kumar and Singh, 2018a,b). Some of the microbes isolated from the contaminated site produce biosurfactants. These microbial biosurfactants can entrap the heavy metals and hydrocarbons and degrade them into products, which can be used

[†] Authors equally contributed.

Bioremediation of Pollutants. DOI: https://doi.org/10.1016/B978-0-12-819025-8.00016-8

by the environment or be degraded further by soil microbes (Karlapudi et al., 2018; Kumar et al., 2018a,b,c).

In the late 1960s, biosurfactants replaced the synthetic surfactants like carboxylate, sulfonate, and sulfate acetic acid as they impose toxicity and cause damage to the environment (Makkar and Rockne, 2003). Hence, biosurfactant attained the attention of food, oil, and pharmaceutical industries as it can dissolve both heavy metals and hydrocarbons (Santos et al., 2016). Biosurfactants are a structurally unique group of surface-active molecules synthesized by microbes. They have amphiphilic nature as they contain both polar and nonpolar part in its structure. A hydrophobic portion includes monosaccharide, polysaccharide, proteins, or peptides group, whereas, the hydrophilic portion contains saturated fat, unsaturated fat, fatty alcohol, or hydroxylated fatty acid (Saharan et al., 2011).

A distinct feature of biosurfactant is a hydrophilic–lipophilic balance which separates the portion of the hydrophobic and hydrophilic constituent within surface-active molecules which allow the increase of surface area for hydrophobic compounds, increase of the bioavailability of water for hydrophilic compounds, and can alter the surface properties of bacterial cells. These surface properties of biosurfactant make it an excellent dispersing, emulsifying, and foaming agent (Pacwa-Płociniczak et al., 2011). Comparing to synthetic surfactants, biosurfactants have many advantages as it is biodegradable, ecofriendly, nontoxic, and less hazardous. Moreover, they are stable at high temperature, pH, salinity, and have better selectivity (De Almeida et al., 2016). Nowadays, they are synthesized using industrial waste or their byproducts as carbon source, which allows the utilization of waste and reduce their toxicity effect at the same time (Freitas et al., 2016).

These advantages make the biosurfactant the targeted molecules and prompt us to find new types of bioemulsifiers and biosurfactants. This chapter aims to describe the microbial biosurfactant, their properties, and potential in bioremediating environmental contaminants.

16.2 Biosurfactants: surface-active compounds

Biosurfactants are amphiphilic compounds containing both hydrophobic and hydrophilic domains. Generally, hydrophobic moiety is a hydrocarbon chain, whereas hydrophilic moiety may be amphoteric, positively or negatively charged (ionic), or nonionic (Otzen, 2017). These hydrophobic as well as hydrophilic moieties aggregate themselves at the interface between liquid phase and reduce the surface tension in liquids (Lombardo et al., 2015). The most commonly used ionic surfactants are ammonium salt, ethylene, ethoxylate, fatty acids, ester sulfonates or sulfates, sorbitan ester, and propylene oxide copolymers (Satpute et al., 2018). Generally, surfactants form the film at the interface, which reduces the IFT and empowers unique properties to these surfactants (Burlatsky et al., 2013). The CMC (critical micelle concentration) signifies the aggregation of surfactant moieties within aqueous phase. At the phases, where the value of CMC is significantly high, these amphiphilic molecules form a supramolecular structure called micelles (Campos et al., 2013) (Fig. 16.1).

The currently used surfactants are synthetically synthesized via petroleum (Silva et al., 2014). These synthetic surfactants are highly toxic and problematic for degradation by microbes (Tian et al., 2016). This has motivated the researchers to explore surfactants that are ecofriendly; this is achieved by microbial produced surfactants, stated as biosurfactants (Marchant and Banat, 2012).

On comparing the physiochemical properties like reduced IFT, temperature, and pH stability of synthetic surfactant and biosurfactant showed a large similarity. But the advantages like greater biodegradability, high stability at varied physical environmental factors (pH, temperature, and salinity), higher specificity, lower CMC value, and less toxicity have drawn the attention of various industries and made biosurfactant as a material of interest (Naughton et al., 2019). Even government have developed rules and regulations, which motivate to explore and develop natural surfactants as an alternative to synthetic surfactants in order to save the environment for future generations (Freitas et al., 2016). Hence, exploitation of microbial

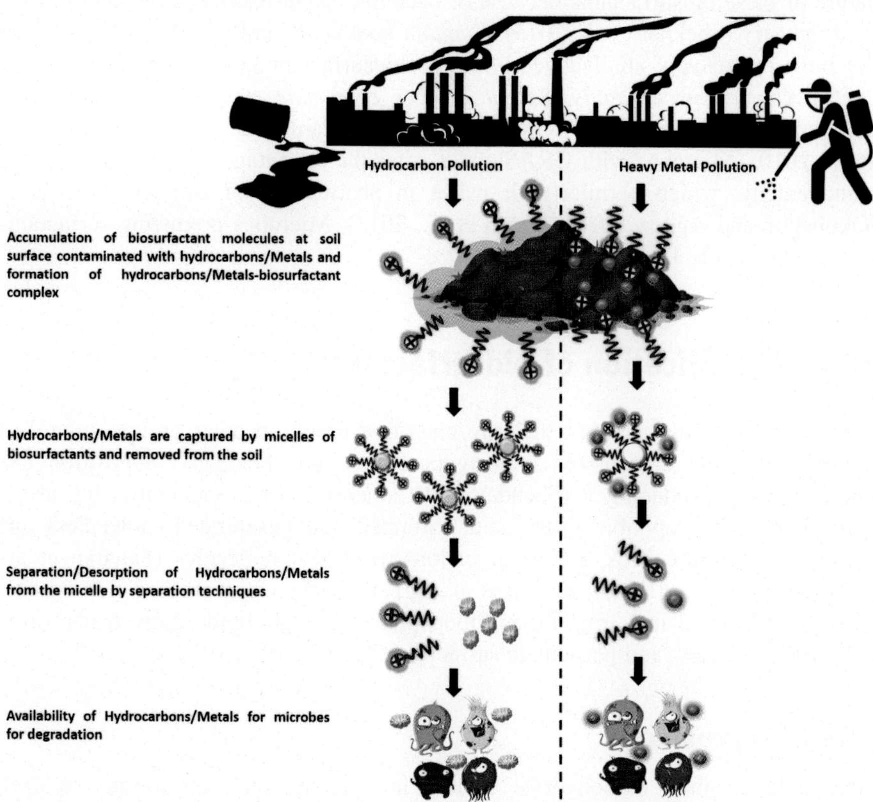

Figure 16.1 Schematic representation of surfactant molecule and micelle formation.

community is required for discovering novel or new biosurfactants with unique properties.

16.3 Biosurfactant-producing microbes

Microbes grow on different kinds of substrate as they require a carbon source for proliferation. The blend of insoluble substrate with carbon sources facilitates the production of different substances and intracellular diffusion (Lynd et al., 2002; Usman et al., 2016). Microbes like bacteria, filamentous fungi, and yeast have the ability to synthesize biosurfactant of varied molecular structure and surface properties (Santos et al., 2018; Datta et al., 2018; Singh et al., 2018; Kaur et al., 2018). Recently, scientists have started isolating microbes which synthesize tensioactive molecules with desired surfactant features like high emulsifying activity, low CMC value, and low toxicity (Sobrinho et al., 2013).

The literature survey has revealed that the genera *Bacillus* and *Pseudomonas* are the major producers of biosurfactants (Loiseau et al., 2018). However, pathogenic nature of these biosurfactants because of bacterial origin restricts their usage in the food industry (Shekhar et al., 2015). *Candida bombicola* and *Candida lipolytica* are the most extensively studied yeast for the biosurfactant production (Silva et al., 2018). *Kluyveromyces lactis*, *Saccharomyces cerevisiae*, and *Yarrowia lipolytica* are the species granted with GRAS (generally regarded as safe) status (Amaral et al., 2010). Microbes with GRAS status offer the advantages of less toxicity and pathogenicity, which permits their usage in pharmaceutical and food industries (Okoliegbe and Agarry, 2012; Shoeb et al., 2013). Microbes producing surfactants are listed in Table 16.1.

16.4 Classification of biosurfactants

Unlike synthetic surfactants which are classified based on polar groups, microbial biosurfactants are categorized majorly based on the chemical composition and microorganism producing it (Sachdev and Cameotra, 2013). Generally, biosurfactants involve hydrophobic (fatty acid, saturated and unsaturated molecules) and hydrophilic (amino acids, anions or cations, peptides) molecules (Saharan et al., 2011; Singh et al., 2017a,b; Kumar et al., 2017). Hence, major microbial biosurfactants are classified into glycolipids, lipopeptides, phospholipids, fatty acids, polymeric biosurfactant, and particulate surfactants.

16.4.1 Glycolipids

Glycolipids contain carbohydrate moiety in combination with a fatty acid or hydroxy fatty acid (Mnif and Ghribi, 2016). Due to the ability to utilize the cheap

Table 16.1 Classification of the major types of biosurfactants synthesized by microbes.

Class	Type	Source	References
Glycolipids	Rhamnolipids	*Pseudomonas aeruginosa*	Maier and Soberon-Chavez (2000)
		Nocardia erythropolis	Park et al. (1998)
		Arthobacter sp.	Li et al. (1984)
		Pseudomonas putida	Wittgens et al. (2011)
		Serratiarubidaea	Nalini and Parthasarathi (2014)
		Burkholderia thailandensis	Funston et al. (2016)
		Burkholderia pseudomallei	Irorere et al. (2017)
		Burkholderia plantarii	Hörmann et al. (2010)
	Trehalose lipids	*Rhodococcus erythropolis*	Kurane et al. (1995)
		Rhodococcus wratislaviensis	Tuleva et al. (2008)
		Mycobacterium sp.	Murphy et al. (2005)
		Arthrobacter paraffineus	Duvnjak et al. (1982)
		Torulopsis bombicola	Göbbert et al. (1984)
		Torulopsis apicola	Hommel et al. (1987)
	Sophorolipids	*Torulopsis petrophilum*	de Oliveira et al. (2014)
		Candida bombicola	Felse et al. (2007)
		Candida apicola	Konishi et al. (2016)
		Candida rugosa	Prabhakar et al. (2017)
		Rhodotorula mucilaginosa	Chandran and Das (2011)
		Rhodotorula bogoriensis	Zhang et al. (2011)
		Pichia anomala	Thaniyavarn et al. (2008)

(*Continued*)

Table 16.1 (Continued)

Class	Type	Source	References
	Mannosylerythritol lipid	*Candida antartica*	Adamczak and Bednarski (2000)
	Cellobiolipids	*Ustilago zeae*	Boothroyd et al. (1955)
		Ustilago maydis	Frautz et al. (1986)
Lipopeptides and lipoproteins	Surfactin	*Nocardiopsis alba*	Gandhimathi et al. (2009)
		Bacillus subtilis	Ghribi and Ellouze-Chaabouni (2011)
	Fengycin	*Bacillus subtilis*	Ghribi and Ellouze-Chaabouni (2011)
	Viscosin	*Pseudomonas fluorescens*	Bonnichsen et al. (2015)
	Lichenysin	*Bacillus licheniformis*	Konz et al. (1999)
	Antibiotic TA	*Myxococcus xanthus*	Thakur et al. (2018)
	Serrawettin	*Serratia marcescens*	Matsuyama et al. (2011)
	Subtilosin	*Bacillus subtilis*	Sutyak et al. (2008)
	Gramicidin	*Bacillus brevis*	Marahiel et al. (1979)
	Polymixin	*Paenibacillus polymyxa*	Choi et al. (2009)
Phospholipids, fatty acids, and neutral lipids	Spiculisporic acid	*Penicillium spiculisporum*	Suzuki et al. (1988)
	Corynomycolic acid	*Corynebacterium lepus*	Cooper et al. (1979)
	Phosphatindyl-ethanolamine E	*Acinetobacter* sp.	Boll et al. (2015)
		Rhodococcus erythropolis	De Carvalho et al. (2014)
		Myxococcus sp.	Kearns et al. (2002)

(*Continued*)

Table 16.1 (Continued)

Class	Type	Source	References
Polymeric surfactants	Alsan	*Acinetobacter radioresistens*	Toren et al. (2001)
	Biodispersan	*Arethrobacter calcoaceticus*	Kosaric and Sukan (2010)
	Carbohydrate–protein–lipid	*Pseudomonas fluorescens*	Desai and Banat (1997)
		Debaryomyces polymorphus	Al-Araji et al. (2007)
	Emulsan	*Acinetobacter calcoaceticus*	Choi et al. (1996)
	Liposan	*Candida lipolytica*	Cirigliano and Carman (1985)
		Yarrowia lipolytica	Csutak et al. (2015)
	Mannoprotein	*Saccharomyces cerevisiae*	Cameron et al. (1988)
		Kluyveromyces marxianus	Lukondeh et al. (2003)
	Protein PA	*Pseudomonas aeruginosa*	Gilboa-Garber et al. (2000)
Particulate biosurfactant	Vesicles	*Acinetobacter* sp.	Käppeli and Finnerty (1979)
		Pseudomonas marginalis	Burd and Ward (1997)
	Whole microbial cells	*Cyanobacteria* sp.	Karanth et al. (1999)

substrate and produce high-yield of the commercial product, the interest of scientists and researchers toward glycolipids has increased (Elazzazy et al., 2015). The best-known glycolipids are trehalolipids, rhamnolipids, and sophorolipids (Bustamante et al., 2012). In 1949 rhamnolipids were first introduced, which are an exproduct of *Pseudomonas aeruginosa*, and are formed by the linking of rhamnose sugar with one or two molecules of β-hydroxydecanoic fatty acid (Jarvis and Johnson, 1949). Modern analytical techniques have enabled us to discover different rhamnolipid homologues synthesized by various bacterial and *Pseudomonas* species (Rikalović et al., 2015). For instance, rhamnolipids synthesized by *Burkholderia glumae* have long alkyl chains in comparison to those synthesized by *Pseudomonas putida* (Wittgens et al., 2018). Rhamnolipids have the ability to decrease the water surface tension from 72.80 to 27 mN/m (Zhao et al., 2018). Most of rhamnolipids studies involve environmental application like petroleum hydrocarbon degradation, control oil spillage, and soil remediation (Das and Chandran, 2011). Increased dissipation of some specific contaminants has been reported on supplementing rhamnolipids, whereas reduced biodegradation ability or no effect has also been recorded in which rhamnolipids were supplemented (Liu et al., 2018). Hence, the presence of these biosurfactant molecules in soil induces changes in the microbial community, as a result, they alter the degradation pathway (Chrzanowski et al., 2012; Wani et al., 2017; Singh et al., 2016; Mishra et al., 2016). Rhamnolipids are considered to be toxic to natural vegetation, but they also decrease the toxicity of specific contaminants by enhancing hydrocarbon solubilization, thus aiding biodegradation (Mulligan, 2005).

Various microbial species like *Mycobacterium* sp., *Corynebacterium* sp., and *Rhodococcus* spp. synthesize glycolipids involving trehalose sugar (Wang et al., 2011). *Rhodococcus erythropolis* is the most extensively studied bacteria synthesizing trehalose dimycolate (Tischler et al., 2013). *Arthrobacter* spp. and *R. erythropolis* synthesize trehalolipids which lowers IFT as well as surface tension to 1–5 and 25–40 mN/m, respectively, in the culture broth (Vijayakumar and Saravanan, 2015). Genus *Candida* is the major producer of sophorolipids. Sophorolipids are composed by linking dimeric sophorose sugars with long hydroxyl-fatty acid chains via glycosidic bond (de Oliveira et al., 2015). *C. bombicola*, yeast species is most extensively exploited for obtaining sophorolipids biosurfactant (Elshafie et al., 2015). Sophorolipids have been found to reduce the surface tension of water and *n*-hexadecane from 40 to 5 mN/m (De Rienzo et al., 2015). Moreover, sophorolipids maintain its stability at varied temperature and pH (Gautam and Tyagi, 2006). The potential and applicability of these glycolipids prompts their usage for better prospects in bioremediation.

16.4.2 Lipopeptides and lipoproteins

These types of biosurfactants comprise cyclic peptides linked with long fatty acid chains (Płaza et al., 2014). *Bacillus subtilis* has been recognized as the potent producer of lipoproteins and lipopeptides. The biosurfactant obtained from *B. subtilis* at very low concentration reduces the surface tension from 72.8 to 27.9 mN/m (Arima et al., 1968). Moreover, *B. subtilis* derived biosurfactant possess the antiadhesive, antibacterial, antimycoplasma, antiviral, and hemolytic activity. Serratamolide, an aminolipid biosurfactant has recently been isolated from *Serratia marcescens* (Gudiña et al., 2016). This type of biosurfactant enhances hydrophilicity of cell via blocking the hydrophobic molecules present on the cell surface (Mao et al., 2015). Another biosurfactant, surfactin is made up of seven amino acid ring structure linked to long fatty acids chain through lactone linkage (Karlapudi et al., 2018). Surfactin is known for its ability to reduce interfacial and surface tension in water. On the other hand, *Bacillus licheniformis* produces lichenysin which is highly stable under extreme environmental conditions like surfactin (Coronel-León et al., 2016).

16.4.3 Fatty acids, phospholipids, and neutral lipids

Various bacterial and yeasts species are capable of growing on different substrates like *n*-alkanes and produce a huge amount of fatty acids, neutral lipids, and phospholipids (Joy et al., 2017). *Acinetobacter* spp. and *R. erythropolis* produce phosphatidylethanolamine vesicles which are optical clear microemulsion of alkanes in water (Käppeli and Finnerty, 1979). Phospholipids synthesized by *Thiobacillus thiooxidans* have been found to be responsible for reducing elemental sulfur from soil (Campos et al., 2013). A mixture of corynomycolic acids has been obtained from *Corynebacterium lepus*, which significantly lowers the surface as well as IFT at varied pH (Cairns et al., 2017).

16.4.4 Polymeric biosurfactant

The most extensively studied polymeric biosurfactants are alasan, emulsan, lipomanan, and liposan (Shekhar et al., 2015). Emulsan comprises heteropolysaccharide backbone which is covalently linked with fatty acids. This polymeric biosurfactant, emulsan is secreted by *Acinetobacter calcoaceticus* RAG-1, and it emulsifies the hydrocarbons at very low concentrations ranging from 0.001% to 0.01% in water (Uzoigwe et al., 2015). *C. lipolytica* synthesizes water-soluble emulsifier "liposan," which comprises carbohydrates (83%) and proteins (17%) (Panjiar et al., 2017).

16.4.5 Particulate biosurfactant

Particulate biosurfactants are involved during the partition of extracellular membrane of vesicle to develop a microemulsion which effect uptake of alkane in microbial cells (Gharaei-Fathabad, 2011). *Acinetobacter* spp. synthesizes vesicles are comprised of phospholipids, proteins, and lipopolysaccharides with 1.158 g/cm^3 buoyant density and 20–50 nm diameter (Bezerra et al., 2018).

16.5 Parameters regulating the properties of biosurfactants

There are different parameters that regulate the properties of biosurfactants, which are discussed below:

16.5.1 Surface and interface activity

Biosurfactants are effective molecules that lower the surface tension of water. Surfactin synthesized by *B. subtilis* is known for lowering surface tension of liquids even at extreme conditions (Gudiña et al., 2015). Rhamnolipid surfactant produced by *P. aeruginosa* has the potential to effectively decrease the surface tension of water in comparison to other surfactants (El-Sheshtawy and Doheim, 2014).

Similarly, sophorolipids also reduce the surface tension. Biosurfactants are found to be more effective than synthetic surfactants as their CMC value is 10- to 40-fold lower, which provides evidence why a very low amount of biosurfactant is needed to reduce the surface tension (Andersen et al., 2016).

16.5.2 Temperature, ionic strength, and pH tolerance

Biosurfactants remain unaffected and sustain their functionality to changing environmental conditions especially temperature and pH (Mahanti et al., 2017). It has been found that lichenysin produced by *B. licheniformis* has reduced activity at temperature 50°C with a pH in the range of 4.5–9.0 (Muthusamy et al., 2008). Lipopeptides synthesized by *B. subtilis* have been found to be stable even after 180 days on assessing it at both high (121°C) and low (15°C) temperature. Moreover, it was also stable at 15% NaCl concentration with pH in range of 4–12 (Mukherjee et al., 2006).

16.5.3 Biodegradability

Biosurfactants are considered to be nontoxic which makes it the elite option for their use in cosmetic, food, and pharmaceutical fields (Vecino et al., 2017). Recently, emulsan has shown LC50 against *Photobacterium phosphoreum*, but its activity is less than rhamnolipids synthesized by *Pseudomonas* sp. (Vijayakumar and Saravanan, 2015). If we compare the commercially available surfactants on the basis of toxicity, synthetic surfactants will hold the most number as most of the biosurfactants are degradable in nature (Vigneshwaran et al., 2018). Due to the nontoxicity of biosurfactants synthesized by *Pseudomonas* sp., they are widely used in various industries like food and pharmaceutical industries (Akbari et al., 2018). Different researches have been done to assess the toxicity of synthetic surfactant and biosurfactant (Satpute et al., 2010). To which the toxic effect of biosurfactant was found to be less than synthetic surfactants. Breaking and formation of emulsion can be achieved in a month, and emulsion can be destabilized or stabilized by biosurfactants (Mouafo et al., 2018). Emulsifiers generally belong to high molecular weight biosurfactants in contrast to low molecular weight biosurfactants (Lamichhane et al., 2017). Biosurfactant sophorolipids are synthesized by *Torulopsis bombicola*. For stable emulsion, polymeric biosurfactants are commonly used as they have oil coat droplets that form oil/water emulsion (Sharma et al., 2016). *C. lipolytica* produces biosurfactant liposan which is unable to reduce the surface tension but does emulsify the edible oil (Rufino et al., 2014). Biosurfactant is comprised of hydrophobic (made up of fatty acids or fatty alcohols) and hydrophilic (made up of protein or sugar) group. Several functions are performed by biosurfactants like increased surface area, enhanced bioavailability of water-insoluble compounds, and entrapped heavy metals for remediation (Usman et al., 2016). Additionally, biosurfactants also exhibit antimicrobial, antiinflammatory, and antioxidant activity (Jemil et al., 2017). Further, different biosurfactants have their unique function but have one common function, that is, decrease liquid surface tension (Chander et al., 2012).

16.5.4 Biofilm formation

Biosurfactants have a property that enables it to create a suitable environment for bacterial adhesion (Geys et al., 2014). Biofilms are the extracellular matrix produced by the microbial community (Nadell et al., 2015). Exopolymeric substances like bioemulsifiers aid in formation of biofilm, which further enhances their survival rate and protects them from severe conditions like predator attack and water-loss from cell (Costa et al., 2018). Adhesion of bacterial cells can take place in both stagnant and mobile phase. Biofilm production, a complex process in which transition of free-floating cells to surface adhering microbial community takes place (Dang and Lovell, 2016). The different types of biofilms are dependent on planktonic cells (Donlan, 2002). Single cells synthesizing biofilm regulate the process by communicating with related microbial species with quorum sensing (Li and Tian, 2012). On the other hand, biofilm formed by involving multiple species is triggered by specific signals which aggregates the various surface adhering planktonic cells (especially bacterial) (Demuyser et al., 2014). Factors like flow velocity, nutrient, smoothness, surface area are related with biofilm formation as it provides favorable condition for bacterial proliferation and attachment (Chrzanowski et al., 2012). Different crucial steps are involved in biofilm formation like adhering organic molecule with submerged material in water and neutralizing its surface charge which was repelling the bacteria (Garnett and Matthews, 2012). Temporary attachment of bacterial planktonic take place by electrostatic forces. Due to the formation of extracellular polymeric substances (EPS) the permanent attachment of bacterial planktonic takes place. This attachment forms an ion-exchange system which entraps nutrients, as it doubles the reproduction rate of organisms (Jayathilake et al., 2017). The high amount of EPS and water are the chief factors responsible for slime nature of biofilm. The primary colonies synthesize secondary metabolites which are consumed by secondary colonies for their growth as a result biofilm formation take place (Decho and Gutierrez, 2017). The genes regulated by quorum sensing that are triggering the EPS secretion have become an important asset for microbes. The quorum sensing allows the microbes to survive against antibacterial compounds and aids in availing nutrients under nutrient limited condition (Abisado et al., 2018). Dental plaques are the most extensively studied biofilms involving multiple species. Cooperative and competitive interaction of oral bacteria exhibited complex communication for releasing metabolites. This is the reason why biofilm formed on medical devices and manufacturing surface are difficult to eliminate (Marsh and Zaura, 2017). Different biofilm resistance mechanisms have been proposed and believed to synergistically work to decrease biofilm susceptibility. The presence of microbial cells within biofilms generates high intensity community response in comparison to single planktonic cell (Hall and Mah, 2017). Moreover, three-dimensional structure protects the persisting cell, which on disruption will more become susceptible to antimicrobial agents (Singh et al., 2017a,b). Recently, the use of magnetic field and ultrasound has been found useful in biofilm eradication (Sadekuzzaman et al., 2015). Even, chemical biocides like detergents, disinfectants, and sanitizers are now commonly used for controlling biofilm formation (Maillard, 2005). These agents are divided in two different categories, that is, oxidizing and nonoxidizing. Chlorine, hydrogen peroxide, iodine, and ozone are the most commonly used oxidizing agents, which perform depolymerase action on EPS matrix and disrupts the integrity of biofilm. Whereas, anionic and nonionic surface-active molecules, formaldehyde, and quaternary ammonium compounds are the nonoxidizing agents widely used for the same purposes (Karlapudi et al., 2018).

16.6 Biosurfactant for heavy metal remediation

Anthropogenic activities like manufacturing, mining, and using of chemical products (batteries, industrial waste, production of domestic or industrial sludge, paints, and pesticides) has elevated the contamination of heavy metals in the soil (Ayangbenro and Babalola, 2017). Now, Ba, Cd, Hg, Ni, Pb, Sr, and Zn have become major toxic metals which contaminated water, urban soil, and agricultural soil. These polluted sites may be landfill sites for industrial waste, old insecticide treated orchards, fields used for treating waste water and sewage sludge, area around mines and chemical industries (Wuana and Okieimen, 2011; Kumar et al., 2014, 2015, 2016). Microbial consortium and the secondary metabolites such as surfactants have emerged as the solution to resolve this environmental issue (Biniarz et al., 2017). Nowadays, biosurfactants are used for remediating heavy metal polluted site as illustrated in Fig. 16.2. The counter ion binding, electrostatic interactions, ion exchange and precipitation–dissolution are the probable interactions used for heavy metal bioremediation by biosurfactants (Franzetti et al., 2014). This biosurfactant mediated heavy metal remediation mechanism involves the chelation of metal ion with opposite charged site of surfactant monomer. Then chelated monomers and metal ions complex form the micelle (Sarubbo et al., 2015). This micelle formation helps in recovering the metal from soil by facilitating easy

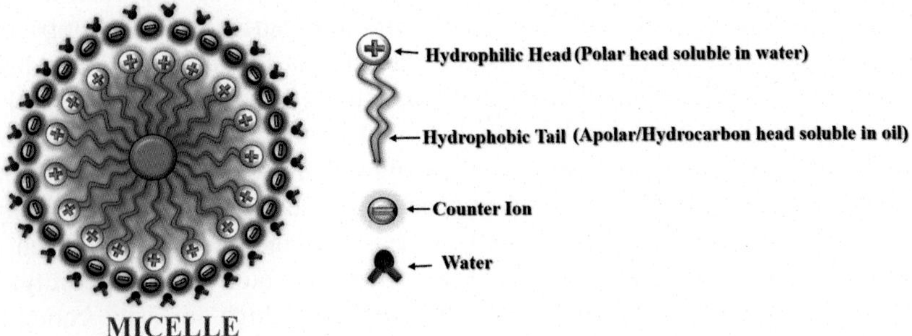

Figure 16.2 Illustration of action mechanism of biosurfactant for hydrocarbons and heavy metals remediation.

transportation to the solution and aiding the recovery of metals via flushing mechanism (Hashim et al., 2011). Metal ions being positively charged enable chelation of anionic surfactants like surfactin, rhamnolipids, sophorolipids, which increase their bioavailability for both plant and microbes in soil (Olaniran et al., 2013). It has been noticed that under moderate IFT condition, biosurfactant directly sorbed the metal and accumulated it at solid solution interface (Usman et al., 2016). Hence, increase in association of metals is observed for anionic surfactants, whereas, it is opposite in the case of cationic surfactants (Upadhyaya et al., 2007).

Miller (1995) reported about the usage of biosurfactant for remediating toxic heavy metal (Ba, Cd, Ni, Pb, and Zn) from a contaminated site. In a study conducted by Sandrin, rhamnolipid is found to be effective in reducing the Cd(II) toxicity. This was validated due to the formation of rhamnolipid and Cd(II) complex, which induces the lipopolysaccharide removal from cell surface, thus altering Cd (II) uptake and reducing its toxicity (Sandrin et al., 2000). In another study, high removal efficiency of about 7% for Strontium (Sr) has been recorded on the usage of 80 ppm of rhamnolipid biosurfactant (Elouzi et al., 2012). Moreover, 1 mg of emulsan has been found to remove 90% of uranium from 0.9 μm of U (Zosim et al., 1983). The lipopeptide synthesized by *C. lipolytica* has also been found effective in removing 96% of Cu as well as Zn (Rufino et al., 2012). Additionally, lipopeptides are also effective in reducing other heavy metals like Cd, Fe, and Pb. Removal efficiency of lipopeptide biosurfactant at different pH has also been assessed and pH 9 has been found to be optimum as well as effective in removing Cd, Cu, Co, Ni, Pb, and Zn (Singh and Cameotra, 2013).

16.7 Biosurfactants for hydrocarbon remediation

The amphipathic nature of biosurfactant increases the solubility of both inorganic and organic compounds by reducing their surface tension. This property of biosurfactant can be used to resolve the worldwide issue of crude oil spillage from oil industries (Patowary et al., 2017). One global issue is oil spillage from oil industries which is affecting both soil and water bodies including aquatic plants and animals (Saadoun, 2015). The probable mechanism of hydrocarbon degradation involves improved hydrophobicity of bacterial cell surface for entrapping hydrophobic compounds. The emulsification of oils or hydrocarbons eases their transportation and increases their bioavailability for hydrocarbon-degrading microbes (Xu et al., 2018). The detailed mechanism is illustrated in Fig. 16.2. The major benefit of the biosurfactant is its ability to reduce interfacial as well as surface tension of water and oil (Sáenz-Marta et al., 2015).

Various biosurfactants have been tested for removing hydrocarbons products from contaminated water and soil. *Pseudomonas* sp., *Bacillus* sp., and *Candida* sp. are among the major biosurfactant producers used in soil remediation (Sachdev and Cameotra, 2013). Out of all the surfactants, rhamnolipid has gained great success in decontamination processes (Müller et al., 2012). In 2009 a study reported that the degradation of 75% of hexadecane by rhamnolipid was produced by *Pseudomonas* sp. (Cameotra and Singh, 2009). Another study reported that the 90% degradation of diesel was by surfactin (40 mg/L concentration) (Bustamante et al., 2012). Even, coculturing of both *P. aeruginosa* (biosurfactant-producing bacteria) and *P. putida* (hydrocarbon-degrading bacteria) have been reported for diesel degradation (Kumar et al., 2006). The BATH (bacterial adhesion to hydrocarbons) assay has enlightened the two important roles of biosurfactant synthesized by one microbe. One is hydrocarbons emulsification and the second is the alteration of biosurfactant molecules for easy attachment of hydrocarbons to cell surface aiding hydrocarbons degradation (Pradhan and Pradhan, 2015). Biosurfactant emulsan synthesized by *A. calcoaceticus* RAG-1 has also been reported to remediate 55%–90% of crude oil from soil (Foght et al., 1989). Naphthalene and phenanthrene are the major polycyclic aromatic hydrocarbons contaminating the soil (Ghosal et al., 2016). Recently, researchers have started exploring biosurfactant which can degrade Polycyclic aromatic hydrocarbons (PAH). About 91% of phenanthrene degradation has been recorded by *P. putida* ATCC 17484 on supplementing rhamnolipid (Zhang et al., 2010). Hence, biosurfactants hold the potential in remediating hydrocarbon representatives like ethanol, hexadecane, indole, naphthalene, phenanthrene, phenol, pyrene, and toluene. Moreover, it can serve as an ecofriendly approach for remediation.

16.8 Biosurfactants production through genetic modification

To increase the production and surface activity of surfactants, bacteria and other microbial populations were optimized using genetic modification methods along with their fermentation processes (Adrio and Demain, 2010). Only a few studies using genetic modification methods of wild *Bacillus strains* have been reported to date, and the obtained resultant is surfactin (Tsuge et al., 2001). To enhance its production, species of bacillus are engineered via promoter exchanges of the operon (srfA) or overexpression of exporter YerP. The expression of this operon is challenging due to complex biosynthetic regulation and attribution to its large genetic sequence (Hu et al., 2019). It is a lipopeptide composed of a circular hepta-peptide and fatty acid tail containing leucine, valine, aspartate, and glutamate. It is a powerful surfactant that has an extraordinary surface activity. It has broad application prospects in various fields such as biopesticides, oil recovery, pharmaceuticals, and cosmetics (Yang et al., 2015).

However, modification using genetic engineering methods all resulted in a few or single gene modifications, and the production of surfactin in the commercial market still has not been achieved (Sekhon et al., 2011). Therefore experimentations-based optimizations are still in progress to synthesize biosurfactin, and other regulatory characters still need to be explored.

16.9 Conclusion and future perspectives

Biosurfactants have enabled themselves as an effective approach for bioremediating both hydrocarbons and heavy metals from the contaminated soil. Still, we need to acquire more knowledge about biosurfactant structures, novelty of biosurfactant, metabolic route, cellular metabolism, cost, and scale-up of biosurfactant synthesis. If hydrocarbon or heavy metal remediation is to be done on-site, then biosurfactant synthesis can also be achieved on-site. However, less information is available regarding the large-scale production of biosurfactant from remediation purpose because of the high-cost of the production process. Nowadays, cheap substrates are used for the fermentation process to obtain high yield and product recovery. Further, optimization of the fermentation process is required at biotechnology in order to reduce the downstreaming and recovery cost of biosurfactant. Ultrafiltration can be additionally used to decrease the utilization amount of surfactant as it concentrates the recovered biosurfactant and makes it available for reuse. Moreover, foam technology of biosurfactant can be further used in in-situ pump and treatment method because it reduces the cost as well as the usage of biosurfactant and other chemicals. On using foam technology, continuous kinetics are studied to enhance its performance for heavy metal removal and recovery. Even, development of equilibrium and kinetic models will enable us to develop a prediction model to predict heavy metal removal and recovery efficiency. Extensive studies have been conducted that are using biosurfactants for removing hydrocarbons, single or multiple metals from contaminated soil. In the near future, focus might shift on developing a more effective system that can efficiently remove the heavy metal and mixed organic contaminants from polluted sites. As biosurfactants are also involved during biofilm formation, a hybrid approach can be developed in which metal-remediating microbes producing biofilm as well as biosurfactant will be cultured. Biofilm can also be used for entrapping hydrocarbons and adsorbing metals as it will reduce the cost of the downstream process. Therefore there is a need for development of innovative research techniques which focus on the usage of biosurfactant for hydrocarbon and heavy metal remediation from contaminated soils.

References

Abisado, R.G., Benomar, S., Klaus, J.R., Dandekar, A.A., Chandler, J.R., 2018. Bacterial quorum sensing and microbial community interactions. MBio 9 (3), e02331–17.

Adamczak, M., Bednarski, W., 2000. Properties and yield of synthesis of mannosylerythritol lipids by *Candida antarctica*, Progress in Biotechnology, Vol. 17. Elsevier Publications, pp. 229–234.

Adrio, J.L., Demain, A.L., 2010. Recombinant organisms for production of industrial products. Bioeng. Bugs 1 (2), 116–131.

Akbari, S., Abdurahman, N.H., Yunus, R.M., Fayaz, F., Alara, O.R., 2018. Biosurfactants—a new frontier for social and environmental safety: a mini review. Biotechnol. Res. Innov. 2 (1), 81–90.

Al-Araji, L., Rahman, R.N.Z.R.A., Basri, M., Salleh, A.B., 2007. Microbial surfactant. Asia Pac J Mol Biol Biotechnol 15 (3), 99–105.

Amaral, P.F., Coelho, M.A.Z., Marrucho, I.M., Coutinho, J.A., 2010. Biosurfactants from yeasts: characteristics, production and application. Biosurfactants. Springer, New York, pp. 236–249.

Andersen, K.K., Vad, B.S., Roelants, S., van Bogaert, I.N., Otzen, D.E., 2016. Weak and saturable protein–surfactant interactions in the denaturation of apo-α-lactalbumin by acidic and lactonic sophorolipid. Front. Microbiol. 7, 1711.

Arima, K., Kakinuma, A., Tamura, G., 1968. Surfactin, a crystalline peptidelipid surfactant produced by *Bacillus subtilis*: isolation, characterization and its inhibition of fibrin clot formation. Biochem. Biophys. Res. Commun. 31 (3), 488–494.

Ayangbenro, A., Babalola, O., 2017. A new strategy for heavy metal polluted environments: a review of microbial biosorbents. Int. J. Environ. Res. Public Health 14 (1), 94.

Bezerra, K.G.O., Rufino, R.D., Luna, J.M., Sarubbo, L.A., 2018. Saponins and microbial biosurfactants: potential raw materials for the formulation of cosmetics. Biotechnol. Progr. 34 (6), 1482–1493.

Bhati, S., Kumar, V., Singh, S., Singh, J., 2019. Synthesis, biological activities and docking studies of piperazine incorporated 1,3,4-oxadiazole derivatives. J. Mol. Struct. 1191, 197–205.

Biniarz, P., Łukaszewicz, M., Janek, T., 2017. Screening concepts, characterization and structural analysis of microbial-derived bioactive lipopeptides: a review. Crit. Rev. Biotechnol. 37 (3), 393–410.

Boll, J.M., Tucker, A.T., Klein, D.R., Beltran, A.M., Brodbelt, J.S., Davies, B.W., et al., 2015. Reinforcing lipid A acylation on the cell surface of *Acinetobacter baumannii* promotes cationic antimicrobial peptide resistance and desiccation survival. MBio 6 (3), e00478–15.

Bonnichsen, L., Svenningsen, N.B., Rybtke, M., de Bruijn, I., Raaijmakers, J.M., Tolker-Nielsen, T., et al., 2015. Lipopeptide biosurfactant viscosin enhances dispersal of *Pseudomonas fluorescens* SBW25 biofilms. Microbiology 161 (Pt 12), 2289.

Boothroyd, B., Thorn, J.A., Haskins, R.H., 1955. Biochemistry of the ustilaginales: X. the biosynthesis of ustilagic acid. Can. J. Biochem. Physiol. 33 (3), 289–296.

Burd, G., Ward, O.P., 1997. Energy-dependent accumulation of particulate biosurfactant by *Pseudomonas marginalis*. Can. J. Microbiol. 43 (4), 391–394.

Burlatsky, S.F., Atrazhev, V.V., Dmitriev, D.V., Sultanov, V.I., Timokhina, E.N., Ugolkova, E.A., et al., 2013. Surface tension model for surfactant solutions at the critical micelle concentration. J. Colloid Interface Sci. 393, 151–160.

Bustamante, M., Durán, N., Diez, M.C., 2012. Biosurfactants are useful tools for the bioremediation of contaminated soil: a review. J. Soil Sci. Plant Nutr. 12 (4), 667–687.

Cairns, W.L., Gray, N.C., Kosaric, N., 2017. Introduction: biotechnology and the surfactant Industry. Biosurfactants and Biotechnology. Routledge, pp. 1–19.

Cameotra, S.S., Singh, P., 2009. Synthesis of rhamnolipid biosurfactant and mode of hexadecane uptake by *Pseudomonas* species. Microb. Cell Fact. 8 (1), 16.

Cameron, D.R., Cooper, D.G., Neufeld, R.J., 1988. The mannoprotein of *Saccharomyces cerevisiae* is an effective bioemulsifier. Appl. Environ. Microbiol 54 (6), 1420–1425.

Campos, J.M., Montenegro Stamford, T.L., Sarubbo, L.A., de Luna, J.M., Rufino, R.D., Banat, I.M., 2013. Microbial biosurfactants as additives for food industries. Biotechnol. Progr. 29 (5), 1097–1108.

Chander, C.R., Lohitnath, T., Kumar, D.M., Kalaichelvan, P.T., 2012. Production and characterization of biosurfactant from *Bacillus subtilis* MTCC441 and its evaluation to use as bioemulsifier for food bio-preservative. Adv. Appl. Sci. Res. 3 (3), 1827–1831.

Chandran, P., Das, N., 2011. Characterization of sophorolipid biosurfactant produced by yeast species grown on diesel oil. Int. J. Sci. Nat. 2 (1), 63–71.

Choi, J.W., Choi, H.G., Lee, W.H., 1996. Effects of ethanol and phosphate on emulsan production by *Acinetobacter calcoaceticus* RAG-1. J. Biotechnol. 45 (3), 217–225.

Choi, S.K., Park, S.Y., Kim, R., Kim, S.B., Lee, C.H., Kim, J.F., et al., 2009. Identification of a polymyxin synthetase gene cluster of *Paenibacillus polymyxa* and heterologous expression of the gene in *Bacillus subtilis*. J. Bacteriol. 191 (10), 3350–3358.

Chrzanowski, Ł., Ławniczak, Ł., Czaczyk, K., 2012. Why do microorganisms produce rhamnolipids? World J. Microbiol. Biotechnol. 28 (2), 401–419.

Cirigliano, M.C., Carman, G.M., 1985. Purification and characterization of liposan, a bioemulsifier from *Candida lipolytica*. Appl. Environ. Microbiol 50 (4), 846–850.

Cooper, D.G., Zajic, J.E., Gracey, D.E., 1979. Analysis of corynomycolic acids and other fatty acids produced by *Corynebacterium lepus* grown on kerosene. J. Bacteriol. 137 (2), 795–801.

Coronel-León, J., Marqués, A.M., Bastida, J., Manresa, A., 2016. Optimizing the production of the biosurfactant lichenysin and its application in biofilm control. J. Appl. Microbiol. 120 (1), 99–111.

Costa, O.Y., Raaijmakers, J.M., Kuramae, E.E., 2018. Microbial extracellular polymeric substances: ecological function and impact on soil aggregation. Front. Microbiol 9.

Csutak, O., Corbu, V., Stoica, I., Ionescu, R., Vassu, T., 2015. Biotechnological applications of *Yarrowia lipolytica* CMGB32. Agric. Agric. Sci. Procedia 6, 545–553.

Dang, H., Lovell, C.R., 2016. Microbial surface colonization and biofilm development in marine environments. Microbiol. Mol. Biol. Rev. 80 (1), 91–138.

Das, N., Chandran, P., 2011. Microbial degradation of petroleum hydrocarbon contaminants: an overview. Biotechnol. Res. Int. 2011, 941810.

Datta, S., Singh, J., Singh, J., Singh, S., Singh, S., 2018. Assessment of genotoxic effects of pesticide and vermicompost treated soil with *Allium cepa* test. Sustainable Environ. Res. 28 (4), 171–178.

De Almeida, D.G., Soares Da Silva, R.D.C.F., Luna, J.M., Rufino, R.D., Santos, V.A., Banat, I.M., et al., 2016. Biosurfactants: promising molecules for petroleum biotechnology advances. Front. Microbiol. 7, 1718.

De Carvalho, C.C., Costa, S.S., Fernandes, P., Couto, I., Viveiros, M., 2014. Membrane transport systems and the biodegradation potential and pathogenicity of genus *Rhodococcus*. Front. Physiol. 5, 133.

de Oliveira, M.R., Camilios-Neto, D., Baldo, C., Magri, A., Celligoi, M.A.P.C., 2014. Biosynthesis and production of sophorolipids. Int. J. Sci. Technol. Res. 3 (11), 133–146.

de Oliveira, M.R., Magri, A., Baldo, C., Camilios-Neto, D., Minucelli, T., Celligoi, M.A.P.C., 2015. Sophorolipids a promising biosurfactant and it's applications. Int. J. Adv. Biotechnol. Res. 6, 161–174.

De Rienzo, M.A.D., Banat, I.M., Dolman, B., Winterburn, J., Martin, P.J., 2015. Sophorolipid biosurfactants: possible uses as antibacterial and antibiofilm agent. New Biotechnol. 32 (6), 720–726.

Decho, A.W., Gutierrez, T., 2017. Microbial extracellular polymeric substances (EPSs) in ocean systems. Front. Microbiol. 8, 922.

Demuyser, L., Jabra-Rizk, M.A., Van Dijck, P., 2014. Microbial cell surface proteins and secreted metabolites involved in multispecies biofilms. Pathog. Dis. 70 (3), 219–230.

Desai, J.D., Banat, I.M., 1997. Microbial production of surfactants and their commercial potential. Microbiol. Mol. Biol. Rev 61 (1), 47–64.

Donlan, R.M., 2002. Biofilms: microbial life on surfaces. Emerging Infect. Dis. 8 (9), 881.

Duvnjak, Z., Cooper, D.G., Kosaric, N., 1982. Production of surfactant by *Arthrobacter paraffineus* ATCC 19558. Biotechnol. Bioeng. 24 (1), 165–175.

Elazzazy, A.M., Abdelmoneim, T.S., Almaghrabi, O.A., 2015. Isolation and characterization of biosurfactant production under extreme environmental conditions by alkali-halo-thermophilic bacteria from Saudi Arabia. Saudi J. Biol. Sci. 22 (4), 466–475.

Elouzi, A.A., Akasha, A.A., Elgerbi, A.M., El Gammudi, B.A., 2012. Using of microbial biosurfactants (rhamnolipid) in heavy metals removal from contaminated water. J. Sebha Univ. Pure Appl. Sci. 11, 85–95.

Elshafie, A.E., Joshi, S.J., Al-Wahaibi, Y.M., Al-Bemani, A.S., Al-Bahry, S.N., Al-Maqbali, D., et al., 2015. Sophorolipids production by *Candida bombicola* ATCC 22214 and its potential application in microbial enhanced oil recovery. Front. Microbiol. 6, 1324.

El-Sheshtawy, H.S., Doheim, M.M., 2014. Selection of *Pseudomonas aeruginosa* for biosurfactant production and studies of its antimicrobial activity. Egypt. J. Pet. 23 (1), 1–6.

Felse, P.A., Shah, V., Chan, J., Rao, K.J., Gross, R.A., 2007. Sophorolipid biosynthesis by *Candida bombicola* from industrial fatty acid residues. Enzyme Microb. Technol. 40 (2), 316–323.

Foght, J.M., Gutnick, D.L., Westlake, D.W., 1989. Effect of emulsan on biodegradation of crude oil by pure and mixed bacterial cultures. Appl. Environ. Microbiol 55 (1), 36–42.

Franzetti, A., Gandolfi, I., Fracchia, L., Van Hamme, J., Gkorezis, P., Marchant, R., et al., 2014. Biosurfactant use in heavy metal removal from industrial effluents and contaminated sites. In: Biosurfactants: Production and Utilization—Processes, Technologies, and Economics, CRC Press, vol. 159, p. 361.

Frautz, B., Lang, S., Wagner, F., 1986. Formation of cellobiose lipids by growing and resting cells of *Ustilago maydis*. Biotechnol. Lett. 8 (11), 757–762.

Freitas, B.G., Brito, J.M., Brasileiro, P.P., Rufino, R.D., Luna, J.M., Santos, V.A., et al., 2016. Formulation of a commercial biosurfactant for application as a dispersant of petroleum and by-products spilled in oceans. Front. Microbiol. 7, 1646.

Funston, S.J., Tsaousi, K., Rudden, M., Smyth, T.J., Stevenson, P.S., Marchant, R., et al., 2016. Characterising rhamnolipid production in *Burkholderia thailandensis* E264, a non-pathogenic producer. Appl. Microbiol. Biotechnol. 100 (18), 7945–7956.

Gandhimathi, R., Kiran, G.S., Hema, T.A., Selvin, J., Raviji, T.R., Shanmughapriya, S., 2009. Production and characterization of lipopeptide biosurfactant by a sponge-associated marine actinomycetes *Nocardiopsis alba* MSA10. Bioprocess Biosyst. Eng. 32 (6), 825–835.

Garnett, J.A., Matthews, S., 2012. Interactions in bacterial biofilm development: a structural perspective. Curr. Protein Pept. Sci. 13, 739–755.

Gautam, K.K., Tyagi, V.K., 2006. Microbial surfactants: a review. J. Oleo Sci. 55 (4), 155–166.

Geys, R., Soetaert, W., Van Bogaert, I., 2014. Biotechnological opportunities in biosurfactant production. Curr. Opin. Biotechnol. 30, 66–72.

Gharaei-Fathabad, E., 2011. Biosurfactants in pharmaceutical industry: a mini-review. Am. J. Drug Discovery Dev. 1 (1), 58–69.

Ghosal, D., Ghosh, S., Dutta, T.K., Ahn, Y., 2016. Current state of knowledge in microbial degradation of polycyclic aromatic hydrocarbons (PAHs): a review. Front. Microbiol. 7, 1369.

Ghribi, D., Ellouze-Chaabouni, S., 2011. Enhancement of *Bacillus subtilis* lipopeptide biosurfactants production through optimization of medium composition and adequate control of aeration. Biotechnol. Res. Int. 2011, 653654.

Gilboa-Garber, N., Katcoff, D.J., Garber, N.C., 2000. Identification and characterization of *Pseudomonas aeruginosa* PA-IIL lectin gene and protein compared to PA-IL. FEMS Immunol. Med. Microbiol. 29 (1), 53–57.

Göbbert, U., Lang, S., Wagner, F., 1984. Sophorose lipid formation by resting cells of *Torulopsis bombicola*. Biotechnol. Lett. 6 (4), 225–230.

Gudiña, E.J., Fernandes, E.C., Rodrigues, A.I., Teixeira, J.A., Rodrigues, L.R., 2015. Biosurfactant production by *Bacillus subtilis* using corn steep liquor as culture medium. Front. Microbiol. 6, 59.

Gudiña, E., Teixeira, J., Rodrigues, L., 2016. Biosurfactants produced by marine microorganisms with therapeutic applications. Mar. Drugs 14 (2), 38.

Hall, C.W., Mah, T.F., 2017. Molecular mechanisms of biofilm-based antibiotic resistance and tolerance in pathogenic bacteria. FEMS Microbiol. Rev. 41 (3), 276–301.

Hashim, M.A., Mukhopadhyay, S., Sahu, J.N., Sengupta, B., 2011. Remediation technologies for heavy metal contaminated groundwater. J. Environ. Manage. 92 (10), 2355–2388.

Hommel, R., Stiiwer, O., Stuber, W., Haferburg, D., Kleber, H.P., 1987. Production of water-soluble surface-active exolipids by *Torulopsis apicola*. Appl. Microbiol. Biotechnol. 26 (3), 199–205.

Hörmann, B., Müller, M.M., Syldatk, C., Hausmann, R., 2010. Rhamnolipid production by *Burkholderia plantarii* DSM 9509T. Eur. J. Lipid Sci. Technol. 112 (6), 674–680.

Hu, F., Liu, Y., Li, S., 2019. Rational strain improvement for surfactin production: enhancing the yield and generating novel structures. Microb. Cell Fact. 18 (1), 42.

Irorere, V.U., Tripathi, L., Marchant, R., McClean, S., Banat, I.M., 2017. Microbial rhamnolipid production: a critical re-evaluation of published data and suggested future publication criteria. Appl. Microbiol. Biotechnol. 101 (10), 3941–3951.

Jarvis, F.G., Johnson, M.J., 1949. A glyco-lipide produced by *Pseudomonas aeruginosa*. J. Am. Chem. Soc. 71 (12), 4124–4126.

Jayathilake, P.G., Jana, S., Rushton, S., Swailes, D., Bridgens, B., Curtis, T., et al., 2017. Extracellular polymeric substance production and aggregated bacteria colonization influence the competition of microbes in biofilms. Front. Microbiol. 8, 1865.

Jemil, N., Ayed, H.B., Manresa, A., Nasri, M., Hmidet, N., 2017. Antioxidant properties, antimicrobial and anti-adhesive activities of DCS1 lipopeptides from *Bacillus methylotrophicus* DCS1. BMC Microbiol. 17 (1), 144.

Joy, S., Rahman, P.K., Sharma, S., 2017. Biosurfactant production and concomitant hydrocarbon degradation potentials of bacteria isolated from extreme and hydrocarbon contaminated environments. Chem. Eng. J. 317, 232–241.

Kapoor, D., Singh, S., Kumar, V., Romero, R., Prasad, R., Singh, J., 2019. Antioxidant enzymes regulation in plants in reference to reactive oxygen species (ROS) and reactive nitrogen species (RNS). Plant Gene 19, 100182.

Käppeli, O., Finnerty, W.R., 1979. Partition of alkane by an extracellular vesicle derived from hexadecane-grown *Acinetobacter*. J. Bacteriol. 140 (2), 707–712.

Karanth, N.G.K., Deo, P.G., Veenanadig, N.K., 1999. Microbial production of biosurfactants and their importance. Curr. Sci. 77 (1), 116–126.

Karigar, C.S., Rao, S.S., 2011. Role of microbial enzymes in the bioremediation of pollutants: a review. Enzyme Res. 2011 (7), 805187.

Karlapudi, A.P., Venkateswarulu, T.C., Tammineedi, J., Kanumuri, L., Ravuru, B.K., Ramu Dirisala, V., et al., 2018. Role of biosurfactants in bioremediation of oil pollution-a review. Petroleum 4 (3), 241–249.

Kaur, P., Singh, S., Kumar, V., Singh, N., Singh, J., 2018. Effect of rhizobacteria on arsenic uptake by macrophyte *Eichhornia crassipes* (Mart.) Solms. Int. J. Phytorem. 20 (2), 114–120.

Kearns, D.B., Bonner, P.J., Smith, D.R., Shimkets, L.J., 2002. An extracellular matrix-associated zinc metalloprotease is required for dilauroyl phosphatidylethanolamine chemotactic excitation in *Myxococcus xanthus*. J. Bacteriol. 184 (6), 1678–1684.

Konishi, M., Fujita, M., Ishibane, Y., Shimizu, Y., Tsukiyama, Y., Ishida, M., 2016. Isolation of yeast candidates for efficient sophorolipids production: their production potentials associate to their lineage. Biosci. Biotechnol. Biochem. 80 (10), 2058–2064.

Konz, D., Doekel, S., Marahiel, M.A., 1999. Molecular and biochemical characterization of the protein template controlling biosynthesis of the lipopeptide lichenysin. J. Bacteriol. 181 (1), 133–140.

Kosaric, N., Sukan, F.V., 2010. Biosurfactants: Production: Properties: Applications. CRC Press.

Kumar, V., Singh, S., 2018a. Kinetics of dechlorination of atrazine using tin (SnII) at neutral pH conditions. Appl. Chem. Eng. 1 (4).

Kumar, V., Singh, S., 2018b. Interactions of acephate, glyphosate, monocrotophos and phorate with bovine serum albumin. Indian J. Pharm. Sci. 80 (6), 1151. - + .

Kumar, M., Leon, V., De Sisto Materano, A., Ilzins, O.A., 2006. Enhancement of oil degradation by co-culture of hydrocarbon degrading and biosurfactant producing bacteria. Pol. J. Microbiol. 55 (2), 139–146.

Kumar, V., Upadhyay, N., Kumar, V., Kaur, S., Singh, J., Singh, S., et al., 2014. Environmental exposure and health risks of the insecticide monocrotophos—a review. J. Biodivers. Environ. Sci. 5, 111–120.

Kumar, V., Singh, S., Singh, J., Upadhyay, N., 2015. Potential of plant growth promoting traits by bacteria isolated from heavy metal contaminated soils. Bull. Environ. Contam. Toxicol. 94 (6), 807–814.

Kumar, V., Kaur, S., Singh, S., Upadhyay, N., 2016. Unexpected formation of *N*′-phenyl-thiophosphorohydrazidic acid *O,S*-dimethyl ester from acephate: chemical, biotechnical and computational study. 3 Biotech. 6 (1), 1.

Kumar, V., Singh, S., Singh, R., Upadhyay, N., Singh, J., 2017. Design, synthesis, and characterization of 2,2-bis (2,4-dinitrophenyl)-2-(phosphonatomethylamino) acetate as a herbicidal and biological active agent. J. Chem. Biol. 10 (4), 179–190.

Kumar, V., Singh, S., Singh, A., Dixit, A.K., Shrivastava, B., Kondalkar, S.A., et al., 2018a. Determination of phytochemical, antioxidant, antimicrobial, and protein binding qualities of hydroethanolic extract of *Celastrus paniculatus*. J. Appl. Biol. Biotechnol. 6 (06), 11–17.

Kumar, V., Singh, S., Singh, A., Dixit, A.K., Srivastava, B., Sidhu, G.K., et al., 2018b. Phytochemical, antioxidant, antimicrobial, and protein binding qualities of hydroethanolic extract of *Tinospora cordifolia*. J. Biol. Act. Prod. Nat. 8 (3), 192–200.

Kumar, V., Singh, S., Singh, R., Upadhyay, N., Singh, J., Pant, P., et al., 2018c. Spectral, structural and energetic study of acephate, glyphosate, monocrotophos and phorate: an experimental and computational approach. J. Taibah Univ. Sci. 12 (1), 69–78.

Kumar, V., Singh, S., Singh, R., 2019a. Phytochemical constituents of guggul gum and their biological qualities. Mini-Rev. Org. Chem. 16, . Available from: https://doi.org/10.2174/1570193X16666190129161757.

Kumar, V., Singh, S., Singh, A., Subhose, V., Prakash, O., 2019b. Assessment of heavy metal ions, essential metal ions, and antioxidant properties of the most common herbal drugs in Indian Ayurvedic hospital: for ensuring quality assurance of certain Ayurvedic drugs. Biocatal. Agric. Biotechnol. 18, 101018.

Kumar, V., Singh, S., Srivastava, B., Bhadouria, R., Singh, R., 2019c. Green synthesis of silver nanoparticles using leaf extract of *Holoptelea integrifolia* and preliminary investigation of its antioxidant, anti-inflammatory, antidiabetic and antibacterial activities. J. Environ. Chem. Eng. 7, 103094.

Kurane, R., Hatamochi, K., Kakuno, T., Kiyohara, M., Tajima, T., Hirano, M., et al., 1995. Chemical structure of lipid bioflocculant produced by *Rhodococcus erythropolis*. Biosci. Biotechnol. Biochem. 59 (9), 1652–1656.

Lamichhane, S., Krishna, K.B., Sarukkalige, R., 2017. Surfactant-enhanced remediation of polycyclic aromatic hydrocarbons: a review. J. Environ. Manage. 199, 46–61.

Li, Y.H., Tian, X., 2012. Quorum sensing and bacterial social interactions in biofilms. Sensors 12 (3), 2519–2538.

Li, Z.Y., Lang, S., Wagner, F., Witte, L., Wray, V., 1984. Formation and identification of interfacial-active glycolipids from resting microbial cells. Appl. Environ. Microbiol 48 (3), 610–617.

Liu, G., Zhong, H., Yang, X., Liu, Y., Shao, B., Liu, Z., 2018. Advances in applications of rhamnolipids biosurfactant in environmental remediation: a review. Biotechnol. Bioeng. 115 (4), 796–814.

Loiseau, C., Portier, E., Corre, M.H., Schlusselhuber, M., Depayras, S., Berjeaud, J.M., et al., 2018. Highlighting the potency of biosurfactants produced by *Pseudomonas* strains as anti-*Legionella* agents. BioMed Res. Int. 2018, 8194368.

Lombardo, D., Kiselev, M.A., Magazù, S., Calandra, P., 2015. Amphiphiles self-assembly: basic concepts and future perspectives of supramolecular approaches. Adv. Condens. Matter Phys. 2015.

Lukondeh, T., Ashbolt, N.J., Rogers, P.L., 2003. Evaluation of *Kluyveromyces marxianus* FII 510700 grown on a lactose-based medium as a source of a natural bioemulsifier. J. Ind. Microbiol. Biotechnol. 30 (12), 715–720.

Lynd, L.R., Weimer, P.J., Van Zyl, W.H., Pretorius, I.S., 2002. Microbial cellulose utilization: fundamentals and biotechnology. Microbiol. Mol. Biol. Rev. 66 (3), 506–577.

Mahanti, P., Kumar, S., Patra, J.K., 2017. Biosurfactants: an agent to keep environment clean. Microbial Biotechnology. Springer, Singapore, pp. 413–428.

Maier, R.M., Soberon-Chavez, G., 2000. *Pseudomonas aeruginosa* rhamnolipids: biosynthesis and potential applications. Appl. Microbiol. Biotechnol. 54 (5), 625–633.

Maillard, J.Y., 2005. Antimicrobial biocides in the healthcare environment: efficacy, usage, policies, and perceived problems. Ther. Clin. Risk Manage. 1 (4), 307.

Makkar, R.S., Rockne, K.J., 2003. Comparison of synthetic surfactants and biosurfactants in enhancing biodegradation of polycyclic aromatic hydrocarbons. Environ. Toxicol. Chem. 22 (10), 2280–2292.

Makombe, N., Gwisai, R.D., 2018. Soil remediation practices for hydrocarbon and heavy metal reclamation in mining polluted soils. Sci. World J. 2018, 7.

Mao, X., Jiang, R., Xiao, W., Yu, J., 2015. Use of surfactants for the remediation of contaminated soils: a review. J. Hazard. Mater. 285, 419–435.

Marahiel, M.A., Danders, W., Krause, M., Kleinkauf, H., 1979. Biological role of gramicidin S in spore functions. Studies on gramicidin-*S*-negative mutants of *Bacillus brevis* ATCC9999. Eur. J. Biochem. 99 (1), 49.

Marchant, R., Banat, I.M., 2012. Biosurfactants: a sustainable replacement for chemical surfactants? Biotechnol. Lett. 34 (9), 1597–1605.

Marsh, P.D., Zaura, E., 2017. Dental biofilm: ecological interactions in health and disease. J. Clin. Periodontol. 44, S12–S22.

Matsuyama, T., Tanikawa, T., Nakagawa, Y., 2011. Serrawettins and other surfactants produced by Serratia. Biosurfactants. Springer, Berlin, Heidelberg, pp. 93–120.

Miller, R.M., 1995. Biosurfactant-facilitated remediation of metal-contaminated soils. Environ. Health Perspect. 103 (Suppl. 1), 59–62.

Mishra, V., Gupta, A., Kaur, P., Singh, S., Singh, N., Gehlot, P., et al., 2016. Synergistic effects of *Arbuscular mycorrhizal* fungi and plant growth promoting rhizobacteria in bioremediation of iron contaminated soils. Int. J. Phytorem. 18 (7), 697–703.

Mnif, I., Ghribi, D., 2016. Glycolipid biosurfactants: main properties and potential applications in agriculture and food industry. J. Sci. Food Agric. 96 (13), 4310–4320.

Mouafo, T.H., Mbawala, A., Ndjouenkeu, R., 2018. Effect of different carbon sources on biosurfactants' production by three strains of *Lactobacillus* spp. BioMed Res. Int. 2018, 15.

Mukherjee, S., Das, P., Sen, R., 2006. Towards commercial production of microbial surfactants. Trends Biotechnol. 24 (11), 509–515.

Müller, M.M., Kügler, J.H., Henkel, M., Gerlitzki, M., Hörmann, B., Pöhnlein, M., et al., 2012. Rhamnolipids—next generation surfactants? J. Biotechnol. 162 (4), 366–380.

Mulligan, C.N., 2005. Environmental applications for biosurfactants. Environ. Pollut. 133 (2), 183–198.

Murphy, H.N., Stewart, G.R., Mischenko, V.V., Apt, A.S., Harris, R., McAlister, M.S., et al., 2005. The OtsAB pathway is essential for trehalose biosynthesis in *Mycobacterium* tuberculosis. J. Biol. Chem. 280 (15), 14524–14529.

Muthusamy, K., Gopalakrishnan, S., Ravi, T.K., Sivachidambaram, P., 2008. Biosurfactants: properties, commercial production and application. Curr. Sci. 94 (6), 736–747.

Nadell, C.D., Drescher, K., Wingreen, N.S., Bassler, B.L., 2015. Extracellular matrix structure governs invasion resistance in bacterial biofilms. ISME J. 9 (8), 1700.

Nalini, S., Parthasarathi, R., 2014. Production and characterization of rhamnolipids produced by *Serratia rubidaea* SNAU02 under solid-state fermentation and its application as biocontrol agent. Bioresour. Technol. 173, 231–238.

Naughton, P.J., Marchant, R., Naughton, V., Banat, I.M., 2019. Microbial biosurfactants: current trends and applications in agricultural and biomedical industries. J. Appl. Microbiol. 127 (1), 12–28.

Okoliegbe, I.N., Agarry, O.O., 2012. Application of microbial surfactant (a review). Sch. J. Biotechnol. 1 (1), 15–23.

Olaniran, A., Balgobind, A., Pillay, B., 2013. Bioavailability of heavy metals in soil: impact on microbial biodegradation of organic compounds and possible improvement strategies. Int. J. Mol. Sci. 14 (5), 10197–10228.

Otzen, D.E., 2017. Biosurfactants and surfactants interacting with membranes and proteins: same but different? Biochim. Biophys. Acta 1859 (4), 639–649.

Pacwa-Płociniczak, M., Płaza, G.A., Piotrowska-Seget, Z., Cameotra, S.S., 2011. Environmental applications of biosurfactants: recent advances. Int. J. Mol. Sci. 12 (1), 633–654.

Panjiar, N., Sachan, S.G., Sachan, A., 2017. Biosurfactants: a multifunctional microbial metabolite, Microbial Applications, Vol. 2. Springer, Cham, pp. 213–229.

Park, A.J., Cha, D.K., Holsen, T.M., 1998. Enhancing solubilization of sparingly soluble organic compounds by biosurfactants produced by *Nocardia erythropolis*. Water Environ. Res. 70 (3), 351–355.

Patowary, K., Patowary, R., Kalita, M.C., Deka, S., 2017. Characterization of biosurfactant produced during degradation of hydrocarbons using crude oil as sole source of carbon. Front. Microbiol. 8, 279.

Pichtel, J., 2016. Oil and gas production wastewater: soil contamination and pollution prevention. Appl. Environ. Soil Sci. 2016.

Płaza, G., Chojniak, J., Banat, I., 2014. Biosurfactant mediated biosynthesis of selected metallic nanoparticles. Int. J. Mol. Sci. 15 (8), 13720–13737.

Prabhakar, S., et al., 2017. A fully enzymatic esterification/transesterification sequence for the preparation of symmetrical and unsymmetrical trehalose diacyl conjugates. Green Chem. 19 (4), 1–21.

Pradhan, A.K., Pradhan, N., 2015. Microbial biosurfactant for hydrocarbons and heavy metals bioremediation. Environmental Microbial Biotechnology. Springer, Cham, pp. 91–104.

Rikalović, M.G., Vrvić, M.M., Karadžić, I.M., 2015. Rhamnolipid biosurfactant from *Pseudomonas aeruginosa*—from discovery to application in contemporary technology. J. Serb. Chem. Soc. 80 (3), 279.

Rufino, R.D., Luna, J.M., Campos-Takaki, G.M., Ferreira, S.R., Sarubbo, L.A., 2012. Application of the biosurfactant produced by *Candida lipolytica* in the remediation of heavy metals. Chem. Eng. 27, 61–66.

Rufino, R.D., de Luna, J.M., de Campos Takaki, G.M., Sarubbo, L.A., 2014. Characterization and properties of the biosurfactant produced by *Candida lipolytica* UCP 0988. Electron. J. Biotechnol. 17, 34–38.

Saadoun, I.M., 2015. Impact of oil spills on marine life. Emerging Pollutants in the Environment-Current and Further Implications. IntechOpen.

Sachdev, D.P., Cameotra, S.S., 2013. Biosurfactants in agriculture. Appl. Microbiol. Biotechnol. 97 (3), 1005–1016.

Sadekuzzaman, M., Yang, S., Mizan, M.F.R., Ha, S.D., 2015. Current and recent advanced strategies for combating biofilms. Compr. Rev. Food Sci. Food Saf. 14 (4), 491–509.

Sáenz-Marta, C.I., de Lourdes Ballinas-Casarrubias, M., Rivera-Chavira, B.E., Nevárez-Moorillón, G.V., 2015. Biosurfactants as useful tools in bioremediation. Advances in Bioremediation of Wastewater and Polluted Soil. Intech Open.

Saharan, B.S., Sahu, R.K., Sharma, D., 2011. A review on biosurfactants: fermentation, current developments and perspectives. Genet. Eng. Biotechnol. J. 2011 (1), 1–14.

Saichek, R.E., Reddy, K.R., 2005. Electrokinetically enhanced remediation of hydrophobic organic compounds in soils: a review. Crit. Rev. Environ. Sci. Technol. 35 (2), 115–192.

Sandrin, T.R., Chech, A.M., Maier, R.M., 2000. A rhamnolipid biosurfactant reduces cadmium toxicity during naphthalene biodegradation. Appl. Environ. Microbiol. 66 (10), 4585–4588.

Santos, D., Rufino, R., Luna, J., Santos, V., Sarubbo, L., 2016. Biosurfactants: multifunctional biomolecules of the 21st century. Int. J. Mol. Sci. 17 (3), 401.

Santos, A.P.P., Silva, M.D.S., Costa, E.V.L., Rufino, R.D., Santos, V.A., Ramos, C.S., et al., 2018. Production and characterization of a biosurfactant produced by *Streptomyces* sp. DPUA 1559 isolated from lichens of the Amazon region. Braz. J. Med. Biol. Res. 51 (2), e6657.

Sarubbo, L.A., Rocha Jr, R.B., Luna, J.M., Rufino, R.D., Santos, V.A., Banat, I.M., 2015. Some aspects of heavy metals contamination remediation and role of biosurfactants. Chem. Ecol. 31 (8), 707–723.

Satpute, S.K., Zinjarde, S.S., & Banat, I.M. (2018). Recent updates on biosurfactants in the food industry. Microb. Cell Fact., 1-20, doi:10.1201/b22219-1.

Satpute, S.K., Banat, I.M., Dhakephalkar, P.K., Banpurkar, A.G., Chopade, B.A., 2010. Biosurfactants, bioemulsifiers and exopolysaccharides from marine microorganisms. Biotechnol. Adv. 28 (4), 436–450.

Sekhon, K.K., Khanna, S., Cameotra, S.S., 2011. Enhanced biosurfactant production through cloning of three genes and role of esterase in biosurfactant release. Microb. Cell Fact. 10 (1), 49.

Sharma, D., Saharan, B.S., Kapil, S., 2016. Structural properties of biosurfactants of lab. Biosurfactants of Lactic Acid Bacteria. Springer, Cham, pp. 47–60.

Shekhar, S., Sundaramanickam, A., Balasubramanian, T., 2015. Biosurfactant producing microbes and their potential applications: a review. Crit. Rev. Environ. Sci. Technol. 45 (14), 1522–1554.

Shoeb, E., Akhlaq, F., Badar, U., Akhter, J., Imtiaz, S., 2013. Classification and industrial applications of biosurfactants. Acad. Res. Int. 4 (3), 243.

Sidhu, G.K., Singh, S., Kumar, V., Dhanjal, D.S., Datta, S., Singh, J., 2019. Toxicity, monitoring and biodegradation of organophosphate pesticides: a review. Crit. Rev. Environ. Sci. Technol. 49, 1–53.

Silva, R., Almeida, D., Rufino, R., Luna, J., Santos, V., Sarubbo, L., 2014. Applications of biosurfactants in the petroleum industry and the remediation of oil spills. Int. J. Mol. Sci. 15 (7), 12523–12542.

Silva, A.C.S.D., Santos, P.N.D., Silva, T.A.L., Andrade, R.F.S., Campos-Takaki, G.M., 2018. Biosurfactant production by fungi as a sustainable alternative. Arq. Inst. Biol. 85.

Singh, A.K., Cameotra, S.S., 2013. Efficiency of lipopeptide biosurfactants in removal of petroleum hydrocarbons and heavy metals from contaminated soil. Environ. Sci. Pollut. Res. 20 (10), 7367–7376.

Singh, S., Singh, N., Kumar, V., Datta, S., Wani, A.B., Singh, D., et al., 2016. Toxicity, monitoring and biodegradation of the fungicide carbendazim. Environ. Chem. Lett. 14 (3), 317–329.

Singh, S., Kumar, V., Upadhyay, N., Singh, J., Singla, S., Datta, S., 2017a. Efficient biodegradation of acephate by *Pseudomonas pseudoalcaligenes* PS-5 in the presence and absence of heavy metal ions [Cu(II) and Fe(III)], and humic acid. 3 Biotech. 7 (4), 262.

Singh, S., Singh, S.K., Chowdhury, I., Singh, R., 2017b. Understanding the mechanism of bacterial biofilms resistance to antimicrobial agents. Open Microbiol. J. 11, 53.

Singh, S., Kumar, V., Chauhan, A., Datta, S., Wani, A.B., Singh, N., et al., 2018. Toxicity, degradation and analysis of the herbicide atrazine. Environ. Chem. Lett. 16 (1), 211–237.

Singh, S., Kumar, V., Sidhu, G.K., Singh, J., 2019a. Kinetic study of the biodegradation of glyphosate by indigenous soil bacterial isolates in presence of humic acid, Fe(III) and Cu(II) ions. J. Environ. Chem. Eng. 7, 103098.

Singh, S., Kumar, V., Sidhu, G.K., Datta, S., Dhanjal, D.S., Koul, B., et al., 2019b. Plant growth promoting rhizobacteria from heavy metal contaminated soil promote growth attributes of *Pisum sativum* L. Biocatal. Agric. Biotechnol. 17, 665–671.

Sobrinho, H.B., Luna, J.M., Rufino, R.D., Porto, A.L.F., Sarubbo, L.A., 2013. Biosurfactants: classification, properties and environmental applications. Recent Dev. Biotechnol. 11, 1–29.

Sutyak, K.E., Wirawan, R.E., Aroutcheva, A.A., Chikindas, M.L., 2008. Isolation of the *Bacillus subtilis* antimicrobial peptide subtilosin from the dairy product-derived *Bacillus amyloliquefaciens*. J. Appl. Microbiol. 104 (4), 1067–1074.

Suzuki, S., Ishigami, Y., Tsuji, Y., 1988. Synthesis and properties of rhodamine-type dyes derived from spicrispolic acid. J. Soc. Color Mater. 61 (6), 327–330.

Thakur, P., Chopra, C., Anand, P., Dhanjal, D.S., Chopra, R.S., 2018. Myxobacteria: unraveling the potential of a unique microbiome niche. Microbial Bioprospecting for Sustainable Development. Springer, Singapore, pp. 137–163.

Thaniyavarn, J., Chianguthai, T., Sangvanich, P., Roongsawang, N., Washio, K., Morikawa, M., et al., 2008. Production of sophorolipid biosurfactant by *Pichia anomala*. Biosci. Biotechnol. Biochem. 72 (8), 2061–2068.

Tian, W., Yao, J., Liu, R., Zhu, M., Wang, F., Wu, X., et al., 2016. Effect of natural and synthetic surfactants on crude oil biodegradation by indigenous strains. Ecotoxicol. Environ. Saf. 129, 171–179.

Tischler, D., Niescher, S., Kaschabek, S.R., Schlömann, M., 2013. Trehalose phosphate synthases OtsA1 and OtsA2 of *Rhodococcus opacus* 1CP. FEMS Microbiol. Lett. 342 (2), 113–122.

Toren, A., Navon-Venezia, S., Ron, E.Z., Rosenberg, E., 2001. Emulsifying activities of purified alasan proteins from *Acinetobacter radioresistens* KA53. Appl. Environ. Microbiol. 67 (3), 1102–1106.

Tsuge, K., Ohata, Y., Shoda, M., 2001. Gene yerP, involved in surfactin self-resistance in *Bacillus subtilis*. Antimicrob. Agents Chemother. 45 (12), 3566–3573.

Tuleva, B., Christova, N., Cohen, R., Stoev, G., Stoineva, I., 2008. Production and structural elucidation of trehalose tetraesters (biosurfactants) from a novel alkanothrophic *Rhodococcus wratislaviensis* strain. J. Appl. Microbiol. 104 (6), 1703–1710.

Upadhyaya, A., Acosta, E.J., Scamehorn, J.F., Sabatini, D.A., 2007. Adsorption of anionic–cationic surfactant mixtures on metal oxide surfaces. J. Surfactants Deterg. 10 (4), 269–277.

Usman, M.M., Dadrasnia, A., Lim, K.T., Mahmud, A.F., Ismail, S., 2016. Application of biosurfactants in environmental biotechnology; remediation of oil and heavy metal. AIMS Bioeng. 3 (3), 289–304.

Uzoigwe, C., Burgess, J.G., Ennis, C.J., Rahman, P.K., 2015. Bioemulsifiers are not biosurfactants and require different screening approaches. Front. Microbiol. 6, 245.

Vecino, X., Cruz, J.M., Moldes, A.B., Rodrigues, L.R., 2017. Biosurfactants in cosmetic formulations: trends and challenges. Crit. Rev. Biotechnol. 37 (7), 911–923.

Vigneshwaran, C., Sivasubramanian, V., Vasantharaj, K., Krishnanand, N., Jerold, M., 2018. Potential of *Brevibacillus* sp. AVN 13 isolated from crude oil contaminated soil for biosurfactant production and its optimization studies. J. Environ. Chem. Eng. 6 (4), 4347–4356.

Vijayakumar, S., Saravanan, V., 2015. Biosurfactants-types, sources and applications. Res. J. Microbiol. 10 (5), 181–192.

Wang, C., Mahrous, E.A., Lee, R.E., Vestling, M.M., Takayama, K., 2011. Novel polyoxyethylene-containing glycolipids are synthesized in *Corynebacterium matruchotii* and *Mycobacterium smegmatis* cultured in the presence of Tween 80. J. Lipids 2011.

Wani, A.B., Chadar, H., Wani, A.H., Singh, S., Upadhyay, N., 2017. Salicylic acid to decrease plant stress. Environ. Chem. Lett. 15 (1), 101–123.

Wittgens, A., Tiso, T., Arndt, T.T., Wenk, P., Hemmerich, J., Müller, C., et al., 2011. Growth independent rhamnolipid production from glucose using the non-pathogenic *Pseudomonas putida* KT2440. Microb. Cell Fact. 10 (1), 80.

Wittgens, A., Santiago-Schuebel, B., Henkel, M., Tiso, T., Blank, L.M., Hausmann, R., et al., 2018. Heterologous production of long-chain rhamnolipids from *Burkholderia glumae* in *Pseudomonas putida*—a step forward to tailor-made rhamnolipids. Appl. Microbiol. Biotechnol. 102 (3), 1229–1239.

Wuana, R.A., Okieimen, F.E., 2011. Heavy metals in contaminated soils: a review of sources, chemistry, risks and best available strategies for remediation. ISRN Ecol. 2011.

Xu, X., Liu, W., Tian, S., Wang, W., Qi, Q., Jiang, P., et al., 2018. Petroleum hydrocarbon-degrading bacteria for the remediation of oil pollution under aerobic conditions: a perspective analysis. Front. Microbiol. 9, 2885.

Yang, H., Li, X., Li, X., Yu, H., Shen, Z., 2015. Identification of lipopeptide isoforms by MALDI-TOF-MS/MS based on the simultaneous purification of iturin, fengycin, and surfactin by RP-HPLC. Anal. Bioanal. Chem. 407 (9), 2529–2542.

Zhang, J., Yin, R., Lin, X., Liu, W., Chen, R., Li, X., 2010. Interactive effect of biosurfactant and microorganism to enhance phytoremediation for removal of aged polycyclic aromatic hydrocarbons from contaminated soils. J. Health Sci. 56 (3), 257–266.

Zhang, J., Saerens, K.M., Van Bogaert, I.N., Soetaert, W., 2011. Vegetable oil enhances sophorolipid production by *Rhodotorula bogoriensis*. Biotechnol. Lett. 33 (12), 2417–2423.

Zhao, F., Shi, R., Ma, F., Han, S., Zhang, Y., 2018. Oxygen effects on rhamnolipids production by *Pseudomonas aeruginosa*. Microb. Cell Fact. 17 (1), 39.

Zosim, Z., Gutnick, D., Rosenberg, E., 1983. Uranium binding by emulsan and emulsanosols. Biotechnol. Bioeng. 25 (7), 1725–1735.

Engineered bacteria for bioremediation

17

Gaurav Sanghvi[1,*], *Arti Thanki*[2], *Siddhartha Pandey*[3] *and Nitin Kumar Singh*[2]
[1]Department of Microbiology, Marwadi University, Rajkot, India, [2]Department of Environmental Science and Engineering, Marwadi University, Rajkot, India, [3]Department of Civil Engineering, Chalapathi Institute of Technology, Guntur, India
*Corresponding author

17.1 Introduction

Pollution, especially in the environment, is a severe worldwide problem. With people shifting to urban areas and committing to high life standards and luxury, there is risk associated with human health. Since the dawn of civilization, the use, overuse, and misuse of natural sources and diversity has led to depletion/lessening of various natural resources to an extent that currently most of our natural wealth is either extinct or on the verge of extinction. Pollution at all levels has adversely affected many lives leading to health risks and deaths. An unstoppable growth in terms of infrastructure, industrial development, and modernization has significantly increased the pollution to a dangerous level. At present, both developing and developed countries are facing huge problems in many facets of environmental pollution (e.g., untreated sewage water, industrial waste, solid waste, and portable water contaminated with nondegradable/xenobiotic compounds). The major contributors of pollution are oil spills, fertilizers, garbage, sewage disposals, and toxic chemicals. Pollutants have contaminated major sources viz soil, air, and water (Adriano, 2003). In an effort to improve crop yield and industrial product quality, many synthetic compounds have been produced with the desired quality but had detrimental effects in terms of generating toxic compounds. Release of these toxic contaminants led to scarcity of clean water and loss of soil fertility (Zhao and Kaluarachchi, 2002). The impact of the polluted sites and associated risks compelled to plan for huge bulk waste management and develop technology to remediate the contaminated sites. The problem of pollution is now spreading rapidly and contaminating many ecological niches.

An early method of waste remediation involved the disposal of this waste by digging a hole in terrestrial space. It is a slow process thereby had the problem of contaminants leaking into the nearby environment. Consequently, other methods were used. Broadly, remediation of waste was carried out by three principal methods; physical, chemical, and biological.

Bioremediation of Pollutants. DOI: https://doi.org/10.1016/B978-0-12-819025-8.00017-X

In physical methods, conventional methods like sparging, soil turn over, excavation, and thermal treatment were used for the treatment of the soil. The physical techniques were based on the principle of either removal of contaminated site or covering of the polluted/contaminated site by forming layer above it. The expense, risk associated with excavation, leakage, and handling of the material during transportation are prime disadvantages associated with usage of this technique. The other method for remediation is chemical methods. It includes mostly degradation of the contaminants by processes like adsorption, catalysis by oxidation, or reduction reactions. These techniques have significantly proved to be worthy for reducing hazards of the compounds mostly falls in xenobiotics of recalcitrant nature. Onsight decrease in the concentration of the compounds like polycyclic aromatic hydrocarbon (PAHs), ketones, BTEX (benzene, toluene, ethylbenzene and xylenes), trinitrotoluene is well documented using these techniques. The expense, final product and its complexity, chemically reactive nature, and exposure to the contact persons, are a few disadvantages of this method (Vidali, 2001). The combination of the physical and chemical methods is also developed but none of the processes could lower the contaminants at desirable levels for the bulk waste management. Also, this process has generated large amounts of undesirable byproducts (Mehta and Chavan, 2009) (Fig. 17.1).

An alternate biological method of treating waste was developed as bioremediation. Bioremediation is a process in which complex hazardous contaminants get converted into non/less hazardous chemicals using plants and microbes.

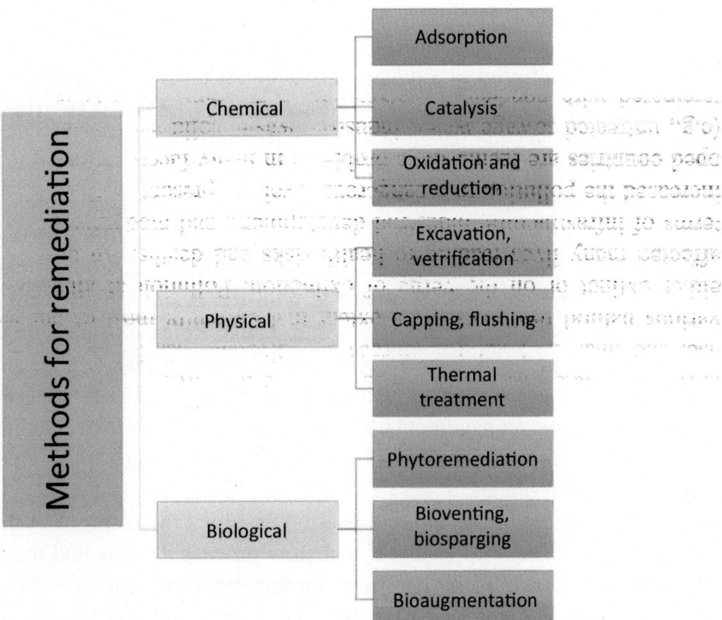

Figure 17.1 Different methods for remediation of polluted or contaminated waste.

17.2 Why microbes?

In context to evolutionary and adaptation point of view, microorganisms are the only entity present in the world that have well adapted in all ecological niches. They play a major role in many ecological cycles, in regulating the biogeochemical cycles and health of animals. Enzyme complexity and diverse metabolic pathways make the microorganisms very special as they can degrade these compounds using this as energy sources for their metabolism (Dash et al., 2014). Due to the versatile nature of microorganisms, they are becoming the model organism in studying pathways for remediation of various xenobiotic/recalcitrant compounds.

Mostly, bioremediation process depends on the degrading efficiency of the cultured/isolated microbes. Generally, it is found that the microbes who are indigenous have better degradation efficiency than the microbes cultured/isolated in in vitro or controlled laboratory conditions (Díaz-Ramírez et al., 2003; Venosa and Zhu, 2003). The organisms that have been continuously exposed to the pollutant become acquainted and develop the capacity to either use this pollutant as energy sources or have genetic adaptation for degrading the pollutant (Atlas and Bartha, 1998). Also, the time frame to remediate this pollutant decreases significantly. In reality, it is found that microorganisms who are found at the polluted sites are the most active degraders (Atlas and Bartha, 1998). Among the diversity of microbes, bacteria are the dominant species found in nature, which makes them an excellent candidate to be used for biodegradation/bioremediation. The few special characteristics like metabolic pathways, presence of toxins in a few species, helps bacteria to easily degrade/breakdown the substrate (pollutant), when provided with growth favorable conditions. Bacteria even can perform well in aerobic as well as anaerobic conditions. The layout for basic bioremediation techniques are as shown in Fig. 17.2. There are more reports with perspectives of biostimulation and bioattenuation. The current chapter portrays the diverse approaches of engineered bacteria for the bioremediation (Fig. 17.2).

17.3 Metabolic engineering

Many pollutants can be degraded by the isolated species/strain. It's the natural capacity of the strain to degrade the pollutant in optimal conditions. However, if we

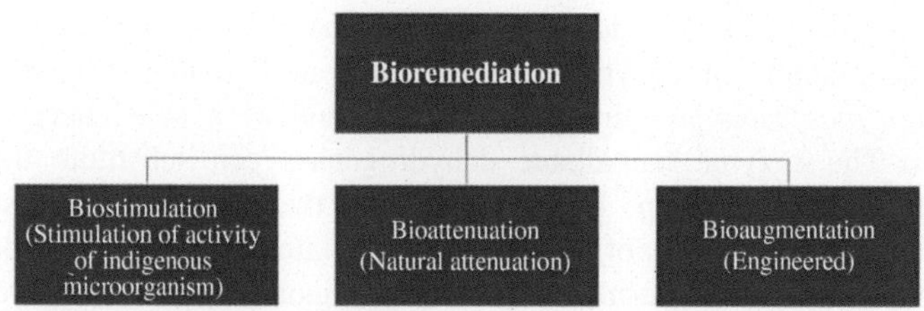

Figure 17.2 Basic bioremediation technology.

want to use the same strain in scale up studies for bulk degradation, it becomes difficult as its productivity is less. Metabolic engineering of biodegradative pathway in microorganisms can potentially offer the best way for expanding the capabilities of the host microorganism to remediate the generated large-scale waste (Chen et al., 1999).

For constructing/designing the metabolic pathway the following points need to be taken care of:

1. Identification of genes encoding for the secretion of enzymes involved in the catabolic pathway;
2. Genes responsible for transport for up-taking the compound and;
3. Identification of the regulatory network regulating the central metabolic pathways.

Extensive information, important for the GEM development is based on the catabolic and co-metabolic pathways (Stapleton and Sayler, 1998). Bacteria living in any experimental/environmental conditions are exposed to the signals which needs to be processed to gain physiological response. For the remediation of the pollutant, there is a need to develop a physiological response adapting/adjusting to the specific regulatory catabolic operons which helps to maintain the growth conditions of the cells (Sato et al., 1997). The transposons and extrachromosomal DNA facilitate the horizontal transfer of genes. Henceforth, thorough understanding is required of the physiological, biochemical, morphological, genetic, and catabolic pathways of organisms and the regulatory gene network associated with it.

There are many reports of designing the catabolic pathway for remediation of hazardous compounds. A unique example is the use of *Rhodococcus erythropolis* for sulphur removal at the sites polluted with fossil fuels. The catabolic pathway of the bacteria is designed in such a way that the bacteria can cleave the sulphur without breaking the ring and also maintain the content of the fuel (Izumi et al., 1994). Many reports of the genetic engineering in *Pseudomonas* are also reported for e.g. cloning and insertion of dsz clusters of genes in the *Pseudomonas* strains which can desulfurize the compound more significantly than the *Rhodococcus* sp. (Gallardo et al., 1997). By using the same strain having the dsz cluster and changing the EGSOX pathway, the ability to remove the sulphur increases four times as compared to the *Rhodococcus* strain.

The design of novel biocatalysts by understanding the structure, function, and control mechanism can offer new opportunities for changing metabolic pathway by altering the genetic makeup. Enzymes in the metabolic pathways can be tailored by random and site-specific mutagenesis. The classic example of site-directed mutagenesis is alteration in the cytochrome P450 (Stevenson et al., 1998) to increase the binding site of degrading enzyme haloalkane dehydrogenase (Holloway et al., 1998).

The other example of enzyme engineering is in *Xanthobacter autotrophicus*. *Xanthobacter autotrophicus* can use dichloroethane as a sole energy source in metabolism. The enzyme haloalkane dehydrogenase can substitute the chlorine atom to OH hydroxyl group. By understanding the structural chemistry of the enzyme, many combinations of enzyme intermediates and enzyme product complexes were designed. The resulting modifiers/variants were found more active in substitution of chlorine in the structure. On the contrary, not a single mutant was able to use more complex compound as substrate like TCE. This led to inference that sometimes rationale design may fail for degradation of complex compounds. The limitation of the site-directed approach is that it allows only a few sequences to change at a single point of time. Nonspecific or irrational approaches like DNA shuffling, random priming, stretched extension can be preferable and leads to significant results for degradation of heavier/complex molecules (Harayama, 1998; Kuchner and Arnold, 1997; Stemmer, 1994; Shao et al., 1998). Among these, DNA shuffling for developing genes, cross breeds seem to be most efficient technique as extended sequences can be explored for the different sites of change (Crameri et al., 1998). These nonspecific strategies were also used to increase the homologous sites for enzymes used in degradation pathway and to insert these modified enzymes in the host microorganisms. For e.g. this approach was used in modifying the operon for arsenic resistance. The variants showed a 40-fold increase in the resistance to the arsenic and also a 12-fold surge in the arc gene product in absence of any physical modification. Such results are also expected in the specific modification/site directed modification but sometimes the natural/nonspecific change in the process parameters helps microorganisms to adapt to those conditions making the system more efficient for the bulk work. The nonspecific approach gives freedom for rapid in-detailed analysis of sequence diversity present in the environment and also this led to the development of novel hybrid enzymes or pathways having desired features.

17.4 Recombinant DNA technology

Novel techniques for alteration of genetic content in microorganisms are developed for remediation/degradation of xenobiotic/recalcitrant pollutants. These techniques include construction of novel expression vectors for gene expression, modification of gene regulatory network for controlled persistence, and methods to report the genetically modified organisms.

For developing an efficient molecular tool, the following points are needed to be considered.

1. Careful designing the synthetic and energy production pathways responsible for the cell growth.
2. The size of the genetic vectors and the systems in the host organisms.
3. Maintenance of the extrachromosomal genetic vector which can be used for possible genetic alterations.

17.5 Plasmids

In any organism, the specific gene sets are linked with specific pollutant degradation pathway. It can be found in both at chromosomal as well as extrachromosomal level. The extrachromosomal genetic material mostly used for the cloning and expression systems are plasmids. Plasmids possess many genes associated for partial or complete degradation of the pollutant. Even few sets of genes are identified for its usage in the chemical specific degradative pathway. The characteristics/attributes like simple molecular structure with simple screening methods have made plasmids a readily assessable system for the introduction/modification of genes in the pathways (Sayler et al., 1990). Many plasmids are linked with the substrate metabolism occurring naturally in the environment. An example of the best used plasmids for degradation is gene characterized in toluene and naphthalene degradation. These plasmids were found to have entire genetic operons to covert xylene and toluene (TOL) to intermediate metabolites. The genetic operon of the naphthalene 7 plasmid has both the upper and lower pathway for salicylate oxidation. The benzoate is the common inducer for both pathways. In case of the toluene degradation pathway, the xylene is converted to toluene and then to toulate and benzoate. In the case of lower pathway, the meta cleavage pathway is followed for degradation of toluate and benzoate. The R and S, specific regulatory genes are used to react with the inducers for increase expression of operons. An overview of important techniques in rDNA technology is depicted in Fig. 17.3.

17.6 Expression systems

17.6.1 Inducible promoters

Most of the natural/engineered microorganism systems require the inducible expression. Expression systems like Lac have the advantages of versatility and complete elucidation. The promoters like Pm from TOL plasmid are available for the genes responsible in the cleavage for the toluene and naphthalene degradation (de Lorenzo et al., 1993). XylS regulatory protein is activated during transcription in catabolic pathway for toluene degradation. Promoter gene Pu is also activated in the upper pathway during degradation of toluene. XylR and NahR regulatory proteins are also activated during transcription. Soon the expression system will be able to respond to the sites contaminated with the metals and other pollutants contaminating the environment (Fig. 17.3).

17.6.2 Post transcriptional processing

Expression of the genes linked with parts of enzymes are controlled using the cistronic mRNA by post transcriptional modification (Carrier and Keasling, 1997a,b). By using this controlled system, the need of many strengths providing promoters will no longer be needed. This technique is also called mRNA stability in which the foreign DNA is in the 5′ untranslated region of specific gene of interest (Carrier and Keasling, 1997a,b). To make the system stable, DNA cassette was designed for hairpin structure formation in mRNA at 5′ end. The reason for introducing the hairpins structure was formation of AG in the secondary structure. The structures, which has large AG formation, shows excellent mRNA stability and elevated

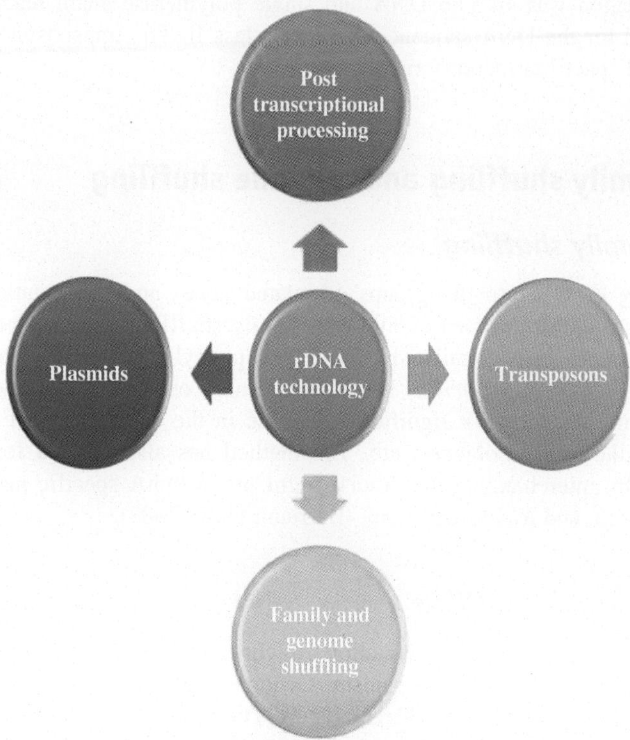

Figure 17.3 Important methods in rDNA technology.

protein levels, which is for the regulation of genes. These techniques will raise a chance to generate the various enzymes at different levels in different pathways using the single promoter as the control point.

17.6.3 Transposons

Transposons area unique DNA sequence catalyzing its movement to any random locations in a chromosome The type of modifications by transposons is nonspecific in nature but it is reported as the best working model for genetically modified organism and also more clones are stable than the plasmids (Davison et al., 1987; Pérez-Martin and de Lorenzo, 1996; Ramos et al., 1994). The exact mechanism of these transposon mediated remediation is yet too characterized. There is also a lack of knowledge in terms of frequency of this movement of transposable elements, nature of the products associates with it, and its phenotypic consequences. There are reports of using these transposable elements in sites contaminated with heavy metals. For example, the findings of Huang et al. (2010) gave the idea of a transposon-based system for mercury contaminated sites. They have identified the transposon from chromosome of *Bacillus megaterium* MB1. The region of transposon region was 14.5 kb DNA and single polymerase chain reaction primer was designed for the DNA sequence of repeat class II. The transposon was encoding the broad-spectrum mercury resistance.

17.7 Family shuffling and genome shuffling

17.7.1 Family shuffling

The shuffling of DNA to the groups of related genes and combination of those genes that accelerates directed evolution is family shuffling (Crameri et al., 1998). The best example is of family shuffling in Biphenyl dioxygenase (bphA) from *Pseudomonas* strain and *Bacillus* strain (Kumamaru et al., 1998). This modified/hybrid enzyme has shown a significant increase in the degradation of recalcitrant compounds like PCBs, toluenes, etc. The method has also applied for the PCBs degradation by interchanging the crucial segments of bphA specific gene set from *Burkholderia* sp. and *Rhodococcus* sp. (Barriault et al., 2002).

17.7.2 Genome shuffling

Genome shuffling is the recombination of chromosomes in several bacteria to improve activity of the whole organism. Genome shuffling generates the mutated strain having better phenotypic characteristics. The mutant is isolated after the multiple rounds of protoplast fusion for optimal combination in between the genomes. Genome shuffling is useful for modification of multitrat phenotypes which are difficult to modify directly. These techniques can be used exclusively for those phenotypes, which is impossible to anticipate that all mutations are needed to develop a complex trait that maintains robust growth. These techniques were used for bioremediation of pentachlorophenol by the mutant strains created by continuous series of protoplast fusions. It was reported that by using this technique 10-fold rise in the pentachlorophenol degradation was achieved and also the tolerance level of the organism was also raised significantly (Dai and Copley, 2004). This technique has worked as a measurement tool for observing the fitness of the mutants generated after repeated cycle of protoplast fusion and then it can be among the best procedures when ideal/specific conditions for which the bacteria was used in the process.

17.8 Omics and bioremediation

Advent of metabolic engineering and recombinant DNA technology had led the construction of novel microorganisms with better implications in the designing of the bioremediation strategy. With combination of molecular biology, "omics" has days become an important tool for designing a strategy for the onsite bioremediation strategy. Genomics had helped to analyze complete genome of many microorganism like

Pseudomonas, *Rhodococcus*, etc. which were isolated mostly from the contaminant sites. Genomics arc also helpful in better understanding the community of the microbes present at the site, and also, its onsite distribution. Other approach in genomics is the use of metagenomics; it is one of the best techniques used for the overall microbial community genomes persistent in nature (Martín et al., 2006). Further, development of DNA microarrays has enabled to analyze and characterize the functional genes of the microbes present at the polluted site (He et al., 2007). mRNA expression profile of both (upregulated and down regulated) gene sets in microbes by transcriptomics profile has helped to analyze the real time situation when microbes are exposed to the pollutants (Jennings et al., 2009). Also, proteomic analyses have led to development of a tool to track the proteins present in microbes residing in the polluted/contaminated site (Kim et al., 2004).

17.9 Genomics

Genomics is the technique for the analysis of total genetic information in a microbial cell. Nowadays, researchers are relying on the genomics data as it also helps in identifying specific sets of genes in the operons which can be helpful in the bioremediation. The complete gene set of many microorganisms have been completely sequenced. For example, the species belonging to the Pseudomonas family is mostly characterized (Nelson et al., 2002). The genome of *Pseudomonas putida* strain KT2440 has revealed the characterization of important enzymes like; *oxygenases*, *ferredoxins*, *dehydrogenases*, and glutathione *S*-transferases which are linked with the defense against the toxic substances and metabolites. Further, it possesses operons coding for the genes clusters which are involved in metabolism of the compounds which are unnatural (Nelson et al., 2002). Genome of Mycobacterium PYR-1 species involved mostly in degradation of the complex and heavy molecular compounds like polyaromatic compounds are also characterized. The genome is comprised of the genes linked for the degradation of the aromatic compounds. One of the interesting examples is of the strain *Polaromonas* sp. (Mattes et al., 2008). These species are found at the sites polluted/contaminated with the heavy metals or hydrocarbons. The genome of the species reveals that it contains specific gene sets which are involved in the catabolism of xenobiotic compounds and resistance to the metals. The genome also shows unique and important features like presence of putative transposons, insertion sequence of transposase 14 families, plasmids, and bacteriophage. The phenotypes were also experimentally proven to be surviving on various concentrations of organic compounds like heptane, catechol, cyclohexane, and salicylate.

17.10 Metagenomics

This approach helps in collecting all the bacterial communities (cultured and uncultured) found at the contaminated sites. The approach is considered as a potential tool to get rid of contaminants found in the environment (Röling, 2015; Bell et al., 2014; Chemerys et al., 2014). This technique is a combination of many techniques under a single niche. Techniques involve screening, assembling, and sequencing, Fluorescence activated cell sorting (FACS), high throughput screening technology.

17.11 Screening, assembling, and sequencing

For screening DNA two approaches are generally followed, in the first protocol the cells are separated from the contaminated soil followed by cell lysis and DNA isolation and purification. In the second a direct environmental sample is taken and lysed for DNA isolation and purification. Once DNA is purified, community DNA is cloned and inserted in the host cell using the best suitable vector for construction of metagenomics library. This is further followed by the metagenome screening. After the screening, the DNA sequences are annotated and assembled to identify the gene of interest. The identification of gene of interest is searched based on the lower protein similarity against the hits which are nearby in the databases. This technique gives a brief of nucleic acid hybridization as an efficient technique for finding novel genes from a large pool of clone libraries.

17.12 Metagenome sequencing techniques

For sequencing the assembled sequences of metagenomes, various technologies are used and also upgraded to increase the sensitivity and detection of gene of interest. Previously, only sangers sequencing technology was used for metagenomics analysis and even in sequencing of the human genome (Sanger et al., 1977; Qin et al., 2010). With the upgradation in technology, the NGS (next generation sequencing) are used for the metagenome sequencing. The advantage of all these technologies is that they give the bulk well defined data or high throughput reads in comparatively less time (Wall et al., 2009; Claesson et al., 2010; Niu et al., 2010).

17.13 Florescence-activated cell sorting

FACS is the cell sorting technique applied for sorting/identifying the microbial cells based on florescence during the metagenomics screening. The accuracy rate of this technique is approximately 5000 cells per second (Herzenberg et al., 1976). This high-through put screening helps to identify direct phenotypes which are readily visible by the attached florescence tags. Further, FACS can also detect the gene expression which is regulated under a florescent biosensor detected in same cell as the metagenome DNA (Handelsman, 2004; Rinke et al., 2014). Therefore, this technique is the best tool for rapid selection/screening of cell from the constructed metagenomics libraries.

17.14 DNA microarrays

The technique is based on hybridization of the targeted DNA molecule to the detected array probes by perceiving the change in florescence signal. Each fluorescence signal is spots of probe-target hybridization and can be individually quantified by signal to noise ratio against its local background. The spot can be quantified by the image analysis software supplied by the manufacturing company. Among the many microarrays; phylogenetic oligonucleotide array (POA), whole genome arrays (WGAs), and functional gene arrays (FGAs) are used for bioremediation analysis. POA is assembled by using the short-stretched fragments of identified nucleotide sequences. The identification of the short nucleotide sequences is based on the rRNA genes of various microbial phylum and are mostly universal/commercially used in the in microarrays to predict the microbes present in the experimental samples. The disadvantage of this method is that short stretches failed to resolve the bacterial lineages at species-level (Bae and Park, 2006). These short stretches were replaced in the technique developed called RHC PhyloChip, where more than 70 probes were used for sample analysis (Loy et al., 2005). This technique was used for detection uncultured bacterial lineages like *sterolibacterium*, *ferri* bacterium from the sludge samples. These uncultured organisms were later found to play a major role in the bioremediation process of waste water. The advance version of this is isotope arrays where the radiolabeled rRNA is used in the microbes. This technique allows simultaneous detection of microbial diversity and its activity using fluorescence tags and radiolabeling. For detection of key enzymes in metabolism and gene products, FGAs are developed. This technique will identify the unique genes linked with bioremediation at polluted sites. Bacterial strains/environmental clones are amplified using PCR to construct large fragments FGAs. The advance version of this is GeoChip which contains more than 24,000 probes covering almost 4 lac gene segments and 4000 functional group. Geochip was successful in the tracking of bacteria involved in remediation sites contaminated with the heavy metals like mercury and uranium. Further, the most advanced technique is WGA. This technique is used to compare or correlate the genomes of the same/related microorganism. Using this technique West et al. (2008) developed a whole genome microarray of *Dehalococcoides* sp. which contains more than 99 of the predicted protein coding sequences having 190 genomes. This WGA was used as a reference data set to compare the genes involved in dehalogenation of compounds such as tetracholorethane and tirchloroethane.

17.15 Transcriptomics approach

Transcriptomics or meta transcriptomics is the study of mRNA transcriptional profiles to understand functional in sights of activities allied with the microbial community.

1. For transcriptomics analysis; the following steps are followed:

2. Extraction, purification, and enrichment of mRNA;
3. Synthesis of complementary DNA;
4. Hybridization (microarray)/complete cDNA transcriptome.

Currently, many researchers have reported the study of complete genome using transcriptome studies. For example, the transcriptome of the *Geobacter* sp. dwelling in the uranium contaminated sites was reported by Holmes et al. (2009). After comparison it was found that 34 types of cytochrome c genes were upregulated in the cells grown in sediments compared to cells grown in normal conditions. Recent studies on meta transcriptomics with cDNA microarrays give useful information of the microbial communities dwelling in the environment. Transcriptomics has a wide approach for the identifying and linking of the structure and function of microbial communities present in the environment.

17.16 Novel approaches

17.16.1 Biofilms assisted bioremediation

17.16.1.1 Biofilms

Biofilms are the association of microbial communities to the surface embedded in the self-made exopolysaccharides. Communities of microbes in the biofilms reside in the mutualism where they benefit each other by forming the favorable niche around each other. The main characteristics of the biofilms include the movement of nutrition and physiological activity of the cells (Costerton et al., 1995). Due to the presence of exopolysaccharides, biofilms show resilience to the environmental stress and have showed high tolerance to physical, chemical, and biological stresses.

17.16.2 Physiological state of cells

Bacterial cells switch from the planktonic phase to the sessile form. However, phenotypic characteristics of the cells show a distinct form of planktonic growth (Hall-Stoodley et al., 2004). Transcription profile of genes and metabolic requirements are totally different compared to the planktonic state. Due to changes in gene expression; metabolic requirement, nutrients, and oxygen availability and signaling molecules inside the layer give a gradient of different niches. This led to the phenotype that remains embedded in the layer of EPS having decreased respiration and metabolic activity (Cogan et al., 2005).

17.16.3 Quorum sensing

Quorum sensing (QS) is a phenomenon which is cell density dependent and it is mediated by the autoinducers (AIs) or the signaling molecules. Normally three types of the AI are found in bacteria namely *N*-acyl homo serine lactones, auto inducing peptides, and AI-2. All three AIs are specific for the specific bacterial groups. The QS regulates important functions like biofilm formation, virulence of the species, and sporulation. Mainly, the data of the mutant of various phases in biofilm development suggests that there are specific quorum sensing regulatory units in the biofilm formation and maturation (Kjelleberg and Molin, 2002).

17.16.4 Biofilms for PAHs remediation

Many recalcitrant/xenobiotic compounds like polyaromatic hydrocarbons (PAHs) are very difficult to degrade and raised concerns about the long-term environmental effects (Pandey and Jain, 2002). Therefore, many processes were developed for the removal of these compounds. Biofilm-based remediation is one of the best alternatives for degradation of PAHs. Recent findings indicate that due to high concentration of calcium present in the exopolysaccharides of biofilm, significant degradation of the PAHs is found in marine strain of *Pseudomonas mendocina* (Mangwani et al., 2014).

17.17 Future prospects for bioremediation

Considering the importance of engineered microbes in enhancing the degradation and detoxification of compounds which are recalcitrant in nature, detailed studies needed to be carried out for increasing the survival rate of the microbes when released in the real site/polluted ecological niches. Although continuous efforts are stressed on using molecular tools for the identification of exact gene regulatory networks, using different approaches but the simple approaches with definite mechanisms are required for the degradation of the pollutant found in the bulk. More efforts are still required in terms of the metabolic engineering of modified microorganisms to ascertain their side effects at the remediation site and nearby ecological niches to the site. The combinatorial effect of plants and microbes are known but still more emphasis needs to be given in this area as both can easily work in the symbiotic relationship. This process needs to be stretched out to make the process economical. Furthermore, the omics data needs to be more annotated so that specific gene sets can be assembled for site directed process. Transcriptomics at mRNA level needs to reach to the small RNA level to check the expression of genes during the resistance or the degradation of compounds. The designing of synthetic and natural media for the exploration/isolation of indigenous microbial population will add success to the bioremediation process.

References

Adriano, D.C., 2003. Trace Elements in Terrestrial Environments: Biogeochemistry, Bioavailability and Risks of Metals, second edition Springer, New York.

Atlas, R.M. and Bartha, R., 1998. Microbial Ecology: Fundamentals and Applications, fourth edition Pearson Education, India.

Bae, J.W., Park, Y.H., 2006. Homogeneous versus heterogeneous probes for microbial ecological microarrays. Trends Biotechnol. 24 (7), 318–323.

Barriault, D., Plante, M.M., Sylvestre, M., 2002. Family shuffling of a targeted bphA region to engineer biphenyl dioxygenase. J. Bacteriol. 184 (14), 3794–3800.

Bell, T.H., Joly, S., Pitre, F.E., Yergeau, E., 2014. Increasing phytoremediation efficiency and reliability using novel omics approaches. Trends Biotechnol. 32 (5), 271–280.

Carrier, T.A., Keasling, J.D., 1997a. Controlling messenger RNA stability in bacteria: strategies for engineering gene expression. Biotechnol. Prog. 13 (6), 699–708.

Carrier, T.A., Keasling, J.D., 1997b. Engineering mRNA stability in *E. coli* by the addition of synthetic hairpins using a 5′ cassette system. Biotechnol. Bioeng. 55 (3), 577–580.

Chemerys, A., Pelletier, E., Cruaud, C., Martin, F., Violet, F., Jouanneau, Y., 2014. Characterization of novel polycyclic aromatic hydrocarbon dioxygenases from the bacterial metagenomic DNA of a contaminated soil. Appl. Environ. Microbiol. 80 (21), 6591–6600.

Chen, W., Brühlmann, F., Richins, R.D., Mulchandani, A., 1999. Engineering of improved microbes and enzymes for bioremediation. Curr. Opin. Biotechnol. 10 (2), 137–141.

Claesson, M.J., Wang, Q., O'Sullivan, O., Greene-Diniz, R., Cole, J.R., Ross, R.P., et al., 2010. Comparison of two next-generation sequencing technologies for resolving highly complex microbiota composition using tandem variable 16S rRNA gene regions. Nucleic Acids Res. 38 (22), e200.

Cogan, N.G., Cortez, R., Fauci, L., 2005. Modeling physiological resistance in bacterial biofilms. Bull. Math. Biol. 67 (4), 831–853.

Costerton, J.W., Lewadowskim, Z.M., Caldwellm, D.E., Korber, D.R., Lappin-Scott, H.M., 1995. Microbial biofilms. Ann. Rev. Microbiol. 49, 711–745.

Crameri, A., Raillard, S.A., Bermudez, E., Stemmer, W.P., 1998. DNA shuffling of a family of genes from diverse species accelerates directed evolution. Nature 391 (6664), 288.

Dai, M., Copley, S.D., 2004. Genome shuffling improves degradation of the anthropogenic pesticide pentachlorophenol by *Sphingobium chlorophenolicum* ATCC 39723. Appl. Environ. Microbiol. 70 (4), 2391–2397.

Dash, H.R., Mangwani, N., Das, S., 2014. Characterization and potential application in mercury bioremediation of highly mercury-resistant marine bacterium *Bacillus thuringiensis* PW-05. Environ. Sci. Pollut. Res. 21 (4), 2642–2653.

Davison, J., Heusterspreute, M., Chevalier, N., Ha-Thi, V., Brunei, F., 1987. Vectors with restriction site banks V. pJRD215, a wide-host-range cosmid vector with multiple cloning sites. Gene 51 (2–3), 275–280.

de Lorenzo, V., Fernández, S., Herrero, M., Jakubzik, U., Timmis, K.N., 1993. Engineering of alkyl- and haloaromatic-responsive gene expression with mini-transposons containing regulated promoters of biodegradative pathways of Pseudomonas. Gene 130 (1), 41–46.

Díaz-Ramírez, I.J., Ramírez-Saad, H., Gutiérrez-Rojas, M., Favela-Torres, E., 2003. Biodegradation of Maya crude oil fractions by bacterial strains and a defined mixed culture isolated from Cyperuslaxus rhizosphere soil in a contaminated site. Can. J. Microbiol. 49 (12), 755–761.

Gallardo, M.E., Ferrandez, A., De Lorenzo, V.Í.C.T.O.R., García, J.L., Diaz, E., 1997. Designing recombinant Pseudomonas strains to enhance biodesulfurization. J. Bacteriol. 179 (22), 7156–7160.

Hall-Stoodley, L., Costerton, J.W., Stoodley, P., 2004. Bacterial biofilms: from the natural environment to infectious diseases. Nat. Rev. Microbiol. 2 (2), 95.

Handelsman, J., 2004. Metagenomics: application of genomics to uncultured microorganisms. Microbiol. Mol. Biol. Rev. 68 (4), 669–685.

Harayama, S., 1998. Artificial evolution by DNA shuffling. Trends Biotechnol. 16 (2), 76–82.

He, Z., Gentry, T.J., Schadt, C.W., Wu, L., Liebich, J., Chong, S.C., et al., 2007. GeoChip: a comprehensive microarray for investigating biogeochemical, ecological and environmental processes. ISME J. 1 (1), 67.

Herzenberg, L.A., Sweet, R.G., Herzenberg, L.A., 1976. Fluorescence-activated cell sorting. Sci. Am. 234 (3), 108–117.

Holloway, P., Knoke, K.L., Trevors, J.T., Lee, H., 1998. Alteration of the substrate range of haloalkane dehalogenase by site-directed mutagenesis. Biotechnol. Bioeng. 59 (4), 520–523.

Holmes, D.E., O'neil, R.A., Chavan, M.A., N'Guessan, L.A., Vrionis, H.A., Perpetua, L.A., et al., 2009. Transcriptome of *Geobacter uraniireducens* growing in uranium-contaminated subsurface sediments. ISME J. 3 (2), 216.

Huang, C.C., Chien, M.F., Lin, K.H., 2010. Bacterial mercury resistance of TnMERI1 and its' application in bioremediation. Interdisci. Stud. Environ. Chem. 3 (11), 21–29.

Izumi, Y., Ohshiro, T., Ogino, H., Hine, Y., Shimao, M., 1994. Selective desulfurization of dibenzothiophene by *Rhodococcus erythropolis* D-1. Appl. Environ. Microbiol. 60 (1), 223–226.

Jennings, L.K., Chartrand, M.M., Lacrampe-Couloume, G., Lollar, B.S., Spain, J.C., Gossett, J.M., 2009. Proteomic and transcriptomic analyses reveal genes upregulated by cis-dichloroethene in *Polaromonas* sp. strain JS666. Appl. Environ. Microbiol. 75 (11), 3733–3744.

Kim, S.J., Jones, R.C., Cha, C.J., Kweon, O., Edmondson, R.D., Cerniglia, C.E., 2004. Identification of proteins induced by polycyclic aromatic hydrocarbon in Mycobacterium vanbaalenii PYR-1 using two-dimensional polyacrylamide gel electrophoresis and de novo sequencing methods. Proteomics 4 (12), 3899–3908.

Kjelleberg, S., Molin, S., 2002. Is there a role for quorum sensing signals in bacterial biofilms? Curr. Opin. Microbiol. 5 (3), 254–258.

Kuchner, O., Arnold, F.H., 1997. Directed evolution of enzyme catalysts. Trends Biotechnol. 15 (12), 523–530.

Kumamaru, T., Suenaga, H., Mitsuoka, M., Watanabe, T., Furukawa, K., 1998. Enhanced degradation of polychlorinated biphenyls by directed evolution of biphenyl dioxygenase. Nat. Biotechnol. 16 (7), 663.

Loy, A., Schulz, C., Lücker, S., Schöpfer-Wendels, A., Stoecker, K., Baranyi, C., et al., 2005. 16S rRNA gene-based oligonucleotide microarray for environmental monitoring of the betaproteobacterial order "*Rhodocyclales*". Appl. Environ. Microbiol. 71 (3), 1373–1386.

Mangwani, N., Shukla, S.K., Rao, T.S., Das, S., 2014. Calcium-mediated modulation of *Pseudomonas mendocina* NR802 biofilm influences the phenanthrene degradation. Colloids Surf. B: Biointerfaces 114, 301–309.

Martín, H.G., Ivanova, N., Kunin, V., Warnecke, F., Barry, K.W., McHardy, A.C., et al., 2006. Metagenomic analysis of two enhanced biological phosphorus removal (EBPR) sludge communities. Nat. Biotechnol. 24 (10), 1263.

Mattes, T.E., Alexander, A.K., Richardson, P.M., Munk, A.C., Han, C.S., Stothard, P., et al., 2008. The genome of *Polaromonas* sp. strain JS666: insights into the evolution of a hydrocarbon- and xenobiotic-degrading bacterium, and features of relevance to biotechnology. Appl. Environ. Microbiol. 74 (20), 6405–6416.

Mehta, V., Chavan, A., 2009. Physico-chemical treatment of tar-containing wastewater generated from biomass gasification plants. World Acad. Sci., Eng. Technol. 57, 161–168.

Nelson, K.E., Weinel, C., Paulsen, I.T., Dodson, R.J., Hilbert, H., Martins dos Santos, V.A. P., et al., 2002. Complete genome sequence and comparative analysis of the metabolically versatile *Pseudomonas putida* KT2440. Environ. Microbiol. 4 (12), 799–808.

Niu, B., Fu, L., Sun, S., Li, W., 2010. Artificial and natural duplicates in pyrosequencing reads of metagenomic data. BMC Bioinforma. 11 (1), 187.

Pandey, G., Jain, R.K., 2002. Bacterial chemotaxis toward environmental pollutants: role in bioremediation. Appl. Environ. Microbiol. 68 (12), 5789–5795.

Pérez-Martin, J., de Lorenzo, V., 1996. VTR expression cassettes for engineering conditional phenotypes in Pseudomonas: activity of the Pu promoter of the TOL plasmid under limiting concentrations of the XylR activator protein. Gene 172 (1), 81–86.

Qin, J., Li, R., Raes, J., Arumugam, M., Burgdorf, K.S., Manichanh, C., et al., 2010. A human gut microbial gene catalogue established by metagenomic sequencing. Nature 464 (7285), 59.

Ramos, J.L., Díaz, E., Dowling, D., de Lorenzo, V., Molin, S., O'Gara, F., et al., 1994. The behavior of bacteria designed for biodegradation. Bio/Technology 12 (12), 1349.

Rinke, C., Lee, J., Nath, N., Goudeau, D., Thompson, B., Poulton, N., et al., 2014. Obtaining genomes from uncultivated environmental microorganisms using FACS-based single-cell genomics. Nat. Protoc. 9 (5), 1038.

Röling, W.F., 2015. Maths on microbes: adding microbial ecophysiology to metagenomics. Microb. Biotechnol. 8 (1), 21.

Sanger, F., Nicklen, S., Coulson, A.R., 1977. DNA sequencing with chain-terminating inhibitors. Proc. Natl Acad. Sci. U.S.A. 74 (12), 5463–5467.

Sato, S.I., Nam, J.W., Kasuga, K., Nojiri, H., Yamane, H., Omori, T., 1997. Identification and characterization of genes encoding carbazole 1, 9a-dioxygenase in *Pseudomonas* sp. strain CA10. J. Bacteriol. 179 (15), 4850–4858.

Sayler, G.S., Hooper, S.W., Layton, A.C., King, J.H., 1990. Catabolic plasmids of environmental and ecological significance. Microb. Ecol. 19 (1), 1–20.

Shao, Z., Zhao, H., Giver, L., Arnold, F.H., 1998. Random-priming in vitro recombination: an effective tool for directed evolution. Nucleic Acids Res. 26 (2), 681–683.

Stapleton, R.D., Sayler, G.S., 1998. Assessment of the microbiological potential for the natural attenuation of petroleum hydrocarbons in a shallow aquifer system. Microb. Ecol. 36 (3-4), 349–361.

Stemmer, W.P., 1994. Rapid evolution of a protein in vitro by DNA shuffling. Nature 370 (6488), 389.

Stevenson, J.-A., Bearpark, J.K., Wong, L.L., 1998. Engineering molecular recognition in alkane oxidation catalyzed by cytochrome P450 cam. N. J. Chem. 22, 551–552.

Venosa, A.D., Zhu, X., 2003. Biodegradation of crude oil contaminating marine shorelines and freshwater wetlands. Spill Sci. Technol. Bull. 8 (2), 163–178.

Vidali, M., 2001. Bioremediation: an overview. J. Appl. Chem. 73 (7), 1163–1172.

Wall, P.K., Leebens-Mack, J., Chanderbali, A.S., Barakat, A., Wolcott, E., Liang, H., et al., 2009. Comparison of next generation sequencing technologies for transcriptome characterization. BMC Genomics 10 (1), 347.

West, K.A., Johnson, D.R., Hu, P., DeSantis, T.Z., Brodie, E.L., Lee, P.K., et al., 2008. Comparative genomics of "Dehalococcoidesethenogenes" 195 and an enrichment culture containing unsequenced "Dehalococcoides" strains. Appl. Environ. Microbiol. 74 (11), 3533–3540.

Zhao, Q., Kaluarachchi, J.J., 2002. Risk assessment at hazardous waste-contaminated sites with variability of population characteristics. Environ. Int. 28 (1-2), 41–53.

Biofilm in bioremediation

18

*Indra Mani**
Department of Microbiology, Gargi College, University of Delhi, New Delhi, India
*Corresponding author

18.1 Introduction

Biofilms are communities of microorganisms, which are attached to a biological or inert surface and coated in a self-synthesized matrix comprising of carbohydrates, water, proteins, and extracellular DNA (Costerton et al., 1987). Microbial biofilms are an intriguing subject, due to their significant roles in the environment, industry, and health. Advances in biochemical and molecular techniques have helped in enhancing our understanding of biofilm structure and development (Velmourougane et al., 2017). Biofilms have various morphologies, depending on the resident bacteria as well as the conditions under which the biofilm has created (Rabin et al., 2015). Generally, biofilms appear as filamentous and mushroom-like shapes in fast moving and static water, respectively (Edwards et al., 2000; Reysenbach and Cady, 2001). It may anticipate that different microbial species present in consortia of biofilms, and each with different metabolic degradation pathway are capable of degrading several pollutants either individually or collectively (Horemans et al., 2013). Under natural environmental conditions, most bacteria persist in biofilm mode encased in an extracellular polymeric substance (EPS) matrix which also provides a beneficial structure to biofilm forming microbes in bioremediation (Flemming and Wingender, 2001, 2010). The EPS is made up of both bound and secreted form of polysaccharides, proteins, lipids, nucleic acids, and water (Branda et al., 2005).

In biofilm formation a series of steps are involved which is shown in Fig. 18.1 (Vlamakis et al., 2013). Responses such as swimming, swarming, twitching motility, chemotaxis, quorum sensing (QS) in the presence of xenobiotics commonly present in soil and water assist microbes to coordinate movement toward pollutants and improved biodegradation (Lacal et al., 2013). The biofilm matrix offers greater resistance than planktonic cells to microbes from environmental stress, shear stress, acid stress, antimicrobial agents, UV damage, desiccation, predation, biocides, solvents, and a high concentration of toxic chemicals and pollutants (Davey and O'Toole, 2000). However, free floating planktonic cells purify environmental contaminants by metabolic activity and such cells are neither stationery nor modified to persevere under mechanical and environmental stress (Sutherland, 2001). Microbial bioremediation is an emerging in situ technology for the elimination of environmental pollutants using microorganisms. The biological processes for

Bioremediation of Pollutants. DOI: https://doi.org/10.1016/B978-0-12-819025-8.00018-1

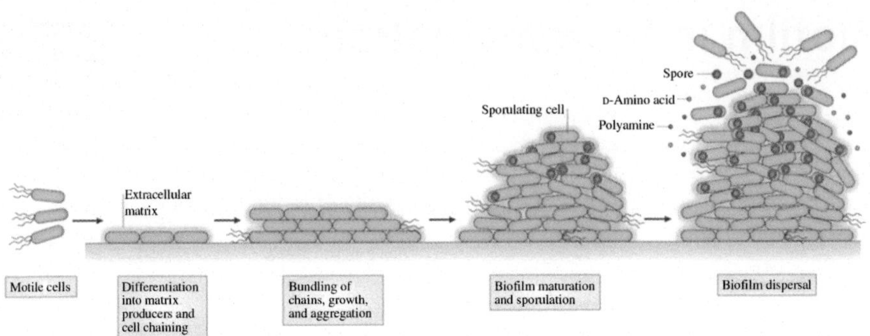

Figure 18.1 Microbial surface colonization; stages in surface attachment and biofilm formation.
Source: Adapted from Vlamakis, H., Chai, Y., Beauregard, P., Losick, R., Kolter, R., 2013. Sticking together: building a biofilm the Bacillus subtilis way. Nat. Rev. Microbiol. 11, 157–168 with permission.

treating toxic effluents are better than chemical and physical methods in terms of their efficiency and economy and the potential of biofilm communities for bioremediation processes has recently been understood (Paul et al., 2005). Biofilm-based bioremediation is an effective substitute to bioremediation with free-floating (planktonic) microorganisms because biofilm forming cells are being protected within the matrix, and it has more ability for adaptation and survival (Decho, 2000). A certain amount of biofilm growth is required for efficient use of biofilms, the ability to control and maintain biofilms at desired thickness is vital (Zhang and Poh, 2018). However, excessive biofilm growth remains a key challenge in the use of biofilms in these applications.

Chloropropham-degrading cultures have been used from sludge and soil samples from two different enrichment cultures like biofilm and planktonic. Denaturing gradient gel electrophoresis fingerprinting indicated that planktonic and biofilm cultures had a different community composition depending on the presence and type of added solid matrix during enrichment. These results have shown that biofilm-mediated enrichment techniques can be used to select pollutant-degrading microorganisms that like to proliferate in a biofilm and that cannot be isolated using conventional shaken-liquid procedures (Verhagen et al., 2011). Though often related with disease and biofouling, biofilms are also significant for engineering purposes, such as bioremediation microbial fuel cells and biocatalysis (Wood et al., 2011). On the other hand, semiconductor trade relies significantly on a procedure known as chemical mechanical planarization, which uses chemical and physical procedures to remove extra substance from the surface of silicon wafers throughout microchip production. These exercise consequences in enormous volumes of wastewater containing dissolved metals including copper (Cu^{2+}) which are essential to be filtered and treated before release into municipal waste systems. Mosier et al. (2015) have examined the potential use of *Lactobacillus casei* and *Pichia pastoris* biomass as a substitute to the presently used ion-exchange resins for the adsorption of dissolved

Cu^{2+} from high-throughput industrial waste streams and which makes this technique an attractive candidate for use in wastewater treatment.

A study has established that thermophilic bacterial communities generate dense biofilm on carbon steel API 5LX and create extracellular metabolic products to accelerate the corrosion procedure in oil pools. In this study, nine thermophilic biocorrosive bacterial strains of *Bacillus* and *Geobacillus* have been isolated. Electrochemical impedance spectroscopy and X-ray diffraction data have shown that these isolates oxidize iron into ferrous/ferric oxides as the corrosion products on the carbon steel surface, as the crude oil hydrocarbon served as a sole carbon source for bacterial growth and development in such extreme environments (Elumalai et al., 2019). Eutrophication of water by nutrient pollution remains a fundamental environmental problem. A study has estimated the nutrient uptake capacity of an algal biofilm as a means to treat polluted water. The algal biofilm has accomplished to remove 99% of phosphorus within 24 hours of P accumulation, with the $PO_4{-}P$ concentration in entering water ranging from 3 to 10 mg/L (Sukačová et al., 2017). The developed biofilm system has been a high efficiency for phosphorus removal and, therefore, it can combine into wastewater treatment processes.

Aminomethyl phosphonic acid (AMPA) and sarcosine are the metabolites and active constituent of glyphosate, which are caused by the river contamination and induced eutrophication of the river. Microorganisms exist as biofilm and are involved in the degradation of glyphosate. The analysis has established the multilateral association among biofilms, glyphosate, and phosphorus in rivers and the results have demonstrated that phosphorus not only was a key driver of river eutrophication but also could reduce complete glyphosate degradation by biofilms and support the accumulation of AMPA in river water (Carles et al., 2019). The major role of biofilms and the trophic status of rivers should be considered in order to better evaluate the outcome and perseverance of glyphosate.

The recent study has focused on the impact of plastic on marine microbial life and on the various functions, which is ensured in the ecosystems (Jacquin et al., 2019; Willis et al., 2017; Gewert et al., 2015). The dynamic features are influencing biofilm progress on plastic surfaces and the possible role of plastic wreckage as a vector for spreading of detrimental pathogen species. It has also given an analytical interpretation of the extent to which marine microorganisms can participate in the decomposition of plastic in the oceans and of the relevance of current standard tests for plastic biodegradability in the seas (Gonda et al., 2000; Krueger et al., 2015; Ghiglione et al., 2016). From the few last decades, it has become clear that plastic contamination offerings are a worldwide public and environmental challenge given its accumulative occurrence in the seas.

18.2 Strategies for use of biofilms in bioremediation

Generally, heavy metals pollution is everywhere in the world, and these are generating from different sources such as domestic, industrial, agricultural, technological, and medical applications. It is critical for human health and the environment (Tchounwou et al., 2012). A magnificent choice for metal contaminated site bioremediation is the use of microbial surfactants, which are considered as complex molecules with a different range of chemical structure (Walter et al., 2013; Nwachi et al., 2016). Now it is essential that the improvement of such compounds since microorganisms have long been found to produce surfactants. Additional alternatives accomplished of demonstrating efficient enhancers include utilizing the chemotactic potential and biofilm forming capability of the relevant microorganisms (Singh and Cameotra, 2004).

The hydrophobic nature of hydrocarbons makes them less bioavailable to microbes, generally leading to low efficiency in biodegradation. The study has determined that the exploitation of naturally formed fungal–bacterial populations for improved biodegradation of hydrocarbons such as hexadecane and synergistic degradation of hexadecane throughout biofilm formation, by a community is comprised of *Bacillus cereus* group and *Aspergillus flavus* complex (Perera et al., 2019). Current bioremediation strategies for hydrocarbon contamination used induced mixed microbial cultures. This in vitro analysis reveals the exploitation of naturally occurring communities in appropriate environments for attaining extremely efficient, synergistic degradation of hydrocarbons in a simple community structure without additives. However, it is not possible to degrade metals into harmless CO_2 like hydrocarbons. But microorganisms are able to remediate or concentrate as a nontoxic form, which are volatilized or precipitated from the contaminated sites (Lovely and Coates, 1997). The behavior, transport, and ultimate fate of contaminants in an aquatic environment may be significantly affected by the sorption and remobilization with biofilms (Headly et al., 1998). Remarkably, biofilm has an ability to sorb the chemical compounds, and cell walls; EPS, cell membrane, and cell cytoplasm work as sorption sites. Due to the sorption properties of biofilms, it can affect the fate of any compounds. For example, after uptake of toluene converted into uronic acids, which increased the sorption capacity of cations molecules (Flemming, 2000). A study has demonstrated that the transition condition between planktonic (free floating) and biofilm formation in the *Pseudomonas putida* KT2440 strain is to be governed by regulatory network controlling signal molecule like cyclic di-GMP (c-diGMP). Synthesis or degradation of c-diGMP can be used to design a genetic device in *P. putida* and other bacterial species for biofilm formation and use in the bioremediation (Benedetti et al., 2016). The study has demonstrated that the morphologies and physical forms of whole-cell biocatalysts can be genetically programmed while knowingly designing their biochemical activity. Biofilm produced by numerous microorganisms possibly postulate a proper microenvironment for effective bioremediation procedures. High cell density and stress resistance properties of the biofilm environment helps in an effective metabolism of a number of hydrophobic and toxic compounds. Bacterial biofilm establishment is frequently regulated by QS, which is a population density-based cell–cell communication process via signaling molecules (Kjelleberg and Molin, 2002). Various signaling molecules such as acyl homoserine lactones, autoinducer-2, peptides, α-hydroxyketones, and diffusion signaling factors have studied in bacteria. For environmental purposes such as biofilm formation, exopolysaccharide synthesis, catabolic gene expression, chemotaxis, motility, and horizontal gene transfer, genetic alteration of QS mechanism can be appreciated to modulate dynamic characters (Mangwani et al., 2016). Thus QS signals can be utilized for the creation of engineered biofilms with enhanced degradation kinetics.

18.3 Types of pollutants remediated by biofilms

Microbial biofilm mediated bioremediation are being progressively used in the removal of different kinds of pollutants including persistent organic pollutants, oil spills, heavy metals pesticides, and xenobiotics. Biofilm remediation has been mainly useful in the treatment of heavy metal contaminated samples from groundwater and soil for often encountered heavy metals such as chromium, cadmium, copper, and uranium (Valls and de Lorenzo, 2002). Phosphatase enzyme in the presence of biofilm matrix helps metal precipitation both for aerobic and anaerobic bacteria (Macaskie et al., 1997). In certain instances, biofilm formation is stimulated by addition of carbon sources in contaminated ground water to create a barrier or reduce flow of pollutants away from the site of contamination to reduce spreading. Heavy metals such as zinc, copper, cadmium, cobalt, and nickel have been remediated by varied biofilm reactors. Biofilm forming sulfate reducing bacteria are especially beneficial in mines for scavenging metals from metal contaminated water into precipitates of metal sulfides (Muyzer and Stams, 2008).

Arsenite-oxidizing bacteria (AOB) have shown an important role in the biogeochemical cycle of arsenic (As) in the environment, and have been used for the bioremediation of As contaminated groundwater. To understand this process, Zeng et al. (2018) have isolated seven different AOB strains from the arsenic-contaminated soils and which can completely oxidize 1.0 mM As(III) in 22–60 hours. Their arsenite oxidase sequences have shown 43%–99% similarities with other known AOB. The biofilm formation considerably promoted the bacterial resistance to arsenic. The study has highlighted the distinct existence of different AOB in arsenic stress and offered essential information for the assessment of useful AOB strains used for manufactures of bioreactors.

Bacterial biofilm shows a vital role in bioremediation of heavy metals from wastewaters. Mosharaf et al. (2018) isolated and identified different biofilm producing bacteria from wastewaters and it has also characterized the biofilm matrix EPS produced by different bacteria. Study has demonstrated that out of 40 isolates from different wastewaters, only 11 (27.5%) isolates (static condition at 28°C) and 9 (22.5%) isolates (agitate and static conditions at 28°C and 37°C) have produced air–liquid (AL) and solid–AL biofilms, respectively. The polysaccharides are known to sequester the heavy metals and consequently, these bacteria might be applied to eliminate these metals from wastewater. Another study has examined biofilm formation on biodegradable plastics in freshwater samples. Poly(3-hydroxybutyrate-*co*-3-hydroxyhexanoate) (PHBH) was to be enclosed by a biofilm after an incubation in freshwater samples. A high throughput sequence analysis of the bacterial communities of biofilms that formed on PHBH films exposed the dominance of the order Burkholderiales. Moreover, Acidovorax, and Undibacterium were the predominant genera in most biofilms. These results have demonstrated that the order Burkholderiales in biofilms functions as a degrader of PHBH films (Morohoshi et al., 2018). *Alcanivorax borkumensis* is a bacterial community that leads hydrocarbon-degrading communities across numerous oil discharges. Actual physicochemical conditions, which speedy bacterial binding to oil/water interfaces, are unclear. It has shown that *A. borkumensis* cells, which attached to the oil/water boundary and not a synthesized biosurfactant, is released into solution and reduces interfacial tension. This study provides key insights into the physicochemical properties that allow *A. borkumensis* to adhere to oil/water interfaces (Godfrin et al., 2018). The study has determined the profusion and viability of biofilm formed on the surface of polylactide (PLA) during its biodegradation in different environments like in lake water, compost, and soil using OxiTop Control. The results have indicated that PLA was sensitive to biodegradation in any environment, particularly in compost. The results have shown that different bacterial species formed biofilm of different abundance and hydrolytic activity levels (Walczak et al., 2015). Naturally, *Pseudomonas* sp. is well known for the production of a wide range of secondary metabolites during late exponential and stationary phases of the growth. Phenazine 1,6-di-carboxylic acid (PDC) is one of such metabolites, which has debated for its origin from *Pseudomonas* sp. (Dasgupta et al., 2015). A study has revealed that interspecific cooperation between *Pseudomonas* sp., which may lead its applicability in bioremediation, has also suggested that the scope of potential investigation on PDC for its therapeutic uses.

18.4 Current status of use of biofilm in bioremediation

Microbial biofilms can be programmed to make living materials with self-healing and evolvable functionalities. Nevertheless, the broader use of nonnatural biofilms has been hampered by restrictions on processability and useful protein secretion capability. Huang et al. (2019) have described a highly flexible and tunable living functional materials platform based on the TasA amyloid machinery of the bacterium *Bacillus subtilis* (Fig. 18.2). This novel tunable platform deals earlier unachievable properties for a variety of living functional materials having future uses in biotechnology, biomaterials, and biomedicine. In another study, the efficiency improvement of three moving bed biofilm reactors has been examined by inoculation of activated sludge cells (R1), mixed culture of eight strong phenol-degrading bacteria consisted of *Pseudomonas* spp. and *Acinetobacter* spp. (R2), and the combination of both (R3). Biofilm development capability of eight bacteria have been evaluated primarily using distinctive methods and media (Irankhah et al., 2018). A study has established that the bacteria with both biofilm-forming and contaminant-degrading capabilities are not only able to support the immobilization of other favorable activated sludge cells in biofilm structure, but also collaborate in pollutant degradation, which all accordingly lead to enhancement of treatment effectiveness.

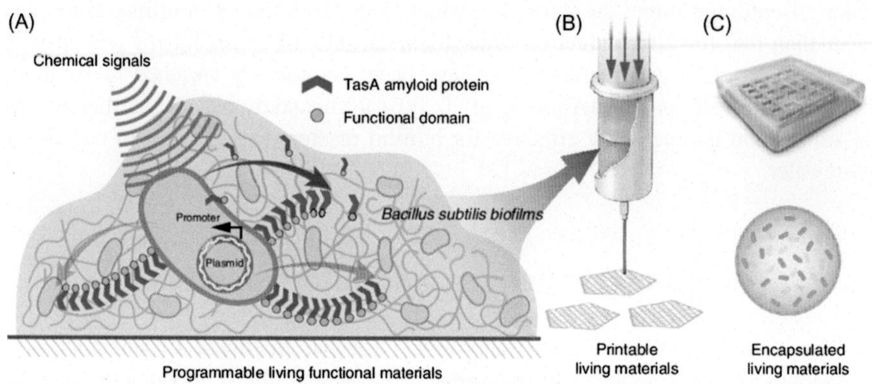

Figure 18.2 Design for a programmable and printable *Bacillus subtilis* biofilm production platform. (A) Schematic of the programmable living biofilm platform based on the TasA amyloid export machinery of *B. subtilis*. In the presence of chemical inducers, *B. subtilis* cells express rationally designed TasA fusion proteins comprising the amyloidogenic TasA domain (blue) and a tunable functional domain (green). Upon secretion, the fusion proteins self-assemble into extracellular fibrous networks that are closely associated with cell surfaces, resulting in programmable biofilms with tunable nonnatural functional properties. (B) Schematic of a printable living biofilm using a 3D printing technique. The living *B. subtilis* biofilms exhibit viscoelastic properties that make them suitable for 3D printing. (C) Schematic showing that the living *B. subtilis* biofilms maintain their natural viability and various cellular capacities such as self-regeneration when trapped in hydrogels or microcapsules.
Source: Adapted from Huang, J., Liu, S., Zhang, C., Wang, X., Pu, J., Ba, F., et al., 2019. Programmable and printable Bacillus subtilis biofilms as engineered living materials. Nat. Chem. Biol. 15(1), 34–41 with permission.

Siderophores are iron chelators, which are produced by several bacteria. Due to this property, there is bacteria fulfillment of their cellular need for iron. There is a class of siderophores, which use catecholate groups to chelate iron. Besides its role in the detoxification of environmental catechols, the catechol 2,3-dioxygenase encoded by *catDE* also protects cells from intoxication by endogenous bacillibactin-derived catechol metabolites under iron-limited conditions (Pi and Helmann, 2018). A continuous release of engineered silver nanoparticles (AgNPs) occurs in the natural water environment. Therefore native biofilms, as the prominent life form of microorganisms in almost all known ecosystems, will be exposed to AgNP contact. Apart from the exponentially growing research activities worldwide, it is still difficult to assess nanoparticle-mediated toxicity in natural environments (Grün et al., 2018). This quantifiable population shift, even after small dose AgNP treatment, causes thoughtful concerns with respect to the broad use of AgNPs and their possibly unfavorable influence on the environmental function of lotic biofilms, such as biodegradation or biostabilization.

In bioelectrochemical reactors, conductive carbon felts (Cf) have been utilized as biofilm carriers to increase the electrical stimulation on treatment of phenol-containing synthetic wastewater. In batch test, phenol biodegradation has accelerated under an optimum direct current (DC), which was 2 mA for Cf biofilm carriers and lower than that for nonconductive white foam carriers. Microorganisms related with genera of *Zoogloea* and *Desulfovibrio* have been distinctively enriched under intermittent applied DC pattern (Ailijiang et al., 2016). This study has shown that electrical stimulation is potentially effective for biofilm reactors treating phenol-containing wastewater.

18.5 Conclusion, challenges, and future perspective

Biofilm activities are particularly appropriate for the treatment of unmanageable compounds because of their high microbial biomass and ability to immobilize compounds. Approaches for successful bioremediation productivity include genetic engineering to improve strains and chemotactic capability, the use of diverse population biofilms, and optimization of physicochemical environments (Singh et al., 2006). Before utilizing any microorganisms for bioremediation purposes, several questions and issues should be assessed and improved. For example, which microorganisms have an ability to form biofilms? Which surfaces are appropriate for biofilm growth? These are some basic questions that appear while designing a biofilm-based method and need to be deliberated. The restrictions of traditional microbiological approaches coupled with the essential nonrepresentative nature of sampling plans delay any full characterization of a microbial community and its specific role in the reduction process. However, the effect of high pollutant loads on bacterial communities is not well understood. Though, by applying an interdisciplinary mode, including bacteriological, molecular, and chemical analyses coupled with the detailed site characterization, the potential for remediation can be assessed by providing more than one association of support.

Acknowledgments

The author appreciates anonymous reviewers of the book for their valuable comments and suggestions to improve the quality of it.

References

Ailijiang, N., Chang, J., Liang, P., Li, P., Wu, Q., Zhang, X., et al., 2016. Electrical stimulation on biodegradation of phenol and responses of microbial communities in conductive carriers supported biofilms of the bioelectrochemical reactor. Bioresour. Technol. 201, 1–7.

Benedetti, I., de Lorenzo, V., Nikel, P.I., 2016. Genetic programming of catalytic *Pseudomonas putida* biofilms for boosting biodegradation of haloalkanes. Metab. Eng. 33, 109–118.

Branda, S.S., Vik, S., Friedman, L., et al., 2005. Biofilms: the matrix revisited. Trends Microbiol. 13, 20–26.

Carles, L., Gardon, H., Joseph, L., Sanchís, J., Farré, M., Artigas, J., 2019. Meta-analysis of glyphosate contamination in surface waters and dissipation by biofilms. Environ. Int. 124, 284–293.

Costerton, J.W., Cheng, K.J., Geesey, G.G., et al., 1987. Bacterial biofilms in nature and disease. Annu. Rev. Microbiol. 41, 435–464.

Dasgupta, D., Kumar, A., Mukhopadhyay, B., Sengupta, T.K., 2015. Isolation of phenazine 1,6-di-carboxylic acid from *Pseudomonas aeruginosa* strain HRW.1-S3 and its role in biofilm-mediated crude oil degradation and cytotoxicity against bacterial and cancer cells. Appl. Microbiol. Biotechnol. 99 (20), 8653–8665.

Davey, M.E., O'Toole, G.A., 2000. Microbial biofilms: from ecology to molecular genetics. Microbiol. Mol. Biol. Rev. 64, 847–867.

Decho, A.W., 2000. Microbial biofilms in intertidal systems: an overview. Cont. Shelf Res. 20, 1257–1273.

Edwards, K.J., Bond, P.L., Gihring, T.M., et al., 2000. An archaeal iron-oxidizing extreme acidophile important in acid mine drainage. Science 287, 1796–1799.

Elumalai, P., Parthipan, P., Narenkumar, J., Anandakumar, B., Madhavan, J., Oh, B.T., et al., 2019. Role of thermophilic bacteria (*Bacillus* and *Geobacillus*) on crude oil degradation and biocorrosion in oil reservoir environment. 3 Biotech 9 (3), 79.

Flemming, H.C., 2000. Sorption sites in biofilms. Water Sci. Technol. 32, 27–33.

Flemming, H.C., Wingender, J., 2001. Relevance of microbial extracellular polymeric substances (EPSs)—Part II: technical aspects. Water Sci. Technol. 43, 9–16.

Flemming, H.C., Wingender, J., 2010. The biofilm matrix. Nat. Rev. Microbiol. 8, 623–633.

Gewert, B., Plassmann, M.M., Macleod, M., 2015. Pathways for degradation of plastic polymers floating in the marine environment. Environ. Sci. Process Impacts 17, 1513–1521.

Ghiglione, J.-F., Martin-Laurent, F., Pesce, S., 2016. Microbial ecotoxicology: an emerging discipline facing contemporary environmental threats. Environ. Sci. Pollut. Res. 23, 3981–3983.

Godfrin, M.P., Sihlabela, M., Bose, A., Tripathi, A., 2018. Behavior of marine bacteria in clean environment and oil spill conditions. Langmuir 34 (30), 9047–9053.

Gonda, K.E., Jendrossek, D., Molitoris, H.P., 2000. Fungal degradation of the thermoplastic polymer poly-ß-hydroxybutyric acid (PHB) under simulated deep sea pressure BT. In: Liebezeit, G., Dittmann, S., Kröncke, I. (Eds.), Life at Interfaces and Under Extreme Conditions. Springer, Dordrecht, pp. 173–183.

Grün, A.Y., App, C.B., Breidenbach, A., Meier, J., Metreveli, G., Schaumann, G.E., et al., 2018. Effects of low dose silver nanoparticle treatment on the structure and community composition of bacterial freshwater biofilms. PLoS One 13 (6), e0199132.

Headly, J.V., Gandrass, J., Kuballa, J., Peru, K.M., Gong, Y., 1998. Rates of sorption and partitioning of contaminants in river biofilms. Environ. Sci. Technol. 32, 3968–3973.

Horemans, B., Breugelmans, P., Hofkens, J., et al., 2013. Environmental dissolved organic matter governs biofilm formation and subsequent linuron degradation activity of a linuron-degrading bacterial consortium. Appl. Environ. Microbiol. 79, 4534–4542.

Huang, J., Liu, S., Zhang, C., Wang, X., Pu, J., Ba, F., et al., 2019. Programmable and printable *Bacillus subtilis* biofilms as engineered living materials. Nat. Chem. Biol. 15 (1), 34–41.

Irankhah, S., Abdi, A.A., Reza, S.M., Gharavi, S., Ayati, B., 2018. Highly efficient phenol degradation in a batch moving bed biofilmreactor: benefiting from biofilm-enhancing bacteria. World J. Microbiol. Biotechnol. 34 (11), 164.

Jacquin, J., Cheng, J., Odobel, C., Pandin, C., Conan, P., Pujo-Pay, M., et al., 2019. Microbial ecotoxicology of marine plastic debris: a review on colonization and biodegradation by the "plastisphere". Front. Microbiol. 10, 865.

Kjelleberg, S., Molin, S., 2002. Is there a role for quorum sensing signals in bacterial biofilms? Curr. Opin. Microbiol. 5 (3), 254–258.

Krueger, M.C., Harms, H., Schlosser, D., 2015. Prospects for microbiological solutions to environmental pollution with plastics. Appl. Microbiol. Biotechnol. 99, 8857–8874.

Lacal, J., Reyes-Darias, J.A., García-Fontana, C., et al., 2013. Tactic responses to pollutants and their potential to increase biodegradation efficiency. J. Appl. Microbiol. 114, 923–933.

Lovely, D.R., Coates, J.D., 1997. Bioremediation of metal contamination. Curr. Opin. Biotechnol. 8, 285–289.

Macaskie, L.E., Yong, P., Doyle, T.C., et al., 1997. Bioremediation of uranium-bearing wastewater: biochemical and chemical factors influencing bioprocess application. Biotechnol. Bioeng. 53, 100–109.

Mangwani, N., Kumari, S., Das, S., 2016. Bacterial biofilms and quorum sensing: fidelity in bioremediation technology. Biotechnol. Genet. Eng. Rev. 32 (1-2), 43–73.

Morohoshi, T., Oi, T., Aiso, H., Suzuki, T., Okura, T., Sato, S., 2018. Biofilm formation and degradation of commercially available biodegradable plastic films by bacterial consortiums in freshwater environments. Microbes Environ. 33 (3), 332–335.

Mosharaf, M.K., Tanvir, M.Z.H., Haque, M.M., Haque, M.A., Khan, M.A.A., Molla, A.H., et al., 2018. Metal-adapted bacteria isolated from wastewaters produce biofilms by expressing proteinaceous curli fimbriae and cellulose nanofibers. Front. Microbiol. 9, 1334.

Mosier, A.P., Behnke, J., Jin, E.T., Cady, N.C., 2015. Microbial biofilms for the removal of Cu^{2+} from CMP wastewater. J. Environ. Manage. 160, 67–72.

Muyzer, G., Stams, A.J., 2008. The ecology and biotechnology of sulphate-reducing bacteria. Nat. Rev. Microbiol. 6, 441–454.

Nwachi, A.C., Onochie, C.C., Iroha, I.R., et al., 2016. Extraction of biosurfactants produced from bacteria isolated from waste-oil contaminated soil in abakaliki metropolis, ebonyi state. J. Biotechnol. Res. 2, 24–30.

Paul, D., Pandey, G., Pandey, J., Jain, R.K., 2005. Accessing microbial diversity for bioremediation and environmental restoration. Trends Biotechnol. 23, 135–142.

Perera, M., Wijayarathna, D., Wijesundera, S., Chinthaka, M., Seneviratne, G., Jayasena, S., 2019. Biofilm mediated synergistic degradation of hexadecane by a naturally formed community comprising *Aspergillus flavus* complex and *Bacillus cereus* group. BMC Microbiol. 19 (1), 84.

Pi, H., Helmann, J.D., 2018. Genome-wide characterization of the Fur regulatory network reveals a link between catechol degradation and bacillibactin metabolism in *Bacillus subtilis*. MBio 9 (5), pii: e01451-18.

Rabin, N., Zheng, Y., Opoku-Temeng, C., et al., 2015. Biofilm formation mech-anisms and targets for developing antibiofilm agents. Future Med. Chem. 7 (4), 493–512.

Reysenbach, A.L., Cady, S.L., 2001. Microbiology of ancient and modern hydrothermal systems. Trends Microbiol. 9, 79–86.

Singh, P., Cameotra, S.S., 2004. Enhancement of metal bioremediation by use of microbial surfactants. Biochem. Biophys. Res. Commun. 319, 291–297.

Singh, R., Paul, D., Jain, R.K., 2006. Biofilms: implications in bioremediation. Trends Microbiol. 14 (9), 389–397.

Sukačová, K., Kočí, R., Žídková, M., Vítěz, T., Trtílek, M., 2017. Novel insight into the process of nutrients removal using an algal biofilm: the evaluation of mechanism and efficiency. Int. J. Phytorem. 19 (10), 909–914.

Sutherland, I.W., 2001. The biofilm matrix—an immobilized but dynamic microbial environment. Trends Microbiol. 9, 222–227.

Tchounwou, P.B., Yedjou, C.G., Patlolla, A.K., Sutton, D.J., 2012. Heavy metals toxicity and the environment. EXS 101, 133–164.

Valls, M., de Lorenzo, Vc, 2002. Exploiting the genetic and biochemical capacities of bacteria for the remediation of heavy metal pollution. FEMS Microbiol. Rev. 26, 327–338.

Velmourougane, K., Prasanna, R., Saxena, A.K., 2017. Agriculturally important microbial biofilms: present status and future prospects. J. Basic Microbiol. 57 (7), 548–573.

Verhagen, P., Gelder, L.D., Hoefman, S., Vos, P.D., Boon, N., 2011. Planktonic versus biofilm catabolic communities: importance of the biofilm for species selection and pesticide degradation. Appl. Environ. Microbiol. 77 (14), 4728–4735.

Vlamakis, H., Chai, Y., Beauregard, P., Losick, R., Kolter, R., 2013. Sticking together: building a biofilm the *Bacillus subtilis* way. Nat. Rev. Microbiol. 11, 157–168.

Walczak, M., Swiontek, B.M., Sionkowska, A., Michalska, M., Jankiewicz, U., Deja-Sikora, E., 2015. Biofilm formation on the surface of polylactide during its biodegradation in different environments. Colloids Surf. B: Biointerfaces 1 (136), 340–345.

Walter, V., Syldatk, C., Hausmann, R., 2013. Screening concepts for the isolation of biosurfactant producing microorganisms. Madame Curie Bioscience Database. Landes Bioscience, Austin, TX, pp. 2000–2013 [Internet].

Willis, K., Denise Hardesty, B., Kriwoken, L., Wilcox, C., 2017. Differentiating littering, urban runoff and marine transport as sources of marine debris in coastal and estuarine environments. Sci. Rep. 7, 44479.

Wood, T.K., Hong, S.H., Ma, Q., 2011. Engineering biofilm formation and dispersal. Trends Biotechnol. 29 (2), 87–94.

Zeng, X.C., He, Z., Chen, X., Cao, Q.A.D., Li, H., Wang, Y., 2018. Effects of arsenic on the biofilm formations of arsenite-oxidizing bacteria. Ecotoxicol. Environ. Saf. 165, 1–10.

Zhang, J., Poh, C.L., 2018. Regulating exopolysaccharide gene wcaF allows control of *Escherichia coli* biofilm formation. Sci. Rep. 8 (1), 13127.

Genetic engineering approaches for detecting environmental pollutants

*Nisarg Gohil, Gargi Bhattacharjee and Vijai Singh**
Department of Biosciences, School of Science, Indrashil University, Rajpur, Mehsana, Gujarat, India
*Corresponding author

19.1 Introduction

Environmental pollution is one of the biggest global threats that the world is facing today. The adverse effects caused by pollutants on the environment, human health and other biotic community are dreadfully hazardous, so it is crucial to detect and degrade them at the earliest. Therefore, the identification and quantification of contaminants remain a foremost priority for bioremediation strategies. A biosensor is an analytical, biological sensing device for easy and rapid detection of analytes that comprises of bioreceptor in the form of microorganisms, antibodies, enzymes, cells, nucleic acids, etc. and a transducer to give out a measurable detection signal in the form of fluorescence, colorimetric, thermal or electrochemical response (Karim and Fakhruddin, 2012).

The advances gained in synthetic biology in the past decade have transformed the design of biosensors by incorporating synthetic genetic parts including riboswitches, promoters, biological gates, etc. (Mahmutoglu et al., 2012; Singh, 2014). Depending upon the biological sensing elements, biosensors are divided into four types:

1. Enzyme biosensors: These are the most commonly used biosensors that use enzyme as a sensing element (Martinkova et al., 2017). In this biosensor, usually the inhibition of specific activity of enzyme caused by the target molecules is measured.
2. Immunobiosensors: These biosensors are highly specific and selectively use antibody–antigen interactions as a biological sensing element.
3. Nucleic-acid based biosensors: These are devices whose sensing elements are based upon DNA or RNA. In DNA biosensors, the target molecules have an affinity toward DNA and resultant hybridization is measured. On the other hand, RNA biosensors measure signals from the process of RNA hybridization or other interactions.
4. Whole-cell biosensors: These types of biosensors use microorganisms as a sensing element. The microbes are designed in a way that reporter elements such as luciferase, green/red fluorescence protein or *lacZ* are attached under the control of an inducible promoter to detect the targets.

In this chapter, various engineered biosensors constructions for environmental pollutants are presented that will provide an insight into the latest biotechnological

Bioremediation of Pollutants. DOI: https://doi.org/10.1016/B978-0-12-819025-8.00019-3

advances that can be used to detect pollutants and develop next-generation biosensors.

19.2 Biosensors for detecting pollutants

Agency for Toxic Substances and Disease Registry (ATSDR, 2017) has provided a priority list of hazardous substances depending upon their significance for public health. This list provides guidance for screening the pollutants that need urgent attention for detection and remediation. For monitoring and identification, numerous microorganisms have been exercised to build biosensors. Some of them are meticulously described here.

19.2.1 Heavy metals detection

As per ATSDR's substance priority list 2017, arsenic (As) has been ranked first because of its widespread significant toxicity toward human health. The International Agency for Research on Cancer has considered arsenic as carcinogenic to humans (IARC, 2004). Therefore it becomes imperative to detect the presence of arsenic conveniently. Mainly, two types of detection platforms are available: (1) recombinant whole-cell based and (2) cell-free based biosensors. In most cases, the microbial biosensors developed for detection of heavy metals (HMs) contain HM resistant gene(s) as receptor molecules (Nigam and Shukla, 2015). In the case of arsenic, most bacteria acquire resistance toward arsenic via *ars* operon which pumps out arsenic from the cell. The arsenic resistant bacteria possess regulatory as well as structural gene(s) wherein regulatory gene codes for transcription repressor and structural gene encodes efflux pump (Chen et al., 1986; Ji and Silver, 1992). A biosensor could be developed in such a way that arsenic binds to the repressor protein and inhibits its functionality. The subsequent synthesis of reporter proteins gives out signals (Kaur et al., 2015). As a detection signal, different transduction mechanisms have been employed such as luciferase (Tauriainen et al., 1997; Petanen, 2001; Stocker et al., 2003; Ranjan et al., 2012; Sharma et al., 2013; Hou et al., 2014), *lacZ* (Stocker et al., 2003; Ivanina and Shuvaeva, 2009; Joshi et al., 2009; Date et al., 2010; Chiou et al., 2011; Cortes-Salazar et al., 2013) and green fluorescent protein (gfp) (Stocker et al., 2003; Kawakami et al., 2010; Siddiki et al., 2012; Chen et al., 2012; Truffer et al., 2014).

Likewise, cell-free based biosensors can be divided into DNA, aptamer and protein-based biosensors. In DNA-based biosensor, the detection signal is generated through the oxidative damage of DNA in presence of As(III) or As(V) (Ozsoz et al., 2003; Solanki et al., 2009). Aptamers are single-stranded DNAs or RNAs used to serve as biosensors because of their ability of selective binding. The affinity of the aptamer for arsenic induces aggregation of gold nanoparticles (AuNPs). As a result, changes in absorption or resonance of AuNPs can be detected (Kaur et al., 2015; Wu et al., 2012, 2013). In protein-based biosensor, some proteins which bear affinity toward arsenic in the form of inhibition, coordination, oxidative coupling or oxidation are used as biosensors (Parker et al., 2005; Cosnier et al., 2006; Sarkar et al., 2010; Fuku et al., 2012).

Lead (Pb, ATSDR 2017 Rank—2) is another lethal HM that poses serious threats by damaging the central and peripheral nervous systems. For detection, numerous research groups have built novel biosensors. For instance, *pbrR* based synthetically developed whole-cell Gram-negative bacterial biosensor. In this biosensor, *pbrR* gene was cloned under the control of a divergent promoter. Additionally, a promoterless, synthetic *gfp* gene was integrated as a detection signal in a plasmid. When the plasmids bearing the *pbrR* and *gfp* gene were transferred via conjugation into different Gram-negative bacteria (such as *Escherichia coli*, *Pseudomonas aeruginosa, Shewanella oneidensis, Enterobacter* sp.), they responded well and detected a concentration as low as 1 μg/mL Pb(II) in different water samples (Bereza-Malcolm et al., 2016). Later, the detection performance of whole-cell lead biosensor was improved by ten times through reconfiguration of six genes of lead resistance operon *pbr* and generating a positive feedback loop. In conventional *pbr* operon-based lead biosensor, *pbrR* lied on one side of promoter whereas *gfp* was located on the opposite side. In the improved biosensor, both *pbrR* and *gfp* were placed on the same side or under the control of two separate promoters (Jia et al., 2018).

Apart from As and Pb, mercury (Hg, ATSDR 2017 Rank—3) is well-known for its toxicity to humans. In an attempt to detect mercury, three different reporter gene-based multifaceted biosensor vectors were constructed successfully by amalgamating mercury inducible promoter (P_{mer}), regulatory gene (*merR*) and reporter gene (*lux*, *lacZ* or *gfp*) (Hansen and Sørensen, 2000). In a similar study, synthetic *merR* gene and promoter were cloned along with *gfp* reporter gene in pUC19 and then transformed into *E. coli* BL21 (DE3) for detecting mercury. The resultant biosensor worked well and detected mercury concentration as low as 10^{-8} M

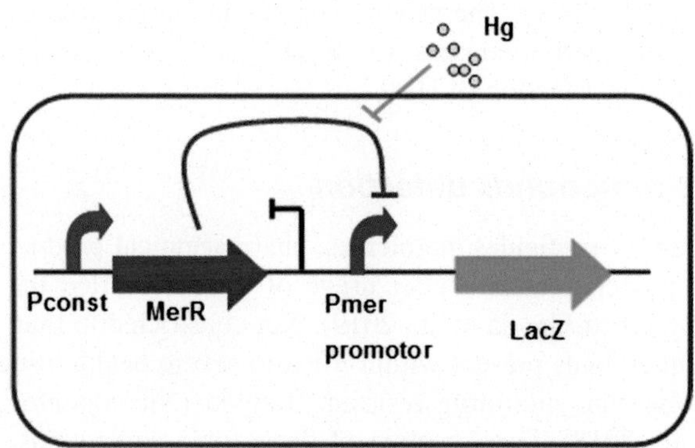

Figure 19.1 The design of mercury responsive biosensor. The *MerR* is expressed under the control of a constitutive synthetic promoter that binds with the operator sequences of the Hg responsive promoter (Pmer), which is inducible with mercury (Hg). When Hg is present in the sample, it will express LacZ which can be detected with color by adding substrate.

(Roointan et al., 2015). In an identical study, *Vibrio fischeri lux* cassette was used and the derived construct was integrated into *E. coli* JM109 (Pepi et al., 2006). Fig. 19.1 shows a design of mercury responsive *lacZ* reporter gene-based biosensor. The shown reporter gene could also be replaced with *gfp* or *lux* gene.

Similarly, a whole-cell biosensor was constructed for sensing toxic cadmium (ATSDR 2017 Rank—7) metal ions by expressing *cadC* gene and promoter *cad* of *Staphylococcus aureus* plasmid pI258 with *gfp* gene in *E. coli* DH5α. The reaction time was 15 minutes with a detection limit of 10 μg/L (Kumar et al., 2017). Additionally, luciferase reporter gene was expressed under the same promoter and resistance determinant in *S. aureus* RN4220 and *Bacillus subtilis* BR151. The resultant luminescent sensor detected cadmium, lead and zinc (ATSDR 2017 Rank—75) (Tauriainen et al., 1998). A novel pigment-based colorimetric biosensor was developed by engineering red pigment-producing *Deinococcus radiodurans*. The engineered strain was developed by introducing the *crtI* gene (carotenoid producing gene) under the control of Cd-inducible promoter DR_0659 in *crtI*-deleted mutant strain. The resultant strain changed from light yellow to red when it came in contact with Cd while giving no response to other metals. The detection range for Cd was 50 nM to 1 mM (Joe et al., 2012). Moreover, a dual reporter-based biosensor has been made for detecting zinc and copper (ATSDR 2017 Rank—118).

Significant efforts have also been taken to detect harmful nickel (ATSDR 2017 Rank—57) and cobalt (ATSDR 2017 Rank—51). For example, *coaLux* and *nrsLux* reporter-based bioluminescent biosensors have been developed for the detection of nickel, cobalt and zinc. In this biosensor, luciferase reporter gene was fused with *coaT* or *nrsBACD* promoter in *Synechocystis* sp. PCC 6803. Among them, the *nrsLux* expressing strain was specific for nickel (detection range 0.2–6.0 μM), whereas the *coaLux* reporter could detect both cobalt and zinc (detection range 0.3–6 and 1–3 μM, respectively). Interestingly, the detection range was enhanced by about four times for *coaLux* when the strain was incubated in dark condition. On the contrary, the *nrsLux* containing strain showed inhibition in the detection in dark condition (Peca et al., 2008). The genetic engineering along with synthetic biology can be applied in microorganisms to develop a more sensitive, specific and portable biosensor for detection of HMs.

19.2.2 Toxic compounds detection

The unpleasant use of pesticides, herbicides, pharmaceutical products and industrial products and improper treatment after usage of thereof baffled the functioning of the ecosystem (Ramírez-García et al., 2019). Not constricted to that, the toxic compounds and antimicrobials present within it led to severe health issues and turn tiny microbes to evolve into multidrug resistant bacteria (Nicolopoulou-Stamati et al., 2016; Gohil et al., 2018). For detection of these toxic compounds, numerous biosensors have been developed.

Pseudomonas putida is known for its environmental adaptability. Even in high concentrations of toxic organic compounds, many strains can easily survive and efficiently metabolize number of pollutants including benzene (ATSDR 2017 Rank

—6), toluene (ATSDR 2017 Rank—74), ethylbenzene (ATSDR 2017 Rank—132), *o,m,p*-xylenes (ATSDR 2017 Rank—65) (BTEX) and other aromatic compounds (Otenio et al., 2005; You et al., 2013). The strain *P. putida* DOT-T1E was engineered to make whole-cell type microbial biosensor for detecting structurally diverse antibiotics and contaminants such as flavonoids and toluene by incorporating *TtgR* (a reporter repressor that has an ability to bind to different antibiotics) regulated TtgABC efflux pump, coupled to a GFP. The resultant engineered strain gained more resistance toward various toxic compounds (Espinosa-Urgel et al., 2015).

Cyanide (ATSDR 2017 Rank—35) is one of the most toxic chemical compounds as it binds with cytochrome *c* oxidase and prevents the transport of electrons to oxygen which eventually inhibits the oxidative respiration and ATP production (Way, 1984). Chemical and mining industries, as well as petroleum refineries, discharge their effluent in the rivers. This effluent contains cyanide as sodium cyanide or free cyanide form making the rivers polluted (Nagy and Kónya, 2003; Sawaraba and Rao, 2015). In order to monitor the cyanide pollution in river water, a microbial whole-cell reactor type cyanide sensor was constructed using *Saccharomyces cerevisiae*. The reactor was immobilized by engineered yeast beads along with two oxygen electrodes. In principle, cyanide present in the water body inhibits yeast's respiration and the activity of respiration is determined by oxygen electrodes (Ikebukuro et al., 1996). A similar type of sensor was also built using cyanide oxidating microbes, specifically *Pseudomonas fluorescens* NCIMB 11764. However, the stability of this biosensor was short (Lee and Karube, 1995).

Like cyanide, organophosphate (OP) is another toxic pollutant used majorly in agriculture fields as a constituent of pesticides. The growing safety concern about toxic OP has stimulated the development of potentiometric microbial biosensor using organophosphorus hydrolase (OPH) expressing recombinant *E. coli*. The biosensor was designed and constructed in a way that the OPH catalyzes the hydrolysis of OP and the released protons are measured using a pH electrode. The amount of hydrolyzed substrate is directly proportional to the released protons (Mulchandani et al., 1998). Similarly, a wild-type naturally endowed OP degrading *Flavobacterium* sp. was also employed to construct a sensor (Gäberlein et al., 2000). It has been proven that the OPH activity of *Flavobacterium* sp. is associated with the plasma membrane (Mulbry and Karns, 1989). As a consequence, the use of membrane fractions of *Flavobacterium* sp. is a better alternative than using the whole cells (Gäberlein et al., 2000). Apart from microbial biosensors, numerous enzymatic biosensors have been developed using an acetyl-cholinesterase enzyme (Ivanov et al., 2010; Choi et al., 2001; Andreou and Clonis, 2002; Law and Higson, 2005; Rekha and Murthy, 2008; Dhull et al., 2013).

Phenolic compounds are also included in prioritized compounds list of many environment concerning agencies. These compounds are predominantly discharged in water bodies from natural sources (such as decomposition of organic matter, synthesis by microorganisms and plants) and/or anthropogenic sources (such as industrial, agriculture, domestic and municipal waste). The exposure of phenol can cause skin irritation, necrosis, dermatitis and may even harm the heart, kidney, liver and central nervous system (Anku et al., 2017). *Moraxella* sp. is recognized to degrade *p*-nitrophenol enzymatically by 4-nitrophenol monooxygenase and convert into hydroquinone. A group of researchers have taken advantage of this property to build a microbial sensor for the detection of *p*-nitrophenol. In their study, the generated electrochemical oxidation current in the process of degradation was measured at carbon paste electrode immobilized by *Moraxella* sp. The minimum detection level for *p*-nitrophenol was 20 nM (Mulchandani et al., 2005). Likewise, other microbial species such as *Rhodococcus* (Riedel et al., 1993), *Trichosporon* (Riedel et al., 1995), *Arthrobacter* (Lei et al., 2003) and *Pseudomonas* sp. (Timur et al., 2003) have also been employed.

In enzymatic biosensors, tyrosinase (Russell and Burton, 1999; Freire et al., 2002; Abdullah et al., 2006; Yildiz et al., 2007; Tembe et al., 2007; Silva et al., 2010) and laccase (Gomes and Rebelo, 2003; Abdullah et al., 2007) are the most commonly used enzymes for development of enzymatic biosensors for detecting phenolic compounds. In principle, the enzymes are first oxidized by oxygen and then reduced by phenolic compounds (Karim and Fakhruddin, 2012). This type of biosensors faces a major issue of reduced signal response and selectivity.

Unicellular green alga has also been employed in detecting many pesticides, herbicides or algicides such as diuron (ATSDR 2017 Rank—254), atrazine, simazine, ioxynil, bromoxynil, dinoseb, terbutryin, prometryne terbuthyl-azine, linuron, and isoproturon (Maly et al., 2005; Koblizek et al., 1998; Koblízek et al., 2002; Giardi et al., 2009; Husu et al., 2013; Scognamiglio et al., 2009, 2013; Masojídek et al., 2011; Avramescu et al., 1999; Bucur et al., 2018). The cyanobacteria are distinct photosynthetic microorganisms having photosystem II (PSII), a specialized membrane protein complex that makes use of light energy to catalyze electron transfer. This electron transfer activity of PSII can be inhibited by numerous pesticides (Bucur et al., 2018), allowing us to construct the biosensors. For example, site-directed mutagenesis in D1 PSII protein of *Chlamydomonas reinhardtii* resulted in atrazine, prometryn and diuron sensitive mutants (Scognamiglio et al., 2009).

19.2.3 Explosives' residues detection

The substantial production and use of explosives have significantly contributed to environmental pollution. In order to detect the residues of explosives many whole-cell based biosensors have been designed (Shemer et al., 2015). In one such attempt, an engineered olfactory yeast strain *S. cerevisiae* WIF-1α was developed by incorporating *gfp* gene with cAMP synthesis. Upon exposure to 2,4-dinitrotoluene (DNT), a dose-dependent fluorescent signal was produced (Radhika et al., 2007). Just as in yeast, soil bacterium *P. putida* was also used for detection of DNT by introducing toluene sensitive XylR regulator along with *lux*AB or GFP as reporting element under the control of Po or Pu promoter, respectively (Garmendia et al., 2008; Behzadian et al., 2011).

A library of around 2000 *E. coli* clones bearing *GFPmut2* gene was fused with different promoters. Among them, two gene promoters (*yqjF* and *ybiJ*) were screened based upon fluorescence response when exposed to 2,4-DNT. Afterwards,

Table 19.1 Important genetically engineered biosensors.

Analytes	Chassis cells	Limit of detection	References
Heavy metals			
Arsenic	*Escherichia coli*	1–100 μg/L	Merulla and van der Meer (2016)
Arsenic	*E. coli*	0.75–3 μg/L	Li et al. (2015)
Arsenic	*Shewanella oneidensis*	40–100 μM	Webster et al. (2014)
Lead	*Enterobacter* sp. NCR3, *Enterobacter* sp. LCR17, *E. coli*, *Pseudomonas aeruginosa* PAO1, or *S. oneidensis* MR-1	0.2–1 μg/mL	Bereza-Malcolm et al. (2016)
Lead	*E. coli* DH5α	0.01 μM	Jia et al. (2018)
Mercury	*E. coli* BL21 (DE3)	0.01 μM	Roointan et al. (2015)
Mercury	*E. coli* JM109	0.002 μg/g d.w.	Pepi et al. (2006)
Cadmium	*E. coli* DH5α	10 μg/L	Kumar et al. (2017)
Cadmium, lead	*Bacillus subtilis*, *Staphylococcus aureus*	Cd^{2+}: 3.3–10 nM, Pb^{2+}: 33 nM	Tauriainen et al. (1998)
Cadmium	*Deinococcus radiodurans*	50 nM	Joe et al. (2012)
Nickel, cobalt, zinc	*Synechocystis* sp. PCC 6803	Ni^{2+}: 0.2–6.0 μM, Co^{2+}: 0.3–6 μM, Zn^{2+}: 1–3 μM	Peca et al. (2008)
Explosives' residues			
2,4-DNT	*Saccharomyces cerevisiae* WIF-1α	Unreported	Radhika et al. (2007)
DNT	*Pseudomonas putida*	Unreported	Garmendia et al. (2008)
DNT, TNT	*E. coli*	2,4-DNT: 5 mg/L	Yagur-Kroll et al. (2014)
TNT	*Escherichia coli* K-12 MG1655	4.75 mg/L	Tan et al. (2015)
DNT	*E. coli*	10 mM	Lönneborg et al. (2012)
DNT	*E. coli*	9.1 mg/L	Davidson et al. (2012)

d.w., Dry weight; *DNT*, dinitrotoluene; *TNT*, trinitrotoluene.

the activity of *yqjF*-based sensor was improved by performing two rounds of random mutagenesis. These uncharacterized promoter genes were cloned in the upstream of *Photorhabdus luminescens lux*CDABE cluster to generate two novel reporter strains, which exhibited a dose-dependent response to 2,4-DNT and 2,4,6-trinitrotoluene (TNT) (Yagur-Kroll et al., 2014). The luminescence intensity was further improved by over 3000-fold and the detection limit was minimized by 75% (Yagur-Kroll et al., 2015). In a similar approach, instead of a single gene promoter, five gene promoters (*yadG*, *yqgC*, *aspC*, *recE* and *topA*) were fused with GFP and the detection threshold of 4.75 mg/L for TNT was achieved (Tan et al., 2015).

DntR, a transcriptional regulator, found within bacteria is known for its capability to degrade 2,4-DNT (de Las Heras et al., 2011). In a study, the DntR was integrated along with GFP in *E. coli* and directed evolution studies were performed. Higher GFP expression providing mutants were selected. The resultant mutants showed a lower detection limit for DNT (10 mM, 25-fold enhancement) referenced to wild-type cells (Lönneborg et al., 2012).

Riboswitch, a tool of synthetic biology, is a small noncoding RNA presents upstream of messenger RNA that regulates downstream or upstream genes in the presence or absence of specific ligand molecule (Patel et al., 2018). This riboswitch could be designed in a way that in the presence of 2,4-DNT or TNT, riboswitch activates expression of a reporter gene (for example, DNT reactive synthetic riboswitch). In this riboswitch, tobacco etch virus (TEV) protease encoding gene was placed in the downstream of the riboswitch. Additionally, fluorescence resonance energy transfer-based fusion protein with the TEV protease cleavage site was also constructed in a plasmid. As a consequence, riboswitch was activated and translated TEV protease in the presence of DNT. This protease cleaved linker in the fusion protein thus inducing the fluorescence signal (Davidson et al., 2012). As riboswitches have the capacity to regulate gene expression, more studies are expected for detection of pollutants.

Some important genetically engineered microbial biosensors developed for the determination of various pollutants are summarized in Table 19.1.

19.3 Conclusion and future remarks

Currently, industrial development is at its peak, considering the high public demand ultimately producing huge amounts of pollutants. Apart from industrial plants, combustion of fossil fuels; synthetic toxic chemicals; improper treatment of effluents; lack of proper waste management and awareness; unpleasant use of fertilizers and pesticides; plastics etc. are also major contributors for environmental pollution (Ramírez-García et al., 2019). Among the generated pollutants, some are highly toxic and need urgent attention for detection and degradation. The ATSDR has released a priority list of noxious substances depending upon their significance on human health. This can be used as guidance for screening the pollutants that need serious attention for detection and remediation. According to the priority list, HMs such as arsenic (As), lead (Pb) and mercury (Hg) are ranked as the top three priority hazardous pollutants. Additionally, numerous toxic pollutants such as cyanide, toluene, OP and phenolic compounds, major constituents of herbicides, pesticides and fertilizers, are also listed in the substance priority list. To curb these pollutants, many novel biological sensors have been developed, in which enzymatic and electrochemical types are the most commonly used biosensors. However, the biosensor comes with a limitation. If the pollutants present in the samples are below the threshold level of biosensors, the pollutants go undetected. Recent advances in synthetic genetic systems have opened up new avenues and boosted the selectivity, reaction time, sensitivity, affordability, and detection threshold of biosensors from time to time. The advent in genome editing technologies, especially CRISPR–Cas9 (clustered regularly interspaced short palindromic repeats-CRISPR associated protein 9) (Jinek et al., 2012; Singh et al., 2017, 2018; Bhattacharjee et al., 2020) and CRISPR-Cas13 technology (Gootenberg et al., 2018; Khambhati et al., 2019), may contribute more for construction of biosensor in the near future.

Acknowledgements

The authors gratefully acknowledge the Gujarat State Biotechnology Mission, Gandhinagar (GSBTM Project ID: 5LY45F) for providing financial support and also Indrashil University for providing infrastructure and facility.

References

Abdullah, J., Ahmad, M., Karuppiah, N., Heng, L.Y., Sidek, H., 2006. Immobilization of tyrosinase in chitosan film for an optical detection of phenol. Sens. Actuators, B 114 (2), 604–609.

Abdullah, J., Ahmad, M., Karuppiah, N., Heng, L.Y., Sidek, H., 2007. An optical biosensor based on immobilization of laccase and MBTH in stacked films for the detection of catechol. Sensors 7 (10), 2238–2250.

Agency for Toxic Substances and Disease Registry (ATSDR), 2017. ASTDR's Substance Priority List, Atlanta. <https://www.atsdr.cdc.gov/spl/index.html> (accessed 23.07.19).

Andreou, V.G., Clonis, Y.D., 2002. A portable fiber-optic pesticide biosensor based on immobilized cholinesterase and sol–gel entrapped bromcresol purple for in-field use. Biosens. Bioelectron. 17 (1–2), 61–69.

Anku, W.W., Mamo, M.A., Govender, P.P., 2017. Phenolic compounds in water: sources, reactivity, toxicity and treatment methods. In: Soto-Hernández, M. (Ed.), Phenolic Compounds-Natural Sources, Importance and Applications. InTech, pp. 419–443.

Avramescu, A., Rouillon, R., Carpentier, R., 1999. Potential for use of a cyanobacterium *Synechocystis* sp. immobilized in poly (vinylalcohol): application to the detection of pollutants. Biotechnol. Tech 13 (8), 559–562.

Behzadian, F., Barjeste, H., Hosseinkhani, S., Zarei, A.R., 2011. Construction and characterization of *Escherichia coli* whole-cell biosensors for toluene and related compounds. Curr. Microbiol. 62 (2), 690–696.

Bereza-Malcolm, L., Aracic, S., Franks, A., 2016. Development and application of a synthetically-derived lead biosensor construct for use in Gram-negative bacteria. Sensors 16 (12), 2174.

Bhattacharjee, G., Mani, I., Gohil, N., Khambhati, K., Braddick, D., Panchasara, H., et al., 2020. CRISPR technology for genome editing. In: Faintuch, J., Faintuch, S. (Eds.), Precision Medicine for Investigators, Practitioners and Providers. Academic Press, London, pp. 59–69.

Bucur, B., Munteanu, F.D., Marty, J.L., Vasilescu, A., 2018. Advances in enzyme-based biosensors for pesticide detection. Biosensors 8 (2), 27.

Chen, C.M., Misra, T.K., Silver, S., Rosen, B.P., 1986. Nucleotide sequence of the structural genes for an anion pump. The plasmid-encoded arsenical resistance operon. J. Biol. Chem. 261 (32), 15030–15038.

Chen, J., Zhu, Y.G., Rosen, B.P., 2012. A novel biosensor selective for organoarsenicals. Appl. Environ. Microbiol. 78 (19), 7145–7147.

Chiou, C.H., Chien, L.J., Chou, T.C., Lin, J.L., Tseng, J.T., 2011. Rapid whole-cell sensing chip for low-level arsenic detection. Biosens. Bioelectron. 26 (5), 2484–2488.

Choi, J.W., Kim, Y.K., Lee, I.H., Min, J., Lee, W.H., 2001. Optical organophosphorus biosensor consisting of acetylcholinesterase/viologen hetero Langmuir–Blodgett film. Biosens. Bioelectron. 16 (9–12), 937–943.

Cortes-Salazar, F., Beggah, S., van der Meer, J.R., Girault, H.H., 2013. Electrochemical As (III) whole-cell based biochip sensor. Biosens. Bioelectron. 47, 237–242.

Cosnier, S., Mousty, C., Cui, X., Yang, X., Dong, S., 2006. Specific determination of As(V) by an acid phosphatase-polyphenol oxidase biosensor. Anal. Chem. 78 (14), 4985–4989.

Date, A., Pasini, P., Daunert, S., 2010. Integration of spore-based genetically engineered whole-cell sensing systems into portable centrifugal microfluidic platforms. Anal. Bioanal. Chem. 398 (1), 349–356.

Davidson, M.E., Harbaugh, S.V., Chushak, Y.G., Stone, M.O., Kelley-Loughnane, N., 2012. Development of a 2,4-dinitrotoluene-responsive synthetic riboswitch in *E. coli* cells. ACS Chem. Biol. 8 (1), 234–241.

de Las Heras, A., Chavarría, M., de Lorenzo, V., 2011. Association of dnt genes of *Burkholderia* sp. DNT with the substrate-blind regulator DntR draws the evolutionary itinerary of 2,4-dinitrotoluene biodegradation. Mol. Microbiol. 82 (2), 287–299.

Dhull, V., Gahlaut, A., Dilbaghi, N., Hooda, V., 2013. Acetylcholinesterase biosensors for electrochemical detection of organophosphorus compounds: a review. Biochem. Res. Int. 2013, 731501.

Espinosa-Urgel, M., Serrano, L., Ramos, J.L., Fernández-Escamilla, A.M., 2015. Engineering biological approaches for detection of toxic compounds: a new microbial biosensor based on the *Pseudomonas putida* TtgR repressor. Mol. Biotechnol. 57 (6), 558–564.

Freire, R.S., Durana, N., Kubota, L.T., 2002. Electrochemical biosensor-based devices for continuous phenols monitoring in environmental matrices. J. Braz. Chem. Soc. 13 (4), 119–123.

Fuku, X., Iftikar, F., Hess, E., Iwuoha, E., Baker, P., 2012. Cytochrome *c* biosensor for determination of trace levels of cyanide and arsenic compounds. Anal. Chim. Acta 730, 49–59.

Gäberlein, S., Spener, F., Zaborosch, C., 2000. Microbial and cytoplasmic membrane-based potentiometric biosensors for direct determination of organophosphorus insecticides. Appl. Microbiol. Biotechnol. 54 (5), 652–658.

Garmendia, J., De Las Heras, A., Galvão, T.C., De Lorenzo, V., 2008. Tracing explosives in soil with transcriptional regulators of *Pseudomonas putida* evolved for responding to nitrotoluenes. Microb. Biotechnol. 1 (3), 236–246.

Giardi, M.T., Scognamiglio, V., Rea, G., Rodio, G., Antonacci, A., Lambreva, M., et al., 2009. Optical biosensors for environmental monitoring based on computational and biotechnological tools for engineering the photosynthetic D1 protein of *Chlamydomonas reinhardtii*. Biosens. Bioelectron. 25 (2), 294–300.

Gohil, N., Ramírez-García, R., Panchasara, H., Patel, S., Bhattacharjee, G., Singh, V., 2018. Book review: quorum sensing vs. quorum quenching: a battle with no end in sight. Front. Cell. Infect. Microbiol. 8, 106.

Gomes, S.A.S.S., Rebelo, M.J.F., 2003. A new laccasse biosensor for polyphenol determination. Sensors 3 (6), 166–175.

Gootenberg, J.S., Abudayyeh, O.O., Kellner, M.J., Joung, J., Collins, J.J., Zhang, F., 2018. Multiplexed and portable nucleic acid detection platform with Cas13, Cas12a, and Csm6. Science 360 (6387), 439–444.

Hansen, L.H., Sørensen, S.J., 2000. Versatile biosensor vectors for detection and quantification of mercury. FEMS Microbiol. Lett. 193 (1), 123–127.

Hou, Q.H., Ma, A.Z., Lv, D., Bai, Z.H., Zhuang, X.L., Zhuang, G.Q., 2014. The impacts of different long-term fertilization regimes on the bioavailability of arsenic in soil: integrating chemical approach with *Escherichia coli arsRp*::*luc*-based biosensor. Appl. Microbiol. Biotechnol. 98 (13), 6137–6146.

Husu, I., Rodio, G., Touloupakis, E., Lambreva, M.D., Buonasera, K., Litescu, S.C., et al., 2013. Insights into photo-electrochemical sensing of herbicides driven by *Chlamydomonas reinhardtii* cells. Sens. Actuators, B 185, 321–330.

IARC, 2004. Arsenic in Drinking-Water, 84. International Agency for Research on Cancer (IARC), Lyon, France, pp. 39–267.

Ikebukuro, K., Miyata, A., Cho, S.J., Nomura, Y., Yamauchi, Y., Hasebe, Y., et al., 1996. Microbial cyanide sensor for monitoring river water. J. Biotechnol. 48 (1–2), 73–80.

Ivanina, A.V., Shuvaeva, O.V., 2009. Use of a bacterial biosensor system for determining arsenic in natural waters. J. Anal. Chem. 64 (3), 310–315.

Ivanov, Y., Marinov, I., Gabrovska, K., Dimcheva, N., Godjevargova, T., 2010. Amperometric biosensor based on a site-specific immobilization of acetylcholinesterase via affinity bonds on a nanostructured polymer membrane with integrated multiwall carbon nanotubes. J. Mol. Catal. B: Enzym. 63 (3–4), 141–148.

Ji, G., Silver, S., 1992. Regulation and expression of the arsenic resistance operon from *Staphylococcus aureus* plasmid pI258. J. Bacteriol. 174 (11), 3684–3694.

Jia, X., Zhao, T., Liu, Y., Bu, R., Wu, K., 2018. Gene circuit engineering to improve the performance of a whole-cell lead biosensor. FEMS Microbiol. Lett. 365 (16), fny157.

Jinek, M., Chylinski, K., Fonfara, I., Hauer, M., Doudna, J.A., Charpentier, E., 2012. A programmable dual-RNA-guided DNA endonuclease in adaptive bacterial immunity. Science 337 (6096), 816–821.

Joe, M.H., Lee, K.H., Lim, S.Y., Im, S.H., Song, H.P., Lee, I.S., et al., 2012. Pigment-based whole-cell biosensor system for cadmium detection using genetically engineered *Deinococcus radiodurans*. Bioprocess. Biosyst. Eng. 35 (1–2), 265–272.

Joshi, N., Wang, X., Montgomery, L., Elfick, A., French, C., 2009. Novel approaches to biosensors for detection of arsenic in drinking water. Desalination 248 (1–3), 517–523.

Karim, F., Fakhruddin, A.N., 2012. Recent advances in the development of biosensor for phenol: a review. Rev. Environ. Sci. Biotechnol. 11 (3), 261–274.

Kaur, H., Kumar, R., Babu, J.N., Mittal, S., 2015. Advances in arsenic biosensor development—a comprehensive review. Biosens. Bioelectron. 63, 533–545.

Kawakami, Y., Siddiki, M.S.R., Inoue, K., Otabayashi, H., Yoshida, K., Ueda, S., et al., 2010. Biosens. Bioelectron. 26 (4), 1466–1473.

Khambhati, K., Bhattacharjee, G., Singh, V., 2019. Current progress in CRISPR-based diagnostic platforms. J. Cell. Biochem. 120 (3), 2721–2725.

Koblizek, M., Masojidek, J., Komenda, J., Kucera, T., Pilloton, R., Mattoo, A.K., et al., 1998. A sensitive photosystem II-based biosensor for detection of a class of herbicides. Biotechnol. Bioeng. 60 (6), 664–669.

Koblízek, M., Malý, J., Masojídek, J., Komenda, J., Kučera, T., Giardi, M.T., et al., 2002. A biosensor for the detection of triazine and phenylurea herbicides designed using photosystem II coupled to a screen-printed electrode. Biotechnol. Bioeng. 78 (1), 110–116.

Kumar, S., Verma, N., Singh, A.K., 2017. Development of cadmium specific recombinant biosensor and its application in milk samples. Sens. Actuators, B 240, 248–254.

Law, K.A., Higson, S.P., 2005. Sonochemically fabricated acetylcholinesterase microelectrode arrays within a flow injection analyser for the determination of organophosphate pesticides. Biosens. Bioelectron. 20 (10), 1914–1924.

Lee, J.I., Karube, I., 1995. A novel microbial sensor for the determination of cyanide. Anal. Chim. Acta 313 (1–2), 69–74.

Lei, Y., Mulchandani, P., Chen, W., Wang, J., Mulchandani, A., 2003. A microbial biosensor for *p*-nitrophenol using *Arthrobacter* sp. Electroanalysis 15 (14), 1160–1164.

Li, L., Liang, J., Hong, W., Zhao, Y., Sun, S., Yang, X., et al., 2015. Evolved bacterial biosensor for arsenite detection in environmental water. Environ. Sci. Technol. 49 (10), 6149–6155.

Lönneborg, R., Varga, E., Brzezinski, P., 2012. Directed evolution of the transcriptional regulator DntR: isolation of mutants with improved DNT-response. PLoS One 7 (1), e29994.

Mahmutoglu, I., Pei, L., Porcar, M., Armstrong, R., Bedau, M., 2012. Bioremediation. In: Schmidt, M. (Ed.), Synthetic Biology: Industrial and Environmental Applications. John Wiley & Sons, Weinheim, pp. 67–101.

Maly, J., Masojidek, J., Masci, A., Ilie, M., Cianci, E., Foglietti, V., et al., 2005. Direct mediatorless electron transport between the monolayer of photosystem II and poly (mercapto-*p*-benzoquinone) modified gold electrode—new design of biosensor for herbicide detection. Biosens. Bioelectron. 21 (6), 923–932.

Martinkova, P., Kostelnik, A., Válek, T., Pohanka, M., 2017. Main streams in the construction of biosensors and their applications. Int. J. Electrochem. Sci. 12 (8), 7386–7403.

Masojídek, J., Souček, P., Máchová, J., Frolík, J., Klem, K., Malý, J., 2011. Detection of photosynthetic herbicides: algal growth inhibition test vs. electrochemical photosystem II biosensor. Ecotoxicol. Environ. Saf. 74 (1), 117–122.

Merulla, D., van der Meer, J.R., 2016. Regulatable and modulable background expression control in prokaryotic synthetic circuits by auxiliary repressor binding sites. ACS Synth. Biol. 5 (1), 36–45.

Mulbry, W.W., Karns, J.S., 1989. Purification and characterization of three parathion hydrolases from Gram-negative bacterial strains. Appl. Environ. Microbiol. 55 (2), 289–293.

Mulchandani, A., Mulchandani, P., Kaneva, I., Chen, W., 1998. Biosensor for direct determination of organophosphate nerve agents using recombinant *Escherichia coli* with surface-expressed organophosphorus hydrolase. 1. Potentiometric microbial electrode. Anal. Chem. 70 (19), 4140–4145.

Mulchandani, P., Hangarter, C.M., Lei, Y., Chen, W., Mulchandani, A., 2005. Amperometric microbial biosensor for p-nitrophenol using *Moraxella* sp.-modified carbon paste electrode. Biosens. Bioelectron. 21 (3), 523–527.

Nagy, N.M., Kónya, J., 2003. Long term effects of cyanide pollution of the river Tisza. In: Barany, S. (Ed.), Role of Interfaces in Environmental Protection. Springer, Dordrecht, pp. 129–134.

Nicolopoulou-Stamati, P., Maipas, S., Kotampasi, C., Stamatis, P., Hens, L., 2016. Chemical pesticides and human health: the urgent need for a new concept in agriculture. Front. Public Health 4, 148.

Nigam, V.K., Shukla, P., 2015. Enzyme based biosensors for detection of environmental pollutants—a review. J. Microbiol. Biotechnol. 25 (11), 1773–1781.

Otenio, M.H., Silva, M.T.L.D., Marques, M.L.O., Roseiro, J.C., Bidoia, E.D., 2005. Benzene, toluene and xylene biodegradation by *Pseudomonas putida* CCMI 852. Braz. J. Microbiol. 36 (3), 258–261.

Ozsoz, M., Erdem, A., Kara, P., Kerman, K., Ozkan, D., 2003. Electrochemical biosensor for the detection of interaction between arsenic trioxide and DNA based on guanine signal. Electroanalysis 15 (7), 613–619.

Parker, K.J., Kumar, S., Pearce, D.A., Sutherland, A.J., 2005. Design, synthesis and evaluation of a fluorescent peptidyl sensor for the selective recognition of arsenite. Tetrahedron Lett. 46 (41), 7043–7045.

Patel, S., Panchasara, H., Braddick, D., Gohil, N., Singh, V., 2018. Synthetic small RNAs: current status, challenges, and opportunities. J. Cell. Biochem. 119 (12), 9619–9639.

Peca, L., Kós, P.B., Máté, Z., Farsang, A., Vass, I., 2008. Construction of bioluminescent cyanobacterial reporter strains for detection of nickel, cobalt and zinc. FEMS Microbiol. Lett. 289 (2), 258–264.

Pepi, M., Reniero, D., Baldi, F., Barbieri, P., 2006. A comparison of mer::lux whole cell biosensors and moss, a bioindicator, for estimating mercury pollution. Water Air Soil Pollut. 173 (1–4), 163–175.

Petanen, T., 2001. Assessment of Bioavailable Concentrations and Toxicity of Arsenite and Mercury in Contaminated Soils and Sediments by Bacterial Biosensors (Doctoral dissertation). University of Helsinki, Helsinki, Finland.

Radhika, V., Proikas-Cezanne, T., Jayaraman, M., Onesime, D., Ha, J.H., Dhanasekaran, D. N., 2007. Chemical sensing of DNT by engineered olfactory yeast strain. Nat. Chem. Biol. 3 (6), 325–330.

Ramírez-García, R., Gohil, N., Singh, V., 2019. Recent advances, challenges, and opportunities in bioremediation of hazardous materials. In: Pandey, V.C., Bauddh, K. (Eds.), Phytomanagement of Polluted Sites. Elsevier, Amsterdam, pp. 517–568.

Ranjan, R., Rastogi, N.K., Thakur, M.S., 2012. Development of immobilized biophotonic beads consisting of *Photobacterium leiognathi* for the detection of heavy metals and pesticide. J. Hazard. Mater. 225, 114–123.

Rekha, K., Murthy, B.N., 2008. Studies on the immobilisation of acetylcholine esterase enzyme for biosensor applications. Food Agric. Immunol. 19 (4), 273–281.

Riedel, K., Beyersdorf-Radeck, B., Neumann, B., Schaller, F., 1995. Microbial sensors for determination of aromatics and their chloro derivatives. Part II1: Determination of chlorinated phenols using a biosensor containing *Trichosporon beigelii* (cutaneum). Appl. Microbiol. Biotechnol. 43 (1), 7–9.

Riedel, K., Hensel, J., Rothe, S., Neumann, B., Scheller, F., 1993. Microbial sensors for determination of aromatics and their chloroderivatives. Part II: Determination of chlorinated phenols using a *Rhodococcus*-containing biosensor. Appl. Microbiol. Biotechnol. 38 (4), 556–559.

Roointan, A., Shabab, N., Karimi, J., Rahmani, A., Alikhani, M.Y., Saidijam, M., 2015. Designing a bacterial biosensor for detection of mercury in water solutions. Turk. J. Biol. 39 (4), 550–555.

Russell, I.M., Burton, S.G., 1999. Development and determination of an immobilized-polyphenol oxidase bioprobe for the detection of phenolic pollutants in water. Anal. Chim. Acta 389, 161–170.

Sarkar, P., Banerjee, S., Bhattacharyay, D., Turner, A.P., 2010. Electrochemical sensing systems for arsenate estimation by oxidation of L-cysteine. Ecotoxicol. Environ. Saf. 73 (6), 1495–1501.

Sawaraba, I., Rao, B.R., 2015. Monitoring of river water for free cyanide pollution from mining activity in Papua New Guinea and attenuation of cyanide by biochar. Environ. Monit. Assess. 187 (1), 4181.

Scognamiglio, V., Raffi, D., Lambreva, M., Rea, G., Tibuzzi, A., Pezzotti, G., et al., 2009. *Chlamydomonas reinhardtii* genetic variants as probes for fluorescence sensing system in detection of pollutants. Anal. Bioanal. Chem. 394 (4), 1081–1087.

Scognamiglio, V., Pezzotti, I., Pezzotti, G., Cano, J., Manfredonia, I., Buonasera, K., et al., 2013. A new embedded biosensor platform based on micro-electrodes array (MEA) technology. Sens. Actuators, B 176, 275–283.

Sharma, P., Asad, S., Ali, A., 2013. Bioluminescent bioreporter for assessment of arsenic contamination in water samples of India. J. Biosci. 38 (2), 251–258.

Shemer, B., Palevsky, N., Yagur-Kroll, S., Belkin, S., 2015. Genetically engineered microorganisms for the detection of explosives' residues. Front. Microbiol. 6, 1175.

Siddiki, M.S.R., Shimoaoki, S., Ueda, S., Maeda, I., 2012. Surface plasmon resonance-based DNA biosensor for arsenic trioxide detection. Sensors 12 (10), 14041–14052.

Silva, L.M.C., Salgado, A.M., Coelho, M.A.Z., 2010. Agaricus bisporus as a source of tyrosinase for phenol detection for future biosensor development. Environ. Technol. 31 (6), 611–616.

Singh, V., 2014. Recent advancements in synthetic biology: current status and challenges. Gene 535 (1), 1–11.

Singh, V., Braddick, D., Dhar, P.K., 2017. Exploring the potential of genome editing CRISPR-Cas9 technology. Gene 599, 1–18.

Singh, V., Gohil, N., Ramírez García, R., Braddick, D., Fofié, C.K., 2018. Recent advances in CRISPR-Cas9 genome editing technology for biological and biomedical investigations. J. Cell. Biochem. 119 (1), 81–94.

Solanki, P.R., Prabhakar, N., Pandey, M.K., Malhotra, B.D., 2009. Surface plasmon resonance-based DNA biosensor for arsenic trioxide detection. Int. J. Environ. Anal. Chem. 89 (1), 49–57.

Stocker, J., Balluch, D., Gsell, M., Harms, H., Feliciano, J., Daunert, S., et al., 2003. Development of a set of simple bacterial biosensors for quantitative and rapid measurements of arsenite and arsenate in potable water. Environ. Sci. Technol. 37 (20), 4743–4750.

Tan, J., Kan, N., Wang, W., Ling, J., Qu, G., Jin, J., et al., 2015. Construction of 2,4,6-trinitrotoluene biosensors with novel sensing elements from *Escherichia coli* K-12 MG1655. Cell Biochem. Biophys. 72 (2), 417–428.

Tauriainen, S., Karp, M., Chang, W., Virta, M., 1997. Recombinant luminescent bacteria for measuring bioavailable arsenite and antimonite. Appl. Environ. Microbiol. 63 (11), 4456–4461.

Tauriainen, S., Karp, M., Chang, W., Virta, M., 1998. Luminescent bacterial sensor for cadmium and lead. Biosens. Bioelectron. 13 (9), 931–938.

Tembe, S., Inamder, S., Haram, S., Karvee, M., D'Souza, S.F., 2007. Electrochemical biosensor for catechol using agarose-guar gum entrapped tyrosinase. J. Biotechnol. 128 (1), 80–85.

Timur, S., Pazarlioğlu, N., Pilloton, R., Telefoncu, A., 2003. Detection of phenolic compounds by thick film sensors based on *Pseudomonas putida*. Talanta 61 (2), 87–93.

Truffer, F., Buffi, N., Merulla, D., Beggah, S., van Lintel, H., Renaud, P., et al., 2014. Compact portable biosensor for arsenic detection in aqueous samples with *Escherichia coli* bioreporter cells. Rev. Sci. Instrum. 85 (1), 015120.

Way, J.L., 1984. Cyanide intoxication and its mechanism of antagonism. Annu. Rev. Pharmacol. Toxicol. 24, 451–481.

Webster, D.P., TerAvest, M.A., Doud, D.F., Chakravorty, A., Holmes, E.C., Radens, C.M., et al., 2014. An arsenic-specific biosensor with genetically engineered *Shewanella oneidensis* in a bioelectrochemical system. Biosens. Bioelectron. 62, 320–324.

Wu, Y., Liu, L., Zhan, S., Wang, F., Zhou, P., 2012. Ultrasensitive aptamer biosensor for arsenic(III) detection in aqueous solution based on surfactant-induced aggregation of gold nanoparticles. Analyst 137 (18), 4171–4178.

Wu, Y., Wang, F., Zhan, S., Liu, L., Luo, Y., Zhou, P., 2013. Regulation of hemin peroxidase catalytic activity by arsenic-binding aptamers for the colorimetric detection of arsenic (III). RSC Adv. 3 (48), 25614–25619.

Yagur-Kroll, S., Lalush, C., Rosen, R., Bachar, N., Moskovitz, Y., Belkin, S., 2014. *Escherichia coli* bioreporters for the detection of 2,4-dinitrotoluene and 2,4,6-trinitrotoluene. Appl. Microbiol. Biotechnol. 98 (2), 885–895.

Yagur-Kroll, S., Amiel, E., Rosen, R., Belkin, S., 2015. Detection of 2,4-dinitrotoluene and 2,4,6-trinitrotoluene by an *Escherichia coli* bioreporter: performance enhancement by directed evolution. Appl. Microbiol. Biotechnol. 99 (17), 7177–7188.

Yildiz, H.B., Castillo, J., Guschin, D.A., Toppare, L., Schuhmann, W., 2007. Phenol biosensor based on electrochemically controlled tyrosinase in a redox polymer. Microchim. Acta 159 (1–2), 27–34.

You, Y., Shim, J., Cho, C.H., Ryu, M.H., Shea, P.J., Kamala-Kannan, S., et al., 2013. Biodegradation of BTEX mixture by *Pseudomonas putida* YNS 1 isolated from oil-contaminated soil. J. Basic. Microbiol. 53 (5), 469–475.

Current status, challenges and future of bioremediation

20

Gajendra Singh Vishwakarma[1], *Gargi Bhattacharjee*[2], *Nisarg Gohil*[2] and *Vijai Singh*[2,*]
[1]Department of Biological Sciences and Biotechnology, Institute of Advanced Research, Gandhinagar, India, [2]Department of Biosciences, School of Science, Indrashil University, Rajpur, Mehsana, Gujarat, India
*Corresponding author

20.1 Introduction

Bioremediation is a collective phenomenon involving processes that use biological systems to either restore or clean-up contaminated sites. The microbial community is consistently reported for bioremediation. Most of the indigenous microbes have the ability to successfully bring up the environmental restoration via oxidizing, immobilizing, or transforming the contaminants (Crawford and Crawford, 2005). It aims to reduce or bring down pollutant levels up to undetectable, nontoxic, or acceptable (i.e., within limits set by regulatory agencies) levels. The concept of bioremediation was first used on a large scale in 1972 for the cleaning of Sun Oil pipeline spill at Ambler, Pennsylvania (Kumar et al., 2015).

However, in laboratory-scale, George M. Robinson was the first to recognize this process during experiments on sewage and oil treatment (Sonawdekar, 2012). Subsequently, in 1992, Environmental Protection Agency (EPA) developed protocols for bioremediation on the basis of different case studies on bioremediation. The modern approaches of bioremediation are to search for a novel microorganism from contaminated sites. The isolated microbes are thought to have a strong potential to remediate pollutants. The use of genetically modified strains and also microbial consortium has been used directly or indirectly to increase the bioactivity of a bioremediant. Various mechanisms including bioaccumulation, biodegradation pathways, and different modes for biosorption have also been investigated for the removal of pollutants (Hlihor et al., 2017; Ramírez-García et al., 2019). In this chapter, we highlight and discuss different bioremediation processes and their classification with future potential.

20.2 Bioremediation process and classification

Bioremediation-based cleaning involves various microbial processes, many of which have been developed and utilized since the advent of this field. The basic

Bioremediation of Pollutants. DOI: https://doi.org/10.1016/B978-0-12-819025-8.00020-X

fundamental principle of all the methods is to stimulate the microbe to yield optimum microbial enzymatic activity. The utilization of microbial communities employs the enzymatic metabolic pathways that have evolved over a long period of time. Hence for the systematic use of microbial communities, the basic requirement is optimization and regulation of process parameters. Under specific conditions, microbes degrade contaminants and gain energy for growth and reproduction by utilizing the contaminants as substrate. Energy is evolved due to the breakdown of chemical bonds and the free electrons released are transferred to an electron acceptor such as oxygen. The basic process involves a redox reaction in which the organic matter is oxidized through loss of electrons while the compound that accepts the electron is said to be reduced. In the anaerobic process, the electron acceptors are nitrate, sulfate, and iron while in the aerobic process it is oxygen. The entire process of oxidation and reduction is regulated by designated gene clusters and for different types of contaminants or heavy metals, the respective degrading gene cluster or pathway lies within either the genome or plasmid of organisms (Ramírez-García et al., 2019).

In the case of mercury (Hg), the *mer* operon allows the conversion of Hg from a toxic form to a less toxic state (Naguib et al., 2018). *Mer* operon is basically a set of structural genes (*merA*, *merB*, *merT* and *merP*) that codes for different enzymes and integral proteins. Among them, *merA* codes for enzyme flavin disulfide oxidoreductase, a mercuric ion reductase. Similarly, *merT* and *merP* codes for the integral proteins that facilitate uptake of Hg^{2+} in the cytosol after which *merA* reduces it into elemental mercury (Hg^0) (Boyd and Barkay, 2012). Arsenic (As) is generally found in two forms, arsenite As(III) and arsenate As(V). Bioremediation of arsenic is performed based on the respiratory reduction of As(V) to As(III) and oxidation AsIII to AsV. Oxidation of AsIII in bacteria is regulated by *aox* operon. The *aox* operon mediated oxidation of arsenic depends on the formation of *aox* operon AoxAB complex. Once the complex is synthesized, it exports the AsIII to the periplasm where arsenite oxidase oxidizes AsIII to AsV (Mitra et al., 2017).

Some bacterial strains including *Pseudomonas* and *Enterobacter* species have been identified to produce exopolysaccharides (EPS) (Kalita and Joshi, 2017). EPS is capable of flocculation and removal of heavy metals ions by forming a biofilm. Cabuk et al. (2006) reported that the EPS of *Bacillus* sp. ATS-2 contains hydroxyl and carboxyl groups and some nitrogen-based bioligands such as amide and sulfonamide that are involved in the binding of Pb(II). A removal efficiency of up to 99.8% Pb(II) using microbial flocculant GA1 was reported (Feng et al., 2013). The bacterial species involved in bioremediation of chromium either use chromate reductase or metabolic end-products such as Fe(II) and HS^- of iron-based catalysis to reduce Cr(VI) to Cr(III). Furthermore, processes such as reducing the uptake of Cr(VI) via mutation in chromosome-encoded sulfate uptake transporters and extracellular reduction of Cr(VI) to Cr(III) and reactive oxygen species (ROS) are a few other approaches used for detoxification (Thatoi et al., 2014).

For the bioremediation of complex substances including pesticides, use of different microbial strains such as *Pseudomonas* sp., *Bacillus* sp., *Klebsiella* sp., *Pandoraea* sp., *Phanerochaete chrysosporium*, *Mycobacterium* sp., *Agrocybe semiorbicularis*,

Auricularia auricula, *Coriolus versicolor*, *Dichomitus squalens*, *Flammulina velutipes*, *Hypholoma fasciculare*, *Pleurotus ostreatus*, *Stereum hirsutum*, and *Avatha discolor* have been reported (Rani, 2014; Wang et al., 2017; Wahla and Shukla, 2018). In addition to that, some composting materials and organic amendments such as corn fermentation byproducts, corn stalks, manure, peat, and sawdust have also been included for the removal of pesticides (atrazine, trifluralin, and metolachlor) (Chen et al., 2015; Moorman et al., 2001).

Bioremediation is classified on the basis of the site of the pollutant. If the pollutants are excavated from the contaminated area and carried elsewhere to remediate other than the site then it comes under the ex situ remediation. Examples of ex situ remediation process include biopiling, windrow, bioreactor, and land-farming. On the other hand, the process that takes place at the site of contaminates is known as in situ remediation. Bioslurping, bioventing, biosparging, phytoremediation are all examples of in situ bioremediation process. These processes are generally based on providing nutrients, oxygen, and stimulating conditions that are necessary for the growth of microbial communities. Bioslurping method involves the extraction of free-floating products from the vadose zone. Bioventing is the controlled stimulation of airflow by delivering oxygen to unsaturated zone in order to increase bioremediation, while biosparging is the process of injecting air into groundwater to provide oxygen for inducing the bioremediation process. Phytoremediation uses plants for removing contaminates while the permeable reactive barrier is the establishment of permeable membrane across the flow of groundwater (Ramírez-García et al., 2019). In addition to the above methods, phytosequestration, rhizodegradation, phytohydraulics, phytoextraction, phytodegradation, and phytovolatilization are also preferred for the removal of pollutants (Garg and Paliwal, 2020). Fig. 20.1 depicts potential approaches to improve bioremediation process.

20.3 Current status of bioremediation

Bioremediation is applied for the removal of a wide range of pollutants. Following laboratory trials, the process is extended to a large contaminated site. In every case, some modifications are carried out into the basic technologies for the optimization of parameters. These parameters depend on the place of contamination, nature of the contaminant, and the source of contamination. Therefore, to understand the current status of bioremediation process in terms of advancement and latest technologies, it is important to look into various insights of this phenomenon. The latest alterations in the process, modification in microbes, and site-specific designing of bioremediation systems are discussed below.

20.3.1 Advanced bioreactor systems

Bioreactor is a closed or open system used to allow maximum contact between the microorganisms and contaminates. In order to design a suitable bioreactor for a

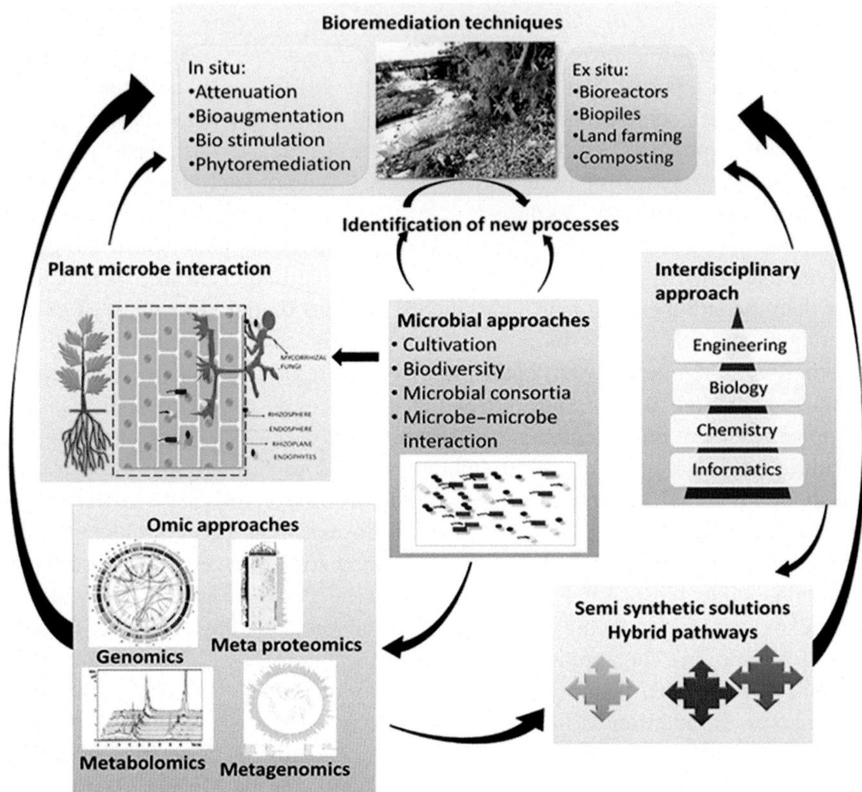

Figure 20.1 Novel bioremediation approaches to facilitate removal of pollutants from the environment.
Source: Adopted from Malla, M.A., Dubey, A., Yadav, S., Kumar, A., Hashem, A., Abd Allah, E.F., 2018. Understanding and designing the strategies for the microbe-mediated remediation of environmental contaminants using omics approaches. Front. Microb. 9, 1132 (Malla et al., 2018).

particular treatment process, extensive research on the biological system (such as growth kinetics, metabolic activity of microbes, and genetic makeup) is required. In the past two decades, different types of bioreactors have been developed. Some of these are based on modifications in conventional technology including stirred-tank bioreactors, airlift bioreactor, fluidized bed bioreactors, wave bioreactors, and many others which are currently being modified with modern technologies including combined or sequential bioreactors and hybrid bioreactors. Combined or sequential bioreactors are designed for the facilitation of combined chemical, physical, and biological processes (Tawfik et al., 2014; El-Gohary and Tawfik, 2009; Hasar et al., 2009; Cui et al., 2014; Hwang et al., 2013).

Combined chemical/biological processes are effective in overcoming the problem of biomass retention time, chemical oxygen demand (COD) reduction, and clogging problems (Su et al., 2016). In the case of azo dyes, degradation takes place in two steps: aerobic and anaerobic, where by in the first step, the dye anaerobically reduces and produces aromatic amines which are further degraded aerobically (Tan et al., 2016). Therefore, complete degradation of azo dye requires a sequential anaerobic and aerobic bioreactor. In this regard, Tan et al. (2016) achieved the degradation of azo dye using sequential anaerobic and aerobic bioreactor. Likewise, Mallick and Chakraborty (2019) developed anoxic–aerobic sequential moving bed reactors for the degradation of automobile service station wastewater and achieved 94% COD removal in combined hydraulic retention time of 24 hours. From their study, they were able to identify *Pseudomonas aeruginosa* SKM2013 and *P. aeruginosa* SC2013 in anoxic biomass which were endowed with the ability to degrade phenol and hydrocarbons, while in aerobic biomass, they found high prevalence of *Lysinibacillus* sp. H200-510, *Stenotrophomonas* sp. LW-34, *P. aeruginosa* LZS8436, and *P. aeruginosa* ISB4 with the potential to utilize NH_4^+-N (Mallick and Chakraborty, 2019).

20.3.2 Cell and plasmid-mediated bioaugmentation

Complete microbial cell bioaugmentation has several limitations such as rapid decrease of bacterial viability and abundance after inoculation, as well as limited dispersal of the inoculated bacteria in the soil matrix. Hence, this method cannot be considered as a successful practice for bioremediation. The genomic analysis of bacterial strains clearly indicates that the genes encoding the enzymes for degradation of organic compounds are often located in plasmids and consequently, they can spread via bacterial conjugation into well-established, ecologically competitive, indigenous bacterial populations. Therefore, the new approach of plasmid-mediated bioaugmentation is based on the spread of contaminant degradation genes among indigenous soil bacteria by the introduction of plasmids isolated from the donor cells, harboring such genes (Garbisu et al., 2017).

20.3.3 Antibiotic resistant microorganisms

Due to the indiscriminate use of antibiotics and biocides, their residues remain persistant in soil and water (Gohil et al., 2018). Thus the selection of antibiotic-resistant or tolerant mutants for bioremediation has increased over the past decade. The resistance to a myriad of antibiotics comes from the co-localization and/or co-migration of genes conferring multiple resistance mechanisms (dos Santos et al., 2015). Antibiotic resistance genes are often naturally present in microorganisms and can show a high level of tolerance against certain antibiotics. These strains show some additional biological advantages such as resistance to other environmental contaminants or interspecies competitive strategies and metabolism of toxic compounds which are structurally similar to antibiotics (Atashgahi et al., 2018).

20.3.4 Use of algal biomass

Remediation in which algae is used for neutralization or removal of pollutants is called phytoremediation. Various algal species have been exercised for the removal of organic, inorganic, heavy metal compounds and pathogens. The species that possess a high growth rate and great potential for biotransformation of pollutants remain the preferred choice for phytoremediation. Among them, *Chlorella* sp. has been reported for remediation of an azo-dye Congo Red (Hernández-Zamora et al., 2015), petroleum effluent, primary and secondary (Znad et al., 2018) as well as tannery wastewater (Das et al., 2017). The strain completely removed nitrate-nitrogen and chromium from tannery wastewater by the 6th and 12th day of treatment, respectively. It also removed excess nutrients such as phosphates, sulfates, and reduced COD and biochemical oxygen demand (Das et al., 2017). Marine macroalgae (seaweeds) can be used for the removal of zinc after treatment with sodium hydroxide. Up to 115.198 mg/g zinc biosorption capacity has been achieved in zinc-containing synthetic wastewater by using chemically modified coastal seaweed community of *Chaetomorpha* sp., *Polysiphonia* sp., *Ulva* sp., and *Cystoseira* sp. (Deniz and Karabulut, 2017). Apart from that, heavy metals such as Cd, Hg, Cu, Ni, Pb, Cr have also been removed by different algal species through bioaccumulation, biotransformation, or bioaugmentation process (Chabukdhara et al., 2017). Phytoremediation of wastewater is a challenging task as parameters such as temperature, light intensity, type of waste, and operation time play a very crucial role in the process (Gupta et al., 2017).

20.3.5 White-rot fungi

Fungi are saprophyte organisms reported for their higher tolerance to toxic environments. White-rot fungus or commonly known as wood rot fungus belongs to the *Basidiomycetes* and is capable of degrading lignocellulose substrates. It produces an extracellular enzymes system (manganese peroxidase, lignin peroxidase, and laccase) which gets involved in the degradation of lignin and xenobiotics (Ellouze and Sayadi, 2016). The enzyme system has a low structural specificity and so acts on a range of similar structured compounds such as synthetic dyes and aromatic hydrocarbons. The major strains of white-rot fungi, such as *P. chrysosporium*, *P. ostreatus*, *C. versicolor*, *Cyathus stercoreus* are widely used for the degradation of different contaminants (Yesilada et al., 2018; Tortella et al., 2015). The fungal mycelia have an additive advantage over single-cell organisms which is through solubilization of the insoluble substrates by producing extracellular enzymes (He et al., 2017).

20.4 Advancement in phytoremediation

Plant-based soil and water remediation (phytoremediation) is a technique that can prove to be ecologically beneficial and capable of overcoming the environmental problems. Phytoremediation uses the abilities of green plants to take up, stabilize, and metabolize the pollutants. It is a cost-effective and environmentally safe approach as compared to conventional methods to solve the problems of soil and water pollution. Currently, investigation is under process for exploring new ways to improve the phytoremediation process. Phytoremediation is in its early stages and many technical problems appear that need to be addressed for its development and further applications. Moreover, the use of applied molecular biology techniques and development of transgenic plant with extended phytoremediation capability are gaining more attention (Ramírez-García et al., 2019).

Genetic engineering is expected to play a vital role in boosting the applicability of phytoremediation technologies (Perpetuo et al., 2011; Singh et al., 2017, 2018; Jaiswal et al., 2019). In this regards, the in vitro studies on transgenic plants have shown successful transformations and efficient remediation of metal. Amongst them, some have progressed from in vitro experiments to field testing. The only concern with phytoremediation is restricting the chances of releasing transgenic genes into the environment. It may be overcome by choosing a very specific gene pool from sterile or vegetative propagated species, or a self-pollinated plant species (Dale et al., 2002).

20.5 Challenges in bioremediation

The major constraint of bioremediation is that not all compounds are biodegradable and this limits the bioremediation process. In some cases, if at all the material is biodegradable, its downstream processing and degradation further generate some toxic material. Sometimes a particular bacterial strain functional at one site may not successfully work on different sites due to the prevalence of some site related limiting factors. The complexity of biological processes is attributed to the microbial metabolic processes, nature of pollutants and presence of an appropriate level of nutrients. As the process demands excavation of soil, construction and customization of specialized site layout, it makes it more time-consuming and laborious. Mostly, the process is carried out in underground and isolated places away from the residential area. However, the use of heavy machineries and pumps may create some noise and disturbances. Sometimes, ethical issues concerning the use of certain bacterial strains also make the process questionable in terms of their effects on another local microflora (Ramírez-García et al., 2019).

20.5.1 Challenges in soil bioremediation

The major challenges that should be minimized during soil bioremediation include the following:

1. Bioavailability or accessibility of pollutants: The accessibility of pollutants to the microbes in soil is determined by its adsorption, toxicity, and utilization. Many organic compounds including various pesticides and polycyclic aromatic hydrocarbons (PAHs)

are present in the soil in a strongly sorbet state. Thus bioremediation of soil that is contaminated with PAHs and pesticides is more affected due to reduced bioavailability rather than the poor density of PAH degrading microorganisms. Bioavailability of contaminants changes with time. This process is known as aging. Aging of pollutants is one of the greatest challenges encountered during remediation of soil pollution. Hence, a need arises to find an appropriate solution to maximize the bioavailability of pollutants in soil.

2. Reversion of pollutant stability: Composting organic matter obtained after waste treatment is one of the potential and sustainable approaches to tackle accumulation of organic products. However, mineralization of organic matter in the environment changes its pH which in turn reverses the pollutant stability.
3. Microbial adaptability: Poor compliance of exogenous microorganisms with the contaminated soils slows down their adaptability to act on a particular site. Thus bioaugmentation becomes ineffective in enhancing the degradation of pollutants. Appropriate methods may be needed to explore and address this problem.
4. Metabolic pathways: Proper understanding of microbial metabolic pathways during bioremediation of organic pollutants and heavy metals is still limited. So, for genetic modification of exogenous microorganism, in-depth exploration of metabolic pathways and dynamics of microbial communities are the need of the hour. Further, the application of molecular biology such as high-throughput sequencing should also be utilized in order to understand the genomic organization of indigenous microbes. Synthetic biology approaches and technology are needed to overcome the aforementioned limitations and clean the environmental pollutants (Dvořák et al., 2017; de Lorenzo et al., 2018; Gohil et al., 2017).
5. Enzymatic studies: The basic catalytic action of different enzymes is responsible for degradation of pollutants which actually is hidden and therefore, extensive research is required for the study of several enzymatic aspects like kinetics, molecular structure, activity, and inhibition process.
6. Pollutant interactions: The synergistic and antagonistic interaction between pollutants from the same soil/compost mixture on their degradation is still unclear. This demands further analysis to rise over this challenge.
7. End product quality: It is obligatory to ensure that the quality of the end product released following bioremediation is free of any kind of toxic organic compounds and metals beyond a threshold limit.

20.5.2 Challenges in groundwater bioremediation

The bioremediation of groundwater is dependent more on the underground conditions and parameters such as geochemistry, geology, hydrology, contaminant concentration, etc. Hence understanding the geological setting and its heterogeneity is critical for developing useful bioremediation technology. The low baseline pH and temperature affect the biological remedies while porosity of soil and strength of bedrock may influence the plume migration. An appropriate prediction of groundwater flow direction and velocity is again a key factor that influences the transfer of contaminants. Groundwater flows preferentially through more permeable zones compared to less permeable zones. Bioremediation is more effective in more permeable zones because liquid and gaseous infiltrates seep much more quickly through high permeability zones (Leeson et al., 2013).

20.5.3 Challenges in phytoremediation

Several challenges such as long duration, low-biomass production of hyperaccumulator plants, climatic interference in the accumulation capacity of plants, and invasiveness of the hyperaccumulator plant species affect the indigenous flora diversity. The above challenges need to be overcome to increase the use of phytoremediation for waste treatment (Odoh et al., 2019).

20.6 Conclusion and future remarks

Bioremediation has emerged as a popular option for cleaning up the contaminated aquifers and soil. By now, the ability of certain microbes to engage in the degradation of environmental pollutants has been well established. The credibility of the failure arises from the thought of transferring the said ability of the microorganisms from lab-to-field level. So far, three major constraints of employing bioremediation have been envisaged which restricts the spread of bioremediation techniques: lack of comprehensive knowledge about how a microbe would react when left in the field, its handling and glitches in stimulating the microbe, and difficulty in ensuring proper contact or engagement between the microbe and contaminant. Continuous efforts by the scientific community working in this genre have been successful in pushing the boundaries of the existing techniques. Scientists across the world have been looking for innovative engineering ideas to incorporate materials that are able to stimulate the microbes. An example of this is the recent development of gas sparging that has readily upgraded the aerobic degradation of petroleum products. The rapidly developing technologies in the area of bioaugmentation are expected to allow a better hold over the genetic manipulation of the cell, thus overcoming limitations arising due to microbiological factors. With an advanced understanding of biotransformation at the ecological and genetic level, novel methods for bioremediation can be made available. This knowledge can also be carried forward to design protocols for treating noxious pollutants such as polychlorinated biphenyls and chlorinated solvents, which were once thought to be nondegradable or could escape biodegradation. Many pollutants tend to escape the degrading action of the microbes predominantly because the microbes fail to interact or get attached to the surface of the contaminants. Newer ways are being devised to enhance the bioavailability of the microbes to add value to the efficiency of bioremediation. Techniques such as solubilization of waste by heat injection through means of hot air, steam or flushing hot water, intense-pressure subsurface matrix fracture, and addition of surfactants are a few of the preferred ways to improve existing technologies.

While progressing with bioremediation, it is not just important to discover efficient yet feasible strategies but also come up with ways to evaluate the functioning and efficacy of current methods. Hence, several protocols have been designed that revolve around three basic aspects: (1) documenting the net reduction of contaminants from the polluted site, (2) estimating the biodegradation potential of the microbes isolated from the site through laboratory tests, and (3) a few other possible information factors that stress on the fact that the methods employed at the field-level actually work. The principle aim is to determine the practicality of theoretically-proposed solution so as to ensure that the cleanup goals are met. Researchers are also showing an inclination toward designing modified protocols based on the knowledge of molecular biology which are expected to be reliable, rapid, and cost-effective. In situ characterization of the physiochemical parameters also seems promising which in the future may revolutionize field assessment tests. Bioremediation is a widely accepted alternative for cleaning up the environment especially because it promises to reach the clean-up goals at a very economical rate with lesser chances of spreading contaminants to other media. However, the coalescence of the fact that it is necessary to ensure cautious handling of the microbes and their continuous monitoring at the subsurface leaves many clients hesitant to accept bioremediation as a preferred choice for waste and pollutant management. The future awaits the discovery of expeditious technologies that would rise above the current challenges and lead the world toward a cleaner and greener environment.

Acknowledgments

GSV gratefully acknowledges the Gujarat State Biotechnology Mission (GSBTM Project ID: L1Y5SU) for financial support. GB, NG, and VS thank Indrashil University for providing infrastructure and support.

References

Atashgahi, S., Sánchez-Andrea, I., Heipieper, H.J., van der Meer, J.R., Stams, A.J., Smidt, H., 2018. Prospects for harnessing biocide resistance for bioremediation and detoxification. Science 360, 743–746.

Boyd, E., Barkay, T., 2012. The mercury resistance operon: from an origin in a geothermal environment to an efficient detoxification machine. Front. Microbiol. 3, 349.

Cabuk, A., Akar, T., Tunali, S., Tabak, Ö., 2006. Biosorption characteristics of *Bacillus* sp. ATS-2 immobilized in silica gel for removal of Pb (II). J. Hazard. Mater. 136, 317–323.

Chabukdhara, M., Gupta, S.K., Gogoi, M., 2017. Phycoremediation of heavy metals coupled with generation of bioenergy. In: Gupta, S.K., Malik, A., Bux, F. (Eds.), Algal Biofuels. Springer, Cham, pp. 163–188.

Chen, M., Xu, P., Zeng, G., Yang, C., Huang, D., Zhang, J., 2015. Bioremediation of soils contaminated with polycyclic aromatic hydrocarbons, petroleum, pesticides, chlorophenols and heavy metals by composting: applications, microbes and future research needs. Biotechnol. Adv. 33, 745–755.

Crawford, R.L., Crawford, D.L. (Eds.), 2005. Bioremediation: Principles and Applications. Cambridge University Press, Cambridge.

Cui, J., Wang, X., Yuan, Y., Guo, X., Gu, X., Jian, L., 2014. Combined ozone oxidation and biological aerated filter processes for treatment of cyanide containing electroplating wastewater. Chem. Eng. J. 241, 184–189.

Dale, P.J., Clarke, B., Fontes, E.M., 2002. Potential for the environmental impact of transgenic crops. Nat. Biotechnol. 20, 567–574.

Das, C., Naseera, K., Ram, A., Meena, R.M., Ramaiah, N., 2017. Bioremediation of tannery wastewater by a salt-tolerant strain of *Chlorella vulgaris*. J. Appl. Phycol. 29, 235–243.

de Lorenzo, V., Prather, K.L., Chen, G.Q., O'Day, E., von Kameke, C., Oyarzún, D.A., et al., 2018. The power of synthetic biology for bioproduction, remediation and pollution control. EMBO Rep. 19, e45658.

Deniz, F., Karabulut, A., 2017. Biosorption of heavy metal ions by chemically modified biomass of coastal seaweed community: studies on phycoremediation system modeling and design. Ecol. Eng. 106, 101–108.

dos Santos, D.F.K., Istvan, P., Noronha, E.F., Quirino, B.F., Krüger, R.H., 2015. New dioxygenase from metagenomic library from Brazilian soil: insights into antibiotic resistance and bioremediation. Biotechnol. Lett. 37, 1809–1817.

Dvořák, P., Nikel, P.I., Damborský, J., de Lorenzo, V., 2017. Bioremediation 3.0: engineering pollutant-removing bacteria in the times of systemic biology. Biotechnol. Adv. 35, 845–866.

El-Gohary, F., Tawfik, A., 2009. Decolorization and COD reduction of disperse and reactive dyes wastewater using chemical-coagulation followed by sequential batch reactor (SBR) process. Desalination 249, 1159–1164.

Ellouze, M., Sayadi, S., 2016. White-rot fungi and their enzymes as a biotechnological tool for xenobiotic bioremediation. In: Saleh, H. (Ed.), Management of Hazardous Wastes. IntechOpen, London, pp. 103–120.

Feng, J., Yang, Z., Zeng, G., Huang, J., Xu, H., Zhang, Y., et al., 2013. The adsorption behaviour and mechanism investigation of Pb (II) removal by flocculation using microbial flocculant GA1. Bioresour. Technol. 148, 414–421.

Garbisu, C., Garaiyurrebaso, O., Epelde, L., Grohmann, E., Alkorta, I., 2017. Plasmid-mediated bioaugmentation for the bioremediation of contaminated soils. Front. Microbiol. 8, 1966.

Garg, S., Paliwal, R., 2020. Green technologies for restoration of damaged ecosystem. In: Meena, R.S. (Ed.), Soil Health Restoration and Management. Springer, Singapore, pp. 357–380.

Gohil, N., Panchasara, H., Patel, S., Ramírez-García, R., Singh, V., 2017. Book review: Recent advances in yeast metabolic engineering. Front. Bioeng. Biotechnol. 5, 71.

Gohil, N., Ramírez-García, R., Panchasara, H., Patel, S., Bhattacharjee, G., Singh, V., 2018. Book review: quorum sensing vs. quorum quenching: a battle with no end in sight. Front. Cell. Infect. Microbiol. 8, 106.

Gupta, S.K., Sriwastav, A., Ansari, F.A., Nasr, M., Nema, A.K., 2017. Phycoremediation: an eco-friendly algal technology for bioremediation and bioenergy production. In: Bauddh, K., Singh, B., Korstad, J. (Eds.), Phytoremediation Potential of Bioenergy Plants. Springer, Singapore, pp. 431–456.

Hasar, H., Unsal, S.A., Ipek, U., Karatas, S., Cınar, O., Yaman, C., et al., 2009. Stripping/flocculation/membrane bioreactor/reverse osmosis treatment of municipal landfill leachate. J. Hazard. Mater. 171, 309–317.

He, K., Chen, G., Zeng, G., Huang, Z., Guo, Z., Huang, T., et al., 2017. Applications of white rot fungi in bioremediation with nanoparticles and biosynthesis of metallic nanoparticles. Appl. Microbiol. Biotechnol. 101, 4853–4862.

Hernández-Zamora, M., Cristiani-Urbina, E., Martínez-Jerónimo, F., Perales-Vela, H.V., Ponce-Noyola, T., Montes-Horcasitas, M.C., et al., 2015. Bioremoval of the azo dye

Congo Red by the microalga *Chlorella vulgaris*. Environ. Sci. Pollut. Res. 22, 10811–10823.

Hlihor, R.M., Apostol, L.C., Gavrilescu, M., 2017. Environmental bioremediation by biosorption and bioaccumulation: principles and applications. In: Anjum, N.A., Gill, S.S., Tuteja, N. (Eds.), Enhancing Cleanup of Environmental Pollutants. Springer, Cham, pp. 289–315.

Hwang, G., Dong, T., Islam, M.S., Sheng, Z., Pérez-Estrada, L.A., Liu, Y., et al., 2013. The impacts of ozonation on oil sands process-affected water biodegradability and biofilm formation characteristics in bioreactors. Bioresour. Technol. 130, 269–277.

Jaiswal, S., Singh, D.K., Shukla, P., 2019. Gene editing and systems biology tools for pesticide bioremediation: a review. Front. Microbiol. 10, 87.

Kalita, D., Joshi, S., 2017. Study on bioremediation of Lead by exopolysaccharide producing metallophilic bacterium isolated from extreme habitat. Biotechnol. Rep. 16, 48–57.

Kumar, A., Govil, M., Singh, S., Sharma, K., Tripathi, S., Tiwari, R., et al., 2015. Role of micro-organisms in bioremediation: a comprehensive model using *Trichoderma* spp. In: Singh, S., Srivastava, K. (Eds.), Handbook of Research on Uncovering New Methods for Ecosystem Management Through Bioremediation. IGI Global, pp. 29–50.

Leeson, A., Stroo, H.F., Johnson, P.C., 2013. Groundwater remediation today and challenges and opportunities for the future. Groundwater 51, 175–179.

Malla, M.A., Dubey, A., Yadav, S., Kumar, A., Hashem, A., Abd Allah, E.F., 2018. Understanding and designing the strategies for the microbe-mediated remediation of environmental contaminants using omics approaches. Front. Microb. 9, 1132.

Mallick, S.K., Chakraborty, S., 2019. Bioremediation of wastewater from automobile service station in anoxic-aerobic sequential reactors and microbial analysis. Chem. Eng. J. 361, 982–989.

Mitra, A., Chatterjee, S., Gupta, D.K., 2017. Potential role of microbes in bioremediation of arsenic. In: Gupta, D.K., Chatterjee, S. (Eds.), Arsenic Contamination in the Environment. Springer, Cham, pp. 195–213.

Moorman, T.B., Cowan, J.K., Arthur, E.L., Coats, J.R., 2001. Organic amendments to enhance herbicide biodegradation in contaminated soils. Biol. Fertil. Soils 33, 541–545.

Naguib, M.M., El-Gendy, A.O., Khairalla, A.S., 2018. Microbial diversity of mer operon genes and their potential rules in mercury bioremediation and resistance. Open Biotechnol. J. 12, 56–77.

Odoh, C.K., Zabbey, N., Sam, K., Eze, C.N., 2019. Status, progress and challenges of phytoremediation—an African scenario. J. Environ. Manage. 237, 365–378.

Perpetuo, E.A., Souza, C.B., Nascimento, C.A.O., 2011. Engineering bacteria for bioremediation. In: Capri, A. (Ed.), Progress in Molecular and Environmental Bioengineering—From Analysis and Modeling to Technology Applications. IntechOpen, London, pp. 605–632.

Ramírez-García, R., Gohil, N., Singh, V., 2019. Recent advances, challenges, and opportunities in bioremediation of hazardous materials. In: Pandey, V.C., Bauddh, K. (Eds.), Phytomanagement of Polluted Sites. Elsevier, Amsterdam, pp. 517–568.

Rani, D., 2014. Bioremediation and biodegradation of pesticide from contaminated soil and water—a novel approach. Int. J. Curr. Microbiol. Appl. Sci. 3, 23–33.

Singh, V., Braddick, D., Dhar, P.K., 2017. Exploring the potential of genome editing CRISPR-Cas9 technology. Gene 599, 1–18.

Singh, V., Gohil, N., Ramirez Garcia, R., Braddick, D., Fofié, C.K., 2018. Recent advances in CRISPR-Cas9 genome editing technology for biological and biomedical investigations. J. Cell. Biochem. 119, 81–94.

Sonawdekar, S., 2012. Bioremediation: a boon to hydrocarbon degradation. Int. J. Environ. Sci. 2, 2408–2424.

Su, C.X.H., Low, L.W., Teng, T.T., Wong, Y.S., 2016. Combination and hybridisation of treatments in dye wastewater treatment: a review. J. Environ. Chem. Eng. 4, 3618–3631.

Tan, L., He, M., Song, L., Fu, X., Shi, S., 2016. Aerobic decolorization, degradation and detoxification of azo dyes by a newly isolated salt-tolerant yeast *Scheffersomyces spartinae* TLHS-SF1. Bioresour. Technol. 203, 287–294.

Tawfik, A., Zaki, D.F., Zahran, M.K., 2014. Degradation of reactive dyes wastewater supplemented with cationic polymer (Organo Pol.) in a down flow hanging sponge (DHS) system. J. Ind. Eng. Chem. 20, 2059–2065.

Thatoi, H., Das, S., Mishra, J., Rath, B.P., Das, N., 2014. Bacterial chromate reductase, a potential enzyme for bioremediation of hexavalent chromium: a review. J. Environ. Manage. 146, 383–399.

Tortella, G., Durán, N., Rubilar, O., Parada, M., Diez, M.C., 2015. Are white-rot fungi a real biotechnological option for the improvement of environmental health? Crit. Rev. Biotechnol. 35, 165–172.

Wahla, V., Shukla, S., 2018. Role of microorganisms in bioremediation of pesticides. In: Pathak, V.M. (Ed.), Handbook of Research on Microbial Tools for Environmental Waste Management. IGI Global, pp. 164–189.

Wang, B., Wang, Q., Liu, W., Liu, X., Hou, J., Teng, Y., et al., 2017. Biosurfactant-producing microorganism *Pseudomonas* sp. SB assists the phytoremediation of DDT-contaminated soil by two grass species. Chemosphere 182, 137–142.

Yesilada, O., Birhanli, E., Geckil, H., 2018. Bioremediation and decolorization of textile dyes by white rot fungi and laccase enzymes. In: Prasad, R. (Ed.), Mycoremediation and Environmental Sustainability. Springer, Cham, pp. 121–153.

Znad, H., Al Ketife, A.M., Judd, S., AlMomani, F., Vuthaluru, H.B., 2018. Bioremediation and nutrient removal from wastewater by *Chlorella vulgaris*. Ecol. Eng. 110, 1–7.

Engineered microbes and evolving plastic bioremediation technology 21

*Alka Kumari and Doongar R. Chaudhary**
Biotechnology and Phycology Division, CSIR-Central Salt and Marine Chemicals Research Institute, Bhavnagar, India
*Corresponding author

21.1 Introduction

Plastic was discovered in 1940 by an accidental reaction due to an instrumental fault and nobody had thought it would be a revolutionary as material element besides this it has given the boon to many sectors primarily to packaging, construction, and medical applications as well (Thompson et al., 2009). Industrialization and urbanization have given many more discoveries by using plastic material as a base to ease human lives and spread endemically everywhere in the world. The long carbon backbone, high molecular weight, and hydrophobic property made plastics nondegradable. Globally plastic waste generation increased to 300 million tons, obviously due to nondegradability and unlimited usage of plastic products (Urbanek et al., 2018). Central Pollution Control Board of India reported that more than 6.2 million tons of plastic waste was generated by 2018, which is 1707.9 tons of plastic waste generated per day basis, however, data excluding few states due to unavailability of the records (C.P.C.B., 2018). Polymer like polyvinyl chloride found to be intact in the marine environment for a long time moreover morphologically damaged and leached microsized particles by the thermal and physical damage (Tang et al., 2018). There is a continuity of negligence by government and public sectors as well as the general public toward the impact of plastic waste on the environmental adversity. Plastics in the environment also adversely affect the soil fertility, releasing toxic, and volatile chemicals in the environment, as well affect lives by causing ingestion of plastic particles and entanglement of the animal in different ecosystem (Hammer et al., 2012; Diepens and Koelmans, 2018) (Fig. 21.1). Many conventional ways had been developed for reducing plastic waste which involve landfilling and incineration, although these methods provided a temporary relief, they also resulted in the release of various toxic end products and contributed to rising levels of global warming distressing the climate (National Research Council, 2000; Rhodes, 2018).

Now, the rapid ecofriendly approaches need to be evolved for reducing the plastic waste load from the environment. The rise of biodegradable plastics from available renewable resources are expected to be a reliable substitute to the synthetic polymers, but they have application limitations due to product instability; and

Bioremediation of Pollutants. DOI: https://doi.org/10.1016/B978-0-12-819025-8.00021-1

Figure 21.1 Plastic in the environment.

production complexity could not substitute the petroleum-based plastics, additionally some bioplastics tricking degradability as synthetic ones as not all the bioplastics are degradable (Vroman and Tighzert, 2009; Babu et al., 2013). Plastic waste remediation through biological approach is an attractive and environmentally safe alternative to conventional practices (Shah et al., 2008). Microbes have been serving human life since after Pasteur's established their role in fermentation and, today microbes and their products largely rely on us in daily life. Their quick duplication rate made them capable of evolving quickly which provided them numerous genomic diversities. Recently, many microorganisms from various sources were found to be capable of degrading different plastics at different rate and extent (Pometto et al., 1993; Ghosh et al., 2013; Yoshida et al., 2016). The general scheme of polymer degradation is microbial adherence to the surface, fragmentation of polymeric chain into oligomers or monomers, assimilation, and mineralization (Lucas et al., 2008; Glaser, 2019).

The advancement in biotechnology allows rational engineering of microbial genome and metabolism for specific degradation of pollutants. However, most of the studies lack fundamental knowledge and comparable experimental design as well as their field application rendition. Emergence of microbial genetic engineering tools for bioremediation technology would augment more efficacy to the area. The initial excitement generated in this area should continue to pave the way for ecofriendly design of microbes, enzymes, or perfect composition of genes with novel remedial properties and maximum efficacy (Kulshreshtha, 2013). The microbiological and genetic engineering amalgamation would provide a solution to develop new generation bioremediation techniques encompassing the expression and cellular genetic typing of microorganisms. The current systemic biology approaches help the researchers to broaden the bioremediation agents from the core and exploit cellular catabolic edifice. The systemic biology technologies offer possibilities for not only renaissance of bioremediation using anthropogenically enhanced microbial degraders, but also countenances of translational implications (Ahmad and Ahmad, 2014). The available databases and pathway prediction systems direct to select of suitable chemically and energetically feasible traits for a desired polymeric substrate. Multiomics analyses provide valuable information which facilitates the choice of a suitable host for specific or broad range substrate (Dangi et al., 2019). Genome-scale and kinetic models along with genetic engineering enable tailoring of gene expression, reduction of metabolic burden, and pathway optimization in context of host metabolism (Lieder et al., 2015). General work schemes and standards for design, building, and fine-tuning of selected biochemical routes, networks and whole cell biocatalysts are simultaneously becoming better defined and accepted (Sin et al., 2009; Chubukov et al., 2016).

Such new approaches (systemic biology technologies) can make a difference by developing an efficient ability in the candidate strains owing to depolymerize broad range of synthetic polymers under ambient conditions, and the identified depolymerases and hydrolases genes offer great promise for developing feasible microbial recycling of plastics. The results of testing any new microorganism in a laboratory cannot automatically be extrapolated to large scale applications. The change in magnitude was further complicated by the lack of rigid controls. Unlike the natural environment, costumed large fermenter under controlled environment with careful regulation could be applied. Nevertheless, despite the formidable obstacles, the potential value of the products in these areas assures continuing efforts.

21.2 Stages of polymer biodegradation

The structural transformation of any chemical present in the environment by the action of microorganisms is known as biodegradation process (Joutey et al., 2013). The nondegradable nature of plastics is leading its accumulation everywhere in the environment, nonetheless a few microbes including fungi, bacteria, and yeasts are reported to degrade as well assimilate plastics (Kale et al., 2015; Yoshida et al., 2016; Gajendiran et al., 2016; Skariyachan et al., 2018). Such reports cued a solution to remediate plastics via greener way.

Executing an exact natural condition in the laboratory is unconceivable particularly, the diversity and efficiency of microbial communities, and their catalytic abilities to use and transform various plastics is difficult to be anticipated. The indepth knowledge of microbe and material interaction and the microbial cellular mechanism is crucial to find a context to develop proficient field application. Basically, the microbial degradation of polymeric material may be outlined into four phases as shown in Fig. 21.2 (surface adherence, fragmentation, assimilation, and mineralization) (Sen and Raut, 2015).

21.2.1 Microbial adherence

The instigation of microbial degradation is contingent by the substrate and microbe interface followed by a sequence of events. This colonization and film formation over the surface of plastics is followed by specific mechanisms and a highly synchronized process (Mohan, 2011). Some microorganisms are reported to change their morphology under harsh conditions, which allows them to exist in a hostile environment (Marshall, 2006). The unavailability of any carbon source other than the plastics in the vicinity of microbial cells accords them

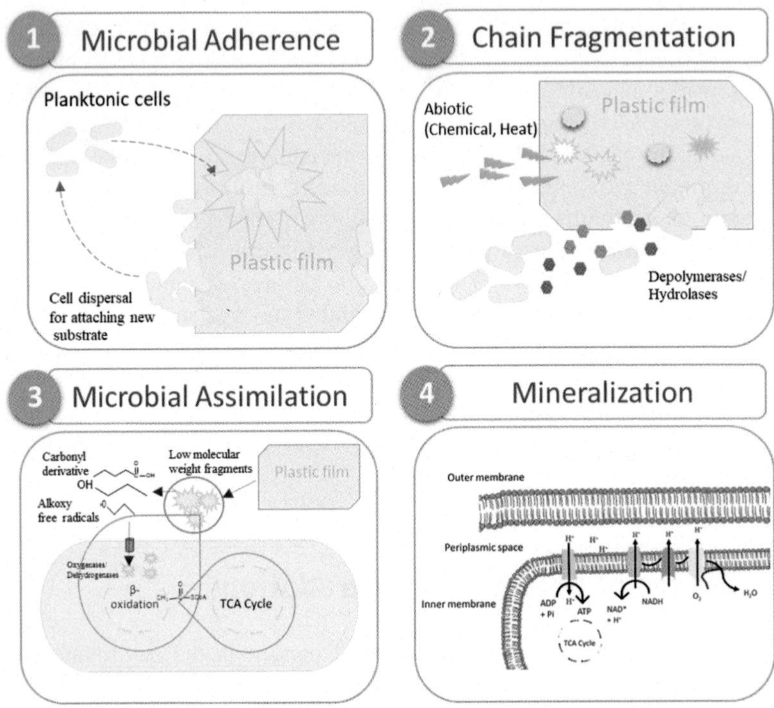

Figure 21.2 Stages of microbial plastic degradation.

toward the hydrophobic plastic surface in quest for carbon assimilation. Microbial cells to plastic surface interaction is a multistage process influenced by many abiotic and biotic factors (Shah et al., 2008). Abiotic factors allies are on surrounding conditions such as surface roughness, topography, hydrophobicity, pH, salinity, temperature, and presence of oxygen (Nauendorf et al., 2016). The morphological loss of polymer due to abiotic elements allows colonization of microbial cells (Pathak and Navneet, 2017). The biotic factors may be exogenous or endogenous. Endogenous factors encompass based on the cell depending on the nutrient source facilitate adherence and formation of film, starting with the initial physical interaction of bacteria toward substrate and ending with the irreversible settling of cell clusters on the surface (Alshehrei, 2017). These attachment forces ultimately determine the properties of biofilm which occurs in three stages: (1) deposition of cells on surface, (2) attachment by the formation of polymeric bridges between cells and substrate surface, and (3) colonization of organisms followed by growth and division of cells in the form of film over the surface (Garrett et al., 2008) leading to the biofilm formation is a pioneering step. The firmness of adherence is time and physical properties of surface dependent process which indicates the induction of suitable bridging between polymer surface and associated microorganisms (Christensen et al., 1995). Also, specific gene regulation events at the individual cell or population levels may play important roles in microbial surface colonization, modification of surface physicochemical properties, structured biofilm development, establishment and maturation of functional colonies (Dang and Lovell, 2016). The thermal properties also play a central role in regulating the binding of bacteria to any surfaces. Surface energy, charge, and roughness have a vital impact on cell to surface adhesion and provide a base for elucidating cellular functions for the biodegradation process (Tokiwa et al., 2009). These events may include the formation of an initial conditioning film by pioneer cells to promote growth and establishment of the biofilm forming microbial community. Variation in any of the factors is expected to affect the degradation process and rate which may be due to acceleration or inhibition of any cellular key step involved. However, not all microbial colonization or biofilm-forming bacteria are involved in plastic degradation. The plasticizers or pretreatments such as heat or chemical to plastic affect the microbial cell adherence and colonization capability (Marini et al., 2007). Advanced microscopic and analytical tools enabled the measurement of forces driving the cell toward the substrate. *Rhodococcus ruber* adheres and colonizes to form biofilm on polyethylene surface that gradually deteriorate the surface and reduces the surface hydrophobicity (Orr et al., 2004). The complete conversion of the material components or only change in the chemical structure results from drastic decrease in molecular weight of polymer resulting from biodegradation (Gu, 2003). The microbial adherence and colonization over any polymeric surface depend not only on microbial abilities but also on type of polymer properties, pretreatments, and surrounding properties. Particularly, for increasing the degradation rate by condition optimization would be crucial. The plastic structural transformations include bond stretching to oxidative functional group formation, resulting in alteration in mechanical properties of the polymer and further fragmentation lowering the molar mass (Corti et al., 2010; Esmaeili et al., 2013).

21.2.2 Fragmentation

Several surrounding dynamic and chemical forces primarily break the large polymeric structure into smaller fragments. The augmentation of the hydrophilicity to the plastic surface due to pretreatments or abiotic course of events favors microbial attack (da Luz et al., 2014). The extracellular depolymerizing enzymes primarily initiate the breaking of terminally available branches to the smaller molecular size that could be ingested intracellularly (Mueller, 2006). Molecular weight reduction and change in carbonyl bond indexes indicate chain scission and oxidation-reactions by the extracellular enzymes catalyzes polymer chains principally at the edges. The polymer fragmentation is mainly concerned about enzymes belong to oxidoreductases and hydrolases (Müller et al., 2005; Mueller, 2006). Hydrolases include esterases, lipases, and cutinases synthesized by microorganisms to hydrolyze the polymers by attacking specifically carboxylic linkages (Fig. 21.3) (Lucas et al., 2008; Kawai, 2010). The presence of aspartate, histidine, and serine residues at the active site of hydrolases is a very common feature of their reaction

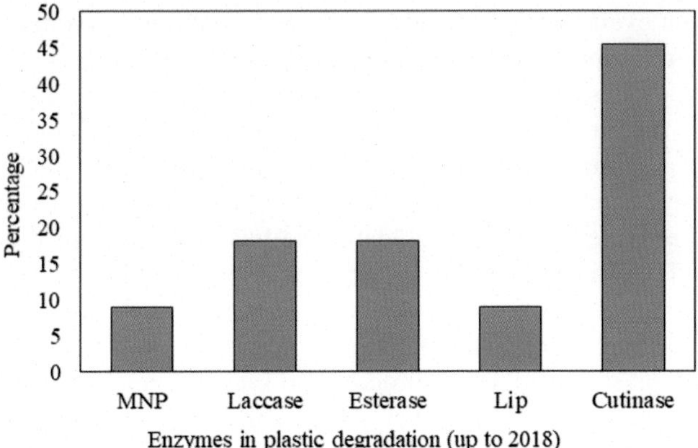

Figure 21.3 Enzymes reported to involved in the degradation of different plastics up to 2018.

mechanism. Aspartate interacts with the histidine ring to form a hydrogen bond. The ring of histidine is thus oriented to interact with serine. Histidine acts as a base, deprotonating the serine to generate a very nucleophilic alkoxide group (−O) (Joo et al., 2018). This group that attacks the ester bond (alkoxide group) leads to the formation of an alcohol end (−OH) group and an acyl enzyme complex. They secrete specific enzymes to generate free radicals. For instance, monooxygenases, dioxygenases, and oxidoreductases add one or two oxygen atoms respectively, forming alcohol or peroxyl groups that are more easily fragmentable (Mishra et al., 2001). Further, peroxidases catalyze reactions between a peroxyl molecule (H_2O_2) and to the electron acceptors. Another group of oxidoreductase, laccase, are metalloproteins copper atoms containing enzyme. The scission reactions of crystalline and highly organized branched polymeric chain are difficult due to inaccessibility of internal structures to the enzymatic attack. Over-all two types of oxidation reactions take place, first hydroxylation reactions which increase the polarity of the molecule and the other one is involved in catalyzing free radicals conducing to chain reactions that accelerate polymer transformations. This could lead to the formation of free radicals and consequently low molecular weight fragments with oxidative ends available for microbial cell ingestion. The oxidation of plastic structure are also determined by length of exposure and the type of additives used in the film (Singh and Sharma, 2008; Tokiwa et al., 2009). Also, metabolizing of low molecular weight fragments depends on the microbial species. For instance, actinomycetes have a high potential for the depolymerization of polyesters, but they are not able to metabolize the formed products (Mor and Sivan, 2008; Sivan, 2011).

The analytical techniques are used to quantify the formation and destabilization of functional groups in the chemical structure of plastics by Fourier transform infrared spectroscopy and separate oligomers of different molecular weight be separated by gel permeation chromatography, high pressure liquid chromatography (LC), and are identified by mass and structural verification by nuclear magnetic resonance.

21.2.3 Assimilation

Formation of low molecular weight fragments on plastic surface resulting from the fragmentation process by exogenous microbial activity allows small oligomers through cell membranes and enters central metabolic pathway (Sen and Raut, 2015; Alshehrei, 2017). Plastic biodegradation is primarily an electron transfer process where energy obtained through the oxidation of reduced molecules resulted from chain fragmentation (Kyrikou and Briassoulis, 2007). Most monomers fragmented as an outcome of degradation are analogs of naturally occurring substrates such as organic acids, alcohol, and amide compounds. Enzymes like laccase, lipases, cutinases, and esterases from microbes are reported to hydrolyze several xenobiotic plastics although their original substrates are phenolic, lipids, and ester compounds (Kawai, 2010). Enzymes catabolizing molecules act as a sole source of carbon and energy for cellular metabolism (Banerjee et al., 2014). Most of research reports stated that the degradation of polyolefins and polyethers are by exogenous depolymerization and oxidative events by a series of oxidative steps followed by β oxidation (Kawai, 2010; Wilkes and Aristilde, 2017). The electrons are released from the substrate material to absorb energy available through the oxidation process. The electrons are moved through the respiratory route via series of compounds to the terminal electron acceptors which are oxygen for aerobic bacteria and for anaerobic organisms, nitrate, and sulfate are electron acceptors, but aerobic biodegradation is typically more efficient (Gottschalk, 2012). Assimilation process for the carbon mobility from the plastic can be estimated by the production of metabolites such as carbon dioxide production or the development of microbial biomass (Skariyachan et al., 2018). This method is possible, if the polymer is a sole carbon source available to the cell. The biodegradation of plastics is typically a surface erosion process due to exertion of extracellular enzymes into the polymer structure and so act only on the polymer surface. Plastic degradation carried out when the pro-oxidants catalyze the formation of free radicals in polymeric chains, which react with molecular oxygen for further oxidation (Chiellini et al., 2006; Eubeler et al., 2010). However, in soil or compost method, the released carbon dioxide cannot be measured. The polymeric carbon labeling with stable isotope would provide the exact mobility of the carbon from stable structure to carbon dioxide conversion. In a study of polystyrene degradation, carbon was tracked using ^{14}C-polystyrene where FTIR analysis showed that ozonation of polystyrene generated carbonyl groups had been decreased and also reduced molecular weight after incubation with *Penicillium variabile* (Tian et al., 2017).

21.2.4 Mineralization

Mineralization is defined as the conversion of biomass to gaseous form, water, salts, and minerals, and residual biomass. Mineralization is complete when all the solid carbon are converted to carbon dioxide and water, and methane resulting from aerobic or anaerobic degradation process, respectively (Vert et al., 2012; Sen and Raut, 2015). The microbial carbon and energy cycle are solely coming from substrate assimilated into biomass for growth and respiration whereas the carbon is released to the atmosphere as carbon dioxide and water.

The simple framework offers an opportunity to study the microbial contribution to the global climate system. Yang et al. (2015) found mealworms are 95% efficient in styrofoam mineralization into biomass and carbon dioxide (*Tenebrio molitor*) verified by feeding ^{13}C labeled Styrofoam. In another study polystyrene was labeled with ^{14}C to investigate its mineralization by *P. variabile* after 16 weeks. In another study, fungi mineralized labeled polymers with a lower molecular weight led to a higher mineralization rate suggesting lower molecular weight being more accessible to microorganisms (Tian et al., 2017). Improving understanding and model representation of microbial community structure and its impacts on carbon and nutrient cycling coming from plastics is important for modeling studies and field experiments.

21.3 Biotechnological intrusion in bioremediation technology

The developing modern world demanded discovery of new chemicals to ease life, however many of them are xenobiotic in nature. Now there is an aim to develop new technologies for maintaining disturbed sustainability by application of xenobiotic substances. Subsequent use of these chemical products ending at the dumpsites persists in the environment affecting indigenous species. Plastic is a xenobiotic material and now it is accumulating with alarming rate in the environment. Plastic wastes continued to be resistant to natural degradation due to lack of proficient catabolic traits in the autochthonous microbial species (Andrady, 2011). Increasing plastic waste and its impact in the environment need new innovations in the area of bioremediation research and public concerns to show interest in it. The solution to these generated plastic wastes should be such that all plastics conveniently go back and integrate into the environment through the microbial elemental cycling.

Microbes are found to degrade many toxic chemicals, mineralize completely, or cometabolize them under suitable physiochemical conditions. Moreover, the 90s has been recognized as the golden era of biodegradation research, Chakrabarty (1981) revolutionized the area by discovery multiplasmid *Pseudomonas putida* strains from four *Pseudomonas* sp. capable of degrading oil which was called as oil eating bacteria. He engineered the plasmid for novel catabolic efficiency from *Pseudomonas* strain to degrade different hydrocarbon. Such metabolic diverse superbugs were expected to be economically feasible and environment-friendly technologies. The *Pseudomonas* sp. are metabolically diverse group of bacteria also found to be active to degrade various types of plastics differentially (Wilkes and Aristilde, 2017). Balasubramanian et al. (2010) found *Pseudomonas* sp. to be the most efficient in degrading high density polyethylene (HDPE) structure followed by *Arthrobacter* sp. Some polyethylene treated with chemical-like nitric acid found the formation of carbonyl group and around 50% weight reduction in 2 months by *Pseudomonas aeruginosa* (Rajandas et al., 2012). Yoon et al. (2012) shown *Pseudomonas* species in a compost condition, low molecular weight polyethylene (PE) was lost 28.6% in dry weight within 40 days without any pretreatment. Microbial consortium composed of thermophilic *Pseudomonas* sp., *Stenotrophomonas* sp., *Bacillus* sp., and *Paenibacillus* sp. isolated from cow dung showed weight loss reduction up to 75% and 55% after 120 days of incubation of low density polyethylene (LDPE) and HDPE, respectively (Skariyachan et al., 2017). Yamada Onodera et al. (2001) isolated a fungus strain *Penicillium simplicissimum* capable of degrading polyethylene without any additives.

The testing methods for plastic biodegradability has many cannons in interpreting the degradation mechanism augmenting uncertainty of authenticity to microbial action. The progress in molecular biology and engineering seemed to be a partial solution for such challenges as expressive conquests in the metabolic networks of the host cells capable to act on plastics aiming to engineer whole cells to understand respective catabolic pathways and corresponding enzymes toward plastics mineralization (Cases and de Lorenzo, 2005; Dvořák et al., 2017). The understanding over catalytic mechanism of these enzymes provides valuable information for engineering aimed to improve biocatalytic applications. The verge of bioinformatics tools and databanks remained limited to genomic and protein sequence data although the core molecular biology demanding needs to enhance and simplify data management and sharing in various area of biotechnology to simplify data mining. To handle huge data would have been impossible without the availability of computers and computer science expertise (Gligorijević and Pržulj, 2015; Karahalil, 2016). Several accessible software had been developed which are crucial for solicitation of many data mining researches. Moreover, quality check, aligning, assembly, completeness, and comparative analysis on the basis of algorithm runs from the ocean of data have been no longer a limitation. However, like all the previous technologies, next-generation sequencing also precincts. The mapping of the numerous and large structural variations remains an issue.

Very often complementing the sequence constrained with additional biological information through experimental or computational derivation can be featured in the sequence databanks. In a number of instances, such knowledge from secondary data processing can be found in other reference data resources. The computational tools play a crucial role in the upgradation of data, analysis tools, accessibility, and data representation creating a huge demand (Chou and Voit, 2009). Some of these protein databases remain very general whereas, many other specialized databases focus on a limited set of objects. The microbes have been engineered from a few decades back for environmental implications, however still legally restricted for ex situ field exposure. The regulations of genetically engineered microbes have relaxed and now plenty of genetic tools are applied for constructing genetically modified organisms for in situ bioremediation. The major challenge to the entire field is inevitable but rather difficult to get popularization; and public environmental concerns and regulatory constraints have not just restricted field tests for engineered microbes, but also affect the quality of fundamental research and overall progress.

Such changes are vital because the ability to minimize plastic wastes production and remove existing waste or use it in bioprocesses to form value added products will shape the fate at global scale.

21.4 Genes in plastic degradation: an overview

There are many reports on microbial plastic biodegradation but a few of them confirmed genes involved in the process and illustrated pathway according to the enzymatic reaction. Genes involved in degradation of the most commonly used PET (polyethylene terephthalate) plastic have been reported from few bacteria, actinomycetes, yeast, and fungi where some are cloned by traditional molecular biology approaches. The gene(s) encoding enzymes from different microbial species like *Thermobifida* sp., *Pseudomonas* sp., *Bacillus* sp., *Fusarium* sp., and *Sacharomonospora* sp., are reported to mainly belongs to class hydrolases where some of them named based on the substrate, for example, PETase, MHETase, and BHETase (Figs. 21.4 and 21.5). Cutinases are major reported enzymes to act on PET and related polymers and increasing surface hydrophilicity. Herrero-Acero

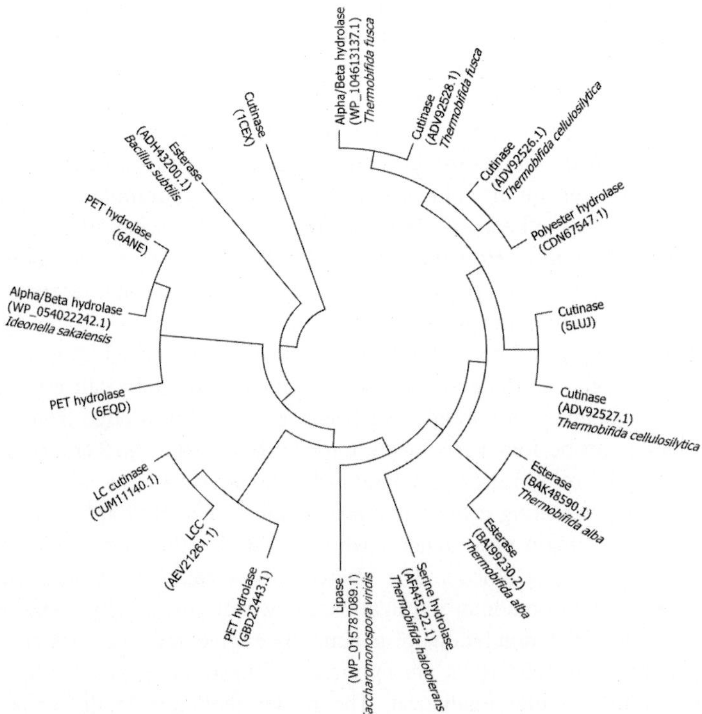

Figure 21.4 The phylogenetic closeness of enzymes in polyesters degradation based on the neighbor joining algorithm.

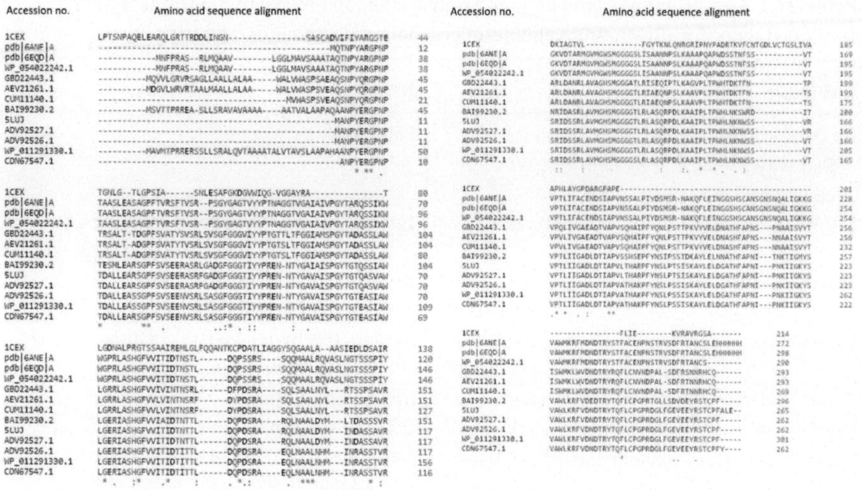

Figure 21.5 Multiple sequence alignment amino acid sequence of cutinase, lipase, and esterase enzymes reported to be actively degrading PET. *PET*, Polyethylene terephthalate.

et al. (2013) identified two cutinases (Thc_Cut1 and Thc_Cut2) and cloned genes from *Thermobifida cellulosilytica* to hydrolyze PET. Cutinase (Tha_Cut1) from *Thermobifida alba* is identical to cutinase from *T. cellulosilytica* with four amino acid difference at active site, perhaps considerably more active in hydrolyzing PET which was also cloned and expressed in *Escherichia coli* (Ribitsch et al., 2012). Yoshida et al. (2016) reported to degrade and mineralize PET polymer by PETase enzyme from *Ideonella sakaiensis*, PETase closely related to cutinase and esterases. *I. sakaiensis* had demonstrated to be capable of degrading PET into its dimer and monomer (ethylene glycol and terephthalic acid) in 6 weeks by sequential enzyme-mediated reaction. Further, MHETase hydrolyzes ethylene glycol and terephthalic acid. The cutinase gene (cut190) was cloned from the thermophile *Saccharomonospora viridis* and expressed in *E. coli* for hydrolyzing polyester films, moreover active site modeling had been performed by mutational substitution of amino acid yielding thermostability and enhanced activity (Kawai et al., 2014). The knowledge over the key genes encoding plastic degrading enzymes belongs to class lipases, carboxylesterases, cutinases, and laccases and pertains to the hydrolysis result and product formation (Table 21.1) (Wierckx et al., 2015). Genomic sequences of many other microorganisms with biodegradation capabilities can be searched through comprehensive databases such as NCBI genome database, a microbial genome annotation and analysis platform provide a complete pipeline for annotations and comparative analyses of microbial genomes. Examination of transcriptomic data uncovers the significant portion of the genome that is differentially expressed when the cells are grown on diverse substrates.

Table 21.1 List of online pathway prediction system and respective institute which developed them for illustrating xenobiotic compound degradation and toxicity.

Database	Institute
Biocatalysis/Biodegradation Database and Pathway Prediction System (http://eawag-bbd.ethz.ch/)	University of Minnesota
EAWAG-BBD/PPS (http://eawagbbd.ethz.ch)	Eawag, the Swiss Federal Institute of Aquatic Science and Technology
enviPath (https://envipath.org/)	Environmental Contaminant Biotransformation Pathway Resource
Eawag-Soil (https://envipath.org/)	Eawag, the Swiss Federal Institute of Aquatic Science and Technology
MetaCyc (https://metacyc.org/)	SRI International
BioCyc (https://biocyc.org/)	SRI International
ATLAS (http://lcsb-databases.epfl.ch/atlas/)	Laboratory of computational systems Biotechnology, Ecole polytechnique fédérale de Lausanne
TOX21 (https://ntp.niehs.nih.gov/results/tox21/index.html)	National Center for Advancing Translational Sciences, US Food and Drug Administration, National Center for Computational Toxicology
EcoCyc (https://ecocyc.org/)	National Institute of General Medical Sciences
UniProt (www.uniprot.org/)	European Bioinformatics Institute and Swiss Institute of Bioinformatics and the Protein Information Resource
RAPID (http://rapid.umn.edu/rapid/)	University of Minnesota
BNICE.ch (Biochemical Network Integrated Computational Explorer)	Swiss Federal Institute of Technology
KEGG (https://www.genome.jp/kegg/)	Bioinformatics Center, Institute for Chemical Research, Kyoto University and Human Genome Center, Institute of Medical Science, University of Tokyo

The integrative study of transcriptomic, proteomic, and metabolomic uncovers the cellular information and their impact. This information could be used to modify the gene expression of host for improving final activity.

21.5 Synthetic biology approaches

The wide range of molecular biological techniques are now accessible for identifying and analyzing various aspects of microorganism such as gene and protein, metabolic, and regulating pathways. The application of genomic, proteomic and metabolomic data would provide a whole understanding of the processes in any living system and could be rewired to address the target. Synthetic biology is an emerging area combining biological and engineering techniques for designing complex biological systems. Synthetic biology attempts to overcome the traditional methods of molecular biology and instead provide computational and conceptual tools for tackling biological hitches comprehensively. Synthetic biology does not exactly stand for importing metabolic pathways and regulatory networks but provides prospects to design proficient pathways from various available genomic and metabolic databases (Oldham et al., 2012). It opens the opportunity for exploring all possibilities such as detailed sequence, comparative genomics, structure, enzymatic reactions, and protein function of any cell or organism.

For the bioremediation of a substance like plastic exploiting metabolic pathways and enzymes can specifically be done for making direct application possible. The efficiency of microbial system can be increased for degrading plastics by applying precise and systemic molecular approaches (Bano et al., 2017). Omics technique contributed markedly for rationally selecting suitable model through systemic study. The integration of specific enzymes into a pathway of interest in order to drive a particular metabolic flux toward the product of interest is normally approached for biosynthetic pathway design and can be applied in plastic biotransformation drives (Jouhten, 2012). Recently, synthetic biologists implemented logic gates to design orthogonal genetic circuits that would establish the skeleton of selected microbe and robust utility of the minimal availability to cells (Brophy and Voigt, 2014). The integration of big data to implement them in a directed way resembling in situ bioremediation encompass the development of test schemes that will expose engineered organisms to simulated environmental conditions of limiting carbon and other nutrient requirements (Austin et al., 2018). These new strains must simultaneously be reliable, amenable to genetic manipulation, have mapped genomes, and ideally metabolic models at hand. The *in-situ* bioremediation quests will be complicated due to the presence of different polymers, their mixtures, and resulting transformation products at a single site (Macdonald and Rittmann, 1993; Azubuike et al., 2016).

21.5.1 Genomic

The traditional bioremediation approaches involve generalized biotransformation study of substrate under laboratory conditions by determining the degradation rates and monitoring microbial growth pattern. Genome of an organism is the complete set of genetic information required for function of the cell. Moreover, the genomic field is subdivided into number of areas, involving various investigation of interactions between genes and information molecules (Thiele and Palsson, 2010). Research on bioremediation technology mainly focused on the reporting of microbial diversity and biotransformation pathways. Microorganisms acquire genes from other organisms present in the vicinity by the process of recombination which enables them to adapt and survive in the new environment. The new combination of genes is commonly accompanied by a sudden change in physicochemical environment. The coupling of DNA with ^{13}C stable isotope biomarkers to trace environmental function countenances identifies microorganisms actively involved in specific metabolic processes under in situ condition. The movement of the ^{13}C-DNA produced during the growth of metabolically active microbial community enriched with ^{13}C carbon source can be separated from ^{12}C-DNA by density gradient centrifugation. The microbial diversity involved in any specific process can be identified (Sathyamoorthy et al., 2018).

Isolation and characterization of an efficient microbe from the natural sources uses traditional techniques though appears to be cumbersome and limited to culturable microbes. The 16S rRNA gene sequencing and microarrays profiling from direct environment sample provide phylogenic characteristics and maps gene expression respectively of single microorganism or communities (Yarza et al., 2014). The high throughput metagenome sequencing and data assembly from the environment for data analysis to delineate the interactions among different microbial populations based on the data are promising for understanding the interactions among different species (Klindworth et al., 2013; Zhou et al., 2015). Microarray-based approaches identify microbial community analysis, especially for complex microbial communities whose comprehensive sampling remains infeasible. The functional gene microarrays could be applied to monitor microbial community dynamics and track microbial activities in a particular environment (Liang et al., 2009). DNA microarray could be effectively applied in determining microbial communities and genes involved in the course of polymeric substrate deterioration with genome-wide transcriptional profiles. Such type of genome expression analysis provides important information for the identification of regulatory pathway of microorganisms (Maphosa et al., 2012; Van Hamme et al., 2003). Complementation of traditional isolation techniques with modern molecular tools for correct identification of potential organism remains an approach of choice.

Moreover, genetic engineering approaches appear to be fascinating but are also associated with technical and ethical difficulties. The enormous amount of genetic data together with robust algorithms and computational tools aided with data mining provide an understanding of dynamic interactions inside and in between cells and the environments (Carr and Church, 2009; Weissenbach, 2016). For the manipulation of microbial system, the presence of robust genetic tools is important where plasmids are routinely used for genetic engineering vehicle although unstable for large scale applications involves the deletion, insertion, and overexpression or mutation of desired gene(s). The enzyme-based pathways can be regulated and altered at various level such as promoter and ribosome binding strength, messenger RNA (mRNA), and protein synthesis.

Moreover, in cases where gene copy number is the limit for protein expression, copy number can be amplified for enough enzyme production involved during the pathway. However, knowledge about the activity of each enzyme in the process is often absent and the quantification of each metabolite in the pathway will provide an idea for determination of pathway is bottlenecks. In addition, mRNA expression or protein stability of the limiting steps could be possible to adjust accordingly.

The discovery of enzyme recombinases allowed deletion, inversion, and insertion of specific sequence depending on the orientation and architecture of the DNA recognition sites. The Cre/lox recombinase system is required for the targeted insertion of the specific genc at the recognition sites (Gaj et al., 2014). Recombinases can also be used to efficiently control gene expression levels when stop cassettes consisting of transcriptional terminator sequences flanked by respective DNA recognition sites are placed into the 5′-untranslated region of a target gene (Gaj et al., 2014). Other site-specific genome engineering tools include zinc finger nucleases and transcription activator like effector nucleases (TALEN) which are programmed to specifically bind to DNA sequence. They are programmed to bind DNA sites in proximity within the genome and each coupled to the catalytic domain of the endonuclease FokI in local proximity of two zinc-finger nucleases (ZFN) or TALEN proteins induces FokI dimerization and activation leading to the double strand DNA break at the target site (Pattanayak et al., 2014). Endogenous DNA repair pathways recruited to break site, results in either error prone nonhomologous end joining causing insertions or deletions, or recruitment of the homology directed repair machinery, which enables the introduction of new genetic sequences with reduced error rates (Chandrasegaran and Carroll, 2016).

The RNA guided endonucleases-based gene editing is a new approach where CRISPR–Cas (clustered regularly interspaced short palindromic repeats associated) systems make programmed RNA bind to the complementary nucleic acid sequence to allow introducing genes into nearly any location in the genome. CRISPR–Cas system works with single chain guide RNA that binds to Cas protein to form a complex located to the target site by recognizing specific target sequence where nucleic acid sequence is excised (Tarasava et al., 2018; Choi and Lee, 2016). Since, CRISPR/Cas is a highly specific system for sequence targeting it has become an excellent choice for precision genome editing. The ability to manipulate the genome of any system by using RGEN technology triggered diverse ethical and regulatory debates (Alper and Beisel, 2018; Tarasava et al., 2018; Ausländer et al., 2017).

Engineering of the enzyme active sites for improving the sorption and catalysis dynamics through specific amino acid substitutions, truncation of particular domain or addition of binding modules, is in trends (Biundo et al., 2018).

Two cutinases from *T. cellulosilytica* found to be dissimilar electrostatically and surface hydrophobicity of active site shows differential efficiency to hydrolyze PET, moreover activity toward PET increased considerably by site directed mutagenesis in *E. coli* BL21-Gold expression vector. In contrast, enzyme active site modeling by amino acid substitution was found to increase in the hydrolytic activity for PET (Herrero-Acero et al., 2013). The functional analysis of indigenous pelagic community grown with naturally weathered plastics. The orders *Rhodobacterales*, *Oceanospirillales*, and *Burkholderiales* dominated the distinct plastisphere communities and genera *Bacillus* and *Pseudonocardia*. The functional analysis predicts overrepresentation of adhesive cells carrying xenobiotic and hydrocarbon degradation genes (Syranidou et al., 2019). The application of genetic engineering tools in bioremediation area has not got so much attention due to lack of their complete understanding. Identifying the pathways and corresponding genetic framework responsible for the biotransformation is of extreme value in order to knock out unnecessary genes and pathways framework (Ramírez García et al., 2019).

Eliminating unnecessary pathways from the highly regulated systems of host cells that might be obstructing the target activity would be promising. However, some of the extended pathways might not be necessarily linked with the biodegradation but indirectly plays significant role in the survival of organism under deficient conditions. The development of strains with enzyme specificity and affinity for desirable substrate is done through pathway assemblage and proteomic-based models. This makes an excellent paradigm for laboratory conditions and enables building a perfect setup for testing imported metabolic pathways genes for the selective degradations process (Pandey and Mann, 2000; Liu et al., 2019).

21.5.2 Proteomics

The protein expression of organisms helps to identify its physiological state at any particular time. Proteomics are post-translational modification which involves protein functionality, protein–protein interaction, and protein sequence analysis. The phenotype of organisms or cells can be mapped by proteomic analysis rather than genome study. The basic of proteomic is molecular weight identification of peptides through polyacrylamide gel electrophoresis rather two-dimensional electrophoresis (2-DE) separates protein based on the isoelectric focusing and molecular weight efficiently in conjunction with mass spectrometry (MS) and by peptide mass obtained from MS can be identified from mass-dependent protein library (Shevchenko et al., 2006). The subtle MS approach has revolutionized the environmental proteomics by countenancing identifying the small peptides and protein molecules within a short time. The coupling of databases search with MS technique played a crucial role in protein identification (Pandey and Mann, 2000).

The LC–MS technique has also opened a new analytical window for direct detection and identification of potential protein and metabolites produced as a result of degradation taken into account to assess the fate during the processes. The improvements in 2-DE for use in compartmental proteomics have been made by introducing an alternative approach via multidimensional protein identification technology. Further, matrix associated laser desorption/ionization time-of-flight MS is the most commonly used MS approach for protein identification in combination of 2-DE (Singhal et al., 2015). The objective of protein profiling is to identify upregulated and downregulated proteins during the experimental condition by comparing expression between control and treatment (Pandey and Mann, 2000). Most of the regulated proteins are involved in different categories such as stress response, energy metabolism, transcription regulation and transport molecule, nucleotide biosynthesis, and cell motility. Proteomics playing an important role in complex physiology of single or microbial consortium where genomic studies are not enough to understand the functional variation. The proteomics analysis could be confirmed by validating the data with complementary technique like microarray or transcriptome profiling. The forthcoming applications of modern proteomic techniques in microbial ecology are identification of novel functional genes, metabolic pathways, and novel proteomes in plastic degrading organisms.

21.5.3 Metabolomics

Metabolomic focuses on whole metabolic pathways instead of specific gene organization or enzyme. All cellular processes have systematic and unique chemical fingerprints involving substrate, intermediated to end products which is called metabolism (Dangi et al., 2019). Microorganisms have quick metabolic adaptation than any other life on Earth due to their rapid multiplication rate (Singh et al., 2010). They can use diverse metabolic approach and often be differentiated from its species based on the nutrient and condition availability. Plastics are supposed to be enzymatically hydrolyzed by the exogenous response of microbes, and subsequent formation of many oxidizing side groups also reduces surface hydrophobicity and depolymerizes the polymer chains into low molecular weight fragments making it available for microbial assimilation (Mc Arthur, 2006).

Specifically, metabolomics is the cellular fingerprints of physiochemical processes during a particular condition (Kosmides et al., 2013). Despite being such a new area, the utility of this metabolomics approach is represented by many industrially relevant portfolios. The beginning of synthetic biology area the demand for de novo designing of whole or manipulation of the existing pathway from the available data and algorithmic tools predict all possibility and their corresponding enzymatic parts based on online biological databases, according to various properties of enzyme reaction (Moreno-Sánchez et al., 2008).

Engineering of the metabolic system includes altering either genomic or heterologous pathway of microbe for maximum synthesis of the desired molecule encompassed with production and regulating enzyme, specificity, and cellular tolerance system which is too often poorly understood. Without detailed experimental evidence, there is little hope for rational mutagenesis strategies to substantially improve yields and pathway efficiencies. When coupled with high throughput screening, random mutagenesis can be a powerful method to explore sequence space with little to no prior knowledge (Choi et al., 2019). For the vast majority of small molecule targets, screening throughput stands as the rate-limiting step with low throughput liquid and gas chromatography being the modus operandi today. The traditional approaches for identifying the flux or of the particular plastic consumption through microbial metabolism and locating the rate-limiting step. The possibility to explore the full understanding of pathway properties under a variety of conditions for successful manipulation of enzyme activity and product formation was limited.

The engineering of metabolism could be achieved through synthetically assembling or engineering existing enzyme network from the single or various host by applying computational and omics techniques form deeper insight. The engineering circuits and computational tools are applied for data mining to the available data for finding out the most suitable pathway for maximum degradation and mineralization proficiency of the engineered new ones (Nielsen and Keasling, 2016). Although, there are many advantages of developing engineered microbes, there are still concerns for introducing them to the environment due to adverse effects such as gene transfer, may affect local microbial community, contaminate natural bodies, and can be invasive to autochthonous species. In the meantime, many of the early efforts for engineering host cells for exploiting them for bioremediation sector, influencing unusual reactions that form parts of many degradative pathways could serve as one of the bases of modern green chemistry.

21.5.4 Pathway prediction systems for various xenobiotic

Pathway information is available through a large number of databases ranging from high-quality databases created by professional curators to massive databases, covering a vast number of putative pathways created through natural processing and text mining. The available pathway databases (Table 21.1) display pathway diagrams in combination of metabolic, genetic, and signal networks based on the available literatures. Some software can produce, edit, and analyze the required pathways. EAWAG-BBD is a remarkable tool providing the idea for possible biotransformation pathway of a new compound resulting from central catabolic pathway. Recently, EAWAG-BBD/PPS rebuilt as an enviPath tool particularly designs biochemical routes and personal databases with biotransformation data as well as for prediction of new catabolic pathways (Latino et al., 2017; de Lorenzo, 2018; Ramírez García et al., 2019). MetaCyc is huge metabolism data repositories which serve primarily to predict metabolic routes from available sequenced genomes based on the free energy of the reaction. BioCyc is a collection of specific pathway/genome databases (PGDB) that provides several useful tools for navigating, visualizing, and analyzing the omics data. Both MetaCyc and BioCyc are linked to numerous other databases. Another extension is EcoCyc, which is also pathway/genome databases (PGDM)-based, integrates genes, proteins, metabolic, and regulatory networks (Keseler et al., 2017). The software performs multiple computational inferences including prediction of metabolic pathways, whole fillers and operons. The software also supports comparative analyses of PGDBs and provides systems biology analyses including interactive metabolic network (Karp et al., 2009). UniProt supplies missing information on function, sequence, or taxonomy of proteins from metabolic or signaling pathways and is further interconnected with more specific protein databases such as RCSB protein data bank or BRENDA, thus completing the picture with structural or kinetic data. These tools consider many factors including substrate specificities, binding sites or reaction mechanisms of enzymes, structural changes in substrate-product pairs, and pathway distance from the substrate to product. RAPID is a complementary prediction algorithm of public database, BNICE.ch (Biochemical Network Integrated Computational Explorer) tools for uncovering new enzymatic reactions following broader reaction rules of the enzyme commission classification system. The method is manually curated KEGG database complete repositories of metabolic data. ATLAS database evaluates on the basis of thermodynamic feasibility of de novo generated reactions. SensiPath algorithm stems from the concept of retro synthesis introduced previously by the same laboratory (Carbonell et al., 2014). Quantitative structure activity relationship models estimate the physical and structural characteristics of any compound other models have recently been developed to predict toxicity based on chemical–chemical interactions. TOX21 is a logarithmic prediction system for defining the toxicity of compound or reaction networks. EcoliTox web server predicts toxicity. EcoliTox dataset is a chemical library of toxic compounds in the most common prokaryotic host *E. coli* for predicting metabolites toxicity throughout the metabolome (Dvořák et al., 2017). A powerful virtual screening is currently applied predominantly to design synthetic pathway and avoid selection of enzymes which might inhibit at any point of degradation pathway in the host cell.

21.6 Conclusion and future remarks

Short life cycle and quick adaptation rate constrained microorganisms to adapt and evolve rapidly with the surrounding conditions. Subsequently, naturally occurring microbes found to readily degrade many pollutants including plastics though the degradation rate may not be adequate for large-scale implementation. Many hydrolytic enzymes isolated from microorganisms from diverse environment sources are found to be capable of degrading various plastics differently. Howsoever, implication of enzymes for degrading plastics is still limited with the current approaches. The research over identification and characterization of enzymes for elaborating degradation kinetics and substrate specificity with various polymer is now in drift. Hence, there is a huge demand in exploring such microbes that can grow in different conditions and under specific stress condition, may be directed to grow and use those carbon polymers as their energy source up to maximum.

The modern biotechnology interventions in molecular engineering techniques enables to derive networking between the genes and pathways. The accessibility to big data made it possible to elaborate and correlate pathways structure and functions of the key proteins and sequence of cell metabolism offers base to expand the naturally existing capabilities. Computational biology resources play a vital role in the prediction of probable mechanisms and metabolic pathways followed by the microbial consortia during the degradation. Among various cutting-edge molecular engineering implements efficient catabolic pathways can be designed with precise genomic map and well-ordered protein and metabolic cycles. This may involve altering the regulation of carbon fluxes such that the specific plastic is directed as a carbon source to preferentially get the product of interest. Though, it has been possible to construct a completely synthetic genome and translation of the result for designing new efficient cell factories depending on both knowledge of biological systems and the identification of a target pathway. Hence, scientists are looking for an effective eco-friendly solution where microbial derived degradation of the long-chain synthetic polymers into their respective monomers proficiently could be possible as a reverse dogma. Bypassing the threat of plastic wastes in the environment to go back to their simple monomers would be an alternative for reusing its monomers for reproducing polymers rather from petroleum fraction.

References

Ahmad, M., Ahmad, I., 2014. Recent advances in the field of bioremediation, Biodegradation and Bioremediation, 11. pp. 1–42.

Alper, H.S., Beisel, C.L., 2018. Advances in CRISPR technologies for microbial strain engineering. Biotechnol. J. 13, e1800460. Available from: https://doi.org/10.1002/biot.201800460.

Alshehrei, F., 2017. Biodegradation of synthetic and natural plastic by microorganisms. J. Appl. Environ. Microbiol. 5, 8–19. Available from: https://doi.org/10.12691/jaem-5-1-2.

Andrady, A.L., 2011. Microplastics in the marine environment. Mar. Pollut. Bull. 62, 1596–1605. Available from: https://doi.org/10.1016/j.marpolbul.2011.05.030.

Ausländer, S., Ausländer, D., Fussenegger, M., 2017. Synthetic biology—the synthesis of biology. Angew. Chem. Int. Ed. 56, 6396–6419. Available from: https://doi.org/10.1002/anie.201609229.

Austin, H.P., Allen, M.D., Donohoe, B.S., Rorrer, N.A., Kearns, F.L., Silveira, R.L., et al., 2018. Characterization and engineering of a plastic-degrading aromatic polyesterase. Proc. Natl. Acad. Sci. USA 115, E4350–E4357. Available from: https://doi.org/10.1073/pnas.1718804115.

Azubuike, C.C., Chikere, C.B., Okpokwasili, G.C., 2016. Bioremediation techniques-classification based on site of application: principles, advantages, limitations and prospects. World J. Microbiol. Biotechnol. 32, 1–18. Available from: https://doi.org/10.1007/s11274-016-2137-x.

Babu, R.P., O'Connor, K., Seeram, R., 2013. Current progress on bio-based polymers and their future trends. Prog. Biomater. 2, 8. Available from: https://doi.org/10.1186/2194-0517-2-8.

Balasubramanian, V., Natarajan, K., Hemambika, B., Ramesh, N., Sumathi, C.S., Kottaimuthu, R., et al., 2010. High-density polyethylene (HDPE)-degrading potential bacteria from marine ecosystem of Gulf of Mannar, India. Lett. Appl. Microbiol. 51, 205–211. Available from: https://doi.org/10.1111/j.1472-765X.2010.02883.x.

Banerjee, A., Chatterjee, K., Madras, G., 2014. Enzymatic degradation of polymers: a brief review. Mater. Sci. Techol. 30, 567–573. Available from: https://doi.org/10.1179/1743284713Y.0000000503.

Bano, K., Kuddus, M.R., Zaheer, M., Zia, Q., Khan, M.F., Gupta, A., et al., 2017. Microbial enzymatic degradation of biodegradable plastics. Curr. Pharm. Biotechnol. 18, 429–440. Available from: https://doi.org/10.2174/1389201018666170523165742.

Biundo, A., Ribitsch, D., Guebitz, G.M., 2018. Surface engineering of polyester-degrading enzymes to improve efficiency and tune specificity. Appl. Microbiol. Biotechnol. 102, 3551–3559. Available from: https://doi.org/10.1007/s00253-018-8850-7.

Brophy, J.A., Voigt, C.A., 2014. Principles of genetic circuit design. Nat. Methods 11, 508–520. Available from: https://doi.org/10.1038/nmeth.2926.

C.P.C.B., 2018. Material on Plastic Waste Management. Central Pollution Control Board, New Delhi.

Carbonell, P., Parutto, P., Baudier, C., Junot, C., Faulon, J.L., 2014. Retropath: automated pipeline for embedded metabolic circuits. ACS Synth. Biol. 3, 565–577. Available from: https://doi.org/10.1021/sb4001273.

Carr, P.A., Church, G.M., 2009. Genome engineering. Nat. Biotechnol. 27, 1151–1162. Available from: https://doi.org/10.1038/nbt.1590.

Cases, I., de Lorenzo, V., 2005. Genetically modified organisms for the environment: stories of success and failure and what we have learned from them. Int. Microbiol. 8, 213–222.

Chakrabarty, A.M., 1981. Microorganisms Having Multiple Compatible Degradative Energy-Generating Plasmids and Prcparation Thereof. General Electric Co, US Patent 4,259,444.

Chandrasegaran, S., Carroll, D., 2016. Origins of programmable nucleases for genome engineering. J. Mol. Biol. 428, 963–989. Available from: https://doi.org/10.1016/j.jmb.2015.10.014.

Chiellini, E., Corti, A., D'Antone, S., Baciu, R., 2006. Oxo-biodegradable carbon backbone polymers—oxidative degradation of polyethylene under accelerated test conditions. Polym. Degrad. Stab. 91, 2739–2747. Available from: https://doi.org/10.1016/j.polymdegradstab.2006.03.022.

Choi, K.R., Jang, W.D., Yang, D., Cho, J.S., Park, D., Lee, S.Y., 2019. Systems metabolic engineering strategies: integrating systems and synthetic biology with metabolic engineering. Trends Biotechnol. 1744, 1–21. Available from: https://doi.org/10.1016/j.tibtech.2019.01.003.

Choi, K.R., Lee, S.Y., 2016. CRISPR technologies for bacterial systems: current achievements and future directions. Biotechnol. Adv. 34, 1180–1209. Available from: https://doi.org/10.1016/j.biotechadv.2016.08.002.

Chou, I.C., Voit, E.O., 2009. Recent developments in parameter estimation and structure identification of biochemical and genomic systems. Math. Biosci. 219, 57–83. Available from: https://doi.org/10.1016/j.mbs.2009.03.002.

Christensen, G.D., Baldassarri, L., Simpson, W.A., 1995. Methods for studying microbial colonization of plastics. Methods Enzymol. 253, 477–500.

Chubukov, V., Mukhopadhyay, A., Petzold, C.J., Keasling, J.D., Martín, H.G., 2016. Synthetic and systems biology for microbial production of commodity chemicals. NPJ Syst. Biol. Appl. 2, 16009. Available from: https://doi.org/10.1038/npjsba.2016.9.

Corti, A., Muniyasamy, S., Vitali, M., Imam, S.H., Chiellini, E., 2010. Oxidation and biodegradation of polyethylene films containing pro-oxidant additives: synergistic effects of sunlight exposure, thermal aging and fungal biodegradation. Polym. Degrad. Stab. 95, 1106–1114. Available from: https://doi.org/10.1016/j.polymdegradstab.2010.02.018.

da Luz, J.M.R., Paes, S.A., Bazzolli, D.M.S., Tótola, M.R., Demuner, A.J., Kasuya, M.C.M., 2014. Abiotic and biotic degradation of oxo-biodegradable plastic bags by *Pleurotus ostreatus*. PLoS One 9, e107438. Available from: https://doi.org/10.1371/journal.pone.0107438.

Dang, H., Lovell, C.R., 2016. Microbial surface colonization and biofilm development in marine environments. Microbiol. Mol. Biol. Rev. 80, 91–138. Available from: https://doi.org/10.1128/MMBR.00037-15.

Dangi, A.K., Sharma, B., Hill, R.T., Shukla, P., 2019. Bioremediation through microbes: systems biology and metabolic engineering approach. Crit. Rev. Biotechnol. 39, 79–98. Available from: https://doi.org/10.1080/07388551.2018.1500997.

de Lorenzo, V., 2018. Biodegradation and bioremediation: an introduction. Consequences of Microbial Interactions With Hydrocarbons, Oils, and Lipids: Biodegradation and Bioremediation 1–21. Available from: https://doi.org/10.1007/978-3-319-44535-9_1-1.

Diepens, N.J., Koelmans, A.A., 2018. Accumulation of plastic debris and associated contaminants in aquatic food webs. Environ. Sci. Technol. 52, 8510–8520. Available from: https://doi.org/10.1021/acs.est.8b02515.

Dvořák, P., Nikel, P.I., Damborský, J., de Lorenzo, V., 2017. Bioremediation 3.0: engineering pollutant-removing bacteria in the times of systemic biology. Biotechnol. Adv. 35, 845–866. Available from: https://doi.org/10.1016/j.biotechadv.2017.08.001.

Esmaeili, A., Pourbabaee, A.A., Alikhani, H.A., Shabani, F., Esmaeili, E., 2013. Biodegradation of low-density polyethylene (LDPE) by mixed culture of *Lysinibacillus xylanilyticus* and *Aspergillus niger* in soil. PLoS One 8, 71720. Available from: https://doi.org/10.1371/journal.pone.0071720.

Eubeler, J.P., Bernhard, M., Knepper, T.P., 2010. Environmental biodegradation of synthetic polymers II. Biodegradation of different polymer groups. Trends Anal. Chem. 29, 84–100. Available from: https://doi.org/10.1016/j.trac.2009.09.005.

Gaj, T., Sirk, S.J., Barbas III, C.F., 2014. Expanding the scope of site-specific recombinases for genetic and metabolic engineering. Biotechnol. Bioeng. 111, 1–15. Available from: https://doi.org/10.1002/bit.25096.

Gajendiran, A., Krishnamoorthy, S., Abraham, J., 2016. Microbial degradation of low-density polyethylene (LDPE) by *Aspergillus clavatus* strain JASK1 isolated from landfill soil. 3 Biotech. 6, 52. Available from: https://doi.org/10.1007/s13205-016-0394-x.

Garrett, T.R., Bhakoo, M., Zhang, Z., 2008. Bacterial adhesion and biofilms on surfaces. Prog. Nat. Sci. 18, 1049–1056. Available from: https://doi.org/10.1016/j.pnsc.2008.04.001.

Ghosh, S.K., Pal, S., Ray, S., 2013. Study of microbes having potentiality for biodegradation of plastics. Environ. Sci. Pollut. Res. Int. 20, 4339–4355. Available from: https://doi.org/10.1007/s11356-013-1706-x.

Glaser, J.A., 2019. Biological degradation of polymers in the environment. In: Gomiero, A. (Ed.), Plastics in the Environment. IntechOpen Limited, London, UK. Available from: https://doi.org/10.5772/intechopen.85124.

Gligorijević, V., Pržulj, N., 2015. Methods for biological data integration: perspectives and challenges. J. R. Soc. Interface 12, 20150571. Available from: https://doi.org/10.1098/rsif.2015.0571.

Gottschalk, G., 2012. Bacterial Metabolism. Springer Series in Microbiology Springer, New York.

Gu, J.D., 2003. Microbiological deterioration and degradation of synthetic polymeric materials: Recent research advances. Int. Biodeterior. Biodegrad. 52, 69–91. Available from: https://doi.org/10.1016/S0964-8305(02)00177-4.

Hammer, J., Kraak, M.H., Parsons, J.R., 2012. Plastics in the marine environment: the dark side of a modern gift. Rev. Environ. Contam. Toxicol. 220, 1–44. Available from: https://doi.org/10.1007/978-1-4614-3414-6_1.1-44.

Herrero-Acero, E., Ribitsch, D., Dellacher, A., Zitzenbacher, S., Marold, A., Steinkellner, G., et al., 2013. Surface engineering of a cutinase from *Thermobifida cellulosilytica* for improved polyester hydrolysis. Biotechnol. Bioeng. 110, 2581–2590. Available from: https://doi.org/10.1002/bit.24930.

Joo, S., Cho, I.J., Seo, H., Son, H.F., Sagong, H.Y., Shin, T.J., et al., 2018. Structural insight into molecular mechanism of poly(ethylene terephthalate) degradation. Nat. Commun. 9, 382. Available from: https://doi.org/10.1038/s41467-018-02881-1.

Jouhten, P., 2012. Metabolic modelling in the development of cell factories by synthetic biology. Comput. Struct. Biotechnol. J. 3, e201210009. Available from: https://doi.org/10.5936/csbj.201210009.

Joutey, N.T., Bahafid, W., Sayel, H., Ghachtouli, N.E., 2013. Biodegradation: involved microorganisms and genetically engineered microorganisms. In: Chamy, R. (Ed.), Biodegradation—Life of Science. IntechOpen Limited, London, UK. Available from: https://doi.org/10.5772/56194.

Kale, S.K., Deshmukh, A.G., Dudhare, M.S., Patil, V.B., 2015. Microbial degradation of plastic: a review. J. Biochem. Technol. 6, 952–961.

Karahalil, B., 2016. Overview of systems biology and omics technologies. Curr. Med. Chem. 23, 4221–4230.

Karp, P.D., Paley, S.M., Krummenacker, M., Latendresse, M., Dale, J.M., Lee, T.J., et al., 2009. Pathway Tools version 13.0: integrated software for pathway/genome informatics and systems biology. Brief. Bioinform. 11, 40–79. Available from: https://doi.org/10.1093/bib/bbp043.

Kawai, F., 2010. The biochemistry and molecular biology of xenobiotic polymer degradation by microorganisms. Biosci. Biotechnol. Biochem. 74, 1743–1759. Available from: https://doi.org/10.1271/bbb.100394.

Kawai, F., Oda, M., Tamashiro, T., Waku, T., Tanaka, N., Yamamoto, M., et al., 2014. A novel Ca^{2+}-activated, thermostabilized polyesterase capable of hydrolyzing polyethylene terephthalate from *Saccharomonospora viridis* AHK190. Appl. Microbiol. Biotechnol. 98, 10053–10064. Available from: https://doi.org/10.1007/s00253-014-5860-y.

Keseler, I., Mackie, A., Santos-Zavaleta, A., 2017. The EcoCyc database: reflecting new knowledge about *Escherichia coli* K-12. Nucleic Acids Res. 45, D543–D550. Available from: https://doi.org/10.1093/nar/gkw1003.

Klindworth, A., Pruesse, E., Schweer, T., Peplies, J., Quast, C., Horn, M., et al., 2013. Evaluation of general 16S ribosomal RNA gene PCR primers for classical and next-generation sequencing-based diversity studies. Nucleic Acids Res. 41, e1. Available from: https://doi.org/10.1093/nar/gks808.

Kosmides, A.K., Kamisoglu, K., Calvano, S.E., Corbett, S.A., Androulakis, I.P., 2013. Metabolomic fingerprinting: challenges and opportunities. Crit. Rev. Biomed. Eng. 41, 205–221. Available from: https://doi.org/10.1615/CritRevBiomedEng.2013007736.

Kulshreshtha, S., 2013. Genetically engineered microorganisms: a problem solving approach for bioremediation. J. Bioremed. Biodeg. 4, e133. Available from: https://doi.org/10.4172/2155-6199.1000e133.

Kyrikou, I., Briassoulis, D., 2007. Biodegradation of agricultural plastic films: a critical review. J. Polym. Environ. 15, 125–150. Available from: https://doi.org/10.1007/s10924-007-0053-8.

Latino, D.A., Wicker, J., Gütlein, M., Schmid, E., Kramer, S., Fenner, K., 2017. Eawag-Soil in enviPath: a new resource for exploring regulatory pesticide soil biodegradation pathways and half-life data. Environ. Sci. Process. Impacts 19, 449–464. Available from: https://doi.org/10.1039/c6em00697c.

Liang, Y., Li, G., Van Nostrand, J.D., He, Z., Wu, L., Deng, Y., et al., 2009. Microarray-based analysis of microbial functional diversity along an oil contamination gradient in oil field. FEMS Microbiol. Ecol. 70, 324–333. Available from: https://doi.org/10.1111/j.1574-6941.2009.00774.x.

Lieder, S., Nikel, P.I., de Lorenzo, V., Takors, R., 2015. Genome reduction boosts heterologous gene expression in *Pseudomonas putida*. Microb. Cell Fact. 14, 23. Available from: https://doi.org/10.1186/s12934-015-0207-7.

Liu, L., Bilal, M., Duan, X., Iqbal, H.M., 2019. Mitigation of environmental pollution by genetically engineered bacteria—current challenges and future perspectives. Sci. Total Environ. 667, 444–454. Available from: https://doi.org/10.1016/j.scitotenv.2019.02.390.

Lucas, N., Bienaime, C., Belloy, C., Queneudec, M., Silvestre, F., Nava-Saucedo, J.E., 2008. Polymer biodegradation: mechanisms and estimation techniques—a review. Chemosphere 73, 429–442. Available from: https://doi.org/10.1016/j.chemosphere.2008.06.064.

Macdonald, J.A., Rittmann, B.E., 1993. Performance standards for in situ bioremediation. Environ. Sci. Technol. 27, 1974–1979. Available from: https://doi.org/10.1021/es00047a002.

Maphosa, F., Lieten, S.H., Dinkla, I., Stams, A.J., Smidt, H., Fennell, D.E., 2012. Ecogenomics of microbial communities in bioremediation of chlorinated contaminated sites. Front. Microbiol. 3, 351. Available from: https://doi.org/10.3389/fmicb.2012.00351.

Marini, M., De Niederhausern, S., Iseppi, R., Bondi, M., Sabia, C., Toselli, M., et al., 2007. Antibacterial activity of plastics coated with silver-doped organic-inorganic hybrid coatings prepared by sol-gel processes. Biomacromolecules 8, 1246–1254. Available from: https://doi.org/10.1021/bm060721b.

Marshall, K.C., 2006. Planktonic versus sessile life of prokaryotes. In: Rosenberg, E., DeLong, E.F., Lory, S., Stackebrandt, E., Thompson, F. (Eds.), The Prokaryotes. Springer, New York, pp. 3–15.

Mc Arthur, J.V., 2006. Microbial Ecology: An Evolutionary Approach. Academic Press, San Diego, CA.

Mishra, V., Lal, R., Srinivasan, 2001. Enzymes and operons mediating xenobiotic degradation in bacteria. Crit. Rev. Microbiol. 27, 133–166. Available from: https://doi.org/10.1080/20014091096729.

Mohan, K., 2011. Microbial deterioration and degradation of polymeric materials. J. Biochem. Technol. 2, 210–215.

Mor, R., Sivan, A., 2008. Biofilm formation and partial biodegradation of polystyrene by the actinomycete *Rhodococcus ruber*: biodegradation of polystyrene. Biodegradation 19, 851–858. Available from: https://doi.org/10.1007/s10532-008-9188-0.

Moreno-Sánchez, R., Saavedra, E., Rodríguez-Enríquez, S., Olín-Sandoval, V., 2008. Metabolic control analysis: a tool for designing strategies to manipulate metabolic pathways. J. Biomed. Biotechnol. 2008, 597913. Available from: https://doi.org/10.1155/2008/597913.

Mueller, R.J., 2006. Biological degradation of synthetic polyesters—enzymes as potential catalysts for polyester recycling. Process. Biochem. 41, 2124–2128. Available from: https://doi.org/10.1016/j.procbio.2006.05.018.

Müller, R.J., Schrader, H., Profe, J., Dresler, K., Deckwer, W.D., 2005. Enzymatic degradation of poly(ethylene terephthalate): rapid hydrolyse using a hydrolase from *T. fusca*. Macromol. Rapid Commun. 26, 1400–1405. Available from: https://doi.org/10.1002/marc.200500410.

National Research Council, 2000. Waste Incineration and Public Health. The National Academies Press, Washington, DC. Available from: https://doi.org/10.17226/5803.

Nauendorf, A., Krause, S., Bigalke, N.K., Gorb, E.V., Gorb, S.N., Haeckel, M., et al., 2016. Microbial colonization and degradation of polyethylene and biodegradable plastic bags in temperate fine-grained organic-rich marine sediments. Mar. Pollut. Bull. 103, 168–178. Available from: https://doi.org/10.1016/j.marpolbul.2015.12.024.

Nielsen, J., Keasling, J.D., 2016. Engineering cellular metabolism. Cell 164, 1185–1197. Available from: https://doi.org/10.1016/j.cell.2016.02.004.

Oldham, P., Hall, S., Burton, G., 2012. Synthetic biology: mapping the scientific landscape. PLoS One 7, e34368. Available from: https://doi.org/10.1371/journal.pone.0034368.

Orr, I.G., Hadar, Y., Sivan, A., 2004. Colonization, biofilm formation and biodegradation of polyethylene by a strain of *Rhodococcus ruber*. Appl. Microbiol. Biotechnol. 65, 97–104. Available from: https://doi.org/10.1111/j.1365-2672.2005.02553.x.

Pandey, A., Mann, M., 2000. Proteomics to study genes and genomes. Nature 405, 837–846. Available from: https://doi.org/10.1038/35015709.

Pathak, V.M., Navneet, 2017. Review on the current status of polymer degradation: a microbial approach. Bioresour. Bioprocess. 4, 15. Available from: https://doi.org/10.1186/s40643-017-0145-9.

Pattanayak, V., Guilinger, J.P., Liu, D.R., 2014. Determining the specificities of TALENs, Cas9, and other genome-editing enzymes. Methods Enzymol. 546, 47–78. Available from: https://doi.org/10.1016/B978-0-12-801185-0.00003-9.

Pometto, A.L., Johnson, K.E., Kim, M., 1993. Pure-culture and enzymatic assay for starch-polyethylene degradable plastic biodegradation with *Streptomyces species*. J. Environ. Polym. Degrad. 1, 213–221. Available from: https://doi.org/10.1007/BF01458029.

Rajandas, H., Parimannan, S., Sathasivam, K., Ravichandran, M., Yin, L.S., 2012. A novel FTIR-ATR spectroscopy based technique for the estimation of low-density polyethylene biodegradation. Polym. Test. 31, 1094–1099. Available from: https://doi.org/10.1016/j.polymertesting.2012.07.015.

Ramírez García, R., Gohil, N., Singh, V., 2019. Recent advances, challenges, and opportunities in bioremediation of hazardous materials. In: Pandey, V.C., Bauddh, K. (Eds.), Phytomanagement of Polluted Sites. Elsevier, Amsterdam, The Netherlands, pp. 517–568. Available from: https://doi.org/10.1016/B978-0-12-813912-7.00021-1.

Rhodes, C.J., 2018. Plastic pollution and potential solutions. Sci. Prog. 101, 207–260. Available from: https://doi.org/10.3184/003685018X15294876706211.

Ribitsch, D., Acero, E.H., Greimel, K., Eiteljoerg, I., Trotscha, E., Freddi, G., et al., 2012. Characterization of a new cutinase from *Thermobifida alba* for PET-surface hydrolysis. Biocatal. Biotransformation 30, 2–9. Available from: https://doi.org/10.3109/10242422.2012.644435.

Sathyamoorthy, S., Hoar, C., Chandran, K., 2018. Identification of bisphenol A-assimilating microorganisms in mixed microbial communities using ^{13}C-DNA stable isotope probing. Environ. Sci. Technol. 52, 9128–9135. Available from: https://doi.org/10.1021/acs.est.8b01976.

Sen, S.K., Raut, S., 2015. Microbial degradation of low density polyethylene (LDPE): a review. J. Environ. Chem. Eng. 3, 462–473. Available from: https://doi.org/10.1016/j.jece.2015.01.003.

Shah, A.A., Hasan, F., Hameed, A., Ahmed, S., 2008. Biological degradation of plastics: a comprehensive review. Biotechnol. Adv. 26, 246–265. Available from: https://doi.org/10.1016/j.biotechadv.2007.12.005.

Shevchenko, A., Tomas, H., Havli, J., Olsen, J.V., Mann, M., 2006. In-gel digestion for mass spectrometric characterization of proteins and proteomes. Nat. Protoc. 1, 2856–2860. Available from: https://doi.org/10.1038/nprot.2006.468.

Sin, G., Woodley, J.M., Gernaey, K.V., 2009. Application of modeling and simulation tools for the evaluation of biocatalytic processes: a future perspective. Biotechnol. Prog. 25, 1529–1538. Available from: https://doi.org/10.1002/btpr.276.

Singh, B., Sharma, N., 2008. Mechanistic implications of plastic degradation. Polym. Degrad. Stab. 93, 561–584. Available from: https://doi.org/10.1016/j.polymdegradstab.2007.11.008.

Singh, B.K., Bardgett, R.D., Smith, P., Reay, D.S., 2010. Microorganisms and climate change: terrestrial feedbacks and mitigation options. Nat. Rev. Microbiol. 8, 779. Available from: https://doi.org/10.1038/nrmicro2439.

Singhal, N., Kumar, M., Kanaujia, P.K., Virdi, J.S., 2015. MALDI-TOF mass spectrometry: an emerging technology for microbial identification and diagnosis. Front. Microbiol. 6, 791. Available from: https://doi.org/10.3389/fmicb.2015.00791.

Sivan, A., 2011. New perspectives in plastic biodegradation. Curr. Opin. Biotechnol. 22, 422–426. Available from: https://doi.org/10.1016/j.copbio.2011.01.013.

Skariyachan, S., Manjunath, M., Shankar, A., Bachappanavar, N., Patil, A.A., 2018. Application of novel microbial consortia for environmental site remediation and hazardous waste management toward low- and high-density polyethylene and prioritizing the cost-effective, eco-friendly, and sustainable biotechnological intervention. In: Hussain, C.M. (Ed.), Handbook of Environment Materials Management, 48. Springer, Cham, pp. 431–478. Available from: https://doi.org/10.1007/978-3-319-73645-7_9.

Skariyachan, S., Setlur, A.S., Naik, S.Y., Naik, A.A., Usharani, M., Vasist, K.S., 2017. Enhanced biodegradation of low and high-density polyethylene by novel bacterial consortia formulated from plastic-contaminated cow dung under thermophilic conditions. Environ. Sci. Pollut. Res. Int. 24, 8443–8457. Available from: https://doi.org/10.1007/s11356-017-8537-0.

Syranidou, E., Karkanorachaki, K., Amorotti, F., Avgeropoulos, A., Kolvenbach, B., Zhou, N.Y., et al., 2019. Biodegradation of mixture of plastic films by tailored marine consortia. J. Hazard. Mater. 375, 33–42. Available from: https://doi.org/10.1016/j.jhazmat.2019.04.078.

Tang, C.C., Chen, H.I., Brimblecombe, P., Lee, C.L., 2018. Textural, surface and chemical properties of polyvinyl chloride particles degraded in a simulated environment. Mar. Pollut. Bull. 133, 392–401. Available from: https://doi.org/10.1016/j.marpolbul.2018.05.062.

Tarasava, K., Oh, E.J., Eckert, C.A., Gill, R.T., 2018. CRISPR-enabled tools for engineering microbial genomes and phenotypes. Biotechnol. J. 13, 1700586. Available from: https://doi.org/10.1002/biot.201700586.

Thiele, I., Palsson, B.Ø., 2010. A protocol for generating a high-quality genome-scale metabolic reconstruction. Nat. Protoc. 5, 93–121. Available from: https://doi.org/10.1038/nprot.2009.203.

Thompson, R.C., Swan, S.H., Moore, C.J., Vom Saal, F.S., 2009. Our plastic age. Philos. Trans. R. Soc. Lond. B Biol. Sci. 364, 1973–1976. Available from: https://doi.org/10.1098/rstb.2009.0054.

Tian, L., Kolvenbach, B., Corvini, N., Wang, S., Tavanaie, N., Wang, L., et al., 2017. Mineralisation of ^{14}C-labelled polystyrene plastics by *Penicillium variabile* after ozonation pre-treatment. New Biotechnol. 38, 101–105. Available from: https://doi.org/10.1016/j.nbt.2016.07.008.

Tokiwa, Y., Calabia, B., Ugwu, C., Aiba, S., 2009. Biodegradability of plastics. Int. J. Mol. Sci. 10, 3722–3742. Available from: https://doi.org/10.3390/ijms10093722.

Urbanek, A.K., Rymowicz, W., Mirończuk, A.M., 2018. Degradation of plastics and plastic-degrading bacteria in cold marine habitats. Appl. Microbiol. Biotechnol. 102, 7669–7678. Available from: https://doi.org/10.1007/s00253-018-9195-y.

Van Hamme, J.D., Singh, A., Ward, O.P., 2003. Recent advances in petroleum microbiology. Microbiol. Mol. Biol. Rev. 67, 503–549. Available from: https://doi.org/10.1128/MMBR.67.4.503-549.2003.

Vert, M., Doi, Y., Hellwich, K.H., Hess, M., Hodge, P., Kubisa, P., et al., 2012. Terminology for biorelated polymers and applications (IUPAC Recommendations 2012). Pure Appl. Chem. 84, 377–410. Available from: https://doi.org/10.1351/pac-rec-10-12-04.

Vroman, I., Tighzert, L., 2009. Biodegradable polymers. Materials 2, 307–344. Available from: https://doi.org/10.3390/ma2020307.

Weissenbach, J., 2016. The rise of genomics. C. R. Biol. 339, 231–239. Available from: https://doi.org/10.1016/j.crvi.2016.05.002.

Wierckx, N., Prieto, M.A., Pomposiello, P., de Lorenzo, V., O'Connor, K., Blank, L.M., 2015. Plastic waste as a novel substrate for industrial biotechnology. Microb. Biotechnol. 8, 900. Available from: https://doi.org/10.1111/1751-7915.12312.

Wilkes, R.A., Aristilde, L., 2017. Degradation and metabolism of synthetic plastics and associated products by *Pseudomonas* sp.: capabilities and challenges. J. Appl. Microbiol. 123, 582–593. Available from: https://doi.org/10.1111/jam.13472.

Yamada Onodera, K., Mukumoto, H., Katsuyaya, Y., Saiganji, A., Tani, Y., 2001. Degradation of polyethylene by a fungus, *Penicillium simplicissimum* YK. Polym. Degrad. Stab. 72, 323–327. Available from: https://doi.org/10.1016/S0141-3910(01)00027-1.

Yang, Y., Yang, J., Wu, W.M., Zhao, J., Song, Y., Ga, L., et al., 2015. Biodegradation and mineralization of polystyrene by plastic-eating mealworms: Part 1. Chemical and physical characterization and isotopic tests. Environ. Sci. Technol. 49, 12080–12086. Available from: https://doi.org/10.1021/acs.est.5b02661.

Yarza, P., Yilmaz, P., Pruesse, E., Glöckner, F.O., Ludwig, W., Schleifer, K.H., et al., 2014. Uniting the classification of cultured and uncultured bacteria and archaea using 16S rRNA gene sequences. Nat. Rev. Microbiol. 12, 635–645. Available from: https://doi.org/10.1038/nrmicro3330.

Yoon, M.G., Jeon, H.J., Kim, N.M., 2012. Biodegradation of polyethylene by a soil bacterium and alkB cloned recombinant cell. J. Bioremed. Biodegrad. 3, 145. Available from: https://doi.org/10.4172/2155-6199.1000145.

Yoshida, S., Hiraga, K., Takehana, T., Taniguchi, I., Yamaji, H., Maeda, Y., et al., 2016. A bacterium that degrades and assimilates poly(ethylene terephthalate). Science 351, 1196–1199. Available from: https://doi.org/10.1126/science.aad6359.

Zhou, J., He, Z., Yang, Y., Deng, Y., Tringe, S.G., Alvarez-Cohen, L., 2015. High-throughput metagenomic technologies for complex microbial community analysis: open and closed formats. mBio 6. Available from: https://doi.org/10.1128/mBio.02288-14e02288-14.

Index

Note: Page numbers followed by "*f*" and "*t*" refer to figures and tables, respectively.

Q

R

U

V

Printed in the United States
By Bookmasters